W0245769

DYNAMIC PROBABILISTIC MODELS
AND SOCIAL STRUCTURE

THEORY AND DECISION LIBRARY

General Editors: W. Leinfellner (*Vienna*) and G. Eberlein (*Munich*)

Series A: Philosophy and Methodology of the Social Sciences

Series B: Mathematical and Statistical Methods

Series C: Game Theory, Mathematical Programming and Operations Research

Series D: System Theory, Knowledge Engineering and Problem Solving

SERIES B: MATHEMATICAL AND STATISTICAL METHODS

VOLUME 19

Scope: The series focuses on the application of methods and ideas of logic, mathematics and statistics to the social sciences. In particular, formal treatment of social phenomena, the analysis of decision making, information theory and problems of inference will be central themes of this part of the library. Besides theoretical results, empirical investigations and the testing of theoretical models of real world problems will be subjects of interest. In addition to emphasizing interdisciplinary communication, the series will seek to support the rapid dissemination of recent results.

The titles published in this series are listed at the end of this volume.

DYNAMIC PROBABILISTIC MODELS AND SOCIAL STRUCTURE

Essays on Socioeconomic Continuity

by

GUILLERMO L. GÓMEZ M.

University of Erlangen-Nürnberg, Germany

SPRINGER-SCIENCE+BUSINESS MEDIA, B.V.

Library of Congress Cataloging-in-Publication Data

Gómez M., Guillermo L.
    Dynamic probabilistic models and social structure : essays on
socioeconomic continuity / by Guillermo L. Gómez M.
        p.   cm. -- (Theory and decision library.   Series B,
Mathematical and statistical methods ; v. 19)
    Includes bibliographical references and index.
    ISBN 978-94-010-5114-9   ISBN 978-94-011-2524-6 (eBook)
    DOI 10.1007/978-94-011-2524-6
    1. Economics, Mathematical. 2. Probabilities. 3. Statics and
dynamics (Social sciences)   I. Title. II. Series.
HB135.G66 1992
330'.01'51--dc20   92-9042

ISBN 978-94-010-5114-9

1492

These pages are dedicated to the unknown and forgotten survivors of cultures dispoiled of everything, even their lands;

to recall them whose deprivation, humiliation, suffering and poverty were destined to nourish freedom, peace and democracy;

to celebrate the real heroes who unwillingly rised Eurocentrism to power and world leadership;

to commemorate all those around the world who day by day feel the actual sense of incessantly talking of humanity and missionary commitment to progress;

to praise the wisdom of nature and ongoing legacy to posterity to which humankind is critically linked.

Contents

II   Models with Persistent Structures   171

Preface

Mathematical models have been very successful in the study of the physical world. Galilei and Newton introduced point particles moving without friction under the action of simple forces as the basis for the description of concrete motions like the ones of the planets. This approach was sustained by appropriate mathematical methods, namely infinitesimal calculus, which was being developed at that time. In this way classical analytical mechanics was able to establish some general results, gaining insight through explicit solution of some simple cases and developing various methods of approximation for handling more complicated ones. Special relativity theory can be seen as an extension of this kind of modelling. In the study of electromagnetic phenomena and in general relativity another mathematical model is used, in which the concept of classical field plays the fundamental role. The equations of motion here are partial differential equations, and the methods of study used involve further developments of classical analysis.

These models are deterministic in nature. However it was realized already in the second half of last century, through the work of Maxwell, Boltzmann, Gibbs and others, that in the discussion of systems involving a great number of particles, the deterministic description is not by itself of great help, in particular a suitable "weighting" of all possible initial conditions should be considered. This is typically the case in the attempt to understand the thermodynamic behaviour of macrosystems like a gas in terms of a microsystem consisting of molecules. Here we leave the domain of deterministic analysis to enter probabilistic analysis. It was realized later on, starting with Poincaré, that even in the detailed study of the majority of simple mechanical systems, consisting of a small number of "degrees of freedom" over a large range of times, the non linearities manifest themselves in the form of very complex behaviour (called nowadays "chaos"), which requires for its description methods of probabilistic or stochastic analysis.

Finally also in the realm of quantum phenomena probability enters in an essential way, through the interpretation of the solutions of the Schrödinger equation as amplitudes determining probabilities. It appears that probabilistic analysis has established itself as a basic mathematical framework for the description of the physical world. What about mathematical modelling, in particular probabilistic modelling, in other sciences ? Also here this modelling plays an increasing role, and in fact the interaction between mathematics and many sciences has been beneficial to both sides, for instance certain developments of probabilistic analysis, like the theory of branching processes, have been stimulated by questions which arose in the study of biological systems. More specifically, what about social economic and human sciences? Let me concentrate here on economics, although there would be much to say about successful mathematical modelling in other sciences of this group. The motivation for probabilistic modelling of economic and social phenomena dates from the very beginnings of probability theory as a mathematical science in the XVI-XVII century. It was already felt at that time that the complexity of such phenomena makes a probabilistic analysis necessary. Not only the initial conditions but also the interactions in the systems are known only with great uncertainty. A probabilistic approach seemed therefore appropriate, even though some of the basic mathematical tools for probabilistic analysis were yet to be developed. For example, in work by Huyghens around 1670 probabilistic methods are used in connection with demographic statistics and the computation of annuities for insurances. In N. Bernouilli's dissertation (1709) probabilistic considerations enter in juristic matters and J. Bernouilli's "Ars conjectandi" contemplated the application of probability theory "to civic, moral and economic problems". Some of these applications were carried through later on by a number of mathematicians, including Euler, Buffon, Laplace, and Quetelet (whose work can be considered as giving birth to mathematical statistics).

Often however the application of probabilistic methods in social and economic sciences has presented great problems of interpretation. Some of these problems were caused by uncertainties in the foundations of probability and statistics itself, partly clarified only in our century, through the work of Kolmogorov and others. Others problems are connected with the role of modelling, its appropriateness, efficiency, relevance, questions which should always be discussed in a true interdisciplinary and critical spirit, to avoid, in particular inappropriate conclusions or even misuses of mathematics for ideological purposes. Although some of the applications of mathematics of the XVIII-XIX century in human sciences can easily be criticized as being too naive or misplaced, it would be wrong to draw an overall negative picture. In fact modern stochastic methods in social and economic sciences which go under the names of decision sciences, operation research, stochastic control have their roots in the somewhat confuse attempts of last century at a probabilistic description of social and economic phenomena.

As to mathematical economics itself its roots can be traced back to the times of the Encyclopédie (Quesnay, Montesquieu, D'Alembert ...) and the influence the successes of Newtonian physics and analysis had on the whole culture. The whole classical (A. Smith, D. Ricardo ...) and marxist school of economics was strongly influenced by classical deterministic physics. This influence has been continued through the neoclassical school into our century. In particular a general equilibrium theory was developed, which uses, and partly influenced, new methods of convex analysis, optimization theory and the theory of dynamical systems. J. von Neumann, in his work with Morgenstern, started a new approach through game theory which also had a dynamical component and in which certain aspects of economics and mathematics came to a productive interaction. All these approaches are basically deterministic. The interest and even necessity of the introduction of stochastic methods in economics has however been felt for a long time, they are present, for example, in work by Cournot (1838), as a means of dealing with the fact that economic agents operate in a probabilistic environment and inferences about economic behaviour are essential statistical in nature (a fact reflected in the modern praxis of econometrics).

Although in the study of stock markets stochastic processes were brought in already at the beginning of our century by L. Bachelier, it is fair to say that a major systematic use of probabilistic tools in economics has only taken place after the second world war, using the newly developed methods of stochastic differential equations and martingale theory. These methods were later on combined with (stochastic) optimal control methods to yield an impressive interactive picture for the study of certain aspects of economics, with stimulation also in the other direction, namely development of new mathematical tools for the needs of economics.

G. Gómez makes accessible the latest developments in stochastic analysis and control theory and applies them to the study of social and economic systems, in particular is their global features (which also include ecological components).

In addition his book represents an original and highly interesting attempt to understand and master the phenomena of labour surplus and formation of a structure of dual type, as occurring in economies linking "peripheral components" (typically third world countries) and "centre components" (typically western developed capitalistic countries). Gómez develops a dynamical model incorporating neoclassical, post-keynesian and neo-marxist elements, with a careful treatment of the microdynamical basis for the macroeconomic phenomena he investigates. His book is conceived as a critical but constructive contribution to new classical models in which the central role of the concept of labour is reinstated and dynamical (time dependent) features are emphasized. Also it represents a courageous attempt to alter the role orthodox mathematical economics as an academic discipline, having an ancillary role in prevalent economic trends, and open it up again to vital global problems like the detrimental consequences of an ever increasing centre-periphery polarization. The book proposes a new socioeconomic analysis based on stochastic modelling and leads to proposals of policies aimed at bringing about better global economic development. This stochastic modelling incorporates practically all relevant recent developments of stochastic analysis (including jump processes as well as diffusions) and is also greatly inspired by the theory of systems, and learning theory on one hand, and by recent methods used in connection with new developments in physics, including quantum physics, on the other hand.

It is an interdisciplinary attempt linking sociological, psychological, economic, and political issues with the exact sciences. It represents also a pladoyer, supported by many appropriate quotations, against the alleged "neutrality of sciences", often used to hide, consciously or unconsciously, underlying philosophical or ideological choices, concerning the type of problems to be studied, the applications envisaged or to mystify certain social implications of research activities. Gómez' opening up of new directions of investigation and application for economic and social research, in particular to bridge the gap between developed and underdeveloped economies, as well as his passionate concern to make the reader aware of the great scientific interest and urgency of the problems, deserves our great attention.

In a time were all values are apparently being shaken it is good to have enlightened and critical authors like Gómez reminding us of our responsabilities as scientists and men. I hope that readers are influenced by this book to embark on new paths of applications of stochastic analysis in the social and economic sciences.

Bochum, April 1992.   S. Albeverio

Part I

Framework of Analysis

1492: The myth of free market

Most of my students have trouble with the idea that a book – especially a *textbook* – can lie. When I tell them that I want them to argue with, not just read, the printed word they're not sure what I mean. That's why I start my U.S. history class by stealing a student's purse ...

So I begin by stealing a student's purse. I announce to the class that the purse is mine, Obviously, because look who has it. Most students are fair-minded. They saw me take the purse off the desk, so they protest: "That's not yours, it's Nikki's. You took it, we saw you." I brush these objections aside and reiterate that it is mine, and to prove it I'll show them all the things I have inside ...

It's time to move on: "Okay, if it's Nikki's purse, how do you know? Why are you all so positive it's not my purse?" Different answers: We saw you take it; that's her lipstick, we know you don't wear lipstick; there is stuff in there with her name on it. To get the point across, I even offer to help in their effort to prove Nikki's possession: "If we had a test on the contents of the purse, who would do better, Nikki or me?" "Whose labor earned the money that bought the things in the purse, mine or Nikki's?" Obvious questions, obvious answers.

I make one last try to keep Nikki's purse: "What if I said I *discovered* this purse, then would it be mine?" A little laughter is my reward, but I don't get any takers; they still think the purse is rightfully Nikki's.

"So," I ask, "why do we say that Columbus discovered America?" Now they begin to see what I've been leading up to. I ask a series of rhetorical questions which implicitly make the link between Nikki's purse and the Indians' land: Were there people on the land before Columbus arrived? Who had been on the land longer, Columbus or the Indians? Who knew the land better? Who had put their labor into making the land produce? The students see where I'm going – it would be hard not to. "And yet," I continue, "what is the first thing that Columbus did when he arrived in the New World?" Right: he took possession of it. After all, he had discovered the place.

We talk about phrases other than "discovery" that textbooks could use to describe what Columbus did. Students start with the phrases they used to describe what I did to Nikki's purse: he stole it; he took it; he ripped it off. And others: he invaded it; he conquered it.

I want students to see that the word "discovery" is loaded. The word carries with it a perspective, a bias; it takes sides. "Discovery" is the phrase of the supposed discoverers. It's the conquerors, the invaders, masking their theft. And when the word gets repeated in textbooks those textbooks become, in the phrase of one historian, "the propaganda of the winners."

Bill Bigelow
(see [123], pp. 123–125)

Chapter 1

Introduction and outline

Mathematics has actually been used in economic theory, perhaps even in an exaggerated manner. In any case its use has not been highly successful. This is contrary to what one observes in other sciences: There mathematics has been applied with great success, and most sciences could hardly get along without it. ...

In order to elucidate the conceptions which we are applying to economics, we have given and may give again some illustrations from physics. There are many social scientists who object to the drawing of such parallels on various grounds, among which is generally found the assertion that economic theory cannot be modeled after physics since it is a science of social, of human phenomena, has to take psychology into account, etc. Such statements are at least premature. It is without doubt reasonable to discover what has led to progress in other sciences, and to investigate whether the application of the same principles may not lead to progress in economics also. ...

We do not worry at all if the results or our study conform with views gained recently or held for a long time, for what is important is the gradual development of a theory, based on a careful analysis of the ordinary everyday interpretation of economic facts. This preliminary stage is necessarily *heuristic*, i.e. the phase transition from unmathematical plausibility considerations to formal procedure of mathematics. The theory finally obtained must be mathematically rigorous and conceptually general. Its first applications are necessarily to elementary problems where the results have never been in doubt and no theory is actually required. At this early stage the application serves to corroborate the theory. The next stage develops when the theory is applied to somewhat more complicated situations in which it may already lead to a certain extent beyond the obvious and the familiar. Here theory and application corroborate each other mutually. Beyond this lies the field of real success: genuine prediction by theory. It is well known that all mathematized sciences have gone through these successive phases of evolution.

John von Neumann and Oskar Morgenstern
(See [224], pp. 3–8)

1.1 Scope of the essays

This work deals basically with the process of social reproduction as it emerges in an economic system and attempts to understand mechanisms governing the phenomenon of labour-surplus and the subsequent formation of a structure of the dual type. Thus we shall naturally ask how to strengthen and restructure the economy at hand in such a way that the system as a whole begins supporting those processes primarily responsible for the fulfilment of its needs for growth and development associated with social reproduction. At a fairly general level the fundamental principles underpinning social reproduction are linked in an inseparable manner with the ubiquitous law of entropy. Hence in this work understanding social reproduction shall be tantamount to grasping the mechanisms ruling processes along which available potentials are used up, renewed and/or recreated.

Expansion of productive capacity, development of productive forces and generation of economic surplus seem to be the most suitable concepts relying on which one may approach the above question. To this end, we construct a sequence of models that capture essential aspects of the system dynamics and investigate using mathematical techniques, mainly stochastic analysis and control theory, behavioural issues concerning responses, adaptation and learning resulting from alternative policies. Special features are a clear separation of short- and long-term questions, consideration of interaction and interdependence of economic processes, introduction of uncertainty as an inherent outcome of the overdetermination of socioeconomic phenomena and the natural assumption of limited information, intervention of institutions in conflicting social configurations.

Owing to its highly interdisciplinary character, this monograph is intended to serve manifold objectives: a general conceptualisation of essential aspects of economic development and growth; to illustrate in an intuitive way the philosophy underlying modelling design; application of mathematical techniques with the purpose of obtaining development strategies; analysis and interpretation of the original questions and answers on the basis of the obtained results. In this sense the present work develops very much in the spirit of the rapidly advancing theory of systems, uses mathematical reasoning and powerful tools supplied by several fields of (applied) mathematics investigating various pressing socioeconomic problems which by now afflict almost two thirds of the world.

In its aims, framework and methodology our inquiry is of such vitality and dynamism that, while its relevance will certainly endure, its centre of interest, the insight and understanding it provides, the adequacy and accuracy of its setting, the sharpness of its models and questions will progressively shift in the course of time. This transitory character is what the sub-title "essays" is intended to emphasise. At this point we shall mention, we are fully aware our models are dwarfs compared to the dimension of the questions at issue. In this regard it is necessary to say, essential to our approach is setting humankind at the centre of economic considerations and elaborating in mathematical terms on answers to situations representing highly simplified descriptions of reality without loosing economic content and social relevance.

Concerning the genealogy of these essays we shall draw attention further to the fact that they are, almost literally, between the boundaries of various fields of scientific endeavour, providing in challenge and excitement. As the present essays point out, an outcome of the system-theoretical nature of modelling and theorising is that prejudices and peculiarities characterising one's state of existence affect one's vision of social reality. Thus, the higher the degree of interdisciplinarity, the greater the chance of delivering an unbiased picture of reality. Yet this may entail risks since some researchers may feel that certain virtues of their favoured theory are not seen in their full power or are altogether neglected, that their failures or shortcomings are being overplayed, or that our own blend can in no way do as well as school of thought X or Y does; all this according to the fervour, communication-readiness or partisan view of the observer, reader or whatsoever receptive mind gets attracted by these pages. In this sense, the qualities or lack of these essays may also become dependent upon the prejudices and bias inherent in the lenses by means of which readers get their own picture of social reality. As far as we are concerned, one cannot overemphasise that this is primarily an address and an invitation for tolerance, cooperation and solidarity in every respect, an appeal to put sciences and human intelligence at the service of the most dramatic needs of mankind.

1.1.1   On goals and views

These essays intend hence to remind mathematical economists that there are more relevant issues than Hessians in their various economic brands, indifference maps, perfect foresight, the rationality principle and many other venerable notions at the service of orthodoxy. On the other hand, they attempt to suggest to those, who as reaction to the alluded placement of priorities, disapprove the quantification of social phenomena, that the misuse and abuse of mathematics in the social sciences is in no way an intrinsic quality of it. Further, it dares to defy some well-established standards of elegance and purity with the aim of keeping the most relevant issues at the forefront of the research agenda.

Mathematical techniques should not serve to distract from the inappropriateness of the assumptions chosen to describe social reality or to hide the reactionary roots of orthodoxy. Abstract thinking should not be reduced to the role of making the life of scientists easy. Abstraction has to remain a way of approaching complex reality by means of simplified pictures of it, the distorted character of which diminishes as science makes progress in the course of time.

A preliminary statement of views

Needless to say, that our contribution succeeds at most in grasping the surface of the complex and dramatic reality of peripheral societies or the so-called Third World. It is precisely the awareness of this which makes our appeal to solidarity and cooperation more compelling. To avoid any misunderstanding, we shall stress that these essays are not essentially against capitalism, neoclassicism and so on. On the contrary, our criticisms are constructive and in no way partisan, and attempt first of all to reinforce all positive aspects that any theory, system, ideology may possibly have.

With respect to neoclassicism, we shall say that we are glad about the powerful machinery this school of thought has made available to economics and understand the strong degree of abstraction along which it seeks to progress. In this regard, let us recall that, for instance, Joan Robinson granted that there is much to be learned from *a priori* comparisons of equilibrium positions as long as no conclusions are drawn from it. Further, she brought forward plenty of arguments supporting the conclusion that such models cannot be applied to concrete reality, since they are not capable of getting out of equilibrium.

As in any other branch of scientific endeavour, one expects that the neoclassic paradigm finally moves towards the everyday facts-of-life. Since the latter does not seem to enter into the actual research agenda of orthodoxy, we can only disprove its pretension of being "the theory" and with it its ideological bias, teaching absolutism and intolerance. It might strike anybody the astonishing success of prevailing orthodoxy in the face of its lack of social perspective. Even more, if one recalls that this great theory failed to advance any consistent explanation of the most crucial problems of employment, capital, production, distribution, to mention but a few. However, a glance at the history of science rapidly explains how such a thing as the neoclassical revolution takes off apparently out of nothing. We will have the opportunity of giving precise references on this subject, for the time being let us reproduce a few ideas from Pasinetti's line of argumentation on the consolidation of orthodoxy under the label "the baffling 1870s".

> A satisfactory and comprehensive explanation of such a striking turnabout may yet require a long time to emerge clearly. But it seems to me that in the end it cannot be abstracted from the combined effects of two major features of the European environment of the time: the appearance of Marx's critique of economic theory (Marx's *Das Kapital* first appeared in 1867), and the widespread social unrest which characterised those troubled years.
>
> Objectively, Karl Marx was a Classical economist in the full sense of the word. He picked up and pursued the Classical approach to economic reality. This gave a sweeping drive to his analysis, since production – and production with capital – undoubtedly is the central feature of any modern industrial system. Subjectively, however, he used classical theory for purposes which were diametrically opposite to those of the classical economicsts. The Classical economists (in a line descending directly from physiocratic thinking) had accepted the society in which they lived as part of the order of Nature; Marx regarded it as a passing phase in the transition from the feudalism of the past to the socialism of the future. The Classical economists had generally been arguing in terms of harmony of interest among the various sections of society; Marx conceived economic life in terms of conflicts of interest and class struggle.
> . . .
>
> This was upsetting. To many people it sounded preposterous. Yet Marx's overall arguments were not easy to challenge. The obvious procedure to follow would have been to question the premises. But this is precisely what was so difficult. Marx's premises were exactly the same as those of Smith and Ricardo, i.e. of all established economics.
>
> If only one could find an economic theory that made no reference to

labour, no reference to means of production, possibly even to production itself. ... that would surely be the sort of thing that a frightened Establishment could not but most warmly welcome. Marginal utility theory provided precisely that.

Luigi I. Pasinetti
(see [160] pp. 12–13)

Regarding capitalism, viewed as the actual forces underpinning the worldwide expansion of capital which has given rise to a centre/periphery polarisation impossible to overcome within the prevailing state of affairs, we believe we have the right to refuse or to accept it, at least, as a fanatic and unsatiable means of generating social value all around the world and transfering it to the centre of the capitalist system at the expenses of the periphery. The least that one may expect is granting the periphery its legitimate right to build its own independent capitalist class, and to back up its activities with its own multinational and financial institutions.

It may be that real capitalism, with its mighty networks, its operations indeed seems diabolical to common mortals as F. Braudel puts it, but it is not easy to imagine Eurocentric capital assuming human attitudes which enable peripheral capital to come out of its crippled state and to promote its own interests. Let us describe our scepticism with some lines taken from Braudel's *Civilization and Capitalism,*

'Tradition and previous generations', Marx wrote, 'weigh like a nightmare on the minds of the living' – and not only on the minds, on the very existence of the living too, one might add. Jean-Paul Sartre may have dreamed of a society from which inequality would have disappeared, where one man would not exploit another. But no society in the world has yet given up tradition and the use of privilege. If this is ever to be achieved, all the social hierarchies will have to be overthrown, not merely those of money or state power, not only social privilege but the uneven weight of the past and of culture. The experience of the socialist countries proves that the disappearance of a single hierarchy – the economic hierarchy – raises scores of new problems and is not enough on its own to establish equality, liberty or even plenty. A clear-sighted revolution, if such a thing is even possible – and if it were, would the paralysing weight of circumstances allow it to remain so for long ? – would find it very difficult to demolish what should be demolished, while retaining what should be retained: freedom for ordinary people, cultural independence, a market economy with no loaded dice, and a little fraternity. It is a very tall order – especially since whenever capitalism is challenged, it is invariably during a period of economic difficulty, whereas far-reaching structural reform, which would inevitably be difficult and traumatic, requires a context of abundance or even superabundance. And the present population explosion is likely to do little or nothing to encourage the more equitable distribution of surpluses.

Fernand Braudel
(see [30], Vol. 3 pp. 628)

To look at this issue from the point of view of economic theory, one needs to recall that in order for investment operations to call forth the envisaged effects, the principle of effective demand has to come into operation. But this is precisely what the incessantly increasing transfer of surplus value to the centre of the system hinders, not to mention the devastating impact of this transfer on the foreign exchange and debt. Furthermore, the overweight of Eurocentric capital at the periphery deprives peripheral economies of the essential intermediate links of production, rendering the main economic principles of national growth and development fully inoperational, and serving in various ways only the accumulation goals of the centre. Highly detrimental has been the imposition upon the periphery of trading patterns according alone to the needs to support and/or counteract trading and industrialisation phases along which state nations at the centre of the system go through: First crippling the natural development path and the domestic industrialisation process of peripheral economies and second frustrating the process of democratisation and liberalisation that such an economic and social life would necessitate.

However, the most devastating ingredients of Eurocentrism have been the financial and monetary mechanisms introduced on the basis of Bretton Woods institutions for the purpose of post-war reconstruction. In this connection we shall mention, the emergence of the Dollar as a virtual universal currency provides the centre of the system with highly effective, continuously operating and pervasive mechanisms of sucking finance out of the periphery by apparently legal monetary manipulations, the major consequences of which are capital flight, self-reinforcing currency devaluation, indebtness and increased poverty. As if this would not suffice, ingenuity of Eurocentrism brought into being a sophisticated network of all embracing and omnipresent international institutions, indeed the "Brockers of Poverty", where state nations of the periphery do not have any theoretical chance – even if they all would agree to do so – to decide an issue in their favour. See in this regard Chapter 2 and Gómez [91].

A note on disinformation and racism

Development in human society is a manifold process taking place at the social and individual level. Hence, it involves, for instance, unfolding individual skills and know-how, increasing creativity and material well-being, and improving capability of leaving the domain of necessity towards that of freedom. On the other hand, these features of development at the level of the individual should entail for society as a whole an increasing capacity to cope with internal and external relationships resulting from interaction and cooperation.

History of mankind shows that people everywhere have amply demonstrated a capacity of acting upon nature with an increased ability to use and transform it so as to satisfy their needs. In this sense development has been universal. Certain feelings of superiority, some sort of religious legitimacy and material motives for removing the last scruples and cultural barriers, release means of encroaching on other people's fruits of development and/or natural entitlements becoming themselves primarily manifestations of racism. The manifold considerations nourishing racism are often related with needs and/or compelling

interests of extending the own limit of accumulation and these are set in motion
by contradictions inherent in a specifically determined framework, e.g. by national, traditional and cultural circumstances. Following I. Wallerstein, racism
constitutes an efficient mechanism at the service of the only international status
group category, namely race, and refers to actions at a specific level of organisation, i.e. within the world arena. Thus, racism is the act of maintaining the
existing international social structure. See in this regard, Wallerstein [225], pp.
180–181.

Racism is so deeply rooted in European culture and institutions that to the
majority of European people it seems sheerly natural to support and defend a
set of scientific untenable generalisations and assumptions ranging from religion
to biology. Racist rhetoric reflects a certitude of superiority on the basis of the
capacity of subjugating other nations, using their labour power and resources,
and finally enforcing the achievement of material property regardless of those
who suffer to make this possible. The consequences of this European superiority, as an essential manifestation of racism, have been devastating for the
non-European state nations of the periphery, see for instance [13, 49, 78, 186].
It is striking however that orthodoxy confuses these with the causes of the
*under*development of state nations of the periphery. Gramsci's reflections on
the philosophy of praxis may perhaps help us to understand this tendentious
position of conventional wisdom.

> If the philosophy of praxis affirms theoretically that every "truth" be-
> lieved to be eternal and absolute has had practical origins and has repre-
> sented a "provisional" value (historicity of every conception of the world
> and of life), it is still very difficult to make people grasp "practically"
> that such an interpretation is valid also for the philosophy of praxis it-
> self, without in so doing shaking the convictions that are necessary for
> action. This is, moreover, a difficulty that recurs for every historicist
> philosophy; it is taken advantage of by cheap polemicists (particularly
> Catholics) in order to contrast within the same individual the "scientist"
> and the "demagogue", the philosopher and the man of action, and to de-
> duce that historicism leads necessarily to moral scepticism and depravity.
> From this difficulty arise many dramas of conscience in little men, and in
> great men the "Olympian" attitude à la Goethe. This is the reason why
> the proposition about the passage from the reign of necessity to that of
> freedom must be analysed and eloborated with subtlety and delicacy.

> Antonio Gramsci
> (see [96], pp. 406,)

To shed more light on the preceding thoughts, let us recall some words by
means of which Hegel claimed to illustrate the roots of the morality underlying
the *modus operandi* of demagogy when he said "First you have to define the
other as something different, - subhuman -, then there are no problems to kill
or enslave this inhuman object." See Hegel's "Mastery and Servitude" about
which we shall have more to say later.

Looking for alternative action, we will appeal to extend to international cooperation between the centre and periphery but in both directions. This means,

an international division of labour that does not entail cheap labour services and shifting of harmful production branches to poor countries, a free market that covers also raw materials as well as products actually produced by peripheral enterprises. International cooperation in an actual geographical sense and without greedy intermediaries and limitations regarding the actual choice of trading partners and objects of trade, on political or strategic grounds, be they white, black, Christian or whatever they may be. Democratic principles for all, positive freedom and peace everywhere but neither delivered by guns, coup d'état, death squads, multinational blackmailing and retaliation, nor based on negotiations with the international underground of big business nor engineered by intelligence agencies.

Blockades, wars and embargos will never contribute to meeting the pressing needs of mankind, even if they are supposedly run in the name of God, freedom and peace under the approval of holy institutions and various constitutional masks. After all, two thirds of the world's population are violently condemned to increasing poverty and starvation. What we expect of all of you is to refuse to let them die in silence: as a way of example let us just mention eleven millions of children dying every year according to WHO reports. At a first glance, everybody may agree although some may still hesitate. In this regard, we hasten to add, this is not a matter of humanity or generosity, but rather of justice.

From Adam Smith to John Maynard Keynes, to mention just two fathers of prevailing orthodoxy, many theorists, known as supporters of that state of affairs, have called attention to the transfer of wealth from the periphery to the centre of the capitalist system, as the most decisive ingredient of modern capitalism; by way of example, see [122], pp. 139–141. Furthermore let us recall at different periods labour surpluses appearing in the centre of the system have been drained off into the periphery, in this way giving millions of people opportunity to escape starvation; this facilitated their lands a way out of depression towards prosperity.

Today the flow of wealth toward the core of the capitalist world goes on through institutionalised channels on the grounds of international division of labour, unequal exchange concealed in the monetary system and the like. Thus, it is of outmost urgency to realise that the penetration of Eurocentric capital into more and more spheres of production in the periphery is tantamount to further annihilation of working and thus income opportunities there. It is necessary to be aware of the fact that, what may lead to progress in the centre has not necessarily to mean progress to the periphery, unless one values human beings here and there differently.

The events accompanying the end of the Cold War may remove the basis of anti-communist thinking, which dominates the centre and its major institutions, and/or reveal the true interest hidden by their missionary rhetoric. The spectre of communism and the worldwide challenge of bolshevism have offered so far plenty of excuses to the centre for denying any space for political autonomy and blocking any essential initiative towards an economically and financially vital development in state nations of the periphery. However, a glance at documents and reports issued by competent offices of state nations of the centre reveals that the real fear has been the eventual success of independent development on the basis of indigenous nationalist forces. Because the virus of independent nationalism might spread through the demonstration effect and lead in turn to the ultimate loss of control over energy reserves, strategic resources and raw materials.

Rhetoric aside, this perceived danger has brought about the most evident manifestation of political dualism: a sublime commitment to democratic principles and an inevitable alliance with dictatorships of fascist type. This basic structure of Eurocentrism exhibits, in the words of N. Chomsky, the childlike simplicity of a fairy tale: the virtuous champion of all that is good and the absolute evil – that by its very nature must seek total domination of the world – which has therefore to be overcome and eliminated. Hence, any nationalist movement that may in principle deprive the driving forces of Eurocentrism of energy resources, (strategic) raw materials, investment opportunities, markets and cheap labour serves by definition the diabolic designs and needs to be uprooted; this way of thinking coined the most usual sense of the term "bolshevik" and/or "communist" which came to embrace critical minds, social reformers, unionists, priests doubting about this devine will and the like, on the other hand Hitler, Mussolini and others appear in official records of the U.S. State Department among "the moderates" and "fascism as compatible with U.S. economic interests"; for precise references on these points and other issues on facts and fancy behind the Cold War see Chomsky's "Deterring Democracy" as well as Galeano's "Open Veins of Latin America".

> As example after example attests, economic nationalism elicits US hostility. Where possible, the culprit is assigned to the Bolshevik conspiracy to destroy Western civilization. In any event, he must be slain. It is as close to a historical law as a complex world allows.
>
> The essential point was captured in John F. Kennedy's celebrated remark that while we would prefer decent democratic regimes, if the choice is between a Trujillo and a Castro, we will choose the Trujillo. It is necessary only to add three points: (1) the concept of "a Castro" is very broad, extending to anyone who raises problems for the "rich men dwelling at peace within their habitations," who are to rule the world according to Churchill's aphorism, while enjoying the benefits of its human and material resources; (2) the chosen "Trujillo", however monstrous, will be a "moderate" as long as he fulfills his function' (3) the "Trujillo" will make a quick transition from favored friend to another beast to be destroyed if he shows the bad judgement of stepping on our toes. This story has been reenacted time and time again, until today. Saddam Hussein is only the most recent example.
>
> The post-World War I pattern does constitute a departure from US intervention in an earlier period of less self-consciousness and global power. There is every reason to expect that pattern to persist, with whatever adjustments are required, after the Bolshevik challenge has lost its last shreds of credibility.
>
> Noam Chomsky
> (see [39], pp. 45)

A major axiom underlying Eurocentric thinking sets natural resources under the "proper hands" since – as Churchill viewed it – satisfied nations wished nothing more for themselves than what they had. After all, they are committed to the integrity and vitality of free society, and to create and maintain worldwide prosperity.

The centre of the (world) capitalist system contributes instead actively – to say the least, with missionary eagerness – to establishing and maintaining dictatorships and other kinds of substitutes for direct colonial controls, which contrasts greatly with the almost daily renewed commitment to democratic principles of Eurocentric institutions.

It is striking how Eurocentric capital seeks to transform state nations of the periphery into its crisis outlets, in particular for surpluses of the military/industrial complex based on which the centre attempts to counteract the always recurring crises of profit realisation, demand deficiencies and unemployment. The élites of Eurocentrism seem to know that the poors of the world do not like to be poor and the dissatisfied masses are "vulnerable to subversion" so as to expect the basic patterns of resorting to violence will remain as means to bar any pretension of the periphery to independent development: first because the periphery needs to fulfil its role as source of raw materials and market as long as Eurocentrism does not have an alternative to it, second because the military/industrial complex is an essential Keynesian instrument of counteracting realisation crises, and third because human murder by poverty – to borrow Galeano's words – begins to preoccupate even minds serving the establishment.

The agencies of Eurocentric capital do not facilitate the periphery means of straightening out its twisted capital and establishing self-financing and self-propelling economic structures which would stimulate a building up of basic democratic institutions. What is worse, the agencies of Eurocentric capital literally force peripheral economies into so-called development projects which are ill-conceived and in no way conducive to unfolding the productive potentials of the periphery. In forty years of Bretton Woods institutions at the service of Eurocentric capital, they have succeeded in frustrating the efforts of poor nations to gain some control over their own economies and destroying their long-term development prospects; in this connection we refer to [161, 162, 185, 105, 125] and to the additional references given there.

It appears that the centre of the capitalist system cannot free itself from the necessity of keeping a sharp eye on its "vital space" and undertaking military operation against peripheral nations that venture a walk on their own. It remains to hope that the end of the Cold War brings a positive change in this sad picture of the world, so that human needs and dignity truly become essential issues of social discourse. Therefore the success of these essays and economic theorising will necessarily rest on the ability to delineate carefully the questions at issue and to visualise the state of knowledge of the matter on the basis of a faithful descriptive work and plausible schematisation. Because only after having formulated the economic situation seriously, there is a real chance to understand what the problems really are. Economic theory will not advance to the status of a science as long as the need of hiding facts of life prevails. Here too, like in astronomy and physics in the Medieval time, it is necessary to dispell obscurantism and ignorance before a decisive break becomes possible. Thus, we have to put our main issues and goals in a clear language and gradually dispel vagueness and obviousness from our picture of the socioeconomic reality so as to allow orthodox rhetoric as little room as possible. That means a careful, meaningful and truthful description of the matters at issue has to precede our system-theoretical investigation if we are to pursue an adequate and appropriate use of the various powerful instruments at hand.

A quick overview of primary goals

The chief goal of the present monograph is to provide a theoretical framework that makes possible the design and elaboration of conceptually simple socioeconomic policies. These are aimed at starting and at stimulating various processes capable of mobilising potential energies and bringing into operation broad linkage effects, which underlie objectives of development and institutional change in a peripheral economy.

The investigation into the transformation process of the above sort and the forces that may make it self-propelling are, both at an intermediate and final stage, at the heart of the present undertaking. It is important to observe that the goals to be described next are in a sense also instruments of development and they interact in a way that renders one and the other the cause and effect. Roughly, the primary goal can be stated as elucidating the main notions underlying (socioeconomic) processes conducive to

- Elimination of any form of chronic unemployment and underemployment by gradually shifting certain fractions of labour surplus into more highly productive activities on the basis of self-sustaining expansions of trade and productive capacity.

- Piecewise continuous expansion of the system productive capacity and its various components accompanied by increases in the productivity of physical and human capital in a way capable of becoming self-sustaining.

- Improvement of the capability to function and well-being of all members and segments of society, including in a broad sense working and socioeconomic environment, public health, educational and vocational services, nutritional and recreational aspects and the like.

- Gaining insight into and control over various processes associated with the disintegration of primary productive activities and other phenomena associated with industrialisation, e.g. system deterioration – either of an ecological or human kind – and social strains linked to labour migration, since the consequences of these have long-lasting effects on the development of society.

The unifying principle will become the one of social reproduction, so that human capital becomes an asset of society, the motor and beneficiary of progress, making it possible to give foregoing goals a sound theoretical basis able to support economic policy studies and empirical research from the point of view of political economy.

On the interdisciplinary goals

On the way to accomplishing our academic objective of advancing a *new economic policy* approach to matters of long range development planning, we shall combine classic, mainly Ricardian and Marxian views, neoclassic and post-Keynesian ingredients. However, to a great extent we shall follow the most recent guises of these schools of thought. Recent ideas on economic growth and development are also be taken into account and incorporated in our suggestion, providing a heuristic but theoretically sound setup. Based on the latter, one can address questions related to cultural, economic, sectorial and regional growth dissimilarities at the microscopic and social level.

Here the role played by mathematics will be mainly an instrumental one. Thus we shall say that the *mathematical objectives* will remain subordinated to that of economics. However, since economic modelling and various related issues represent a special challenge to mathematics, we expect this work will stimulate the communication and cooperation between economists and mathematicians and induce changes in the mathematical techniques themselves, and broaden the field of successful application of mathematics.

Without intending to discriminate against any field, let us mention our belief that the beauty and fascination of any theory can be better appreciated by showing its usefulness. The theory of martingales, control sciences and other branches of (applied) mathematics are in this regard prominent. Hence a further academic objective is to test the applicability and power of stochastic analysis in modelling and explaining decision processes under uncertainty and limited information, designing feedback and steering mechanisms assumed to occur in a system-theoretical approach to socioeconomic development.

Theory of systems will act as an interdisciplinary link between the economic development process itself and our highly simplified version of it. To begin with we shall model development as a diffusion process with jumps. The latter will be expected to describe the evolution of the most relevant qualitative and quantitative features of the actual system dynamics. Impulse control techniques and large deviations make possible the simultaneous consideration of the effects of systems interacting at different time-scales, as it happens in developmental phenomena. This facilitates the construction of a dynamic model of the economy, where human factors as well as human responses to changes in their environment and institutions actually play a decisive role. Through controlled impulses the economic system becomes able to switch to technologies and socioeconomic strategies not previously considered as suitable, a feature which makes planning a more flexible and realistic but still challenging task.

1.1.2   Science contribution to mastering social problems

It may appear natural that science will assist society in mastering the problems it faces. However in a social system, where needs or use-values not necessarily correspond to aims of (commodity) production or exchange-values this is not naturally the case. Because the dominant productive structures of society are the essential determinants of social agencies including science and individuals.

Therefore in order to ascertain the potential contribution of contemporary science to understanding, assessing and solving present-day social problems, we have to start examining how free and powerful a social system is with regard of generating, diffusing and applying the kind of knowledge the vital requirements of social reproduction necessitate.

The myth of plurality and power of science

At the heart of Eurocentric rhetoric is its innate legitimacy as guardians of free society, first as a natural outgrowth of Christianity and second as representant of true knowledge rooted in science, i.e. rationalism. Anything else is superstition, myth, subculture and subhuman. Radical liberals generously concede survivors of crushed cultures access into free society – as "free" and "equal" individuals – whenever they are willing to accept the ruling principles of Eurocentrism, meaning loosing their identity; in this regard we refer to P. Feyerabend's "Science in a Free Society". As a historical remark let us mention, when Renaissance Europeans headed across the ocean in 1492 they were representants of cultures largely irrelevant compared to that of China, India or Egypt. Their superiority was embodied in their destructive power and lack of scruple abusing this power, an enduring feature currently marking their global superiority. The over 60 million victims of World War II have become a visible consequence of this power and its abuse, just because it affected directly Eurocentrism.

At issue is the inconsistency of science in the servitude of power and science as rationalism, since it derives from a mechanistic model which rests on the postulate of separation of mind and body. Science and scientists are social artefacts and thus loaded with prejudices, moreover they both depend on prevailing political and economic institutions and are therefore susceptible to bias toward the establishment. The following *a priori* claims characterise the orthodox picture of the world capitalist system:

- Its pluralistic foundation and legitimacy as self-regulating market society without alternative. Akin to this presumption is the existence of a ritual framework of peacemeal criticisms and circumscribed margin of manoeuvre within which institutionalised oppositional forces may act.

- Its quality of being the democracy *per se* with voluntary and temporary eventual suspension of the relative autonomy of its decision-making units. This underlines the presumption that problems and contradictions of the established order are strictly *transient* phenomena.

- Its value commitment to neutrality and unchallengeable objectivity resulting from the advancement of scientific knowledge and the systematic application of the latter to production according to the objective requirements of its functioning.

- Its inherent adherence to a timeless state of harmony which no amount of concentration and centralisation of capital can radically alter. This follows from the presumption that capital by its very nature is constituted as the plurality of capitals.

Needless to say that such claims cannot reach very far, see [149]. For instance, they fail to recognise the effective linkage of pluralism to partial interests of competing capitals *vis-á-vis* those of the overwhelming majority of society. They fail to recognise that scientific knowledge and scientists are a result of the ongoing social process, too, and that the roles and functions they assumed correspond very much to powerful socioeconomic and political determinants.

After all the alleged autonomy of science, scientists cannot escape specific manifestations of an increasing alienation and division of labour which divorce science from its social reality. The resulting fragmentation and isolation of intellectual production hinder science and intellectuals as well, who are themselves "sovereign individuals", to obtain a comprehensive picture of society making it virtually powerless with regard to the pressing social challenges facing contemporary society. Needless to say that the overwhelming majority are either lacking appropriate knowledge and unbiased information on symptoms, causes and possible remedies to social phenomena or are in any realistic sense deprived of the power of decision making.

It is hard to understand a civilisation that proudly presents a technology the greatest achievement of which is its capacity of general annihilation and a production machinery whose financial and trading power fail to recognise any reasonable limit of satiety. An Eurocentric civilisation, which under the lead of a powerful minority does not hesitate to exclude the great majority of human beings from the circular flow of income underlying social reproduction. This Eurocentrism that even in the last decades continues supporting crime against humanity, bases its legitimacy in a highly sophisticated intellectual and ideological confrontation.

The fact that science accomodates itself to serve "vital needs", which are characteristic of a socioeconomic context emerging from Eurocentric interests akin to the ruling plurality of capitals, does not however change anything in the yearly starvation tragedy involving millions of human beings. The prevailing international order may be seen as a masterpiece of political, diplomatic and social engineering faithfully reflecting the unending financial, economic and political avidity of power of Eurocentric institutions. It legitimates appropriation of natural resources by a purposeful state fragmentation of former colonies, it institutionalises the transfer of economic surplus and other forms of income drains on the basis of Bretton Woods arrangements, it elevates the military mechanisms of a sham desintegrating colonialism to holy guardians of freedom and interventions, e.g. blockades, invasions, bombing and other forms of genocide against defenseless peoples, as crusades in favour of the free market, pluralism and democracy.

Daily life tells us how peoples in the periphery benefit from the established international order. Criticisms and constructive suggestions from orthodoxy are rare and apologies by contrast abound. This however is the natural consequence of eliminating social agents and awareness from the structural framework underlying the division of labour and relation of production. This avoids thinking of social relations as interaction between direct producers and shifts attention to relations between exchange-values or commodities. From this point on science or as Mészáros puts it " a veritable travesty of science" fragmented and divorced from social reality starts ignoring and disregarding the social implications and its far-reaching intervention in the process of social reproduction. Let us recall Mészáros' vivid picture of alienated science:

> Its methodology, making use almost *ad nauseam* of models, diagrams, formulas, twisted statistical 'evidence', 'mass-observations' and 'mass-intervies' (based on 'scientifically' devised – though in reality derisorily puny – 'representative samples'), etc., reflected a vital need and practical imperative of commodity society. Namely: to secure the reproduction of exchange value on a constantly enlarged scale by means of the wanton *manipulation* of the social processes in every sphere of activity, from material 'demands' generated by 'supply management' to grossly *influencing* public opinion while pretending to represent it objectively, and from cynically 'producing' artificial scarcities in a world of plenty to 'massaging' the facts in the service of eliciting the required ideological and political responses in a systematically miseducated public.
>
> The ideology of neo-positivistic scientism, which continued to idealize a science subservient to the reified technolgical requirements of the prevailing mode of production, was most appropriate to assume the leading role in this process of manipulation, in that it could promise to stamp with the lofty authority of science writ large even the most prosaic of manipulative practices. Indeed, the ideology of scientism – to be sure, not simply on its own, but largely thanks to its inherent linkages to the dominant productive practices – was so powerful that it successfully penetrated not only the citadels of knowledge but virtually all facets of everyday life as well. Its manifestations ranged 'from the sublime to the ridiculous', so long as they could be quantified or turned into models, formulas and 'paradigms'. For a rare example of the 'sublime', we can think of Max Weber's ingenious system of 'value-free ideal types'. As to the ridiculous and often grotesque abuse of science in the service of manipulation, the examples are legion: from the 'Mortuary Science' (read: lucrative funeral undertaking) and 'Apiary Science' (that is: bee-keeping) Departments of some American Universities to the *'technology of the unified field'*unified field!technology of the of the Maharishi Mahesh Yogi and his meditating followers, with their 'scientifically quantifiable' mumbo-jumbo about the *'square root* of the world's population'.

István Mészáros
(see [149], pp. 189–190)

Mészáros' picture is not encouraging and the option of reorienting science toward scientific practices at the service of human needs is definitely not optimistic, because constructive alternatives seem to be conceivable only if the system itself turns to an essentially more human society.

Financial subordination of research objectives

It is stimulating to know that one of the greatest scientists of our century, Norbert Wiener, had a deep preoccupation and distress about the danger and disastrous consequences of the increasing fragmentation of sciences. Wiener's reflections in this regard are convincingly documented in his path-breaking book "Cybernetics: or Control and Communication in the Animal and the Machine", see [229].

At a fairly general level one recognises as a central motive for launching a programme of the nature described in cybernetics, the need to achieve a common vocabulary, which enables scientists from different fields to interact and use conceptual frameworks and ideas better developed in other branches of scientific inquiry, with the aim of easing communication and increasing research efficiency. In this sense cybernetics may be seen as means of counteracting lack of perspective and narrow insight resulting from fragmentation and over-specialisation. In the introduction to his Cybernetics Wiener tells us "we had dreamed for years of an institution of independent scientists, working together in one of these backwoods of science, not as subordinates of some great executive officer, but joined by the desire, indeed by the spiritual necessity, to understand the region as a whole, and to lend one another the strength of that understanding".

The fact that a few lines below he adds "The deciding factor in this new step was the war" is however alarming, in particular knowing that in Chapter VIII, p. 160, he writes, although in another context, "A certain blend of wheedling, bribery, and intimidation will induce a young scientist to work on guided missiles or the atomic bomb" and categorically states somewhere else "Non-cooperation in military matters should be an essential moral principle for all true scientists". See in this regard *Einstein on Peace* and other references given in [149].

At this point we wish to mention that the tribulations and dispair of outstanding figures as Einstein, Oppenheimer and Wiener associated with their critical posture over the militarisation of sciences makes evident the extremely one-sided and illusory tenet of an autonomous development of science and technology which increasingly shape the dominant aspects of every day life. Viewing science as a creative and conscious agency of social emancipation ignores the most essential links of knowledge production with the structural framework underlying the prevailing social division of labour, first at the level of the individual as a relatively autonomous and isolated scientist whose functions are determined socially and instrumentally. Second, at the social level where science acts as a materially, politically and financially subordinate agency. In this sense science cannot scape the structurally determined orientation toward production of exchange-values or commodities in contrast to use-values or human needs. Alienated science is in fact deprived of the social determination of its research objectives.

Let us close these considerations stating that the material forces corresponding to the fundamental structural determinations of society produce the *men they need* in every walk of life, including science, as Mészáros asserts, is closer to what we can see in reality than to say "in the end men will get what they deserve", Einstein's unusual articulation of threat and resignation. See [149], pp. 208–213.

The military/industrial complex

The principle of the effective demand clarifies various misleading notions of conventional wisdom. Let us call attention to the belief of orthodoxy that any autonomous increase in the expenditure on non-consumable commodities can be justified on the basis of the commodity market it generates. In this sense, it becomes customary to deal with the problem of an insufficient demand by increasing the expenses on weaponry. The absurdity of this sort of policy-making measures is a natural outgrowth of divorcing production from the role of serving primarily human needs. Production becomes instead subordinated to the drive of generating an always increasing quantity of exchange value, regardless of its potential usefulness.

The alienation of activities formerly underpinning social reproduction by eliminating their function as means of satisfying human needs, brings about a theorising framework where the notion of work as task disappears and one gets rid of the idea of workers as direct producers. The capitalists become to be seen as producers, the act of purchasing acquires the role of consuming blurring the difference between purchaser and consumer. Thus purchasing or transfering commodities from an owner to the next, opens the possibility of acquiring commodities without consuming in reality anything at all and without being aimed at satisfying any human need. Allocating resources to purposes of purchasing means of destruction becomes also analogous to replenishing or regenerating the consumed capacity to produce which underlines the state of confusion of conventional wisdom, since waste and destruction of (productive) resources are put on the same footing as consumption and production of means of satisfying social needs.

The relevance of the aforementioned misrepresentations resulting from the ill-conceived therapy to demand deficiencies lies in the ability they offer to re-cast and legitimate the expansionary tendencies of the system and in the size and scope of profitable operations, which appear as inexhaustible outlets for the always recurring crises of overproduction and profit realisation. These reflections outline the economic background, the mode of functioning, the specific characteristics and structural determinations of the military/industrial complex which incorporates and represents the most essential socioeconomic, political and ideological manifestation of imperialism.

The fact that the military/industrial complex embraces vast areas of production linked in various ways to structural determinants of social reproduction, brings about an increasing subordination of all fields and branches of knowledge to its imperatives. In this sense one has to understand Wiener's warning on the dangers deriving from the progressive fragmentation of science and the resulting lack of comprehensive vision, an almost natural outgrowth of this fragmentation, as well as Einstein/Oppenheimer/Wiener's struggle against militarism and its impact on science. We refer to [149] for a detailed treatment of this issue and additional references.

1.1.3   Mathematics as means of capturing social reality

The intention of applying mathematical techniques as a way of improving our understanding of social reality leads naturally to viewing society as a dynamic system consisting of interacting agents. One characteristic feature of such a dynamic system is its ability to change its movements, to pass into various states under the effect of the interactions of its agents.

A basic contribution of mathematics to the acquisition of knowledge on this system, may consist in the description of communication, information, observation and organisation processes associated with the system's ability to change and the mechanisms of change themselves. Since assessing the impact of the acquired knowledge needs also to belong to this domain of mathematical activities, we will shortly recall some of Wiener's reflections on human labour which are still of great relevance for the future development of mankind.

Human use of human beings

Human labour constitutes doubtless the most decisive social agency of creativity, development and progress. It has been human labour that over the centuries has taught men to develop the riches and forces of nature to the aims and interests of mankind. No matter how large the achievements of automation, control, communication, computing and information among others may be, human labour will remain the decisive link among machines, man, nature. Along the lines suggested by N. Wiener, cybernetics has the aim to investigate the interaction between man and machine following two major objectives:

(1) to explore constructive developments and the perspective of cooperation between man and machine in labour processes taking into account the ability to operate at different speeds;

(2) to study acquisition and obsolence of skills resulting from an always changing configuration of functions assigned to the man and the machine in processes of labour of varying complexity.

Wiener interprets the first industrial revolution as the devaluation of the human arm by the competition of machinery. The intervention of the market reduces the achievements of mechanisation to the opportunity of saving money and of getting rid of unskilled labour. Wiener says, "There is no rate of pay a United States pick-and-shovel labourer can live which is low enough to compete with the work of a steam shovel as an excavator". And he adds, "The modern industrial revolution is similarly bound to devalue the human brain, at least in its simpler and more routine decisions. Of course, just as the skilled carpenter, the skilled mechanic, the skilled dressmaker have in some degree survived the first industrial revolution, so the skilled scientist and the skilled administrator may survive the second"; see [229], pp. 27–28.

In this connection it is important to recognise that no moral problems derive directly from automation and technological progress as such but only from a practical implementation and use which ignore the short- and long-term implications in any particular type of society. It is precisely at this level when a system-theoretical investigation of the issues under consideration should become compelling with the aim of putting together fragmented knowledge as advanced by fragmented science.

However, it is necessary to recall that particular questions of nature and society arise only in particular societies, since the sort of science a society develops is akin to the sort of society it represents. Therefore, in spite of the vitality and urgency of the questions posed by scientists of the stature of Einstein and Wiener, they remain unfashionable and outside of the conventional domain of scientific inquiry. Their isolated actions could obviously not reverse the ongoing process but just remind us of the social origin and responsibility of science.

Mathematical models as veils for lack of relevance

The contribution of mathematics to the rise of contemporary science and technology is inseparable from the dangerous trend associated with the prevailing determinants of social production. Mathematical sciences have in turn obtained great benefits materially and intellectually from impulses coming about as natural consequences of the spectacular development of productive potentials of society. In this regard we shall stress the fact (applied) mathematics have entered virtually all branches of scientific inquiry. Let us recall in this sense P. Lax's considerations on the flowering of applied mathematics in USA:

> For a few decades, the late 1930s through the early 1950s, the predominant view in American mathematical circles was the same as Bourbaki's: mathematics is an autonomous abstract subject, with no need of any input from the real world, with its own criteria of depth and beauty, and with an internal compass for guiding further growth. Applications come later by accident; mathematical ideas filter down to the sciences and engineering.
>
> Most of the creators of modern mathematics—certainly Gauss, Riemann, Poincaré, Hilbert, Hadamard, Birkhoff, Weyl, Wiener and von Neumann—would have regarded this view as utterly wrongheaded. Today we can safely say that the tide of purity has turned; most mathematicians are keenly aware that mathematics does not trickle down to the applications, but that mathematics and the sciences, mainly but by no means only physics, are equal partners, feeding ideas, concepts, problems, and solutions to each other. Whereas in the not so distant past a mathematician asserting "applied mathematics is bad mathematics" or "the best applied mathematics is pure mathematics" could count on a measure of assent and applause, today a person making such statements would be regarded as ignorant.

Peter Lax
(see [134], pp. 533–541)

Unfortunately as a result of the fragmentation underlining the division of labour in the production of knowledge, mathematics ended up discriminating between pure and applied fields. The applications of mathematical techniques to the investigation of social problems has also a long tradition and has experienced immense technical progress in the last decades. However, to a major extent the contributions of mathematics have not been integrated into the realm of social theorising owing to their obstinated tendency of ignoring rather than clarifying essential issues around which debates revolve, see [197]. It is not easy to discern what hinders a serious scientist, working in a mathematical description of certain social problems regarding human behaviour, to take into account available knowledge and crucial results from cognitive sciences. In particular when the descriptions of human beings focus on fallacies typically attached to orthodox views of rational behaviour.

A great deal of research and debate centre upon such questions as how to compute the numerical magnitude of the (illegitimately assumed) uniform rate of profit that characterises theoretically a certain state of equilibrium and what may be its real source. But a state of equilibrium with a uniform rate of profit is known not to exist in practice and to be a theoretical impossibility, see [66]. Needless to say, the study of social problems on the basis of mathematical techniques will contribute to clarifying and understanding the complexity and intricacy of socioeconomic events, provided the scientific framework on which these techniques will be applied has been freed of fundamental theoretical misconceptions.

There are great differences between simplifying and fallacious assumptions, i.e. the latter negate certain vital, essential and fundamental aspects of a system. Applying mathematical techniques to elucidating social events represented in models devoid of actual relevance, as a consequence of illegitimate assumptions, reveals a pretension of exact knowledge, diverts attention from the real issues and is therefore highly misleading.

The poor success of most known undertakings of mathematics in social sciences and the lack of relevance of their findings can easily be attributed to the fact they have not developed based on honest, faithful and knowledgeable pictures of the social object under investigation. Thus it hardly surprises the application of mathematical techniques to the economic problem of general equilibrium could not achieve additional knowledge of economic relevance, when one realises it started from an account of society characterised by fully unrealistic assumptions of egalitarian nature, utility notions unable to distinguish between individuals according to status, wealth, income, education and the like. A recent study on the fundamental theoretical underpinning of orthodox mathematical economics comes to the following conclusion:

> We feel it is possible to conclude that the mathematical analysis of general economic equilibrium theory in the context of the classical hypotheses as codified by Debreu's axiomatization has led to one clear result. There is a contradiction between the theory's aims and the consequences derived from the system of hypotheses constituting its structure, and hence a contradiction between aims and hypotheses withing the theory. The only way out of this situation is to jettison explicitly the programmatic central core that has so carefully been preserved throughout the many paradigmatic shifts. ... It appears more than questionable to persist in partial adjustments: the merit of Debreu-type axiomatic approach is precisely that of having made possible a complete and exhaustive analysis of all the implications of the classical approach. ...

We should also like to underline Debreu's effective reference to Bacon when he says that "citius emergit veritas ex errore quam ex confusione". It would be a mistake to lower the level of analysis and clarification. The only way possible is a *thorough* reexamination of the theory's basic hypotheses, i.e. a true paradigmatic revolution.

Bruno Ingrao and Giorgio Israel
(see [113] pp. 361–362)

Besides, the information and rationality pattern according to which, in the neoclassical framework, the agents of society are supposed to act is at best a more-or-less coherent body of fictions concerning social reality, lacks any scientific rigour and is therefore open to ideological clashes. One reason may be, for instance, the questions of legitimacy of profits and the definition of individual satisfaction on which the concept of economic rationality relies cannot be separated from the choice of objectives and the determination of ends on the one hand and means of measuring them on the other, see [84]. By the same token, the failure of economic orthodoxy to forward a consistent theory of employment, inflation and crisis shall not surprise anybody.

In this sense, these essays shall contribute to disentangling ignorance and disinformation, to uncovering racism and obscurantism, to clarifying the links between ideology and social reality, with the purpose of providing a truly descriptive vision of society that sheds light on the actual interdependence and interplay configuration in the prevailing (world) capitalist system in a way amenable to mathematical techniques of analysis.

To close these preliminary notes we shall say, our object of study and ultimate aims are fairly modest compared with the wealth of questions that are awaiting deep, proper and honest investigation. Aware of this fact, we shall start in a fairly heuristic manner and at a purely descriptive level, and attempt to move without losing socioeconomic relevance to a more mathematical treatment. Thus, at various stages of our work we shall just make assertions, the validity of which may rely merely upon empirical knowledge. The process of mathematisation is not all obvious, whenever one insists on keeping ties to the real social phenomena under study and avoids abstracting completely from the multiple manifestations and complex structure of mankind's struggle for survival within the perspective of social reproduction.

Thus, at the heart of investigation we have various crucial aspects of development in an economy of state nations in the periphery and then the process of mathematisation of them. In this sense the present inquiry breaks, in the words of Paul Baran, "with the time-honored tradition of academic economics of sacrificing the relevance of the subject matter to the elegance of analytical method", since as he argues, "it is better to deal imperfectly with what is important than to attain virtuoso skill in the treatment of what does not matter". See [11].

1.1.4   Characteristic features of the present work

At this point we should say a few words about certain distinguishing aspects of the present work and illustrate how they build an actual contribution, explaining thereby why we believe our essays are needed. To this end we will make some comments on the existing literature which may also serve to place our studies in this broad spectrum. However, we shall begin presenting a quick overview of our general perspective.

This work is concerned with building mathematical pictures (models) in a way conducive to development of skills helpful to understand, interpret and critically assess certain segments of the world in which we live. At the centre of our enquiry is humankind, as the driving force of society and interacting with nature. That means, we shall be concerned with the process of social reproduction and with labour processes as the main manifestation of human creativity. This places human capacity to function at the centre of investigation which needs to be coupled with the capacity of society, as a system of interacting individuals, to recreate the capacity to function of its members. Capacity to function as worker, capitalist, manager, artist, mother and the like is a notion akin to energy, e.g. an available potential, so that social reproduction becomes a reflection of the principle of maximum entropy which reminds us of the need of living also in harmony with nature. This is the general perspective of our view of an economic system, at any stage of development, embedded in the world capitalist system.

State of the art and the question of method

In order to understand a dynamic system, it is always necessary to understand the inner structure of the system, its environment, the interaction between environment and system, and the corresponding dynamics. However, to initiate an authentic process of knowledge and learning, one necessarily has to distinguish notions and meaning of objects forming a theory (language) and the loaded interpretation of analogous notions and objects (metalanguage) as existing in daily life or in other theories.

Like many workers in the field, we hesitate very much to use established labels like "theory of development," "least-developed countries," "development aids." Instead, we feel it is more appropriate to refer to more plastic notions of the sort "development of *under*development," "periphery" and "centre" of the capitalist system, expansion of the "market for inputs and outputs of Eurocentric production". However, we shall stick to the conventional labels and reveal, in the course of this work, the real meaning and actual practices behind the prevailing state of affairs, its brilliantly erected façade and cunning institutionalisation of its *modus operandi*.

There are numerous theories of economic development available. Some of them are mainly concerned with least developed countries, meaning perhaps countries most in need of development. T. Szentes presents in [215] a thorough and critical review of received theories to which we refer. Quite many studies on economic development look at it from a *historical perspective* and devote great attention to issues related to the genesis of the present state of *under*developed countries. This is unquestionably the first step towards the understanding of development problems and the design of economic policies capable to bring about a deliberate and purposeful change. See [51, 3, 117, 118, 10, 13].

However, most of these works are of an apologetical nature and transpire racism of varying intensity, quite often dressed-up ingeniously and in a great variety of guises. Due to the fact that the main body of orthodox wisdom regarding development owes its existence also to the successive introduction of *ad hoc* hypotheses by means of which one attempted to react to criticisms and counter-instances, this theory becomes unable to distinguish between material causation and effects of development. As a consequence, it greatly confuses the symptoms and the roots of *under*development leading to the neglect of the role played by society and its organisational structures in the way nature affects human beings and vice versa. Therefore, as one may expect, observational consequences deduced from conventional development economics in conjunction with certain specified initial conditions deliver no evidence of any historical validity. See [215].

An objective investigation into developmental aspects of social systems requires instead focusing on internal and external features of processes underpinning the most essential features of social reproduction, e.g. the forces behind the main struggle of mankind in and between interacting modes of production and consequences resulting from interplay of social structures at different levels of development. The notions involved are for instance: necessary means of subsistence, physical organisation of human beings, material conditions of production, creation of new needs as society unfolds and expansion of human needs as human population increases. See [157, 156, 193, 194, 79, 119, 77].

Production is a process based on which any society attempts to fulfil the social needs attached to its stage of development. Changes in production come to bear on the market and vice versa as the latter may induce in the former a deeper division of labour. The interplay of the market and production brings about thus a transformation of both, but also of the structures of distribution which are directly responsible for qualitative changes in society in a broad sense. Because distribution depends on the nature of social relations of productive forces and these in turn depend on the development degree of production and on the market. This feature is of crucial relevance, for it turns out that, in various ways, the same causes that give rise to progress in society put a check on it, forcing the capitalist system and its components to ensure its own equilibrium by generating disequilibrium somewhere else. See [84, 137, 138].

Therefore, any honest investigation on matters of development should concentrate first on the roots of capital accumulation as the driving force of the process of social reproduction, second on the multiple consequences for mankind deriving from it, and look finally for alternative ways of humanising the International Economic Order and all the Agencies representing Eurocentric interests around the world. Since principles of mutual cooperation, self-determination, defense of common interests and bargaining power among others are highly praised by orthodox rhetoric, it is very strange that state nations of the periphery have no realistic chance of decisively influencing policies that emanate from the international agencies of the free world – like the World Bank, IMF or whatever –, even if they all agreed to vote together on a certain issue, say economic and financial policies, which impact upon their destinies.

Several authors forward comprehensive, systematic and quantitative studies on the *present state* and *common economic policy practices* of developing countries. At any rate, from the perspective of the aims, we see in this an unavoidable second step and want to mention in this regard [219, 220, 63, 108] among the prominent. In [97] Griffin presents a detailed analysis of the most often adopted strategies of development and traces the experiences of various countries applying those strategies; [232] is a valuable reference from the neoclassical point of view.

Let us next make a few comments on the question of method. From the foregoing considerations it becomes evident that there is a big difference between giving an explanation for a process and providing a description which usually limits itself to an account of a state or sequence of states of the process at stake. However, this difference may tend to vanish whenever a description is of a causal nature. This observation seems to suggest the need to investigate and to spell out various implications deriving from a separation of the concept of human practice into processes of material and ideal causality, since by identifying these two processes one becomes aware of the role theories play in the construction of knowledge as well as in the observation and interpretation of facts. Mankind's capacity of significantly conscious and purposive actions is linked to an important degree to its ability to recognise real causal relations in its outside world. Further, the intimate connection of action with the struggle for existence stresses heavily the need for *freedom of action* rather than freedom of will. We shall later come back briefly to these issues.

It should not surprise anybody that development economics inherited to a great extent the state of confusion prevailing in neoclassical orthodoxy and the lack of relevance associated with models, and questions resulting from its fictitious picture of social reality. Owing to the fact that in backward economic systems various modes of production coexist and by means of the international division of labour interactions with more advanced economic systems come into existence, the consequences of this state of confusion and the wrongly-obtained body of conventional wisdom have dramatically affected human beings.

At the latest, after these sad facts of life became the regular outcomes, themes like unemployment, effective demand, terms of trade at a domestic and international level, foreign exchange, balance of payments and debt traps should have shifted to the focus of investigation. However, here the centre of attention is misleadingly full of distortions and ideological bias. As customary in the prevailing orthodoxy the economic aspects of human life have been detached from its political and social background, anthropological and historical issues have been fully ignored; see [183].

Of course, we have to grant everybody the right to adhere to the school of thought and to believe in what she or he pleases, but as Hodgson [110] puts it, "there is no such a right to carry millions into the mire of poverty and destitution as a consequence of ideas which are formally 'interesting' but practically disastrous when applied to policy. The state of economic theory today would be farcial if it were not so tragic."

The course of time forces us to think of the world of past as characterised by experience, knowledge, determinism and necessity on the one side and the world of the future as that of freedom and action. This reminds us in some sense of the Kantian dualism of ethical freedom and natural necessity, counterposing causality and finalism, a dualism which reappears in the form of an opposition between factual and value judgement. To close these remarks, let us point out, following Colletti [41], that "value judgements are inevitably present in scientific research itself, but as judgements whose ultimate significance depends on the degree to which they stand up to historical-practical verification or experiment, and hence on their capacity to be converted ultimately into factual judgements." It is precisely in this sense that the link between knowledge and transformation of the external world which in turn links teleology and causality, starting in a sense the differentiation of science and politics, becomes to be known in social sciences as *normative*.

In this regard, it is necessary to keep in mind that many studies on development have been requested by those international institutions that after the Second World War came to act as substitutes for the open colonial system. The so-called development agencies are in many ways the engineers of the prevailing practices of development and in this sense responsible for the prevailing state of affairs. These agencies not only ignore the fact that what furthers the market expansion of the multinational concerns they represent does not necessarily bring about progress and development in the periphery, but actively contribute by means of embargoes and market burdens to blocking and inhibiting the efforts of countries that abandon the way envisaged by the agents of Eurocentric capital. Let us mention also that [141, 120] provide fairly interesting insights into various development questions using Marxist analysis.

Foregoing reflections pave the way for a third step which attracts our attention, i.e. the *normative* and *planning point of view*. This is an outcome of the recognition that the existence of an external world composed of individuals, which are actually overdetermined by their mental and physical activities, and the material conditions determining their existence, give rise to the need of acquiring additional knowledge of that external reality. This need of knowing closer an external reality existing independent of the knowing subject obeys the aim of changing it, in accordance with material conditions one may expect or desire to prevail, see [52]. The main ingredients are the process of material causation, by way of which external nature acts and affects human beings, and the teleological process, final causality, including the action and effect of human beings upon nature.

It is important to realise that the process of material causation always mediates between nature and society, and varies with the form and level of social organisation. While society acts through labour as an externalisation of ideas, ideal motives, plans and designs according to which nature is transformed to meet human needs and aims. It is worth mentioning that an issue requiring a great deal of subtlety and delicacy is the one related to the transition from the domain of necessity to that of freedom which takes place through the society of men and not through nature, see [96], pp. 403–416.

Our approach owes a lot to the works of Marglin, Prebisch, Zarembka,

Chakravarty and Sachs as will be apparent in the course of these essays, see [145, 168, 232, 36, 191]. Likewise the presence of ideas advanced by Robinson, Kalecki, Pasinetti, Steindl and many other scholars will become evident at different stages. For the time being, we like to mention [183, 116, 160, 210].

In regard to various fairly technical aspects [145, 146, 106, 221, 201, 34, 35, 25, 151, 152, 158, 11, 36, 9, 166] are among the most influential. However, our approach is basically different and new in spite of the fact that we make some use of the mentioned works.

Some distinguishing features of our approach

Let us point out a few distinguishing features which identify and characterise our approach, when they are considered together, i.e. it is the interplay and combination of these features which set this approach apart from others.

1. The economic system at stake is seen and considered as a set of learning and interacting individuals and components, among which certain definite and more-or-less stable relations exist. These relations and the socioeconomic environment of the system as well undergo qualitative and quantitative variations in relatively differing (time) scales, which give in turn rise to various responses of individuals.

2. Based on facts of everyday life we shall share the point of view, market mechanisms are unable to solve satisfactorily basic developmental problems of socioeconomic and cultural nature. Thus the government or control board has to play an active role in promoting development and growth, as well as settling various types of social conflicts.

3. Foregoing features make necessary to consider simultaneously spatial and temporal differentials distinguishing phenomena associated with the development process, the microscopic background and macroscopic behaviour of the system. Further, it requires to move into the forefront of inquiry interactions and interplay of various kinds arising in the system.

4. Application of control techniques is aimed at capturing gradual and adaptive features associated with policy changes. In this way we model system efforts to overcome emerging financial, socioeconomic, cultural and political difficulties, i.e. insufficient initial capital-endowment, resistance against new technologies, inefficiency of a big push and a continuous monitoring of systems.

5. Our choice favours approximation by means of a diffusion setting with jumps. This offers an efficient and suitable framework for examining phenomena reflecting a world where responses are brought about as the result of combined actions called forth by a constellation of only partially known and to a great extent only indirectly observable decisions, e.g. signals of interacting individuals and institutions, which often pursue conflicting goals.

A few remarks are appropriate. The more dynamic becomes a model and the closer it approaches reality, the greater the need of using stochastic tools. The foregoing features are aimed at establishing essential links between economic and noneconomic aspects of the system, as well as bringing into play essential elements of microscopic and macroscopic behaviour. Since our modelling endeavour attempts to capture interactions as well as the interplay of system elements and environment, we let them act through a system of social preferences including rewards and value-judgements; however our social preferences and value-judgements rest on a well-defined system of characteristics, i.e. capability to function. This sort of interaction mechanisms represents on the one hand society's incentives to action and motives of individuals, and on the other describes changes induced by the latter in the socioeconomic environment and in individuals themselves.

The intricate features characterising this mechanism necessitate the introduction of performance-functionals that account for subjective and objective aspects of social and individual perception of rewards. This mechanism has to be able to explain behaviour differentials resting upon various aspects of the otherwise theoretically obscure notions of survival and profit motivation. The notions behind our performance functional allow hence a less ambiguous treatment of rationality and may capture various traits of human behaviour based on culture and tradition, paving the way for a more suitable and meaningful application of mathematics to the proxy of "maximum satisfaction".

The aforementioned interventions are conceived as a vital ingredient structurally determined by setting a framework, within which capital – both human and physical – may flourish consistent with coherently designed links of cooperation between the modes of production that reflect the most legitimate interest of interacting elements. These interventions are intended to be the result of institutionally guided negotiations among capitalists and workers; they are by design likely to remain limited in time and magnitude and have to be exercised mostly through the system of social (priorities and) preferences. Finally, these interventions depend to a certain degree upon an *a priori* unknown set of goal-tailored expansions of the productive capacity.

Further, let us call attention to the fact that, for the sake of simplicity, we need to abstract from the smoothness of certain changes. However, precisely this discontinuity of changes at a macroscopic level lets interactions move explicitly to the forefront of our analysis and enables us to construct a dynamic and operational model, where various types of human actions and their integrating processes are analytically described. By way of example, we shall mention learning processes, decisions aimed at fulfilling social and individual goals, and other features where sometimes conflicting interests come into play.

An additional feature of the system evolution emanating from the assumption of jumps is the linearity the drift and diffusion coefficients exhibit between successive jumps, since this provides a great degree of operationality and dynamics in the short- and long-run. Nevertheless, nonlinearity reemerges as soon as the impulses enter our picture of the system and structural changes materialise, which is a characteristic feature of the long-run. It is in this way that the diffusion process with jumps which represents the dynamics driving the (economic) system, become capable of switching along its (development) paths to technologies and strategies which were not previously ranged as feasible. This peculiarity of the dynamics renders long range analysis and planning flexible and realistic.

The drawback is that changes associated with the underlying dynamics and structural jumps pose various problems regarding adaptive behaviour, which rely on observations, measurements and information of a highly stochastic nature. In this connection it seems reasonable to approximate in the short-run the dynamics of changes in the system by the sum of two displacements: the first one as due to the macroscopic drift of the surrounding socioeconomic environment while the second as a random displacement caused by unpredictable system fluctuations and shocks. One may think of these changes as generating among other things, a sort of chaotic behaviour and confusions of an *global-relative* as well as a *permanent-transitory* character, which are typically attached to intertemporal decisions in situations marked with incomplete and asymmetric information.

The confusions just referred to are due to the fact that, in intertemporal decisional processes, individual agents in the face of partial and asymmetric information are not always able to distinguish, to what extent system changes are of a permanent or transitory nature. Neither do they distinguish precisely, in what respect global changes of the system bring about changes in their own immediate socioeconomic environment. These facts may render obsolete some human responses and actions, in spite of the fact they may have been appropriately conceived in time and/or space making thereby human behaviour inconsistent or even contradictory.

As a way of example, we shall mention inherent difficulties marking an appropriate timing of trading and industrialisation phases, which may be stimulated or frustrated by the accompanying timing in the expansion of transportation and means of production. The fact that responses and actions eventually become out of date is mainly due to switchings of policy, structural changes, impulses etc. and capacity expansions on the one side and on the other due to relatively inefficient information mechanisms. Information delays are a consequence of different time-scales at which these mechanisms operate; such delays may bring about uncertainty and a shaking of confidence in expectations about the future. Needless to say these circumstances put a check on the operationality of the market forces.

As soon as the salient characteristics of the long-run become the objects of our analysis, the diffusion model aforementioned is no longer appropriate. In order to accomodate the system dynamics to changes resulting from structural jumps, innovations and other processes driven by a relatively high-number of sudden impulses of a random nature, one needs to resort to more general processes. The theory of semimartingales offers adequate models and instruments of analysis. With the help of Lévy processes one comes to add a third displacement to our model dynamics represented by means of a martingale measure, since we shall deal indeed with diffusion processes with jumps accounting thereby for various types of discontinuity. However, this latter point will be the subject matter of a subsequent monograph dedicated to phenomena featured by discontinuity; see [94].

Moreover, uncertainty also arises in our model due to the fact that some variables are latent and hence only indirectly observable by means of a set of not precisely definable associated factors, e.g. structural changes and various factors related to learning.

1.2 Bringing in the language of mathematics

Let us next present some preliminary ideas on the manner of translating some of the socioeconomic reflections outlined above into the language of mathematics, which offer us an opportunity to be more precise about what will be developed in the essays presented here.

At the outset we shall stress the marked interdisciplinarity characterising various projects of quantification of moral, social and economic behaviour as witnessed by labels like Physiocrats, equilibrium of social forces, social mathematics, political arithmetic, social physics among others linked with the work of Quesnay, Montesquieu, D'Alembert, Lavoisier, Lagrange, Quetelet and many others. These attempts were always accompanied by an awareness of the difficulties inherent in the applications of physicomathematical knowledge to social phenomena. At this point let us mention, the approach of characteristics we use to describe social preferences and value-judgements in the form of capabilities to function goes back at least to Lavoisier and Lagrange.

It was only almost two hundred years later, as the Newtonian picture of the world was getting outdated and the fascist model rised as the most natural reaction of rich and middle classes against dissatisfied masses, when a picture of a national economy devoided of any essential social content – as reflection of the Cartesian separation of mind and body – emerged as an economic paradigm with a scientific format on the basis of its (alleged) analogy with Newton's model of celestial mechanics.

The main value of this model is rather of a rhetoric nature: it gives the illusion of freedom, harmony and justice an apparently scientific flavour attainable at the equilibrium state – lying at infinity – which the economy will reach in due time provided economic agents behave well and in no way disturb this inexorable process.

That this picture is not consistent with the physicomathematical model it supposedly imitates can be seen for instance in [150], concerning its mathematical flaws see [113] and regarding misconceptions of an economic theoretical nature there is a wealth of references going from Keynes, Joan Robinson, Sraffa, Pasinetti, Hollis and Nell, Eichner, Hodgson and many others. In the words of von Neumann and Morgenstern, there is thus no fundamental reason why mathematics should not be used in economics. Therefore, the reason why mathematics has not been more successful in economics shall be found elsewhere.

The usual point of entry of mathematics is mathematical modelling. This is a task accomplished in the lines of the theory of systems which we present briefly in Chapter 7. The following citation reproduces clearly the instrumental role mathematics is intended to play:

The fact is that symbolism is useful because it makes things difficult. What we wish to know is what can be proved from what. Now, in the beginnings, everything is self-evident; and it is very hard to see whether one self-evident proposition follows from another or not. Obviousness is always the enemy of correctness. Hence we invent some new and difficult symbolism, in which nothing seems obvious. Then we set up certain rules for operation on the symbols, and the whole thing becomes mechanical.

Bertrand Russell
(Mathematics and Metaphysicians)

Of course, we do not indeed expect to have things becoming mechanical. The magic of mathematical models rests on the insight they provide and on their improvement potentials. The use of mathematics tends to suggest a sort of causality in system-theoretical representations of objects one subjects to investigation by quantitative methods. This becomes more apparent when one deals with systems governed by differential operators. However, some methodologies in social sciences, we need to stress, reject causality since this seems to presume a causal ordering which may entail reducing a set of social events or aspects of society to a prima causa, or viewing some particular aspects, say scarcity and preferences, as causes which are not also effects. This is seen to contradict complex interactions and interdependence social systems typically exhibit.

The Markov property associated with a great deal of stochastic systems is in this sense not without problems, since it places future and present on the same footing. The nonanticipation principle associated with martingale systems is in this sense less problematic and allows time to appear in a natural way distinguishing future from present. However, a word of caution is in due course to avoid overinterpreting results derived from mathematical models or extending conclusions to a framework incompatible with the relatively partial pictures (mathematical) models provide.

1.2.1   The framework of stochastic analysis

Let $B = (B_t, \mathcal{F}_t)_{t \geq 0}$ be an m-dimensional Brownian motion on the probability space $(\Omega, \mathcal{F}, \mathbf{P})$. We will consider dynamic systems represented by the state process $X = (X_t)_{t \geq 0}$ governed by a stochastic differential equation (SDE) of the type

$$dX_t^x = \sigma(t, X_t^x)dB_t + b(t, X_t^x)dt, \qquad t > s, \qquad (1.1)$$
$$X_s^x = x \in \mathbb{R}^n,$$

endowed with a filtration $(\mathcal{F})_{t \geq 0}$ satisfying the usual properties. Unless stated otherwise, the time variable s will denote in the sequel any instant of time at which the process X may start, with $s \in [0, T]$. Under adequate assumptions (1.1) defines a Markov process (X_t, P_x) for each $x \in \mathbb{R}^n$. More precisely, the process $X = (X_t)_{t \geq s}$ is a Markov process with continuous sample paths and with a differential generator $\mathbf{A}$ given for $f \in C_b^2(\mathbb{R}^n \to \mathbb{R})$ by the expression

$$\mathbf{A}f(x) = \frac{1}{2}\sum_{i,j=1}^{n} a^{(ij)}(t, x)\frac{\partial^2 f}{\partial x^{(i)}\partial x^{(j)}} + \sum_{i=1}^{n} b^{(i)}(t, x)\frac{\partial f}{\partial x^{(i)}}, \qquad (1.2)$$

where the diffusion coefficient σ is taken as the square root of the matrix a, i.e. the relationship

$$a(t, x) = \sigma(t, s)\ \sigma^*(t, x)$$

is assumed to hold with σ^* indicating the transpose of σ.

On the assessment of performance

The Itô formula gives a direct connection between the operator $\mathbf{A}$ and the process X. Under appropriate conditions on $u \in C^{1,2}(\mathbb{R}_+ \times \mathbb{R}^n \to \mathbb{R}_-)$, we obtain Dynkin formula

$$\mathbf{E}_{s,x}\left[u(t, X_t)\right] = u(s, x) + \mathbf{E}_{s,x}\left[\int_s^t \left(\frac{\partial u}{\partial r} + \mathbf{A}u\right)(r, Xr)dr\right], \qquad (1.3)$$

which provides a vital connection between the Markov process X and certain partial differential equations (PDE). Further, it opens various alternatives to the study of functionals of the trajectory associated with X. In particular, we will exploit this relationship in matters related to model-building and various control theoretical considerations of (socioeconomic) systems. Let us mention the following instances:

- In qualitative and quantitative investigation into the evolutionary behaviour of various (socioeconomic) systems. Specially, when one attempts to include essential economic elements like human and physical capital, labour migration, as well as other notions which are not directly observable as learning, motivation and impatience.

- Applying and developing computational methods with the purpose of evaluating the average of certain functionals associated with the process X at a macroscopic level due, however, to the microscopic behaviour of agents of the system and to some features of interaction.

- Looking for a reference by means of which to compare the effect of alternative control strategies upon the performance of the associated diffusion with the purpose of elucidating the impact deriving from certain differentials of economic nature.

To see how we intend to proceed along these lines, let us consider several typical PDE problems related to the operator $\mathbf{A}$ and the stochastic representation of their solutions.

(a) For example, let us consider the solution of the *Cauchy problem* defined by the relationships

$$\begin{aligned}
\frac{\partial u(t, x)}{\partial t} &= \mathbf{A}u + c(t, x)u, \quad t > s, \quad x \in \mathbb{R}^n, \qquad (1.4)\\
u(s, x) &= g(x),
\end{aligned}$$

where $c(t, x)$ and $g(x)$ are assumed to be bounded and continuous in their domains of definition, which is given by the celebrated *Feynman-Kac* formula

$$u(s, x) = \mathbf{E}_{s,x} \left\{ g(X_t) \exp \left[\int_s^t c(r, X_r) dr \right] \right\}. \qquad (1.5)$$

One easily translates the physical content of eqs. (1.4) and (1.5) into economics. Let $g(X_t)$ represent the productive capacity of a plant, or the utility deriving from an economic undertaking, at time t. Further, let $c(t, X_t)$ describe the rate of depreciation, discount factor or time preference at time t. Then, $u(s, x)$ in eq. (1.5) measures the expected productive capacity of the plant at time t, or alternatively the expected utility, as one evaluates it at time s.

About the associated dynamics, let us say that eq. (1.4) reflects changes per unit of time in the productive capacity or in the utility flow, first due to the average time-change rate of the production capacity, i.e. $\mathbf{A}u$, and second, due to the rate of depreciation of the plant, or rate of utility dissipation concerning time preference, i.e. $c(t, x)u$. Here, $c(t, x) < 0$.

(b) Let us assume that G is a bounded domain in $\mathbb{R}^n$ with smooth boundary ∂G and define the stopping time τ as the random variable given by means of the relationship $\tau = \inf\{t : X_t \in \partial G\}$, which is also known as the exit time from G. Then the solution of the following *Dirichlet problem* defined as

$$\begin{aligned} \mathbf{A}u(x) &= 0, & x \in G, & \qquad (1.6) \\ u(x)|_{s \in \partial G} &= g(x), \end{aligned}$$

is given by

$$u(x) = \mathbf{E}_x \{ g(X_\tau) \}. \qquad (1.7)$$

Let us mention that functionals $u(x)$ in eq. (1.7) are extremely important working with asset valuation, option pricing and other risky ventures. Here g represents a randomly varying reward, utility or price. One has then to decide when to stop operations, sell or buy. Eq. (1.7) suggests intuitively a simple computational strategy to determine the value $u(x)$ of an asset with evolution starting at any $x \in G$. One releases diffusions from the point x, then one watches their trajectories until they first hit the boundary ∂G and one records the boundary values in order to build an average of these over all trajectories.

(c) Finally, let us consider the solution of the parabolic PDE defined by the relationships

$$\begin{aligned} \frac{\partial u(t, x)}{\partial t} + \mathbf{A}u(t, x) + f(t, x)u &= 0, & t > s, & \quad x \in \mathbb{R}^n, & \qquad (1.8) \\ u(s, x) &= \phi(s, x), \end{aligned}$$

where f and ϕ are sufficiently smooth functions, which is then given by

$$u(s, x) = \mathbf{E}_{s,x} \left[\int_s^t f(r, X_r) dr + \phi(t, X_t) \right]. \qquad (1.9)$$

Similarly, $u(s, x)$ in eq. (1.9) can be seen as a utility or production cost functional, where $f(t, X_t)$ stands for the rate of current utility or running cost and ϕ for a terminal pay-off.

Functionals of the type (1.5), (1.7) and (1.9) have been used in a variety of problems arising in natural science, engineering and in social sciences. Under additional assumptions we may, for instance, think of $u(s, x)$, in eq. (1.5), as giving the temperature at the point $X_t \in \mathbb{R}^n$ at time t taking into account that heat at X_t dissipates at the rate $k(t, X_t) = -c(t, X_t)$ and that the initial temperature condition is given by $u(s, x) = \phi(s, x)$. Summing up, functionals of the type (1.5), (1.7) and (1.9) offer a way of figuring out how systems that can be described by eq. (1.1) evolve and, by means of a qualitative analysis of their associated PDE's (1.6), (1.4) and (1.8), suggest a method of evaluating and analysing their performance.

Shadow prices and adjoint variables

At this stage we shall add that in physics it is customary to attribute to nature an optimising behaviour. Thus the functionals (1.5), (1.7) and (1.9) are assumed to fulfil certain extremal properties. This is not so in economics, where in order to be able to use fully the interpretations we just pointed out, we need to introduce devices or instruments which enable us to interact with the system evolution and steer it to an optimal state. The development of a philosophy suitable for the type of problem at hand has been taken over by the theory of stochastic control which, using various methods, manages to forward several versions of the Maximum Principle. This relies on a state and adjoint equation describing the evolution of the state of the system and its adjoint. The latter is sometimes referred to as the imputed or shadow (price) variable.

With the purpose of visualising the notion known as shadow price, let us consider the Cauchy problem given by (1.4) and its solution (1.5), and think of the interpretation of the process X, i.e. (1.1), as representing the output of a production plant, with $g(X_t)$ reflecting the productive capacity and $\exp[\int_s^t c(r, X_r) dr]$ the deterioration rate of a plant installed at time s with a capital endowment given by x.

Let us imagine the production management weighs the alternative of starting operations, at time s, with a higher capital endowment $x + \delta x$. To make it simpler, let us assume that the market can absorb the production at any capacity level of the plant, so that the balance between fixed capital, i.e. machinery, equipment and the like, and circulation capital, i.e. raw material, wages and so on, poses the only constraint to be faced by the management, in addition to those deriving from the period of turnover. That means that the higher δx the less the capital available to cover purchasing raw materials and wages, or the greater the credit services associated with eventual liquidity shortages.

Due to the uncertainty assumed in the output X given by (1.1), it becomes apparent that the variable representing changes in u given by (1.5) resulting from changes in x, taking place at time $t = s$, has to be a random one. Here we shall let $p = (p_t)_{t \geq 0}$ denote the time path of these changes, where for any

$t, t \geq s$, the variable p_t, called adjoint state variable, is given by the relation

$$p_t = \frac{\partial}{\partial x} \mathsf{E}_{s,x} \left\{ g(X_t) \exp \left[\int_s^t c(r, X_r) dr \right] \right\}, \qquad (1.10)$$

and the process p has to satisfy a SDE, the so-called adjoint equation, coupled in a certain sense with (1.1). Since the adjoint (state) variable p_t cannot be observed and/or measured directly, it has become usual to call it *shadow price* or *opportunity cost*, reflecting the idea that starting at time $t = s$ from $x + \delta x$ instead of x, entails giving up a certain amount of the otherwise expected reward.

Intuitively, the relationship (1.10) means that p_t conveys information on the average changes attributed to changes in the plant capacity and its deterioration rate, times those changes associated with the production itself as given by (1.1) with the initial state being $X_s^x = x + \delta x$. Formally, we can write (1.10) as

$$p_t = \mathsf{E}_{s,x} \left[\frac{\partial}{\partial x} \left\{ g(X_t) \exp \left[\int_s^t c(r, X_r) dr \right] \right\} \Phi_{t,s}^x(x) \right],$$

where $\Phi_{t,s}^x(x) = \frac{\partial}{\partial x} X_t^x$ stands for the generalised solution associated with (1.1). The preceding formula calls attention to the fact that the evaluation of the shadow price p at any time t assumes knowledge of the alternative realisations from s to t of the process X, as well as of the attached probability measures. Further, it requires information about the time-profiles of the plant capacity and deterioration rate associated with alternative initial states. Since the chosen initial scale of operation may have a profound impact on the subsequent states of the system, an assessment of the opportunity cost that neglects these issues cannot provide an unbiased picture of the states of affairs at any time t.

We shall look, based on techniques associated with the stochastic Maximum Principle, for new ways of dealing with two basic questions of optimal planning of stochastic growth and development of an economic system. Namely:

- Optimal resource allocation under uncertainty, taking into account the most crucial aspects of society related to production, distribution and consumption of material objects.

- Design of a method for the evaluation of the social value attached to the optimal economic policy, taking into account the impact of the former question, often called the *primary problem*, upon the various spheres of activities, even those of noneconomic nature.

One cannot overemphasise that at the heart of our investigation is the process of social reproduction, so that the second question, usually called the *dual problem*, acquires a crucial dimension. Thereby, the primary problem recovers its social character due to the fact that any economic activity relies upon social relations, and as such is involved in the functioning of a social structure of which it is only a part. Hence, the economic activities of society cannot be given priorities independently of its whole structure and their values cannot be properly measured neglecting their consequences with respect to spatial and temporal dimensions.

Ramsey used as early as 1928 the Euler-Lagrange equation to solve a deterministic version of the primary problem. Since von Neumann's remarkable paper on an expanding economy, resource allocation and valuation are fully and explictly recognised, via the duality concept, as part of the same problem. Indeed through this duality, the primary problem responsible for the optimal allocation of resources, i.e. concerning the flow of goods and services, becomes associated in an indissoluble way with the dual problem of allocating social value among the activities under consideration.

In the framework of neoclassical theory both equilibrium states and optima coincide. However this result has been seriously challenged by Joan Robinson and associates, on the ground that there is no physical measure that reflects the properties on which the qualitative behaviour of equilibrium states relies, and that such a notion is not compatible with historical time; See [181].

1.2.2   Processes underlying social reproduction

Central to any dynamic theory of society is the concept of social reproduction, since it underlies in a systematic fashion basic relations and forms of existence in space and time in human society. Understanding the process of social reproduction leads to better knowledge of fundamental phenomena. Therefore it is not surprising that classical economists felt that the purpose of analysis should be to identify the forces in society that promote or hinder development, and hence the forces governing the operation and progress of the economic system, and consequently to provide a basis for policy and action to influence those forces.

Accordingly the first problem every society faces is the explicit recognition of the need for consistency of principles and patterns underlying production, distribution and consumption with the requirements of social reproduction regarding the material conditions of production, i.e. the means of production and labour. These considerations determine the starting point of our investigation. At a further stage, we shall also address various highly sensitive issues of international character related to the International Division of Labour and other institutionalised mechanisms based on which one transfers economic surplus systematically from the periphery to the "affluent" centre of the world capitalist system.

Constructing models of capital accumulation

Fundamental to the process of social reproduction is the consistent renewal of capital and labour as well. In this sense, the construction of a model of capital accumulation becomes a primary goal. Starting from the concept of economic surplus, which includes profits, the process of accumulation will reflect essential features of the reproduction of capital:

- The *drive to accumulate*, i.e. the logic underlying mechanisms by means of which economic surplus is converted into capital with the purpose of generating more surplus which in turn leads to further additions to capital and so on. This process of accumulation brings about fundamental changes in the structure, organisation, in the scale and methods of production. Crucial are changes which render some productive skills and crafts obsolete.

- The *inherent contradictions* associated with the increasing socialisation and concentration of production without always generating consistent changes in the structure of capital as social reproduction requires. The private appropriation of the economic surplus of society aims at extending social relations of production to more and more people, thus invading more and more aspects of people's lives, in a manner not necessarily consistent with the principle of effective demand.

The dynamics resulting from the model of capital accumulation shall support the design of economic policies aimed at shifting into production the reserve army of labour. For convenience we will distinguish between two levels of analysis.

Corresponding to the first level, we shall construct models abstracting from changes in the structure and organisation of production attached to the process of capital accumulation in the wide sense. That means we shall assume production takes place at constant cost in terms of labour, or that the technical methods of production are independent of the scale of operations, and that the structural parameters of production remain constant despite changes in scale. The initial level of physical capital is given, as well as the fraction of the reserve army which is then to be absorbed into production in a given planning horizon.

At the second level of analysis the full impact of capital accumulation on all aspects of social production will be considered, and changes in the structural parameters will be allowed for. Further, the initial levels of physical capital and the lengths of the planning horizon will be explained by means of capital impulses. The latter level of analysis shall be referred to as the long-term, the former as the short-term analysis, in spite of the fact that a sharp separation of this kind is to a great extent fully unrealistic.

Structural changes and development

Starting from the assumption capital impulses generate manifold structural changes – in the form of changes in structural parameters like production elasticities, rate of technical progress, propensities to consume, to save and to invest, and income elasticity, etc. – we will try to capture essential sources of socioeconomic dynamics.

At an intuitive level, let us point out that the application of capital impulses is primarily directed to the increase of the productive capacity, as well as to enhancing growth and development potentials of basic components via structural changes. This calls attention to the need for a deep investigation of the impact of structural changes on the drift and diffusion coefficients, since the latter pass on to the system, i.e. through the dynamics driving it, the effects of capital impulses.

Let us assume thus (capital) impulses of sizes $\theta^1, \theta^2, \ldots$ are applied at the impulsion times $\tau^1, \tau^2, \ldots$. Then, for any fixed but arbitrary j, the stochastic differential equation and associated relationships of the type

$$dX_t = \sigma(\tau^j, X_t^x)dB_t + b(\tau^j, X_t^x)dt, \quad \tau^j \leq t < \tau^{j+1}, \qquad (1.11)$$
$$X_{\tau^j}^x = X_{\tau_-^j}^x + \theta^j \qquad j = 1, 2, \ldots, \qquad (1.12)$$

$$X_0^x = X_{0-}^x = x \quad \text{and} \quad \tau^1 = \tau_-^1 = 0, \qquad (1.13)$$

describe fully the evolution of the system state X, where the new variable τ^j is a random variable defined on $(\Omega, \mathcal{F}, \mathbf{P})$ in a way such that $\tau^{j+1} = \tau(\theta^j)$. Then τ^{j+1} is a stopping time that depends on the size θ^j of the j-th impulse and the initial state $X_{\tau^j}^x$ attached to the j-th time period is given by (1.12) with $X_{\tau_-^j}^x$ defined by the relationship

$$X_{\tau^j}^x = \sup\{X_t^x, \tau^{j-1} \leq t < \tau^j\}.$$

In the simplest case the SDE (1.11) describes the evolution of capital building in the j-th short-term period, where the drift $b(\tau^j, X_t)$ stands for a certain average of net investment and $\sigma(\tau^j, X_t^x)$ for the associated nonanticipative deviation from the above average at time t. Capital impulses bring about in this case changes in the accumulation of capital due, for instance, to changes in structural parameters attached to the production technology, and/or changes in the fraction of profits to be recommitted to the production process, and so on. Further, let us stress that eq. (1.12) stands for a change of the size θ^j which takes place in the level of capital at the impulsion time τ^j. This makes apparent that capital impulses influence the level of capital as well as the pace of accumulation, since for any arbitrary j the relationship (1.12) gives the initial condition for an accumulation régime described by (1.11).

Let us mention that we are fully aware that a concept like "structural change" is vague and wide. Therefore, including it in a meaningful way in a model of development creates some difficulties. Accordingly we shall opt for the use of a little simpler, more manageable and in many respects almost equivalent concept. Namely, that of "changes in the structural parameters", i.e. those parameters of the system the change of which takes place relatively slowly. For this purpose, we introduce the process of structural changes $\Gamma = (\Gamma_t)_{t \geq 0}$, where $t \in [0, T)$, which will be defined in [94]. The transformation of the economy resulting from the application of (capital) impulses calls forth innovations, labour mobility, flow of job openings and various other phenomena of socioeconomic nature underpinning essential processes of development and the growth. Based on these phenomena one can observe indirectly to what extent structural changes take place and gain information about the readiness of people to change working and, to a lesser degree, living habits and environment, as well as the system's ability to adapt to and even to shape a more dynamic production apparatus. This is essential to the design of employment policies, and to working out a rationality principle more akin to problems of real life which will involve looking at development as a learning process.

Investment, employment policies and migrant labour

With regard to eq. (1.11) and the simple case mentioned above, there is a group of associated questions concerning short-term policies of investment and employment that we approach on the basis of a model consisting of a relatively advanced sector and a reserve army of unemployed labour. As we mentioned already, for any arbitrary but fixed j the system evolution represented by (1.11) will not depend directly on the capital impulses, e.g. on the control variables τ^j and θ^j that make up the impulse control. Besides, the drift and diffusion coefficients do not depend on the time t either, although this is not essential.

The j-th time period will depend crucially on the investment and employment policies which make up the continuous control acting in the time interval, $\tau^j \leq t < \tau^{j+1}$, a fact that will be made more explicit by indexation of the coefficients of (1.11) with the (continuous) control process α, i.e. by writing $b^\alpha(\tau^j, X_t)$ and $\sigma^\alpha(\tau^j, X_t)$ instead of $b(\tau^j, X_t)$ and $\sigma(\tau^j, X_t)$. The latter fully characterise, from the economic point of view, the average drift and diffusion coefficient of the (continuous) system dynamics.

Since investment policies do in general neither materialise nor strictly follow the pattern conceived *a priori* and various agents involved base their responses to the actual course of the process X on either incomplete information or indirect observations, a variety of alternative pictures emerges. Special attention will be given to cases corresponding to concentrating uncertainty upon the demand and/or supply of labour power. First considering them separately and second, when both interact, and the interplay of the resulting process of accumulation and employment formation comes to the forefront. The state process describes accordingly capital accumulation in per capita terms, as well as per available labour, both of which convey information of different nature and shed light on socioeconomic configurations of varying perspectives.

Let us finally mention that owing to the fact that our analysis focuses on a peripheral economy, we do not devote much attention to phenomena associated with price adjustments in market for consumption commodities. This is due to our assumption that underdevelopment manifests itself mainly by way of deficits of various kinds. Behind this view lies the painful fact that effective demand does not keep pace with actual social needs, and that realised demand, owing to demonstration effects of various kinds, is to some extent detached from the domestic production, which has as a natural result the fact that the own productive capacity also lags well behind the effective (aggregate) demand; in particular, with respect to luxury and productive consumption for the upper segments of society and entrepreneurs.

Notwithstanding this limitation in our scope, we claim that our analysis extends to the so-called "developed" economies. To cover the latter we shall need to consider in addition various processes of innovation, which we shall do somewhere else.

1.2.3   Toward an objective notion of reward

The process of capital accumulation shall be shown to solve a stochastic differential equation of the diffusion type, on which the process of economic devel-

opment relies. Interpreting development as a learning process, we are able to
reformulate the conflict between the flow of social benefits and costs attached
to it as a race between the average time-rate of increase of an unknown payoff
process, and the attached flow rate of social costs. Based on microeconomic
considerations, we will derive a subjective rate of discount which takes into
account various kinds of uncertainty and maturity, establishing thereby clear
links between development and the learning process.

Since higher accumulation paths may, on average, require heavy social sacrifices as mentioned above, and this in turn may harm the process of development
in various ways, it is natural to direct further efforts to examining from the
point of view of political economy the impact of these economic policies upon
the socioeconomic behaviour of individuals. To capture some issues associated
with social benefits, and eventual damage to development in a way amenable
to socioeconomic and mathematical scrutiny, it is necessary to have a precise
characterisation of the notions of social rewards and costs as represented, for
instance, in (1.9) by ϕ and f.

Benefits versus costs and learning

The process of "giving up current consumption" at the macroscopic level has
its own dynamic related to various socioeconomic factors, going from production and technological aspects over social relations of production to a variety
of issues like individual and social maturity, awareness, motivation, etc. On
the other hand, the process of "benefits to be derived" from the policy under
consideration has also its own dynamics, the effects of which are expected to
mitigate social burdens resulting from the former, and what is more, to stimulate even stronger social solidarity, higher performance standards, etc. From
the interplay of these two processes, which unfold on different time-scales, one
expects it ultimately brings about the fulfilment of requirements of social reproduction entailing benefits overcome the costs.

In order to capture in a realistically objective fashion various aspects concerning social valuation and performance of the development process, that at
the same time accounts for subjective features, we shall assume for any $t \in [0, T]$
and any $y \in \mathbb{R}^n_+$ the relationships

$$\phi(t,y) = \Lambda^s_t \cdot \Phi(y) \quad \text{and} \quad f(t,y) = D^s_t \cdot \psi(y) \tag{1.14}$$

hold. By means of these relationships we shall represent a certain kind of
reward-like index deriving from the development process, i.e. representing social
benefits and costs as perceived by the representative individual.

In (1.14), we introduce a subjective discounting factor due to impatience D^s_t
and a factor Λ^s_t due to motivation as functions of time t. Taking into account
(1.14) the performance function can now be given following more enlightening
expression

$$\mathsf{E}_{s,x} \left[\Lambda^s_{\tau^x_s} \cdot \Phi(X^x_{\tau^x_s}) + \int_s^{\tau^x_s} D^s_t \cdot \psi(X^x_t) dt \right], \tag{1.15}$$

where the function ψ measures deficiency of (physical and social) functionings
attributed, say, to a low level of the minimum consumption per capita $\underline{c}$ result-

ing from policy making purposes, while the function Φ measures on the other hand the improvement in the choice set of alternative functionings. The factors Λ_t^s and D_t^s bring in further subjective aspects of performance. The term D_t^s may reinforce the existing deficiencies of functionings, while Λ_t^s may strengthen the improvement in capability.

Recalling that living conditions are, as states of existence and functionings, the reflections of various aspects of newly emerging reality, we like to think of the parameters involved in $\rho = (\rho_t)_{t \geq 0}$, a process to be specified below, as quantitative indicators of the ability of the representative individual and her or his socioeconomic environment to transform various characteristics of the necessary personal-consumption bundle, in particular those responsible for longevity, health, schooling and maturity, into their corresponding functionings ("states of being and doing"). Thus, the factors D_t^s and Λ_t^s besides providing some information at time s concerning the probability of survival on various grounds and the readiness to postpone rewards up to the future time t, offer us a link of interaction between policy-making and its impact upon individuals. That is to say, the parameters involved in ρ together with our understanding of impatience and motivation, as introduced above, allow us to assess the state of impatience resulting from the economic policy at stake.

Motivation versus impatience and positive freedom

The factors D_t^s and Λ_t^s just introduced correspond to expectations formed by the individual at time s about the state of affairs to prevail at time t, based on reflections concerning future uncertainty that results in a global discounting factor of the type

$$\exp\{-\int_s^t [(d_r + \xi_r) + m_r] dr\}$$

depending on the process X as already stated, where d_r and ξ_r stand respectively for the probability of dying at any time r independent of consumption and dependent on the level of consumption per capita, while m_r represents the rate of discount in accordance to personal maturity associated, for instance with schooling expenditure per capita.

To save notation, let ϱ_t denote the integrand in the expression above, i.e. $\varrho_t = d_t + \xi_t + m_t$ for any t, and $\varphi_t^s = \int_s^t \varrho_r \, dr$. Here again ϱ_t and φ_t^s are assumed to be parametric functions given by the relationships $\varrho_r = \varrho^s(X_r)$ and $\varphi_t^s = \varphi^s(X_t)$, where the upper index s indicates that the underlying expectations are formed at time s. With these notational simplifications, and taking into account we need growing rather than discounting weights, we define the factors Λ_t^s and D_t^s by means of the expressions

$$\Lambda_t^s = \exp\{-\varphi_t^s\} \quad \text{and} \quad D_t^s = 1 - \exp\{-\varphi_t^s\} \tag{1.16}$$

with the usual convention that $\Lambda^s = \Lambda^s(X_t)$ and $D_t^s = D^s(X_t)$ for any t, and any s fixed but arbitrary. We will call Λ_t^s and D_t^s respectively the *sub-jective discounting factor*, the first due to *motivation* and the second due to *impatience*, while $\varrho = (\varrho_r)_{s \leq r \leq t}$ will be termed the *subjective rate of discount*.

Loosely speaking, motivation decreases and impatience increases with uncertainty. High values of Λ_i^s are associated with low levels of motivation and vice versa.

Let us recall, the command over commodities that wage income may give a labourer can be better appreciated through the characteristics identifying these commodities, and the capability and functioning deriving from their consumption. In this sense, the capability to function is what comes closest to the notion of *freedom of action* or *positive freedom*, i.e. the ability to choose and accomplish a set of objective and goal-conducive actions by way of which individuals and systems, in the face of given social and individual configuration, seek to unfold their creative energies on their own.

1.3   Organisation and content

In the foregoing sections we spelled out the most essential features upon which our work relies. Further, we presented our goals and the framework within which we shall formulate questions and attempt to elaborate solutions. The aim of this section is to call attention to the links on the basis of which the disciplines involved are glued together to build a coherent machinery through which we seek to understand the problem at hand and search for viable ways out. This gives a preliminary opportunity to be more specific about the questions we shall ask and the sort of answers we may expect to obtain as well as on the methodology and tools that can effectively support our inquiry.

1.3.1   On the various disciplines involved

To begin with, let us say that within fairly reasonable limits these essays are self-contained. That means, any idea is carefully discussed when it first appears, and its links with the basic aspects of the problem at issue are explained in detail.

Going from the preliminary presentation of concepts and processes of socioeconomic nature to their mathematisation, we make every effort to visualise essential features of the model building philosophy. However, we have to mention that the emphasis lies on understanding certain socioeconomic phenomena based on reality-tailored modelling and application of various mathematical techniques, keeping in mind the need to illuminate soci(ologic)al problems and contribute to their solution. Thus, a full display of technical details, elegance and generality, axiomatisation and rigour is not on the research agenda.

This monograph is organised in three parts, each consisting of three chapters. The last of these, Part III, has the character of an appendix where we introduce briefly basic concepts of the theory of systems and (deterministic) optimal control, stochastic analysis, and control of random systems. Essentials of development economics which help identify distinguishing features of a labour-surplus economy of the peripheral type within the world capitalist system have been disseminated throughout.

Part I examines available theories of political economy and develops a body of theoretical tools appropriate to our undertaking. Part II is dedicated to the elaboration of our own approach using probabilistic and variational techniques.

The models are differentiated according to where uncertainty emerges. For the case where uncertain rates of return on capital prevail, we construct a model of capital accumulation with uncertainty deriving from labour demand. The case uncertainty threatening primary activities motivate uprooted workers to search for jobs in more highly productive activities, giving rise to a model with stochastic accumulation owing to uncertainty in the supply of labour. However, this model is casted in terms of functioning characteristics deriving from jobs in certain sorts of employment. The interplay of the sources of uncertainty that characterise the models just mentioned builds the background of a third one. When economic development is seen as a learning process, uncertainty acts also through behavioural features like impatience and motivation, bringing about a fairly new accumulation model where interaction of individuals plays a crucial role.

The integration of various features of these models in one, where one attempts to capture structural changes and impulsive expansion of production capacity, is a further example of economic model-building and policy design. Since the present essays concentrate on phenomena of social continuity this type of model will be treated in a subsequent work, see [94]. Extension of these models will be also explored using numerical techniques and simulation in the just mentioned work.

The system-theoretical approach

The magic behind good (mathematical) models points to the ability of the model designer to decide what is of primary importance and how to codify the relevant information. This suggests, first of all, that the model-maker needs to be fairly well-acquainted with the subject matter from which the problem emerges in order to be able to know what and when to observe, and the like. Secondly, the model-maker needs to know how to capture her (his) observations and insights in the language of mathematics in a way capable of communicating and improving efficiently the knowledge that such a "formal representation" of reality conveys.

Thirdly, the resulting construction, named a "mathematical model", needs to be amenable at least to the available techniques of analysis, adding further to the requirements to be met by its designer: namely, acquaintance and reasonable skills handling the appropriate set of tools as well as the ability to improve and extend it. Finally, since mathematical modelling is primary a language, it turns out to some extent that the same requirements are to be fulfilled by the model user. Only in this way can communication become effective, and the knowledge and information coded in a mathematical model can be properly deciphered and added to the user's picture of reality.

Aware of the demanding features of a system-theoretical investigation and wanting to reach as many receptive minds as possible, we have added, as already mentioned, a chapter with four concise sections summarising some of the major concepts used in this monograph. Moreover, in every chapter we indicate the main sources, further readings and call attention to some crucially distinctive features of our point of view. Concerning the necessary background in systems theory, Chapter 7 gives a fairly complete overview of introductory notions on systems and modelling in a manner suitable to the needs of these essays. Any other concept or technique shall be explained as soon as it appears or a precise reference is given whenever a further development of it is not of immediate relevance for the subject matter at issue.

Stochastic analysis and time

In the preface to his book on "Dynamic probabilistic systems", [111], R. A. Howard says good ideas are simple and a Markov process is not an exception, and he proceeds to claim that there is no problem in his book that cannot be made clear to a child. His expository device is a pond covered with lily pads among which a frog may jump randomly without falling into water. Using the lily pond analogy, he works out the most interesting questions about dynamic probabilistic systems, including optimisation procedures of dynamic programming that have proved so useful in the control of physical systems.

The same can be said in general, we believe, of the various probabilistic techniques of stochastic analysis as means of dealing with uncertainty, complexity and dynamism which will continue to challenge mankind's quest for understanding and management the manifold processes underpinning social reality. Markovian and semimartingale features provide stochastic models with the ability to capture the essence of interdependent processes typically attached to dynamic and complex systems. They also suggest relatively transparent and effective methods to characterise adaptive and certain optimising properties the behaviour of such systems exhibits in practice as well as in theory. However, filling in the details represents a hard task.

Thus we have decided to remain at a heuristic and intuitive level, concentrating in a few sections the most technical aspects of our approach. These are marked with a * and can safely be omitted. However, in order to assure this without affecting adversely the clarity in the exposition, a certain degree of redundancy has been necessary. In Chapter 8 we have condensed basic notions and tools needed to handle probabilistically the various problems related to uncertainty and randomness emerging in the course of our study. This comprises a relatively complete summary of definitions, classification of processes and interpretations in the sense of our essays. A modern version of stochastic integrals and semimartingales is given, where predictibility is essential.

Part II is dedicated to stochastic analysis of the problems we have outlined, following methods of highly probabilistic nature, and then using analytical techniques of the variational type. Concerning the probabilistic methods we shall say these allow, in spite of a high degree of sophistication in the theory, a fairly intuitive and clear treatment of the main issues. Switching controls, a notion based on a change of probability measure, brings transparency and leads to an appealing introduction of the value function and the most basic concepts of stochastic control.

The variational methods are more elementary and bring about stronger results. However, owing to the fact that analytical techniques are less directly connected with the trajectory of processes, some control concepts remain in the background. On the other hand, working more closely with the differential operator of the process, these techniques seem to offer better alternatives to discretisation procedures and numerical treatment of the problems under consideration.

Control and optimisation

The optimising paradigm continues to pervade more and more fields of scientific endeavour, and social sciences are not immune against the spread of its manifold techniques, despite simultaneously increasing prejudices, displeasure and not rarely unjustified criticisms. The evolutionary ideas come more strongly to the forefront and its methodological machinery advances rapidly, too. In this regard, our option has been to stick to the optimisation methodology without ignoring the efficiency and power of the evolutionary conception of various fundamental processes of society; learning is one of those.

The by-now-classical techniques of optimisation from the point of view of control theory are the object of Chapters 7 and 9. The more modern aspects of control, both deterministic and stochastic, are treated there where they show up. The main emphasis reflects our commitment to describe phenomena under consideration as processes. From this point of view, the methodology underlying the theory of controlled processes as it is known from modern engineering applications attracts our attention. The fact that actual socioeconomic systems do not admit a strictly deterministic analysis directs our efforts to the study of systems subject to random factors which influence their behaviour and exhibit thus fluctuations attributable to random failure of their components.

It is natural in such cases to choose a control strategy on the basis of an average expected behaviour which attempts to account for a large number of possible variants of systems behaviour. Switching from average behaviour to possible variants, we turn in fact from macroscopic to microscopic features of the system. In this sense changes in characteristics at the microscopic level are cause and effect of changes in system characteristics.

Optimisation in general, and control theory in particular, provide here the tools for designing decision principles incorporating, based on certain interaction and learning features, various fundamental links between individuals, social production and institutions.

The natural vehicles of the learning perspective are processes of information, knowledge and change underpinning social reality which are dealt with in Chapter 7 in connection with modelling, and in Chapter 3 concerning time and uncertainty from the point of view of economic theory. In this sense rational choice and optimising behaviour acquire more realism and relevance, although rising thereby further questions and bringing more elements into work.

In Chapters 4 we shall cover various matters related to the ability of the system and individuals to change, adapt to chosen objectives and accomplish envisaged goals. Thereby, issues concerning adaptation, motivation, impatience, culture and the like enter explicitly the analysis illuminating in that connection the socioeconomic character of the objects of economic theorising and stressing the instrumental role played by optimisation.

On the economic theory

Having taken the classical view according to which the central object of economics is the process of social reproduction and its associate processes, e.g. material production, consumption, exchange, distribution, financing and management among others, it becomes obvious that the term "socioeconomic" shall emphasise various social, institutional and anthropological links underlying such processes. In this sense, the notion of (economic) system draws attention to the fact that society consists of individual agents linked by socioeconomic relations and engaged in manifold activities directly and/or indirectly aimed at reproducing the system.

The capitalist system also has a core of activities aimed at reproducing in many ways among which production of commodities by means of commodities may be seen as its basic process. Notions like capital, as means of production and as social relation, and labour, in its prevailing social form of wage labour build the distinguishing categories of this system. We shall look at the capitalist system from the perspective of the world economy and see it as consisting of a centre and a periphery.

A variety of comments surveys the main issues concerning the duality centre/periphery and Chapters 1 and 2 specialise relevant topics of economic theorising to our purpose of understanding the main processes of social reproduction. Chapter 3 deals with various highly neglected aspects concerning time and change as means of visualising the evolutionary power of economic systems. In this sense we will better be prepared to understand technology and techniques of social reproduction; horizontal and vertical structures of production; integration, change and its institutions; in particular, capital accumulation, growth and development from the perspective of political economy. As a whole, Part I is intended to spell out the theoretical framework on which our investigations on the peripheral economy rest and closes with reflections on the ideological underpinning of economic systems.

1.3.2   The design of policies for economic development

The design of policies for economic development belongs to the main themes of our essays. Roughly speaking, we have to distinguish between questions concerning growth and questions about development aspects of evolutionary processes of the economy. In this sense, a great deal of attention is dedicated to problems of capital accumulation given the productive capacity associated with available capital, and one looks for the most efficient accumulation policy using techniques of continuous control. The long-range problems of development are purposefully associated with structural changes and the expansion of productive capacity, which can better be dealt with using discontinuous functions of control, e.g. impulses, jumps and reflections. These topics are the object of a forthcoming monograph, see [94].

Continuous control and accumulation policies

Chapters 5 and 6 centre on the logic behind the problem of continuous control and its associated policy of accumulation for an arbitrary but fixed j. Further, these chapters give a detailed construction of such policies. In order to give an idea of these issues, let us consider a dynamic system with an evolution steered through the SDE

$$
\begin{aligned}
dX_t^x &= \sigma^\alpha(\tau^j, X_t^x)dB_t + b^\alpha(\tau^j, X_t^x)dt, \qquad t > s, \qquad (1.17)\\
X_s^x &= x \in \mathbb{R}^n, \qquad s \geq \tau^j,
\end{aligned}
$$

along the planning horizon $[0, T]$ with $T \in \overline{\mathbb{R}}$ and $s, t \in [0, T]$. Everything else including notation is as before. The characterisation of the coefficients $\sigma^\alpha(\tau^j, X_t)$ and $b(\tau^j, X_t^x)$ varies with the concrete problem at stake and depends on the methodology chosen.

For the purpose of introducing a performance criterion that retains relevance also in long-term analysis, we will consider an objective function which may be described as

$$
\begin{aligned}
u^\alpha(s, x) = \mathbf{E}_{s,x}\Bigg\{ &\int_s^{T \wedge \tau^{j+1}} f^\alpha(t, X_t^x)dt + \phi(T, X_t^x)\ \mathbb{1}_{\{T < \tau^{j+1}\}}(\omega) \\
&+ \psi(\tau_-^{j+1}, X_{\tau^{j+1}}^x)\ \mathbb{1}_{\{\tau^{j+1} \leq T\}}(\omega) \Bigg\}, \qquad (1.18)
\end{aligned}
$$

where the function f^α is the flow rate of instantaneous reward, ϕ and ψ are two alternative pay-off functions, and $\mathbf{1}_A$ the characteristic function of the set A. Here ϕ represents the social cost, i.e. reward foregone, associated with the failure of attaining full employment relative to the j-th time period during the interval of time $[0, T]$, so that no additional shifting of unemployed labour takes place. On the other hand, ψ describes those costs attached to structural changes, capital impulses owing to a further step towards full employment.

Further, let us denote the family of admissible (piecewise) continuous functions of control α by $\mathcal{A}$. Whenever τ^{j+1} is such that $T \leq \tau^{j+1}$ holds, the associated control problem is then of the type referred to in Section 1.2.2, i.e. it consists in the search for an optimal continuous control β among the control functions $\alpha \in \mathcal{A}$ and a function $v(s, x)$, usually known as the value function, such that the relationships

$$
v(s, x) = \operatorname*{ess\,sup}_{\alpha \in \mathcal{A}} u^\alpha(s, x), \qquad \mathbf{P}_\beta \quad \text{a.s.,} \qquad (1.19)
$$

$$
v(s, x) = u^\beta(s, x), \qquad \mathbf{P}_\beta \quad \text{a.s.} \qquad (1.20)
$$

hold. Here $\mathbf{P}_\alpha$ stands for the probability measure solving (1.17) and associated with an arbitrary continuous control α. On the other hand, should $\tau^{j+1} < T$ hold, then an impulse becomes mandatory, a case which loses relevance when T goes to infinity.

Optimisation problems as formulated by (1.17), (1.19) and (1.20) are often studied on the basis of the so-called maximum principles. In Chapters 5 and 6 we study various problems involving continuous controls, where the régime of capital accumulation is governed by (1.17). Various situations are considered dependent of the source of uncertainty and the type of capital accumulation under examination, i.e. accumulation per worker or per available labour.

Probabilistic and analytical methods are used according to the nature of the questions involved and the expected results. Questions concerning assessment and qualitative properties of accumulation paths are also treated and interpreted in the light of the principle of effective demand and social reproduction.

1.3.3   How to read this monograph

Interdisciplinarity is a word one finds in connection with a variety of themes. It seems everybody agrees it is a necessary and adequate tool, but it is very hard to put it into practice. It is not easy to explain why it is so. However, we shall point out a certain similarity concerning the consensus on certain notions and individual prejudices in face of an eventual exercise of activities described by such notions. Thus, let us say notions like egalitarism, frankness, plurality, humanism, flexibility, tolerance, criticisms and many others enjoy a wide consensus and manifold reservations at the individual level.

About what is ahead and how it is linked

Therefore, from the outset we shall stress this work is addressed to readers willing to digest the implications the use of interdisciplinarity entails. Thus, let us emphasise interdisciplinarity is neither a method nor a theory, but rather a statement concerning the framework on which a scientific enquiry takes place. In this sense the magic of interdisciplinarity creates a metaframework characterised by its ability to move to different levels of analysis and to be embedded into the particular framework of disciplines involved. The potential usefulness of interdisciplinarity is linked to a great extend to the flexibility and richness of the conceptual language. These features are dominated by the methodology of systems theory. Hence, those who are not familiar either with the style of interdisciplinary work or theory of systems should start skimming through Chapter 7.

Otherwise the right point to begin is Part I which we recommend to read first quickly with the purpose of acquiring a global perspective, to discover the links between fields and their roles, and to get a feeling of interdisciplinarity before entering to investigate the main points. In order to facilitate a relatively smooth first reading of Part I, we have chosen a terminology with a great deal of intuitive appealing. In particular, this is true with respect to probabilistic concepts and stochastic control questions.

Dealing with the terminology of economic theory is more delicate. This is so not only because the conceptualisation of events, objects and relations between them are forced *ad hoc* into a form conducive to the desired perspective. The difficulty is neither linked to the diverse use of the same concepts nor to the interest to choose notions which maintain their essential meaning within a narrow or an open framework which may serve the need of accounting for certain – say sociological, anthropological or psychological – features of events.

The problem is fundamentally related to the reductionist and rationalist guise of orthodoxy which rises the pretension of universal validity.

This contrasts heavily with our approach which, owing to its system theoretical spirit, attempts to grasp reality on the basis of models. Models, viewed as pictures of reality, have in turn temporary and partial validity due to the passing of time and a naturally limited perception. Meaning and validity of economic concepts are extremely sensible to the characteristics of the observer and/or the agent, a fact that needs to be accounted by clearly and faithfully delineating the domain of possibilities and feasibilities. Therefore, in Chapters 2 and 3 readers particularly interested in the economic conceptualisation and terminology will find the necessary background on which Part II rests.

Since it is extremely important to distinguish a wealth of concepts expressed by symbols appearing in the language of orthodoxy and in everyday life (the metalanguage), we have stressed with some obstination where and why certain notions are ill-conceived and/or misleading. This is unavoidable if we want to give economic theorising a scientific character. To this end we have carefully chosen a good numbers of authoritative citations with precise indication of their source, to which the reader may recur to deepen her (his) understanding on the subject and to seek further evidence or additional references. These citations are an integral part of this monograph and we urge our readers to go through them thoughtfully.

Before going into Part II we advise readers who are not familiar with stochastic control methods and are willing to learn about them to skim over Chapter 9 and the first section of Chapter 8. As a matter of fact Chapters 8 and 9 together with Sections 5.2, 5.3 and 6.3 constitute a good introduction to stochastic control theory. In Part II sections marked with an asterisk * are of mathematical nature and may be skipped out at least in a first reading without loosing the main lines of argumentation. However, since these sections have been written in a relatively intuitive and heuristical manner, we highly recommend those who are interested in interdisciplinarity as presented here not to do so. Stochastic analysis and stochastic control theory are indeed difficult subjects but understanding their basic methodology provides valuable insights and helpful feelings of how a great variety of phenomena unfolds.

The analysis and interpretation of results is in a critical way related to our knowledge on methods and objects under investigation, a point we have carefully explained in Chapter 7 in connection with models, learning and information. In this regard, it is necessary to call attention to two potential sources of methodological bias, i.e. hiding and/or hidden links, as a result of the fact that investigators are also part of the reality about which they are scientifically inquiring. Let us formulate the first as "having the spectacle behind the eyes" which leads to "hiding the findings" while constructing the theory. The other relates more explicitly to links between observation and comprehension, so that lack of knowledge or prejudices come to block our ability to discern the truth behind observations and corresponds to "having the spectacles behind the eyes". Regarding socioeconomic phenomena where experimental interrogations are of limited applicability, history acquires a fundamental role as arbiter of scientific theory. In this sense several citations should help to clarify from a historical point of view many reflections on the framework on which our models rest. The purpose of other citations is to draw attention on the vulnerability of rational science, since no matter how impressive its achievements may be, it remains a part of human discourse.

Readers with an adequate training in economics and mathematics may also go straight to Part II, which consists of three fairly independent chapters, and just consult those sections where some basic assumptions, equations and identities have been collected. Nevertheless the essence of Part II will be better appreciated from the perspective of systems theory which the reader will find scattered all over this monograph.

A survey of chapters in Parts I and II

The aim of **Chapter 1** is to give a preliminary overview about the problems to be treated, about the theoretical and the methodological machinery to be used, and about the results that we are aiming to obtain and questions that may be seen as the most natural objective of future endeavour.

In the first section we present the scope of our undertaking and explain our goals and views. Some reflections on the contribution of science and on the role of mathematics are intended to let the reader know our view. Some characteristic features of the present work should help to identify our aims.

In the second section we tell how to put some of these thoughts in a mathematical fashion using some analogies with classic and probabilistic mechanics. The third section makes preciser statements on our objectives and sketches the construction of accumulation and development policies as well as some questions on future research.

The purpose of **Chapter 2** is to give a brief account of the most basic economic concepts underlying our approach to the allocation of resources in a way that facilitates recasting our views and aims in a manner amenable to mathematical scrutiny.

In the first section we present some key notions of the process of social reproduction based on which we outline our model of a peripheral economy. In the second section the notion of uncertainty and its relations to the labour process are presented briefly for the purpose of explaining the need of accounting for uncertainty in the dynamics driving the system. The third section deals with notions linked to production, social output and distribution of available income in a highly neoclassic fashion and indicates essential points of divergence.

The third section serves also to describe on the basis of notions already introduced steering mechanisms and policy goals, and give a quick idea of the impulsive-control problem and the underlying process of structural changes.

In **Chapter 3** the objective is to understand some aspects of the dynamics of socioeconomic systems, with the aim of explaining essential links between change and time affecting the system as a whole and the interplay of processes underpinning its own reproduction in the course of time.

The first section centres on time and change from the perspective of socioeconomic phenomena. This comprises examining various interacting processes like information, knowledge and organisation which underly basic structures of production and build the heart of social change. An examination of the notions of causality and motion under uncertainty precedes the introduction of various thermodynamic concepts casted in the language of economic dynamics, in the basis of which we pass in the second section to the study of capital accumulation from an eminently dynamical point of view. There we draw attention to the moving forces of capital accumulation and the role of the principle of social reproduction in maintaining the productive power of the system.

Various essentially new aspects of the evolutionary approach to socioeconomic change are the object of the third section. This deals with the operational nature of mechanisms of change, deterioration of structures and organisation, and the pressure of social reproduction permits identifying to some extent the forces behind socioeconomic phenomena exhibiting various discontinuity and continuity features that characterise crucially the main time-structure of a system. The fourth section is dedicated to various issues on political economy deeply related to the reproduction of structures of production and organisation which serve the purpose of maintaining a particular state of affairs.

The purpose of **Chapter 4** is to explore how we can use some stochastic ideas and stopping techniques to support the construction of a simple development policy and obtain insights into accumulation behaviour over time as well as in its interaction with uncertainty. To this end, we shall present in the first section the concept of choice of technique from a probabilistic point of view and derive a simple stochastic model of accumulation susceptible of extension to Pasinetti's picture of structural change and growth. On the basis of this model we introduce concepts, results, reflections and models concerning capital accumulation and development on which we shall rely in Section 4.3.

In the second section control mechanisms which resort to stopping techniques are briefly explained and point out some mathematical and economic features which offer the links between the stopping and development problem. Interpreting development as a learning process we are able to reformulate the conflict between the flow of social benefits and costs attached to it as a race between the average time rate of increase at time t of a process to be sought and the flow rate of social costs f expected to hold at time t, both as viewed at time s. Associated with this race there exists an optimal payoff u that solves the generalized Stefan's problem. Based on microeconomic considerations we derive a subjective rate of discount which takes into account various kinds of uncertainty and maturity establishing thereby clear links between the development and the learning process.

In the third section we discuss also various matters concerning the logic underpinning resource allocation and stress various technological and institutional aspects which may act as bridges between economic theorising and thermodynamic formalism. Then the chapter concludes with a brief analysis of qualitative features associated with benefits and costs on the one hand, and on impatience and motivation on the other.

Chapter 5 deals with the characterisation of an economic policy which leads over a relatively long period of time to the elimination of labour-surplus, i.e. one which directs available labour from less productive or idleness to more highly productive employment.

The crucial asumption in the present essay is that the response of individuals to economic policies at issue is uncertain and only partially known. More precisely, we assume that changes in the labour participation are uncertain and described by a linear stochastic differential equation. This assumption renders the accumulation of capital random and feature reflected in a system dynamics governed by a stochastic differential equation of the diffusion type. We formulate in the first section a stochastic optimisation problem in the framework of social accounting prices or social opportunity costs as forwarded by S. A. Marglin. With the help of semimartingale and stochastic control techniques, we get an optimal strategy that in three-phases eliminates a given labour-surplus.

In the second and third sections we present a fairly detailed account of the main aspects of stochastic control based on semimartingale methods. The aim is a clear and complete understanding of the key steps of the control problem. For that reason we rely on intuitive and formal reasoning, and give precise references regarding a rigorous treatment of various relevant questions.

The final section is dedicated to a thorough application of the presented techniques to various issues of accumulation. A Marglin-Pontryagin accumulation path consisting of three phases is worked out and interpreted. The delicate matter concerning assessment of the resulting economic policy is approached on the basis of a simple and helpful approximation idea.

For the purpose of **Chapter 6**, we shall split the design of an economic policy in one serving the objective of accumulation and another persuing a given employment target. To this end one needs to distinguish between accumulation of capital per available labour and per actually employed worker. From this perspective emerges a variety of issues characterising either a process of accumulation or employment. The analysis of the interplay of these two processes shed light on the structural requirements a model has to meet, from which full employment may result as a lasting outcome of the system evolution. The main task becomes to work out, by means of a stochastic maximum principle development paths along which surplus-labour can be absorbed into production.

In the first two sections, we call attention to the main economic features that are at heart of the present accumulation problem. Relying on stochastic differential equations, we introduce economic concepts and ideas needed for the formulation of an optimisation problem amenable to stochastic control techniques of the variational type. In Section 3 we give a fairly complete derivation of the maximum principle. In the last Section, we deal with questions related to the attainability and reversibility of the optimal labour allocation policy obtained and make detailed economic interpretation of it, in order to keep the links with our original problem alive.

1.3.4   Concluding remarks and acknowledgements

We shall close this introduction and outline with a few remarks on various matters. Some are intended to mention briefly on our forthcoming work, others have supplementary character and the last ones are intended to let readers have a quick view behind our academic and private stage.

Concluding remarks

At this point it should be fairly clear our models of socioeconomic reality have roots in evolutionary thinking and probabilistic physics. From this background one gets an appropriate framework which enables pointing out misconceptions of economic orthodoxy and rationalist science, since the first arose as an imitation of Newtonian mechanics and the second as the attempt to establishing this mechanistic picture as a pattern of universal thinking. The shift of physico-mathematical theorising from its deterministic to its probabilistic counterpart and the emergence of evolutionary theories has a highly instructive methodological value and it still remains a source of inspiration and encouragement in humankind quest for grasping reality. A detailed investigation of central points in this vein regarding development and growth processes of the type outlined here will be given elsewhere, see [5, 6]. There special emphasis is given to notions like knowledge, know-how and information which are viewed from the perspective of learning and evolution. On the other hand, learning and evolution as basic elements of the process of social reproduction are examined with respect to their contribution to anti-entropic processes on the basis of which a system attempts to counteract the natural disintegration of its potentials.

The choice of a probabilistic framework, we shall mention, obeys the recognition that mass phenomena are appropriately characterised by causal relationships assumed to be concealed by the presence of wealth of individual events exhibiting manifold variations of a genuine and relevant nature. In this sense instead of a causal explanation one seeks to assess an average trend and a sort of regular fluctuations about it on the basis of a theoretically calculable probability law. It is extremely important to be aware of the fact that one does not seek to identify a causal explanation of individual events but rather a probabilistic assessment of their consequences; this is a sort of causal explanation at the macroscopic level. Thus stochastic models provide partial pictures of reality which may become progressively more accurate as the flow of information comes to reflect more faithful current observations and observations themselves come to involve more representative points of view.

Our final remarks concern the socioeconomic interpretation of Heisenberg's uncertainty principle. Here we think of the interplay of microscopic and macroscopic events and the implications resulting from this interplay to any attempt of evaluating states of being and states of becoming. States of being and doing defines the capability set of individuals, in the sense of determining an individual capacity to function in a given environment, at a given period of time. Thus, the capability set of an individual is a notion akin to the set of potentials available to an individual. On the one hand, we have that changes in the environment affect and hence call for a reassessment of the capability set of an individual, and on the other hand her (his) actions come to bear likewise upon the environment. In this sense, we see a type of simultaneously indeterminacy between states of being and states of becoming of the sort known in Heisenberg's uncertainty principle which may be captured for the time being by a diffusion approximation.

Acknowledgements

At different stages during the elaboration of the ideas presented in these essays I have benefited from advice, encouragement and discussions with many individuals. It is a pleasure to thank all of them here.

It was in the academic years 1979–1981, while I was at the Division of Applied Mathematics of Brown University as (a student and) post-doctoral fellow, that I initiated a careful and detailed exploration of questions in economic growth and development. Searching for meaningful contributions of mathematics to a basic understanding and unraveling of those forces at work that render virtually fruitless the immense efforts of state nations of the periphery toward progress and mastering of basic needs, I came to outline a research program on the labour-surplus economy which builds essentially the core of the present work.

During this period H. J. Kushner provided encouragement and fruitful critiques in addition to financial support. His incisive questions, continuous guidance and illuminating suggestions helped me to achieve a better understanding of the fascinating world of controlled diffusions, stochastic modelling and approximation tools. During my stay at Providence, I enjoyed a great support and hospitality from the Kushners including the generous professional advice of Linda Kushner that I want to acknowledge with deep gratitude.

In the spring of 1985, having been invited to join the BiBos Research centre, I had the opportunity to continue my research on the labour-surplus economy and formulate clearly various problems describing the unfolding of its productive forces. At BiBos I found a highly stimulating scientific atmosphere that contributed greatly to the shaping and development of these essays. I feel honoured to mention my immense indebtedness to Sergio Albeverio, who in many conversations contributed most to deepen my understanding on intricate matters of stochastic analysis and the art of applying mathematics. He has been a continuous source of ideas, motivation and encouragement. Through the generous guest program of BiBos I got the opportunity to expose parts of my work to well-known experts in the field of stochastic analysis which brought about invaluable feedback. In this manner cooperative work with M. H. A. Davis and J. L. Menaldi came into being. My skills and understanding of the powerful machinery of the variational approach to stochastic control have been decisively improved through various conversations with Alain Bensoussan and his team at INRIA to whom I am very indebted.

In 1986 I came back to the University of Erlangen-Nürnberg, where I had an appointment at the Institute of Applied Mathematics from 1981 to 1985. During this second sojourn in Erlangen, I was first in a joint appointment with the Institute of Mathematics at Erlangen and the Institute of National Economics at Nürnberg, and later as a researcher associated with the Research Group "Stochastic Analysis and its Applications" both supported financially by the DFG (German Research Society).

During the last four years at Erlangen I accumulated a large number of personal debts. First and foremost, I like to single out H. Bauer. Owing to his continuous interest in my research and engagement that cristalises into our joint research project (DFG) by means of which I came to enjoy at the Institute of Mathematics the best possible conditions without which I would have never brought my work to its present stage. Further, I like to thank both H. Bauer and M. Neumann whose patient listening, encouragements and human knowledge helped me to maintain my enthusiasm along the long gestation period of these essays.

At the Computer Station of our Institute, where I spent innumerable days and nights bringing parts of my manuscript into a LaTeX-version, I got helpful suggestions from many individuals. In particular, I like to express my appreciation to H. J. Schmid and K. Mach who provided advice and support with invariable readiness. Thomas Werner, Torsten Füg and Stephan Kaiser processed various parts of the original manuscripts fast and reliable. In the last and decisive phase of the production of the final LaTeX-version of the essays I had the fortune to get the skillful support of Stefan Schmiedl who with great talent and care succeeded in the task of making a heterogeneous and to some extent incoherent set of LaTeX-segments a uniform and complete LaTeX-output. This includes many suggestions regarding editing, booksetting and spelling; further, Stefan produced the index, most of the pictures and extended the available software with the purpose of suppressing the highly frustrating "LaTeX-capacity exceeded, sorry" message.

Before I come to acknowledge the private aspects or shadow price of this undertaking, I like to express my gratitude and profound admiration for two great scientists who are no longer among the living: Gerhard Tintner and A. T. Bharucha-Reid. The first initiated me in the mathematical aspects of economic theorising and the second in the illuminating world of probabilistic analysis, processes and Itô calculus. Both coined deeply in my mind the central role that time and change play in systems of whatever nature and, as a consequence, the dynamism and power that mathematical equations bring in all branches of (applied) mathematical sciences. My way of understanding the function of stochastic ideas in applications has been greatly influenced by their insistence that the most remarkable distinction between the deterministic and the stochastic approach lies mainly in the nature of the questions to which they give rise and the interpretation of the results they provide as well as the flexibility and generality that the corresponding formulations allow and that this in turn characterises essentially the connections between mathematical equations and the real phenomena.

The private indebtness to which I incurred is both immense and manifold. My greatest feeling of thankfulness goes to my parents, who with generosity and self-denial devote themselves to the social, cultural, educational and intellectual unfolding and well-being of their seven offsprings.

Last but not least, I am most deeply appreciative of the support, assistance, understanding and encouragement in many ways and at various levels that I was lucky to get from Renate Schlegel – companion and wife – in spite of her tight academic and medical schedule. She together with Glenna Burckel and Barbara Völkel made many efforts to transform my manuscript into comprehensible English and contributed largely in matters concerning style making the final version easier to read.

Chapter 2

Towards a working model and goal setting

Hunger is not an unavoidable phenomenon like death and taxes. We are no longer living in the seventeenth century when Europe suffered shortages on an average of every three years and famine every ten. Today's world has all the physical resources and technical skills necessary to feed the present population of the planet or a larger one. Unfortunately for millions of people who go hungry, the problem is not a technical one – nor was it wholly so in the seventeenth century, for that matter. Whenever and wherever they live, rich people eat first, they eat a disproportionate amount of the food there is and poor ones rarely rise in revolt against this most basic of oppressions unless specifically told to 'eat cake'. Hunger is not a scourge but a scandal.

The present world political and economic order might be compared to that which reigned over social-class relations in individual countries in nineteenth-century Europe – with the Third World now playing the role of the working class. All the varied horrors we look back upon with mingled disgust and incredulity have their equivalent, and worse, in Asia, Africa and Latin American countries where over 500 million people are living in what the World Bank has called 'absolute poverty'. And just as the 'propertied classes' of yesteryear opposed every reform and predicted imminent economic disaster if eight-year-olds could no longer work in the mills, so today those groups that profit from poverty that keeps people hungry are attempting to maintain the status quo between the rich and poor world.

Susan George
(see [80], pp. 23)

2.1   Introduction and outline of the model

Economic activities emerge and evolve within an institutional framework underpinned by adequate ideological structures. Therefore to understand means and objects of economic actions it is necessary to examine their ideological roots and the moral and ethic background of their institutions. In this sense we begin this chapter with a quick view of several notions like development, growth, employment and institutions, to mention only a few, which are in various ways linked to the present essays. This brings us to touch on some issues of political economy that we comment cursorily.

2.1.1   Preliminary remarks and background

For the purpose of the present essay we assume the most basic political, social and cultural features of development can be represented in a sort of performance functional, which reflects the state of well-being growing out of the underlying *modus operandi* of the system. Further, we assume the ruling party or, for the sake of simplicity, the control board after getting acquainted with the performance functional advances the available information on that state of affairs and reveals its views on the prevailing system of social rewards and value-judgements to the limits of political, cultural and moral strength and civil courage. On the basis of the newly emerging state of information, economic agents reevaluate and update the framework underlying their economic and political choices, see [200, 201]. Thereby, we shall try to draw a separating line, although a subtle one, between economic and noneconomic matters and point out the need to stress the normative function of institutions.

Issues concerning the performance functional are customarily the object of study of welfare economics, which is expected to offer a normative behavioural framework based on criteria of social rationality. These are usually aimed at the maximisation of disposable income, or a functional of it, leading to the optimal allocation of the economic resources of society. However, the fact that most of welfare economics depend on assumptions about the "rational man" and that this postulates a system of individual preferences independently of economic events, i.e. a sort of *prima causa*, renders the whole construction contradictory and irrelevant.

The process of economic development and its components

Economic development is a very complex process involving in addition to economic also natural, social, political and cultural evolutionary components. The latter are usually referred to as the noneconomic aspects of development and reflect a whole system of dynamic relationships linking the economy to society. We shall consider the economic, technological and institutional components as the growth aspects of development.

The interrelation of the above mentioned components is responsible for the interdependence of policy decisions concerning economic growth and social welfare, that in different ways affect the development paths. From this perspective, the theory of economic growth is forced to deal with problems traditionally outside its scope like relationships between the development of the productive forces and the nature of the social relations of production, which together determine both available potentials and incentives of capital accumulation.

Accordingly, we shall be concerned with the introduction of elements that counter the effects of factors inhibiting economic growth by increasing the productive capacity and improving of available labour and productivity of physical resources, skills and know-how in the economy under consideration. Since skills and know-how are mainly a consequence of better trained and organised labour, policy measures aimed at stimulating higher labour efficiency open the way to higher levels of the socioeconomic living environment, further performance motivation and social awareness, and this in turn increases the level of individual expectations, social aspirations and purchasing power breaking vicious circles typically attached to state nations in the periphery.

Let us point out that, when one attempts to take into account some of these relevant aspects of economic development, many limitations become immediately apparent due to the fact that at most one can realistically assume only partial knowledge of them. Think, for instance, of present and future preferences of individuals, attitudes of domestic government and planners towards long-term development issues, vested interests represented by foreign governments, often articulated by international aid agencies, all of which bring to bear upon the performance of the economy in various ways. These weighty aspects of reality call for the consideration of uncertainty, which shall account in this sense for imperfect knowledge of incentives for capital accumulation, motivation, attitudes of fulfilment growing out of developmental processes and different kinds of support for the economic policies promoted.

Social values and unemployment

Our primary objective will be to work out development paths along which surplus-labour can effectively be absorbed into production. Here, further obstacles come to light. To begin with, we shall call attention to the fact that training labour to a level of skill adequate to participation in the productive process entails capital costs as well as resettling and endowing unemployed labour with working tools at the existing standards, so that it can successfully enter production. Further barriers to the elimination of unemployment come from the fact that training and resettlement are associated with human costs, since this may require giving up traditional and cultural elements, customs, habits and the like.

However, even more formidable resistance may come from those already employed, the wealthy and all those getting benefits from the prevailing state of affairs. What renders the matter so delicate is that precisely the latter are in most cases well organised, enjoy political and economic power, and are in a better position to articulate their views and interests in a manner conducive to making national goals out of their individual and class issues.

There are still plenty of reasons why topics in economic policy are so controversial. We shall mention just one more. The principle of rationality on which the whole of policy making rests is extremely fragile and manipulable.

For, in order to decide on the worth of a particular action, it is necessary to weigh advantages and disadvantages associated with that action. As sources of disagreement the following questions arise:

- Which effects will be ranked as advantageous, whose advantages are to be considered and how are chosen mechanisms ruling the sense of advantage to hold?

- Should individual preferences always count, whose preferences again, and who is to decide about what are the relevant preferences?

- How will social benefits be defined, which segments of society are considered and how one prevents mechanisms of power bring about a loaded notion of social benefits?

- Which sense of justice, equity and equality is to prevail, do indeed exist effective mechanisms of social control and fair means of disseminating dissenting ideas?

- Which priorities will be attached *inter alia* to health, education, full employment and poverty, who decides about these issues and about organisations of knowledge, information and communication?

Dealing with these issues, one has to keep in mind that any notion of equality, freedom or freedom of choice remains void of any meaning and relevance unless one clarifies its dependence on the way differential social standings are inherited, on the extent to which individuals are free and able to express their grievances, on the presence or absence of obstacles to the realisation of desires, on the capacity to execute the choice. In particular, we shall point out that any genuine kind of political freedom can flourish only in an atmosphere of economic freedom.

Although it is always possible to disagree about what basic consequences of an action can be judged as advantageous, about the definition of an adequate performance function and so forth, whatever the choice we make, there are always capital costs involved in carrying out an action, in meeting the target, and social costs of diverting resources to other choices, i.e. those opportunity costs measuring how many units, let us say, of education have to be given up in order to allow for an extra unit of defense and the like. Social costs are also known as *accounting* or *shadow prices*, reflecting the fact that they cannot be directly observed and/or that they do not have in general an unambiguously defined market price.

Thus, a further objective of this study is to illustrate, based on adjoint state variables in the language of the theory of stochastic optimal control, how we can obtain accounting prices and, with their help assess in a precise manner the social costs associated with the action in question. Thereby, we shall be able to study some qualitative aspects of the system behaviour and get further optimality conditions.

Some subtle and unpopular issues

At the heart of orthodox neoclassical theorising is the analysis of economic agents' motives independently of their social environment and usually disregarding any social connection between them, see [208]. This may be seen as a consequence of the duality of mind and matter proclaimed by Descartes. However, the need of fully separating mind and matter was forced by the need of avoiding the free will attributed to the mind may jeopardise the conservation of motion if mind and matter are allowed to interact.

We have already mentioned that when orthodoxy rised to dominance, the mechanistic view of the world had undergone profound changes. Notwithstanding this, orthodoxy postulated a full separation of social life and motives of economic agents on the basis of a naturally given system of individual preferences which remain immutable as time passes by as an expression of the free will of mind; in "The Making of Neoclassical Economics" J.F. Henry [107] presents a straightforward analysis of the scientific, social, political and economic background from which neoclassicism came to predominance. In spite of the presumption of free will of mind, adherents of this school cannot escape the logical fallacy of having to add quite often the attribute of social to individual motives of its economic atoms when the need arises to explain a social phenomenon.

Adam Smith in his "Theory of Moral Sentiments" maintains that conscience is a product of social relationships on the ground that our first moral sentiments are concerned with the actions of others, and insists that considerations of utility are the last, not the first, determinants of moral judgements. However, he sees basic judgements of right and wrong as directly concerned with the agent's motive, not with the effect of his action. His notion of sympathy as the sharing of any feeling, is associated with the moral approbation of the agent's motive and to a lesser degree with the feeling of the person affected by the action. The concept of impartial spectator and others like self-love, fellow-feeling, self-command, constitute the basis on which the notion of market forces guided by "The Invisible Hand" rests.

Building upon the assumption that man is found in a social state and within a particular type of social structure, Smith's conception of morality and ethics echoes the principle of gravity, which governs the interactions among particles in Newtonian mechanics. By analogy he sought to give a solid foundation to the principles and rules, that in his view, guide the activities of and interactions among individuals based on a complex of abilities and propensities, which include sympathy, imagination, reason and reflection. The process by means of which individuals come to distinguish between objects of approval or disapproval and to judge their own actions as well as those of others, is however as in the case of gravity depending on distance and hence on the circumstances surrounding the actor and the spectator. Since in the modern society producers and consumers are separated, the markets are complex and rather imperfect, information is asymmetric and scarce, etc., the power of Smith's Invisible Hand rarely attains its expected efficiency; see [33, 222].

The general equilibrium theory that dominates economic theorising of orthodoxy sets out to develop the main issues of Smith's economic analysis taking self-interest and an economic structure where exchange relations prevail as the basic hypotheses. Thus ignoring the essential features of human existence in the social state, in particular the process of production and anything related to it. Despite the high degree of sophistication reached by the theory of general equilibrium and despite the great variety of methods applied, this theory can explain very little and the economic relevance of its content is slight; see [146, 58].

As a consequence, we lack a utility theory of value that in some dynamic sense relates to the most basic processes underpinning social reproduction and reflects judgements about social welfare in terms of capacity to function of individuals and social institutions, that takes into account rights and responsibilities of existing generations toward those not yet born, that adjusts to the evolutionary laws of the system. Hence, we need to bring in a more visible hand that makes up for some imperfections of the market, that provides some guidance about intertemporal decisions and the associated institutional framework and the like. The delicate nature of this task shall be visualised by stressing that in matters of economic policy there may arise some disagreement between private and social objectives, that pursuing social goals one needs to introduce economic policy instruments, the implementation of which creates conflicts related to tradition, the status-quo, etc.

Finally, economic policies have also to take into account the vitality of prevailing market mechanisms and make sure these mechanisms are not used as mask for privilege. Since we do not explicitly consider government, we shall assume the existence of a control board, the role of which will consist on the one hand, in making available adequate information, building up social awareness, solidarity, etc., within society and in promoting socioeconomic incentives to bring about a goal-conducive system of social preferences and value-judgements. On the other hand, this control board will be responsible for the design of an appropriate economic framework within which capitalists and workers operate. The objective function will thus consist of two components. The first reflecting rewards derived by individuals from the entire economic system involving value-judgements about intertemporal allocation, distribution and employment, while the second evaluating the system performance in regard to unemployment and capital to be bequeathed to the next generation.

Myths of freedom and masks for privilege

Examining various links and interactions between the periphery and the centre of the world capitalist system, one can better appreciate the real meaning of a purposefully and carefully elaborated machinery for ideological rhetoric and the actual impact of the "progressive" forces of Eurocentric capital upon millions of people in the periphery. Freedom, peace, international cooperation are among the key words by way of which one pretends to put the system together. In the mean time however the social distance between centre and periphery dramatically increases, while the high priests of orthodoxy still intensify the vacuous arguing in the name of Free Market and for the Anticommunist Crusade.

The economic, financial and political intervention of Eurocentric capital in the periphery and the resulting disintegration of it is then hidden behind the mythical forces of a market that has never been free, and in no way beneficial to the periphery. Think, as a way of example, of the world market for raw materials comprising mineral and agricultural products. A glance through recent history reminds us how the centre uses the world market for raw materials as an instrument for economic, financial and political purposes bringing in the periphery crucial factors of uncertainty and dependency.

Furthermore, changes in prices or quantities in the world market are linked by means of price indexes, capital flight and other tools to exchange rates and a monetary system tailored to suit the interests of the centre, which in this manner frees a great deal of purchasing power, i.e. gains negative entropy, making available means of counteracting its chronic deficiency of effective demand, overproduction and unemployment. A glance into works like [216, 155, 22] offers an insight into facts and fancy of free markets. These facts render the world market for raw materials extremely fragile and expose countries of the periphery to the arbitrariness of the centre. In this regard, international trade and agreements on exchange rates have been the subject of persistent and increasing criticisms which in turn have been suffocated with the ghost of communism. Hopefully, the capitulation of the latter contributes to change the prevailing state of affairs so detrimental in any sense to the periphery of the world capitalist system.

Let us make a comment on how the Invisible Hand might come to regulate the market. Moulding social preferences and value-judgements is no longer necessary since they are supposed to exit as a prima causa. Assume that information becomes available about damaging consequences for individuals and the social environment resulting from expansion processes of Eurocentric capital, in particular those that may lead to inhibiting productive activities of individuals, classes or nations and thus jeopardise their survival. This information would be expected to compensate the one becoming inoperative due to the fact that the market operates on an international scale, so that an opportunity shows up for manifestations of fellow feelings and the like as invisioned by Adam Smith in his Theory of Moral Sentiments. Then, these manifestations and the associated value-judgements, acting again through the market, would enact sanctions against those that violate the rules unless as a natural consequence, economic agents associated with the process under consideration make appropriate adjustments as economic rationality requires.

The weaknesses are obvious. First, the hierarchical structure of the world capitalist system would throw a veil on either the damages attributed to one of those enjoying political or economic power, or hide the perpetrator behind the complex organisation of capitalist production. In this connection let us recall that the admiration or respect for the rich and the great is, following Smith, perfectly natural and contributes to the stability of the social order.

Second, it is a matter of prejudices including racial ones. When the victims are aborigines somewhere in the Amazonia and the perpetrator a fighter for freedom and progress, say McDonald's or Exxon, then the outcome is easy to guess.

However, the high priests of orthodoxy may still argue on the grounds "of the supremacy of free-world and free-market principles over all inferior interests that the wise and virtuous men should at all times be willing to sacrifice their own private interest, even more their own particular order and subordinate society to the greater interest of the state, sovereignty or prosperity of the universe". Smith's notion of self-interest begins to exhibit flexibility, adaptability and other virtues. Self-interest of little man has to surrender to higher priorities which are naturally defined by the rich and the great, since self-interested pursuit of profit ranks highest. They may add, following the great master, "good soldiers, who both love and trust their general, frequently march with more gaiety and alacrity to the forlorn station, from which they never expect to return, than they would to one where there was neither difficulty nor danger. In marching to the latter, they could feel no other sentiment than that of the dulness of ordinary duty: in marching to the former, they feel that they are making the noblest exertion which is possible for man to make. They know that their general would not have ordered them upon this station, had it not been necessary for the safety of the army, for the success of the war. They cheerfully sacrifice their own little systems to the prosperity of a greater system"; see [208], pp. 235–236.

It is remarkable that writing a compendium of virtues and morality did not give his author the opportunity to mention critically genocide, plunder and wars of annihilation against non-European peoples. Adam Smith is of course a vivid and eloquent testimony of a transition period at the end of which old and repudiated passions turn into virtues. See in this regard Hirschman's The Passion and the Interest [109], Max Weber's The Protestant Ethic and the Spirit of Capitalism [227] as well as Sombart's Der moderne Kapitalismus [209]. Smith's long-lasting fame can be explained by the fact that his works provide a gold mine of skillfully masked and attractively written arguments for the subordination of mankind to the self-interested pursuit of profit. Millions of people became victims of the rise of this glorious system.

A few decades ago more millions followed as another great nation under the lead of fascism put forward her claims to supremacy in the world economy. The incessant advance of self-interest's search for profit, meanwhile under the fancy names free-world and free-market, continues increasing the human price for the sort of progress akin to the ideology of "rational man". Starvation amounts to 10.000 people every six hours estimates Susan George [80].

High technology and modernisation incorporated in weaponry as well as asphyxiating debts complete the picture in the periphery. In the meantime privileged nations in the centre of the world system under the sway of self-interest pursuit of profit intensify the struggle for any millimeter of free-market. The success conceived as astronomical figures of profits, national pride personified in the destructive power of their armies and well-being in terms of kilos of consumption to which large masses of society surrendered, swept away any remaining institutional or moral protection of society seen as obstacles for the further advancement of the spirit of self-interest. Let us close these thoughts with a citation of Sartre from his preface to Fanon's The Wretched of the Earth.

Let us look ourselves, if we can bear to, and see what is becoming of us. First, we must face that unexpected revelation, the strip-tease of our humanism. There you can see it, quite naked, and it's not a pretty sight. It was nothing but an ideology of lies, a perfect justification for pillage; its honeyed words, its affection of sensibility were only alibis for our aggressions. A fine sight they are too, the believers in non-violence, saying that they are neither executioners nor victims. Very well then; if you're not victims when the government which you've voted for, when the army in which your younger brothers are serving without hesitation or remorse have undertaken race murder, you are, without a shadow of doubt, executioners. And if you chose to be victims and to risk being put in prison for a day or two, you are simply choosing to pull your irons out of fire. But you will not be able to pull them out; they'll have to stay there till the end. Try to understand this at any rate: if violence began this very evening and if exploitation and oppression had never existed on the earth, perhaps the slogans of non-violence might end the quarrel. But if the whole regime, even your non-violent ideas, are conditioned by thousand-year-old oppression, your passivity serves only to place you in the ranks of the oppressors.

Jean-Paul Sartre
(see [65], pp. 21)

There is a tendency to judge the success of economic policies resulting from economic principles based on self-interest and all the rest, i.e. the so-called free-world economies, and those deriving from principles relying mainly on social objectives and value-judgements subordinated to the essential aims of the process of social reproduction, i.e. social and institutional economics, reducing anything to a comparison of the performance of capitalist and socialist countries. There are plenty of reasons why such a procedure cannot stand any serious scientific scrutiny. Let us just mention one.

The capitalist world consists of a centre and a periphery, i.e. the industrialised and the developing countries. Through a variety of channels, mechanisms and international institutions, i.e. the world market, Multinationals, IMF, BID, etc., a systematic net transfer of economic surplus from the periphery to the centre continues to take place as on the eve of emergence of capitalism, see [13, 74, 78, 141, 161, 162]. Therefore, in order for any comparison to deliver relevant information, if any, about the performance of different economic systems, think of the standard examples FRG and GDR, any net transfer of economic surplus generated in any country of the corresponding periphery shall be netted from the National Accounts of these countries. Unless one is ready to accept neocolonialism as the source of wealth and progress Eurocentric capitalism cannot do without. See [142, 143, 188], a neatly written report and political analysis of events marking the dissolution of GDR – the most astonishing, fast, spontaneous and peaceful popular revolution ever – and the unification of Germany – ironically the most amazing, effective, quick and ingenuous counterrevolution ever. There are too many things to learn from these events. Let us mention one point: discovering the deep differences which have developed between the two Germanies over the past 40 years, should let people think how crippling to Latin America must have been 500 years of Eurocentric domination.

At this point we are not interested in debating whether there may be any reasonable argument explaining why a responsible government should tolerate the continuing existence of surplus labour, whether there may be any economically logical basis for arguing that it may be optimal to leave idle substancial amount of labour with income redistribution undertaken to compensate those unemployed, whether profitability criteria should have priority over those oriented towards social benefits. For the time being, we confine ourselves to pointing out our commitment to the view that democratic principles like individual freedom and equal rights, and concepts like national sovereignty and peaceful international coexistence acquire real meaning only by adhering to each other and by interacting to generate world-wide social and economic development, cultural and technical progress from which no nation should be deprived for any ideological or strategic reason whatsoever.

2.1.2   Basic features of the model

Theories are devices or lenses by means of which one tries, consciously or not, to understand how and why reality exists, works and evolves as it does. We all form judgements and have beliefs, i.e. theorise, concerning the most diverse aspects of the world in which we live to the extent that one may safely conceive our theories as those elements playing the major role in shaping our reality and vice versa. Similarly, different socioeconomic environments result in a great variety of experiences of individuals and these in turn generate alternative economic theorising perspectives.

Hence, only the understanding of the nature, implications and interconnections of alternative social and economic theories may lead us to appreciate better and grasp contemporary pressing social and economic issues. We shall begin by describing briefly the lenses, that we will use, and the object of our analysis.

On the theoretical roots

Our theoretical studies on the labour-surplus economy rely on a highly aggregated and simple dynamic model involving neoclassical, post-Keynesian and neo-Marxian elements. This model captures essential features associated with the development process in peripheral societies of the world capitalist system.

Broadly the most crucial issues of developmental economics can be dichotomised into internal and external, or better national and international. The present essays focus essentially on national or domestic aspects of developments. However, we shall consider international issues whenever it is necessary to emphasise and explain various factors inhibiting and/or stimulating development. There are a variety of reasons for limiting the scope of our analysis. First of all, we want to keep our model as simple as possible, and secondly because these are the most controversial and subtle aspects of development economics, i.e. multinational corporations, international aid agencies, World Bank and the like. Although the activities of institutions of Eurocentric capital cover a wide range, their links to post-war arrangements of Bretton Woods and associate types of operations led to the characterisation of policies, intervention and impacts deriving from them with the general label of neo-colonialism. See [161, 13, 74, 78].

Let us assume a model which consists of three basic sectors obtained by the criterion of mode of production and contains enough structure to allow exploring short-term and long-term implications of intertemporal decisions and development strategies. This model undergoes structural changes and impulsive expansion of capacity; see [94]. To facilitate the identification of the theoretical roots of various ingredients of our model, we shall give a concise survey of those features of neoclassical, post-Keynesian and neo-Marxian economics that are relevant for the economic analysis of the present inquiry.

Neoclassical economics emphasises self-interest as the basic motive that acting through the principle of rationality, characterises the behaviour of individuals and firms as well. It divorces individuals from their social environment, and as a consequence, neglects interaction among individuals and the interplay between man and society. Individuals are free, they do not have needs but rather wants which lack satiation levels; *rational man* as neoclassicists called their protean hero is completely well-informed, perfectly foresighted and uncompromisingly seeks his maximal satisfaction only guided by self-interest and on the basis of exchange operations for the accomplishment of which individuals are exogeneously provided with wealth endowments. Neoclassical theory places major emphasis on the market as the best possible institution capable of accomodating the diverse goals of the most diverse individuals with the most diverse property holdings.

In its extreme version the market is elevated to the form of economic rationality *per se* independently of any framework of social relations. Neoclassical democracy is thereby restricted to the political domain leaving the evolution of society to objective forces of the market. These in turn, being governed mainly by (rationalistic) science and technology, operate beyond the reach of human will. As a consequence, two major tautologies emerge: capitalism equals democracy and freedom equals (free) market including (free) trade at a world scale from which world-wide development results as a natural outgrowth.

Post-Keynesian and neo-Marxian economics focus in contrast, on social relationships which shape and change what human beings are, think and do. Society is seen as consisting of arrangements of individuals in more or less definite socioeconomic relationships from which various patterns of individual behaviour result. Social classes and exploitation are the main issues, while self-interest and free markets are seen merely as the vehicles by means of which one conceals and perpetuates social injustice as it prevails in a class society. Social institutions are given a central and essential role as means of enabling action and moulding preferences in a way conducive to rational operation of the social system. They are regarded as mechanisms for promoting and enhancing the capability of the system to function rationally, e.g. avoid unemployment, centre/periphery polarisation in world development, ecological waste and other manifestations of pseudo-rationalisations attributed to the free market. In this sense democracy recovers its social and cultural dimension. Freedom of action in an ever changing socioeconomic environment substitutes the empty notion of freedom of choice.

In these essays we shall refer to neoclassical economics as orthodoxy or orthodox economics. [146, 230] offer detailed analysis of differences and similarities between these major schools of economic thought. In Chapter 3 we shall have more to say in this regard. As a closing remark, let us say, the fondness for clear ideas – one might attribute to neoclassicists – leads to an unrealistic picture of rational man, which defines – away from the outset – the essence of any decision problem. "Clear ideas are excellent when we are able to conceive them", writes the well-known physicist Polkinghorne, "but it may be that at certain times with certain problems it is better to be content with creative confusion than to strive for an oversimplified solution. Clarity can be purchased at the expense of the complexity of the truth" and later concerning the consistency of mathematical systems, "truth transcends theoremhood. Even in the austere discipline of mathematics there is more than meets the calculating eye". See [167], pp. 2 and 25.

Social reproduction as a primary aim

The most crucial socioeconomic issues surface when we look at society from the perspective of the *fundamental law of social reproduction*, depicted in Fig. 2.1 below; according to this law society engages in the process of material production in order to provide itself with the means of personal and productive consumption. In turn consumption is aimed at meeting the material and social needs of its population and replenishing depleted physical stocks, so that production can be maintained at the existing level.

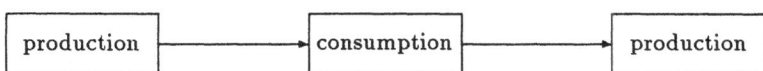

Figure 2.1: Fundamental law of social reproduction

Hence, production is not an end in itself but is the starting point of a social process aimed at providing the means of necessary consumption, i.e. means of livelihood for the direct producers and replacement of means of production used up. Consumption is in turn aimed at making possible a further phase of production at the current pace at least. The *economic surplus of society* that according to our definition amounts, roughly speaking, to the unconsumed part of production and includes the profits of the owners of the means of production, is what allows society to adjust to the pressure of an increasing population and to provide improvements in the standard of living, when ploughed back into the production process.

The relevance of the notion of social economic surplus rests thus on the fact that the growth potential of society is tantamount to its potential for generating economic surplus. This is because it enables the specialisation of a fraction of the labour force, freeing them from working for their survival, and preparing the grounds for an appropriate social division of labour. Once the social division of labour appears, it makes headway as a cumulative process contributing, although with some lag, to the improvement of the techniques of production, and this in turn brings about further increases in production and in economic surplus. More precisely, economic surplus acts as means of financing investments and technological improvements, and once started it may become self-propelling.

Consequently, the barriers to the generation of economic surplus and its purposeful application, interfere with the processes of growth and development. This motivates us to take a closer look at the forces behind it, which are mainly responsible for a variety of socioeconomic processes reflecting conflicting interests.

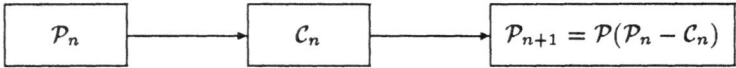

Figure 2.2: The process of social reproduction

Let us denote the processes of production, necessary consumption, formation of economic surplus by $\mathcal{P}$, $\mathcal{C}$ and $\mathcal{S}$ respectively. Then the process of social reproduction as described in Fig. 2.1 can be visualised by Fig. 2.2, which expresses the fact that the scale of production in the $(n+1)$−th period depends on the economic surplus generated in the n−th period. In order to make more apparent the insight behind this process, we need to elaborate these ideas a bit more. Let W, L, δ and K denote respectively the wage rate, units of labour actually employed, the rate of capital depreciation and the stock of capital associated with the n-th period. Then necessary consumption $\mathcal{C}$ in the n-th period can be written as

$$\mathcal{C}_n = W_n L_n + \delta_n K_n,$$

which represents certain amounts of characteristics attached to wage goods and means of production relying on which labour and production can go on. Further, let N be the available labour, so that $N - L$ becomes the reserve army. If production $\mathcal{P} = PQ$ with P and Q denoting respectively a price per unit of output and output expressed in units per period of time, then the corresponding economic surplus $\mathcal{S}$ can roughly be written as

$$\mathcal{S}_n = \mathcal{P}_n - \mathcal{C}_n = P_n Q_n - \{W_n L_n + \delta_n K_n\},$$

from which we can easily see that the process of social reproduction, depicted in Fig. 2.2, really means $\mathcal{P}_{n+1} = \mathcal{P}(\mathcal{S}_n)$. It is necessary to keep in mind this picture of the process of social reproduction in order to avoid the confusion dominating orthodox economics. First, the generation of economic surplus entails an excess of production over consumption as means of promoting development requiring mainly labour-time and other resources usually incorporated in the general notion of means of production. Second, the level of productivity of an economic system depends primarily upon technology, know-how, entrepreneurship and the like, rather than on the amount of wealth that this may represent. It is on this basis that economic surplus comes into being.

Since production is a process that involves time, it requires to distinguish between prices that hold during a period of time and prices without any time dimension. In this sense wage rates are related to that period of time during which one commits the labour-power of workers. Similarly, profit rates per unit of time are the price for committing means of production, which then acquire the form of capital, to production during a period of time. The accumulation process relies on supply of productive equipment to assist workers in producing social output and on availability of finance needed to meet the requirements of necessary consumption. This points to the need for differentiating sharply between factors of production and sources of income, in particular, between capital as stocks of means of production and capital as financial property. From the point of view of national accounting, one gets a better insight into some of these issues; see [23], Chapter 1.

2.1.3 The reserve army of unemployed labour

As a first approximation, we shall assume an economy consisting only of a secondary sector and a reserve army of unemployed labour. A fiction that while reflecting the social costs of the advancement of the capitalist mode of production brings to light essential issues concealed by orthodox economics. To see this, let us recall a few economic concepts of classical political economy.

On the concept of economic surplus

In matters related to developmental planning, the notions of social economic surplus and economic surplus of society play a crucial role. The per capita difference between material production and the consumption necessary to produce it will be called *economic surplus*, while *social economic surplus* stands for the (total) economic surplus of society. Necessary consumption comprises the means of production used up, which requires thus replacement in order to keep production going on, and the personal consumption of the direct producers, aimed at maintaining or regenerating labour power.

Stated in this way, the concept of economic surplus hints at a framework paralleling the theory of social reproduction of classicists and stresses the continuous renewal nature of production, as a material process with the help of which society replenishes its means of production and as a social process that enables the regeneration of labour power of workers. This type of reflections suggests the most natural way of introducing entropy; namely in relation to lack of available potentials the system necessitates to the aim of maintaining the process of social reproduction going. Concerning the concept of economic surplus there are three points which we would like to emphasise:

1. It appears to be more natural for planning and management purposes to deal with *potential* rather than *actual* economic surplus and this is defined as the per capita difference between the potential material production society could bring about using all employable productive resources and that which might be regarded as essential consumption.

2. The assessment of the magnitude of the potential economic surplus involves difficulties of various kinds, since

   - it requires a more-or-less clear goal visualisation, i.e. the picture of a more rationally ordered society,
   - it presupposes an adequate knowledge and appraisal of the performance of the socioeconomic organisation underlying the existing social order which is in general only partially and/or indirectly observable,
   - one has to work with responses of individuals and with goal-dependent notions like rationality, efficiency, etc., which might induce some sort of conceptual circularity.

   Our alternative is to resort to a more objective notion of performance based on the capability approach, i.e. relying on the capability of the system and/or its individual agents to function in a certain way.

3. The mode of utilisation of the economic surplus is roughly characterised by the path of the investment-consumption mix and constitutes the engine of the process of capital accumulation determining, together with the issues mentioned in 1. and 2. , the pace and direction of the development process.

The realisation of the potential economic surplus requires substantial institutional changes and a thorough reorganisation of production, with the market playing a central role as a mechanism of distribution and exchange, which must remain, however, subordinate to the real activities of production and consumption. However, it depends itself on the degree of development of the productive forces, on the corresponding structure of socioeconomic relations, and on attitudes about social welfare as well as on value judgements promoted by the government or control board.

This view of economic surplus shifts the focus from actual to potential production and essential consumption, entails identifying and measuring the productive capacity, excess consumption and output losses. These output losses are due in general to the existence of unproductive workers, irrational and wasteful organisation of the existing productive apparatus, and to the existence of under-, un-, and misemployment of productive resources. A theory of value or system of adjoint variables, which mimics the equations of motion of classical mechanics for conservative and dissipative systems, provides us with a way of balancing present and future consumption claims associated with the economic policy in course and completes the working framework; see [5, 6].

Conflicting interests versus the equilibrium paradigm

An analysis of the two foregoing relationships helps us to draw following enlightening observations on the nature of the conflicts hidden by the orthodox and highly idealistic conception of general equilibrium.

- Increases in the level of employment L or in the wage rate W lessen the growth potential of the economy and act as a check on the profit rate of capital by inhibiting the expansion of the economic surplus. On the other hand, the larger the reserve army, the greater the downward pressure on wages and the upward tendency of the rate of profits. The latter in turn entails higher economic surplus.

- The larger the fraction of the economic surplus dedicated to widening the production process, the larger the demand for labour. On the other hand as a result of a larger demand, the greater the upward pressure on the wage rate, the greater the downward tendency of the rate of profits, which keeps in check a further expansion of the economic surplus.

- Increases in the productivity of labour may have contradictory effects on wages. Because higher productivity of labour implies less labour-time dedicated to the production of the value-equivalent of wages, entailing thereby a reduction in the value and price of the means of livelihood.

- The latter brings about a downward pressure on wages, a higher pace of accumulation, replacement of men by machines, and replenishment of the reserve army slowing down and even stopping the growth of labour productivity by means of idleness of a higher fraction of the available labour force.

- On the other hand, a higher level of labour productivity and the induced price reduction associated with it may effect an expansion in the scale of operations of production, and a widening of the market which requires in turn enlarging the commodity basket on which the notion of subsistence minimum relies, and an increase in the level of employment.

Economic rationality as understood by orthodoxy requires saving working time, i.e. the number of man-hours contained in a unit of social output, based on an increasing efficiency of the means used to produce. At this stage we have to face questions like what prevents capitalists from taking advantage of pressures on the wage rate coming from the reserve army, or of transfering to the centre or other domains of the system a fraction of the economic surplus of the society under consideration in order to counteract a downward pressure on the rate of profits, etc.

A further question is what prevents society from consuming the whole economic surplus jeopardising thereby its development potential by hindering an advancement of the social division of labour and reducing the decision horizon on which provisions for the future are made. Because it is yesterday's economic surplus that ensures today's investments and tomorrow's increases in the social product, partly freeing society from the urgency of necessary consumption which provides room for improvements in available techniques and the prevailing organisation of production.

A labour force limited by the handicaps of a low or inadequate income is deprived of any opportunity for improving its capabilities and functionings, and entails requirements for an efficient participation in the social process of production are not given. The affected fraction of the labour force can neither contribute to generating social output nor to expanding the market, lacking the needed purchasing power, and is likely to become socially and economically a burden on society.

As we already suggested, we shall not define away, as orthodox economics does, social injustice, the irrationality of a misconceived allocation of (human and material) resources, wasteful application of the social product and the like, resulting from decisions based on concerns fairly limited in time and social horizon, to conclude ultimately that the social system so envisaged moves towards and finally settles down to a mythical state of equilibrium. On the contrary, we shall accept these facts of life as a possibility and resort to institutional constraints with the aim of neutralising existing disequilibrium configurations and recreating social potentials of development.

2.1.4   A simple model of social reproduction

The main themes of political economy are dominated by *the principle of social reproduction* and this manifests itself as the rationale of social activities around the process of production by means of which goods and services are provided to satisfy social needs. The latter consist primarily in the continuous reproduction of the factors of production: first, sustaining labour power or the capability of labourers to carry out work and second, renewal of means of production. The history of mankind has been an incessant struggle to increase and improve the social production of goods and services beyond the level of necessary consumption but also the struggle for sharing the surplus of social output.

Thus the evolution of the productive capability of a system is reflected by the unfolding of its capacity to generate economic surplus, and this is intimately connected with the principles governing its appropriation. The relations of production and distribution which are essential to the identification of a society – its individuals, know-how, culture and environment – have to exhibit certain features of coherence to keep them going. These features are stressed by calling such a society a social system. At a fairly general level one identifies three central features of social systems.

- *Self-reproduction*, which refers to interaction of the system with its natural environment according to its know-how, culture and autonomous patterns of organisation. It stresses dependency on the natural base.

- *Self-renewal*, which in addition takes into account a continuous recreation of available potentials including replicative fidelity (order) and completion of replication processes (cleavage).

- *Self-transcendence*, which hints to diversity and mutations with the aim of pushing the system beyond limits associated with the natural base, culture and organisation bonds.

These features contain both the essence of coherence and break, try to maintain the system identity and alter its structure. Therefore, they are at the heart of our analysis of growth and development processes from the perspective of learning, information and evolution; see [5, 6].

Renewal processes associated with production

The aforementioned social needs give rise to two renewal processes underlying production. One process corresponds to the recognition that no society can go on producing without continuously reconverting a part of its social output into means of production. The other process responds to the need to renew the capacity of labour or labour-power of active members of society, i.e. equipping people with the capacity to do given work. This comprises a continuous renewal of nourishment, rest, appropriate leisure facilities, shelter, maintenance of health and skills. The renewal of available labour-power is the result from various processes of biological, social and cultural nature.

Failure to appropriately meet the renewal requirements posed by means of production and labour-power leads to a decline in the capacity to produce and in the incentives associated with social relations of production and other sources of social development. Let us point out the role played in this connection by labour: first, contributing to the conservation of value of factors and means of production, and second, generating economic surplus. The crucial relevance of this observation can only be fully appreciated through an analysis of the potential economic surplus foregone and the opportunity costs deriving from the existence of reserve armies of unemployed. Not to mention social costs in terms of moral deterioration of human capital, social strains, poverty, starvation and other social maladies eventuating even in fascism.

In order to use mathematical techniques to study the capability, e.g. characteristics like functioning, of economic systems for social reproduction, it is necessary to design a model. For this purpose one needs to determine a set of variables which may serve as quantitative characteristics of the system operation, and then establish relations among them which approximately describe those aspects of functioning of the system viewed as essential to the process of social reproduction. A further step in designing a model is the mathematical description of what enters the system and what it produces.

We shall outline next a very simple model that illustrates in an intuitive way various features of modelling and suggests that the information content of any model, as a representation of reality, depends critically on the goals for which it has been constructed and on the theories the model designer endorses. This is not surprising at all, since in different environments people experience life differently and think of its meaning differently as well.

Production, distribution and allocation

Let us present briefly some issues in the production, distribution and allocation of the social output using a static model that takes into account outlays of labour and means of production with the purpose of obtaining certain social output. Let L represent labour, K the means of production and Q the social output. In general L may be a bundle of various kinds of specific labour, e.g. ploughing, sowing, spinning, etc. Similarly, K may be a bundle of various types of means of production, e.g. tools and machines, coal, iron, various kinds of land, etc. Finally, Q may also be a single commodity or a bundle of specific commodities. It is customary to work with vectors of the type

$$
a_l = \begin{bmatrix} \frac{L^{(1)}}{Q} \\ \frac{L^{(2)}}{Q} \\ \vdots \\ \frac{L^{(n)}}{Q} \end{bmatrix} \quad \text{and} \quad a_k = \begin{bmatrix} \frac{K^{(1)}}{Q} \\ \frac{K^{(2)}}{Q} \\ \vdots \\ \frac{K^{(m)}}{Q} \end{bmatrix},
$$

indicating the labour and capital requirements in order to get 1 unit of the output Q (viewed as a single commodity). The components of these vectors are usually called the *technical coefficients* of production, since they reflect the technological features underlying the process of production.

These technical coefficients represent the outlays of the factors of production, i.e. specific kinds of labour and means of production which under the given technical conditions, the production of one unit of output requires. On the other hand, the reciprocals of these coefficients are seen to represent the *productivity of* the given *factors of production*. Needless to say, that a higher level of labour productivity has to be associated with greater outlays of the means of production connected with labour and higher level of social output. In this sense a given production technique or technical process is characterised by the corresponding vectors of productivities of the factors of production which may change in the course of time. As a rule, a product may be produced according to various production techniques.

For the sake of simplicity, let us assume that production can be represented by a simple model in which social output consists only of one commodity produced by means of a technology with fixed coefficients a_l and a_k, where a_l represents the labour requirements in man-years per unit of output and a_k the capital requirements per unit of output. It is intuitively appealing to think of this commodity as corn which helps to visualise the assumption that the only factors of production are capital in the form of seed-corn and labour. All the work is done by labourers in exchange for wage payments from capitalists which are the owners of seed corn.

Suppose that L units of labour measured in workers-year combine with the quantity K of seed-corn to produce by the end of the year, or of the production period, the quantity of corn Q. Let W denote the (annual) wage per worker measured in bushels of food-corn, which is assumed to be paid at the beginning of the (working year) production period. That means, in order to get the output Q the capitalists need to advance corn in the amount $C_n = WL + \delta K$.

It is customary to assume that independent of its age capital, i.e. seed-corn, wears out at the rate d_n per unit of output. For the sake of simplicity, we assume that $d_n = d$, i.e. deterioration or spoilage of seed-corn does not depend on the period of production. It is natural to think of production going as long as the surplus $S_n = Q_n - C_n$ and the rate of profit $r_n = \frac{S_n}{C_n}$ remain positive. It is easy to see that r_n depends on a_l and a_k. Setting $r = r_n$ and corn the unit of account with current price $P = 1$, we obtain the following *distributional relationship* known as the *price equation*

$$1 = Wa_l + (d+r)a_k, \tag{2.1}$$

where Wa_l represents labour costs and $(d+r)a_k$ seed-corn costs involved in producing one unit of corn. W and r are as usual the wage rate and the rate of return. Furthermore, denoting consumption per worker by c and assuming net investment equals gK where g is the rate at which corn output grows, we get a relationship reflecting *allocation of resources* known as the *quantity equation*

$$1 = ca_l + (d+g)a_k, \tag{2.2}$$

which says that the production of one unit of corn requires consumption and investment of corn in the amounts ca_l and $(d+g)a_k$. In algebraic terms, each of these equations contain two unknowns and cannot, therefore, deliver uniquely determined solutions. However, they give information on how one variable changes as the other varies.

Taking a_l and a_k as known, one easily recognises that to each feasible pair (c,g) and (W,r) corresponds respectively a distribution of corn income between wages and profits and an allocation of corn resources between consumption and investment. Under certain assumptions on the substitution ability of labour-power for seed-corn requirements, one can postulate the existence of a *production relationship* denoted by $\mathfrak{h}$ which in a sense characterises a family of technical coefficients of production. A family of technical coefficients of production is usually referred to as a technology, a notion which we will make preciser at a later stage. Thus we have also the *production equation*

$$\frac{1}{a_l} = \mathfrak{h}\left(\frac{a_k}{a_l}\right) \qquad \text{or} \qquad q = \mathfrak{h}(k), \tag{2.3}$$

where $q = a_l^{-1}$ and $k = a_k a_l^{-1}$ which represent respectively units of corn-output and seed-corn per worker.

Moreover, one recognises easily that the *trade-offs* between *growth and consumption* $\frac{dq}{dc}$ and between *profits and wages* $\frac{dr}{dW}$ are given by the rate $-\frac{a_l}{a_k}$, i.e. these are tantamount to trade-offs between labour and capital requirements, as a little reflection shows. This suggests that considering the coefficients a_l and a_k as not purely technological parameters but as dependent on the cultural and institutional structure may lead to a more realistic understanding of the issues associated with the distribution and allocation of resources. A look at eqs. (2.1) and (2.2) indicates that the introduction of mechanisms or rules for the determination of W or r on the one hand, and of c or g on the other, brings about alternative perspectives for the distribution of income and the allocation of social output. Regarding different choices as alternative models, one finds various alternatives of economic theorising, i.e. classic, Ricardian, Marxian, neoclassic, Keynesian, etc. See [146, 53].

Capability issues of development and growth

In a dynamic version of the simple model of the foregoing section, one would need to assume that the technological coefficients a_l and a_k vary in regard to time. In particular, one would assume that the ratio $\frac{a_k}{a_l}$, known as capital intensity, evolves as time passes. This would entail that the patterns of production, distribution and allocation also become time-dependent shifting attention to questions about time-dependent changes in the ability of the system to meet the requirement of social reproduction. The understanding of matters of change and time requires sharpening the way we view the process of production and its components, since this process comprises different operations involving various kinds of labour and working periods.

Thus we have to develop precise notions concerning the overall time needed to produce a given commodity known as the production period, specify used-up means of production in a working and in a production period which leads to considering concepts like working and fixed capital, short-term and long-term issues of production among others. Only in this way we may come to grasp the essentials of the process of social reproduction. In this regard, let us mention once more that a dynamic view of a_l and a_k that incorporates basic elements of the capability approach, i.e. characteristic features of consumption bundles on which "states of being", "states of doing" and "functioning" rely, offers a realistic way to understand and investigate the capability of a system to produce, to generate an economic surplus and to allocate it efficiently.

One major theme in political economy is the analysis of economic growth and development. Considerable efforts are directed to the identification and understanding of the basic processes promoting and hindering development and consequently progress. The ultimate aim of these efforts is, however, to provide a sound basis for the design of economic policies and actions that facilitate steering forces carrying the actual processes of development and growth. Since it is unanimously recognised that accumulation and reinvestment of a part of the social product among the chief driving forces behind economic growth and development, one naturally directs attention to those general principles underlying distributional patterns and governing the allocation of social output, as well as to the main mechanism motivating social behaviour with respect to working and allocation of capital and labour income, i.e. profits and wages, for purposes of consumption and investment. Issues of this sort can also be dealt with adequately from the point of view of capability, since one can directly relate in this way, say, the ability to function efficiently with certain consumption choice. This puts consumer behaviour and policy making in a more solid footing, retaining however essential elements of subjective assessment of rewards.

From a macroeconomic point of view, as we would like to emphasise, it becomes necessary to consider the capital intensity at any time as a random variable. That would mean, at any time there is a variety of values that the ratio a_k/a_l may assume, and all of them with an appropriate rate of occurrence. To this manifold of capital-intensity values there corresponds a manifold of values of the rate of return on capital that reflects realistically the process of capital accumulation, and constitutes a necessary condition for the rise of competition.

As mentioned before the technical coefficients a_l and a_k contain information on productivity or efficiency of the factors of production and on the rate of profit. The assumption that the knowledge involved in a_l and a_k is, broadly speaking, accessible to anyone corresponds to the belief that new knowledge gradually diffuses through the entire economy, contributing in this sense to the uniformity of profit rates. Since the diffusion of technical knowledge, innovation and discovery are rather overdetermined phenomena, it seems more appropriate to see the ratio $\frac{a_k}{a_l}$ as a random variable. This would reflect realistically the manifold values that this ratio may attain as a consequence of an incessantly chasing, catching-up and continuous adoption of improved techniques by capitalists wishing to maximise the return to capital either in the form of means of production or means of finance. Such a probabilistic view of capital accumulation makes possible a macroeconomic analysis of growth and development from a microeconomic perspective. [5, 91] deal with some of these aspects. See also [66].

Hence starting from available labour $N = (N_t)_{t \geq 0}$ and human actions upon nature, processes reflecting participation of labour $L = (L_t)_{t \geq 0}$ and capital K, both understood as forces driving production, come about, since they are means of fulfilling social reproduction. Labour and capital characteristics represent specific potential of directing human labour, codified and natural energy toward rearrangement of materials according to certain aims; see in this regard subsections 2.1.4 and 2.2.1. Human actions are usually visualised as combination of labour and capital characteristics with the purpose of generating artefacts. In an abstract form human artefacts may assume the guise of social output, a concept usually formalised by means of a functional relationship of the kind $Q = \bar{F}(t, K, L)$, which we discuss briefly in Section 2.3.

To understand the real meaning of this relationship let us recall the notion of forces of production includes labour and means of production, in particular the tools and machines which are placed at worker's disposal for production purposes at stake. Instruments of labour are on the other hand themselves the result of human labour and as such they embody human knowledge of a degree of sophistication which depends on prevailing technology and know-how. Knowledge does not evolve by itself but necessitates assistance and this comes in the form of energy and matter, a feature which stresses the relevance of learning as an evolutionary process and accentuates its time dependency as the presence of entropy entails. Since acts of acquiring and codifying knowledge and human information take time and require energy and materials as well, processes of learning are linked in general to the availability of means of finance, the most fundamental form of which is surplus product or economic surplus. A basic feature of production activities is generating surplus product, i.e. a physical excess of output produced over inputs employed, since only in this way can the circular flow of production be maintained.

To find out the proportion of social output that needs to go back into production so as to provide for necessary consumption acquires the utmost importance. In particular this is true, whenever labour and means of production or instruments of labour enter the production process as separate entities owing to property rights inherent to capital. Since the role of necessary consumption is to renew the productive power of society and this includes labour and capital characteristics which are subject to depreciation and obsolence as we already know, putting the decision to allocate economic surplus in the hands of the owners of capital brings in asymmetry and fragility in the process of social reproduction. Usually the owners of capital hire the labour of a large number of individuals at agreed wage rates and organise production directly or through hired managers. The excess of the product over the wage bill or wage income constitutes capital income or income of property. The principle of effective demand provides information on the adequacy of the purchasing power supported by the wage bill and may contribute to properly anticipate expectations of capitalists about the future and to shape the desired rate of accumulation. Therefore one cannot overemphasise the significance of income distribution on issues concerning development planning, improvement of technology, realisation of profits, motivation and expectations.

2.2   Random features of employment

By now the roles and aims of labour in the process of social reproduction should be fairly clear. Thus, we turn our attention to examining how labour itself changes alongside the process of production. This involves analysing technology, machinery and social relations including organisation and control of labour.

However, we need to confine ourselves to some aspects of the evolution of the labour process, comprising occupations and required skills as well as shifts of labour among occupations and acquisition of the corresponding skills. Lack of mobility and flexibility, as well as failure of adapting to and/or shifting among available occupations result in unemployment. Within the context of our model, where workers own no more than their labour power, unemployment means no access to means of labour and as a result no opportunity to gain a livelihood.

The consequences and roots of this problem are manifold. We shall focus mainly on some aspects associated with uncertainty and explore the manner in which they interact and shape the dynamics underlying labour demand and supply as well as migration and mobility. Nevertheless, we shall begin with an account of the labour process, unemployment and their connection with social relations of production.

2.2.1   The transformation of the labour process

The insistence of orthodoxy to disregard social relations leads to a state of confusion and contradictions which are most vividly reflected in current writings about labour and employment.

Some curiosities of orthodox wisdom

To begin with, let us point out that the principle of economic rationality aims at applying means of production in a manner considered as the most efficient. That is, in a way that allows more and better production using fewer working hours and more capital, and hence saves working time across the whole society. This should constitute working time available for production of additional wealth. On the other hand, the essence of a theory of unemployment known under various guises tells us that there is a problem of satiation — people now have all they need. The relatively low social demand that follows from the problem of satiation entails a relatively low level of social output which in turn explains unemployment.

Other theories centre not so much on the system's inability of providing employment opportunities as on the sort of work it provides. Production has become increasingly subdivided into petty operations that fails to sustain the interest or engage the capacity of humans with a standard level of education. On the other hand, economic rationality dictates precisely division of labour to a point that demands ever less skills and training. This not only reduces costs but renders calculable factors, on which rationality and management depend.

Further, there are even some theories that sustain that it is not the collapse of labour processes underlying the capitalist mode of production but rather the appalling success emanating from these that need to attract our attention. Since pressures of poverty, unemployment and wants have been eliminated, the prevailing discontent and distress cannot be touched by providing more prosperity and jobs because these are precisely the source of dissatisfaction. Thus, not additional benefits but rather suppression of them is what is needed to counteract unemployment and other social maladies. American and British experiences in the eighties falsified once again the alleged efficiency of cures tailored according to this brand of orthodox wisdom.

Plenty of empirical and analytical studies looking at this matter from the point of view of anthropology, economic theory, psychology and sociology indicate that with unemployment also physical and mental health problems increase, family and community cohesiveness deteriorate. Drug and alcohol addiction, aggression and delinquency are not rarely accompanying symptoms. However, orthodoxy understands only the language of profits and insists in its straitjacket of rationality according to which "the more the better" and "more of the same" constitute optimal behavioural rules of *homo economicus.* In this one-dimensional world there is thus neither room nor understanding for human despair and pain; even less, as it is mostly the case, when the victims are blacks, women, minorities and people in state nations of the periphery.

These observations and the support statistical information provides reveal the true meaning of the aims of freedom, well-being and defense of vital space behind the rhetoric of big business and its representatives. How people somewhere in the Upper Guinea Coast, Haiti, Bangladesh or Ethiopia may get benefits, when great nations from the centre embark upon new economic, financial or military undertakings, will remain a puzzle. With certainty we can only say that income drains, surplus transfer, terms of trade and other mechanisms of the international economic order will make sure that they too carry the finan-

cial and economic burden of such operations. Bretton Woods institutions are
in this sense highly efficient.

Some features of the human capacity for work

To understand the nature of the process of labour and the development of productive forces, it is necessary to examine carefully the essential links among
social relations of production, technology and organisational structures of society. Since there is no automatic and immediate transformation of a mode
of production as a result of changes in its social relations, overdetermination
hinders a precise characterisation of the historical process any society in transition undergoes. This does not amount to say that processes shaping society
are accidental and/or the outcome of blind forces. On the contrary, they are
the result of a social determinate and gradual evolution of forces and relations
of production. Nevertheless, they are complex historical processes no susceptible to simpleminded reductionism or determinism. We shall deal only with
some fairly limited aspects of the labour process and refer to [31, 95, 175] for
a detailed study.

Putting mankind at the centre of the process of production brings into
play essential anthropological and evolutionary perspectives which enhance our
understanding of labour processes. First of all, let us recall that any form
of life goes on existing by acting upon nature for the purpose of providing
itself with a vital sort of energy. This includes changing the form of natural
products to make them more suitable for its needs. A distinguishing feature of
human labour is that of being a conscious and purposive activity that starts
in the imagination of the labourer before its material realisation. This human
capacity for work following a well-conceptualised model is deeply related to
human culture. Fig. 2.3 depicts the main features of the labour process.

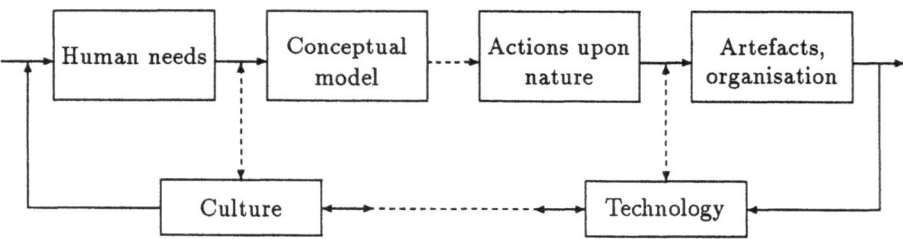

Figure 2.3: Main features underlying processes of labour

Working tools and machinery, which represent stored human labour, witness
the increasing complexity of labour processes and mark the progressive development of mankind along history. An additional feature of human labour, we
would like to stress, is that at a social level the motive force, conception and execution of labour itself build a unity. Further, this unity of motive, conception
and execution can be separated at the individual level, i.e. they may involve
different individuals, because the conception always precedes and governs the

execution of labour. In Fig. 2.3 we call attention to this fact using broken lines.

This separation is mainly own to mankind's capacity to assume a great variety of functions which can further be broken down to the basis of family, group, social and individual assignments. However, this richness and adaptability of human labour have led to confusing labour power, i.e. human capacity to perform work, with other sorts of power as the notion of factors of production demonstrates.

The consequences of this confusion are manifold. Let us mention two of them. First, the separation of workers from the means of production to which they regain access only by selling their labour power. Second, the labour process, originally aimed at creating products suitable to the satisfaction of human needs, has become specifically a process dominated and shaped by the drive to accumulate capital.

From commodity characteristics to the capacity to function

The need to recapture cultural and social links that get lost to the process of labour, as a consequence of the aforementioned separation between conception and execution as well as the division of functions, becomes more apparent when we have to face some issues concerning the supply of and demand for labour. Because at this point we are then required to look at structural features of society related to the capacity of its individuals to do specific types of work.

Thus, based on Sen's capability approach, introduced briefly in Chapter 1, we want to draw attention to those characteristics of consumption commodities that enable individuals to meet labour requirements and to function in a way suitable to a successful participation in the process of social reproduction. Attached to the notion of commodity we consider a function $c(\cdot)$ converting any bundle of commodities x_i into one of characteristics $c(x_i)$ and identifying the commodity bundle at hand. The set X_i of (vector) bundles x_i will be called the *entitlement* associated with the individual i. Individuals are distinguished by a set F_i of *utilisation functions* f_i that characterise the pattern underlying the commodity use that individual i actually makes. In Fig. 2.4 we illustrate diagrammatically the logic behind the capability approach.

Within the described framework, we will assume any individual i is given a unique entitlement set X_i out of which he or she is restricted to choose commodities, and that in a certain way she or he gets associated with a set F_i representing her or his personal features. In a sense, the latter can be thought of describing i's personal conversion pattern of characteristics into her or his achievements, i.e. *functionings*. Then, whenever the person i chooses the utilisation function $f_i(c(\cdot_i))$ and the commodity vector x_i, the achieved functionings will be given by the vector b_i.

Thus, a functioning is an achievement of a person: what she or he manages to do with a given commodity vector, i.e. being well-nourished, knowledgeable, mobile, well-clothed. Hence, the set $Q_i(x_i)$ reveals the various functionings available to the individual i taking into account her or his personal features F_i given x_i, i.e. the "states of being" and "states of doing" characterising individual i. On the other hand, the class $Q_i(X_i)$ shall represent the freedom

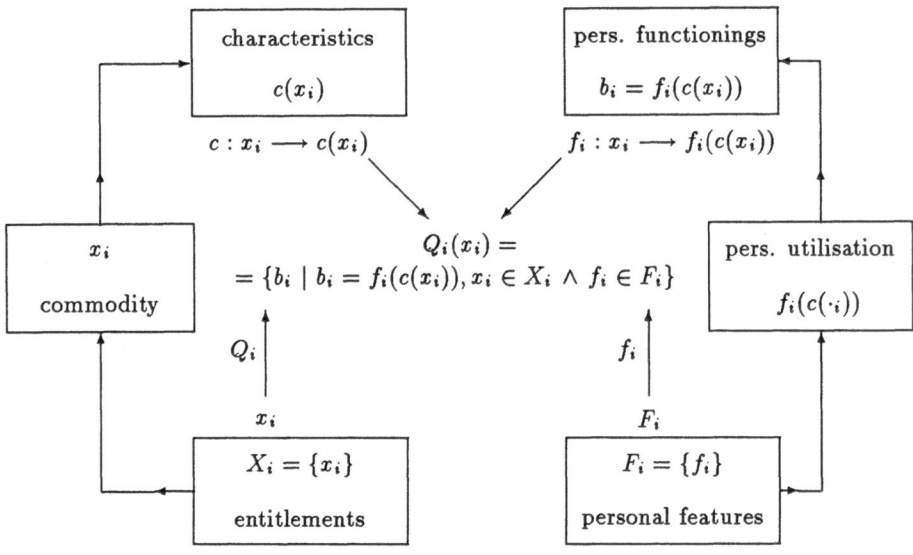

Figure 2.4: Capabilities in terms of entitlements and personal features.

that a person has in terms of the choice of functionings given her or his personal features F_i and entitlements X_i. The Q_i's are called the personal *capabilities* of i given F_i and X_i.

Let us proceed remarking that the conversion of characteristics of commodities into personal achievements – which are called functionings – depends on a variety of factors of individual, cultural and social nature. Thus, it is important to distinguish between choice and nonchoice factors in the determination of the capabilities Q_i. Moreover, it is necessary to realise, the distinction between choice and nonchoice factors is interwoven with that of separating structural and nonstructural elements about which we will have shortly something to say. For complete details see [202, 203].

At this stage we expect it has become clear that it is fully inappropriate to see the labour process merely from a technical point of view and to ignore its cultural and social links, reducing thereby economic analysis to price considerations and drive for profits.

To close these notes on the capability approach, let us say that shifting attention to the consumption of the characteristics (behind commodities) rather than merely (consumption of) commodities gives more sense to the formulation of performance functions at a microscopic and/or macroscopic level. In this sense, a performance criterion can indeed reveal the capacity of a system to contribute to the unfolding of its productive forces. Assessing the system performance will then attempt to measure quality and quantity of "states of being and doing" of its individuals. The set of "functionings" points thus to the more tangible *freedom of action* rather than the vacuous and apologetic notion of *freedom of will*.

The evolutionary perspective of society and labour

Let us now set some evolutionary issues of labour in a context that accomodates the interplay of cultural and socioeconomic elements as well as structural interactions in society. To facilitate an appropriate picture of the evolutionary perspective of society, let us survey some properties of labour processes in connection with human structure.

- With regard to social groups, classes or nations, let *conciousness* be understood as a complex state resulting from the perception of changes in moods, opinions, feelings and sentiments of individuals with some common interest, problems and prospects. Conciousness reflects itself as states of *cohesion* in attitudes, understanding and activities of individuals acting in structured configurations of a type like a group, class or nation. Both conciousness and cohesion are related in various ways to tradition, experiences, education and organisation.

- In the same form as life processes and bodily functions cannot be separated from living organisms and functioning bodies, human labour and labour power cannot be set apart. Thus, the need to apply labour power to a vast range of productive activities brought about the cultural and social development of human labour as an *informed* and *directed understanding* which is the essence of active labour processes. Hence, in potential this understanding resides in labour power which is *flexible* and *adaptable* to the most diverse purposes.

- The peculiar capacity of (human) labour power to produce a surplus, in addition to its intelligent and purposive character, suggests an infinite *potential* inherent in labour power for the expansion of capital accumulation. However, the *realisation* of the greatest profit based on labour power is limited by the subjective state of workers, general social and cultural conditions, specificity and organisation of the working process, and technical setting among others.

- The technical and social features incorporated in the capitalist mode of production, in particular the need of workers to sell their labour power as well as the separation of working into conception and execution, shift to capitalists the full responsability for the fulfilment of the aims of the labour process. This poses various challenging problems. Some go back to the fact that outlays in labour time cannot be evaluated with the same precision as it is used with outlays in building, materials, machinery, etc. Thus the portion of capital expended on labour power constitutes a *variable* quantity of value and is indeterminate. Other problems are related to questions of management.

Equipped with the knowledge encapsulated in the properties sketched above and drawing on Foster's evolutionary approach to economic behaviour [73], we depict in Fig. 2.5 the most salient evolutionary links and structural features of our own distillation of the labour process and its interplay with the human structure of society viewed at a macroscopic level.

From the point of view of capabilities, we want to see consumption commodities, in terms of characteristics $c(x_i)$, as primary and secondary energy inputs that are translated into informed and directed understanding including acquired skills by going through the whole network of evolutionary links and structural relationships. Upper-half of Fig. 2.5 includes the anthropological components and links of the system structure. Thus, they characterise and are responsible to a large extent for the slow-moving elements of the system. The utilisation functions f_i, or states of being and doing, are of this type. See Fig. 2.4.

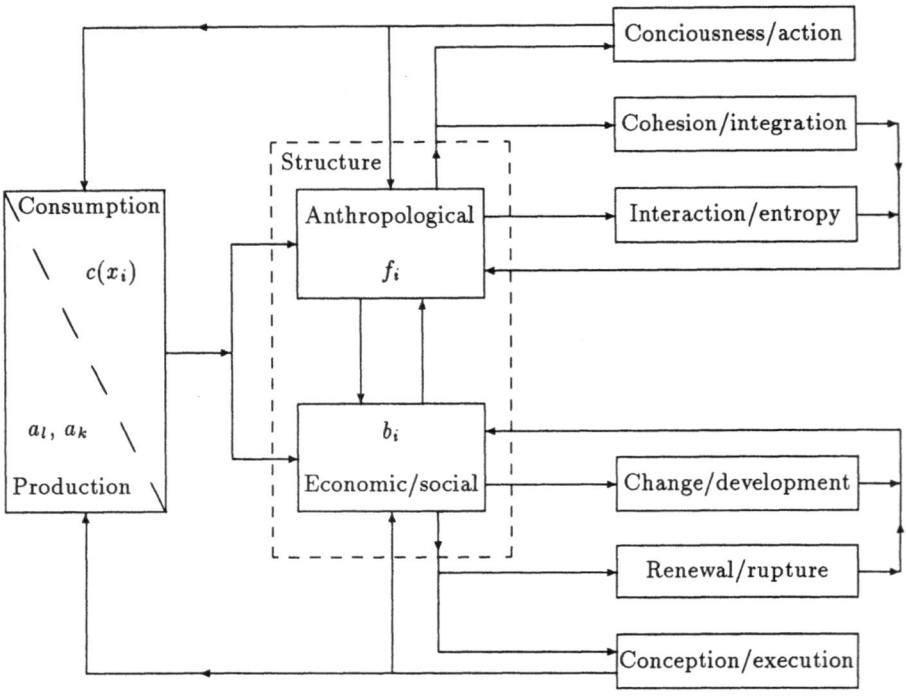

Figure 2.5: Evolutionary links and structure of society

The lower-half of Fig. 2.5, on the other hand, depicts the economic and social components as well as the structural links. There we outline therefore those mechanisms which are believed to contribute most to the system dynamics; indeed they reflect rather some relatively visible and fast-moving elements. The centre of Fig. 2.5, which we have marked as (human) structure, shall be represented by stock variables, e.g. f_i and b_i. These undergo changes of a dialectic nature, i.e. as the result of a progressive accumulation of quantitative and qualitative changes in the system. The arrows indicate the structural links in the system and the paths they follow attempt to visualise how direct or delayed may be the impact upon any element of the interactions going on in the system. At the left- and right-hand side of this diagramm we have to do with flow variables.

At this stage the rest of Fig. 2.5 may be self-explanatory. However, at the risk of boredom, we want to make a few additional remarks. The structure of a system is what binds its elements together, so that it can be understood therefore as consisting of energy bondings. This observation explains the connection of (human) structure to the box cohesion/integration and indicates its relation to its potential energy. Any dynamic interaction involving structural links lowers that potential energy and with it the tightness of the system structure. A consequence of this fact may be a lower capacity of the system to engage effectively in further dynamic interactions and structured motion. This points to a sort of entropy law reflecting the system commitment to its structure.

On the other hand the mechanisms behind the process of social reproduction, which involves conflicting interests, push to change social structures and to develop productive forces as well as to maintain social relations of production and to break unwanted social arrangements. Precisely, these features seem to defy the law of entropy and tend to deteriorate irreversibly its ties to a distinctive structure, bringing in disorder and ground for mutations. At this point, the impact of dynamic interactions and structured motion upon processes underpinning change/development necessitates to be reconciled with the action of less visible and slow-moving elements, a feature linked with the role of forces akin to renewal/rupture mechanisms.

To close these remarks, let us say that in order to see how to keep or to increase the capacity of the system to transform itself, one needs to take into account the forces represented in the boxes at the top and bottom of Fig. 2.5. In regard to conciousness/action, we should say that its role is to make sure that new energy comes to support the structure of the system so as to replenish its potential energy and to sustain its potential of interactions and structured motion. Consciousness, which is itself a manifestation of the potential to comprehend external events and to interact accordingly, is deeply related to entropy and decays at various rates depending on the degree of development and complexity of the structures and links involved. The release and absorption of energy is also supported by transfering energy emanating from a variety of creative and imaginative processes. These build upon stored knowledge and acquired skills as the box conception/execution indicates. These processes constitute the most powerful thrust driving mankind continuously to new areas of motional activity and to mastering new skills.

A thorough and interesting analysis of various aspects concerning the economic behaviour of *homo creativus* is given in [73], to which we refer for a precise description of conciousness, human structure, evolutionary time and other constructs of evolutionary thinking. We just want to point out that in this context economic behaviour manifests a tendency to create structured interactions with the external environment of an innovative character. However, this behaviour depends critically on the existing internal structure and on the potential conciousness attached to it. It evolves in a cumulative sense building upon history and creativity.

2.2.2   A few remarks on the labour market

As we have already announced, our models are differentiated according to where uncertainty emerges. Therefore it is our intention to examine more closely this notion and explain how this enters into the labour market, giving rise to uncertain demand for and/or supply of labour.

For the purpose of the present enquiry we want to mention briefly some basic features of the labour market which we believe may be helpful in the sequel and refer to [165, 164, 189] for detailed information and other references. In the labour market, one usually considers workers offering labour services and on the other hand capitalists who own the means of production, or if you prefer, firms represented by entrepeneurs that buy capital in capital markets and obtain labour at the prevailing wage rate.

Some considerations of firms in planning

For the sake of simplicity, we shall assume that the social supply of labour is infinitely elastic up to the level of full employment, but that the labour supply to individual firms depends on their and others' wage bids. Capital and commodity markets are perfect although random. That means that the choice of technique to be adopted has to be the result of considerations of the availability of labour and capital, and of expected future conditions in the markets for capital and labour, as well as the current position of each firm in these and commodity markets. Associated with this choice are decisions related to various rates of turnover and capital replacement, nature and quantity of capital employed, both human and physical.

Corresponding to the splitting of the global development strategy between short-term and long-term components, see Subsection 2.3.3, we get the two-department dichotomy of the firm, responsible for short-term and long-term planning decisions respectively. In the latter domain falls the choice of technique, optimal allocation of funds to fixed capital, i.e. capital tied up in long-lived plant and equipment, and to circulating capital, i.e. capital that turns over rapidly like the one covering wages and raw materials. Some long-term features of job offers belong also to this domain. Short-run policy making is concerned mainly with recruiting labour, producing, selling and some fine tuning of investment decisions allocation of unproductive capital, like the wage and accumulation fund, etc., which we shall abstract in this research. In the short run, once capital is installed, there is no substitution between labour and other factors and there is an upper bound, imposed by its own capital capacity and market share, on the amount of labour a firm can employ.

Principal criticisms of the conventional view of the labour market

A special feature of human labour relates to the fact that its conception always precedes and drives its execution. Conception and execution build a unity, i.e. a task, which can be dissolved in the individual and to be only recovered at a social level. The many advantages deriving from this special features are very well known. At this point we want to focus on a few drawbacks that constitute crucial sources of uncertainty.

First, concerning the supply of labour or the sort of know-how and skills that those which will carry out the labour process are expected to acquire and supply at a precise point in time and space. Second, in regard to the full utilisation of the labour-power firms get at the market and the realisation of the expected rates of profit. Here, it is necessary to call into mind that workers by selling their labour power surrender to a great extent their interests in the labour process to those for which they carry out operations of production and/or services. It is important to stress, human conciousness remains the driving force of labour. However, the fulfilment of labour processes requires sophisticated processes of information, co-ordination and motivation with a strong cultural origin. These processes take time and are costly in many ways.

This suggests that supply of labour is a process that unfolds in the course of time in the sense described in Fig. 2.5 and cannot be confused with the activity underlying job search and in this sense it is illegitimately reduced to the attached price relations. On the other hand, the demand for labour is a relatively less involved and more flexible process due to the fact that capital, viewed as a stock, has no life on its own as labour does.

The realisation of profits, valuation of capital and its various rates of turn-over nevertheless cannot be separated from changes in the process of labour as a little reflection and application of the principle of effective demand show. These observations point to the superficiality of orthodoxy in matters concerning labour which at most comes to grasp a few symptomatic issues of an ill-posed problem for the treatment of which a huge and highly sophisticated machinery has been developed.

Next, we will make some comments limiting ourselves to the rational underlying system adjustments to any deviation from equilibrium. According to the marginal productivity doctrine, the employment level should not go beyond the point at which a further worker adds as much to revenue as it adds to costs. This finding of conventional wisdom has been challenged on the ground that the major assumptions on which it relies, i.e. capital-labour substitutability, profit maximisation and price-taking behaviour of firms in the labour market, etc., are not valid in general but might become relevant only under fairly restrictive conditions. Criticism has been concentrated on the following points:

- As a condition of equilibrium, the marginal-productivity theory makes requirements like (1) the behaviour of firms will aim at maximising profits and (2) a full Walrasian list of assumptions will be satisfied in the markets for output, capital and labour. They are questionable in the long run and fully irrelevant in the short run.

- Since choice of technique, methods of production, quality and quantity of capital are not freely variable in the short run, firms may not be capable of changing the labour-capital mix whenever the relative prices change. Furthermore, the labour-capital mix being not variable in the short run, the demand for labour cannot always adjust to changes in the wage rate as the marginal productivity theory requires.

In the sense alluded above, one may adhere to the view that, if the Walrasian assumptions do not apply, the marginal productivity conditions will not hold but may still be useful studying the demand for labour. The problem of adjustments involves however finding arrangements and initiating structured motional activities aimed at the most favourable characteristics of the system structure. A point that we explained above.

On the notion of uncertainty

Whenever one has to choose in accordance with the patterns of rational behaviour among alternative courses of action, the consequences of which are not predictable in a realistic way, the need arises to consider *uncertainty*. Besides, the need to consider uncertainty results also as a consequence of partial and/or only indirect observability of state variables of the system or some of its components. Furthermore, in agreement with criteria underlying economic behaviour, the act of selecting an alternative action or policy will be aimed at the most favourable output of structural characteristics in historical time.

However, the desirability of an event is a matter of judgement that involves some subjective notions and beliefs, as well as the assessment ability of the policy- or decision-maker concerning the most likely outcomes and responses associated with the policy in question. In the spirit of intertemporal analysis one is forced to assume that individuals own claims on future production. But since the future is unknown the realisation of these claims turns out to be uncertain. Thus some quantitative aspects of these claims are random and depend on the states-of-nature resulting. Our modelling rationale directs attention to a sort of approximative approach which relies upon our belief that individuals learn and that under adaptive pressure they come to behave in a way that can be described as optimal.

In principle, the above description of human behaviour is simple and corresponds to received views according to which individuals usually act. Moreover, due to perceived and/or expected external events and actions that may, or are believed to, affect their own well-being, individuals may abandon their already established patterns of behaviour and try to adjust and learn anew, so as to keep their structural characteristics at the desired level. This hints at a further source of uncertainty which has to be taken into consideration.

2.2.3   Structural roots of uncertainty in labour dynamics

To account for uncertainty in the labour market, it suffices to bring to mind the interplay of flows and structural changes involved in the transformation of the labour process dealt with at the beginning of this section. Structural characteristics like the technical coefficients a_l and a_k, capabilities related to consumption commodities and individual characteristics of the type f_i and b_i among others are the cornerstones of the supply and demand for labour. The main line of argumentation goes along the ideas introduced with the purpose of motivating and explaining Fig. 2.4 and Fig. 2.5. Further, it is necessary to take into consideration Thesis 7.11 presented in Chapter 7.

One may assume that an arbitrary individual i is able to find out and to assess the income opportunities and other attractivity features of employment associated with the prevailing technical coefficients a_l and a_k characterising firms. Then giving its personal characteristics F_i and X_i, individual i commits himself to acquiring certain functionings $\{b_i\}$ embodying knowledge and skills compatible with certain technical coefficients a_l and a_k. Since this process takes time, it may occur that the distribution attached to the technical coefficients changes so that the achieved functionings become obsolete. Since individuals are different so will be their choices and characteristics.

Therefore the supply of labour to a firm with certain technical coefficients, will be specified also by a probability distribution, as the capital intensity and the (own) rate of return are. These observations and noting any adjustment to changes in the labour market requires time as well as additional efforts and costs, constitute the essence of uncertainty. The demand for labour undergoes processes of similar nature, since it depends on capital intensity and the resulting rate of return which are, for any given time, random variables.

Besides choice of technique, allocation of finance to capital and labour outlays, the realisation of profits is the major source of uncertainty. However, we shall only include uncertainty in a fairly simple way and postpone a detailed study to a later point.

According to the preceding considerations, we shall assume individuals respond to signals from their economic environment, wage or profit differentials for instance, although we may not know to what extent it happens, nor how they build their expectations, nor how they formulate subjective probabilities of future events, even making enough information available. When the need to face a choice of technique arises, firms have to take into account the own requirements and availability of labour and capital in future markets, as well as their own expected future position in factor and commodity markets.

The demand for labour and uncertainty

The frequency of capital replacement, the technique used, the nature and quantity of capital and labour employed are all firm-specific, so that a fairly simple approximation scheme hints at a mean value and some oscillations. Similarly, we may consider distributions of technical coefficients describing the actual configuration of firms and relative rates of change in the stocks of capital which may deviate from the rate at which social output increases.

To begin with let us point out, the demand for labour increases or shrinks according as the rise in output per man-hour goes on at a slower or faster pace than the corresponding increase in social output. The rise in output per man-hour, due to a great extent to technical progress, is mostly related to competition, innovations and population growth. Thus, even in the favourable case of output per man-hour rising at a slower pace than capital is accumulating, a growing unemployment may still come about. As an example, we wish to draw attention to the situations resulting from rising and declining rates of population growth.

First, let us assume a population growing at a declining rate. Then it may occur, that the growth of the disposable income per head associated with an eventually increasing demand for labour cannot keep the effective demand expanding at the same rate as capacity. This question is usually seen in relation with either a system reliance upon an ongoing increase in population or an expanding purchasing power able to keep it going. Because as soon as entrepreneurs anticipate, the market for commodities will cease to expand they may react reducing investment plans, which in turn bring down the rate of profits.

Second, from population growing at a faster rate than the demand for labour, there does not automatically result an increase in the effective demand, despite the fact that human needs increase. Neither eventual contributions to the consumption of those unemployed made in one way or another by institutions and/or individuals do necessarily stimulate effective demand. There are many interesting questions concerning the interaction among labour demand, realisation crises, effective demand and the reserve army which are deeply related to uncertainty in the rate of return on capital. See [184, 182].

To understand the following models of changes in the demand for labour, we shall recall we consider n types of labour $L^{(1)}, L^{(2)}, \ldots, L^{(n)}$ which correspond to labour requirements in terms of characteristics to the production of commodities determining the social output. Form the point of view of firms the demand for labour characteristics is determined by prevailing technical coefficients, expectations about rates of return and expected demand for their own output. All this suggests uncertainty we shall capture assuming changes in $L_t^{(i)}$, the level of actually employed labour of type i at time t, satisfy, for any $t \in [\tau^j, \tau^{j+1})$, the diffusion equation

$$dL_t^{(i)} = a_t^{(i)} L_t^{(i)} dt + \sum_{l=1}^{n} \zeta_t^{(il)} L_t^{(il)} dB_t^{(l)}, \qquad (2.4)$$

where $a_t^{(i)}$ and $\zeta_t^{(i)}$ are real-valued coefficients. These coefficients depend on structural parameters which may remain constant within the j-th time period, as we have already mentioned and will shortly explain. The dependence of drift coefficient $a_t^{(i)}$ on the unknown rate of return appears to us as a natural assumption, however it may also be convenient to assume instead, dependence on the average rate of interest. For more details and an economic motivation of this equation we refer to [92, 91].

Following the conventions established in regard to eq. (2.30), we shall set $a_t = a_{\tau^j}$ and $\zeta_t = \zeta_{\tau^j}$ for any $t \in [\tau^j, \tau^{j+1})$. Thus, whenever our analysis refers only to the j-th time period, the structural parameters remain constant; we may as well write $a_{\tau^j}^{(i)} = a^{(i)}$ and $\zeta_{\tau^j}^{(i)} = \zeta^{(i)}$. Thereby eq. (2.4) becomes

$$dL_t^{(i)} = aL_t^{(i)} dt + \sum_{l=1}^{n} \zeta^{(il)} L_t^{(il)} dB_t^{(l)}. \qquad (2.5)$$

The designation *diffusion* refers to the fact that, under the just-mentioned conditions of uncertainty, and assuming that at time t the level of labour participation is represented by $L_t^{(i)}$, a change of the type $\Delta L_t^{(i)}$ in the labour participation taking place during the incremental time Δt will be assessed from consideration of the following two expressions:

- First of all, $a^{(i)}L_t^{(i)}\Delta t + \delta^{(il)}L_t^{(il)} + o(\Delta t)$ which consists of two terms. The first, $a^{(i)}L_t^{(i)}$ describing a macroscopic change usually called *drift*, which is due to the employment dynamics; this reflects the ability of the system at the given level of employment to absorb idle labour. The second, $\delta^{(il)}L_t^{(il)}$ is an unpredictable change caused by collisions and conflicting goals of interacting agents pursuing mostly their self-interest and acting under conditions of limited, indirect and biased information. Above, $\delta^{(il)}$ denotes a constant.

- Secondly, $\mathsf{E}(\delta^{(il)}L_t^{(il)})^2 = \zeta^{(il)}L_t^{(il)}\Delta t + o(\Delta t)$, where $\zeta^{(il)}$ represents in a certain sense the mean uncertainty or mean energy. This energy is released by the individuals motivated to action by the current socioeconomic environment or by their subjective assessment of the current state of this environment. Here E stands for the expectation operator.

From any process based on human interactions of the sort described above results a sort of hardly predictible or chaotic configuration of fluctuations around a more-or-less systematic average tendency; this is attributable, in general, to difficulties individuals face whenever the need arises to distinguish between lasting and transitory changes. This situation is accentuated by the difficulty in ascertaining the extent to which they themselves or certain segments of society, which are relevant for their considerations, are affected by changes in progress. These facts may render some reactions inappropriate in degree, timing and the like. To give a more precise picture of the sort of changes described in the foregoing SDE's, we present a highly simplified model of this complex phenomenon which rests on Pasinetti's description of production in the short run and the dynamics of demand; see [160] pp. 35–48 and 61–79, full details will be given in [94].

To begin with, let us direct attention to a feature concerning our approximation of the demand for labour. This refers to sharpening focus so as to make the interplay of essential economic processes dependent mainly on structural parameters which within a short period of time remain in the average close to stationary values. This is the distinguishing feature of what we have called persistent structures, an analytical device which allows a great deal of operationality by bringing in component by component a formal structural similarity. For instance, let us consider the i-th component of (2.4), i.e. the level of labour employed in the production of output i given by

$$dL_t^{(i)} = a_t^{(i)}L_t^{(i)}dt + \sum_{l=1}^{n}\zeta_t^{(il)}L_t^{(il)}dB_t^{(l)},$$

where the parameters $a_t^{(i)}$ and $\zeta_t^{(il)}$ are given by

$$a_t^{(i)} = \sum_{\kappa=1}^{n}q_t^{(i\kappa)}(L_t^{(\kappa)}/L_t^{(i)}),$$

$$Z_t^{(il)} = \sum_{\kappa=1}^{n} S_t^{(i\kappa)}(L_t^{(\kappa)}/L_t^{(i)})S_t^{(\kappa l)}(L_t^{(\kappa)}/L_t^{(l)}),$$

$$\zeta_t^{(il)} = Z_t^{(il)} L_t^{(il)}.$$

The parameters $q_t^{(i\kappa)}$ and $S_t^{(i\kappa)}$ represent respectively the normal demand for the output i deriving from labour employed in the production of output κ and the corresponding deviations. By assumption the ratios $(L_t^{(\kappa)}/L_t^{(i)})$ are structural parameters. The notion of normal demand is naturally associated with that of normal prices and can be interpreted as average demand which explains the need of considering deviation from the normal level of demand.

Therefore, we may in the sequel either work with one-dimensional SDE's or operate component by component. However, for notational simplicity we shall suppress the use of the super-indices (i), (l) and (il).

The foregoing reflections prepare the ground for a realistic model of the supply of labour which depends on experience and information. Thus, from the fact that labour characteristics of the type i are known to be demanded in a form represented by $L^{(i)}$, it becomes natural to postulate that given certain circumstances reflecting the current capability of the system, individual decisions of workers will lead to supplying labour characteristics of the type i in a manner condensed in $N^{(i)}$ which we explain next.

The supply of labour and uncertainty

From our foregoing considerations it follows also that an analysis of the supply of labour includes many complicated and important questions. Some of them are connected with the heterogeneity of labour, the firms and other potential employers. By way of illustration, let us mention that the composition of available labour depends on age, class, sex, skills and experience and these in turn vary according to the rates of growth of population the system has experienced in the recent past, as well as in its earlier phases of development.

Let us recall the reserve army of unemployed labour consists of open and disguised unemployed labourers with a great variety of backgrounds, e.g. craftsmen dislodged by competition, people who acquired education but no opportunity to use it, small-scaled capitalists, etc. Thus the responses of individuals in the reserve army to changes in the system cannot be captured in a homogeneous scheme and the cost of integrating a fraction of the unemployed into the production process varies accordingly. As a consequence of the various processes interacting in the system, we want to assume that the supply of labour is also governed by a SDE of the type

$$dN_t^{(i)} = \epsilon_t^{(i)} N_t^{(i)} dt + \sum_{l=1}^{n} \pi_t^{(il)} N_t^{(il)} dB_t^{(l)}, \tag{2.6}$$

where $N = (N_t)_{t\geq 0}$ stands for the total labour available, and $\epsilon_t^{(i)}$ and $\pi_t^{(il)}$ are real-valued coefficients that depend on structural parameters, which remain constant within the j-th time period and $t \in [\tau^j, \tau^{j+1})$.

The characterisation of the SDE above obeys closely the same logic as that of (2.4). However, a fundamental difference has to do with the fact uncertainty in (2.4) is related to firm specific features like rates of return, demand for the output of the firm and the like. Thus, the underlying uncertainty is likely to undergo effects resulting from firm-specific control attempts in terms of capital and product diversification which are without reach of workers.

On the other hand, concerning the dynamics driving the supply of labour, given appropriate information and ability to decode it adequately, workers theoretically have a chance of reducing uncertainty on the basis of diversification. That means, when making decisisons to acquire working skills of the type $i = 1, 2, \ldots, n$, or choosing a corresponding job, they are assumed to conciously consider links among labour characteristics, learning perspectives resulting from work, in job widening of skills and the like, so as to improve in the average the likelihood of long-term employment.

Change and time are two major notions on which most of the process of social reproduction rests. Basic aims as development of the forces of production and generation of economic surplus clearly indicate the pervasive impact of time and change. The complexity resulting from the interplay of a wealth of purposive actions gives rise to formidable requirements on information. Thus, providing effective flows of information and adequate structures able to support acquisition of system potentials should dominate socioeconomic discourse.

Let us think of a policy directed to increasing the efficiency of production through new ideas and improved methods. This would require, for instance, to support appropriately each person involved and provide enough opportunities of on-the-job training and improvements in production methods for a time before qualitatively and/or quantitatively increased contributions materialise, as one may expect. This is a further manifestation of the need of freeing human creativity from certain structural barriers that prevent it to get in motion. This idea appears in different ways in these essays and in particular in this and the last section of the present chapter. These reflections shift attention to various crucial issues concerning the relation between population and economic growth, migration and investment in human capital. Because long before the benefits deriving from a growing population and enriched human creativity come to bear on the well-being of society, one needs to tie up capital in a way furthering structural stocks, e.g. f_i's and b_i's in Fig. 2.5, and linkage effects.

Investments in the structures of society are in no way different from investments in plant, working tools or inventories. However, from a microscopic point of view it is necessary to mention, those generations that provide for today's investment are not the ones that get the benefits. Today's generations are enjoying benefits deriving from contributions of those of yesterday. The understanding and functioning of these aspects of society require the consideration of social and economic institutions as well as the fundamental role played by change and time. See [206, 207].

Labour mobility and migration are in this sense essential aspects of socioeconomic evolution and constitute two major factors of development. Thus, they need to be integrated in any dynamic analysis of structured motional activity. It seems natural to assume that labour mobility and migration are deeply related to structural stocks and personal characteristics available to individuals and potential migrants, there, where they expect or consider to actively participate in the process of social reproduction. Nevertheless, it is necessary to mention that in the process of aggregating structural characteristics of individuals interacting at the microscopic level, they undergo certain quantitative and qualitative changes. This feature leads to the emergence of structures which characterise the evolutionary behaviour of the system at the macroscopic level in a way that may even differ radically from its initial microscopic features. See [98].

2.3   The allocation of social output

Capital accumulation and the generation of economic surplus always involves a qualitative and quantitative transformation of the process of material production, embracing more and more people as well as covering more and more aspects of their lives. This is the result of the increasing pressure of competition under which the need emerges for adopting new methods of production and exploiting further possibilities of large-scale production, establishing wider markets and a deeper social division of labour and so on. It is in this sense that the process of social reproduction may attain the capability of accomplishing the transformation of natural objects into means of satisfying social needs at an increasing level of sophistication.

To facilitate the analysis, we shall make the rather artificial distinction between economic and noneconomic aspects of development. The latter includes the institutional aspects of this transformation. Although this will, as one may expect, not be a sharp one. Under the economic aspects, we shall understand those reflecting fully the impact of accumulation on the process of (material) production, comprising in particular changes in the structural parameters, i.e. the quantitative relationship between factors brought together in the productive process corresponding to the interpretation of technique in the standard economic sense. Institutions are expected to articulate, mediate and resolve social conflicts by way of short- and long-term determination of the range of certain variables.

Thus, in order to be able to grasp the main features of social reproduction and capital accumulation, we need to identify the economic structure and the sort of institutions that should prevail. It is of utmost importance to recognise, the process of (material) production embraces on one side the materialisation or fulfilment of processes of labour, and on the other the transformation of objects of labour with the help of means of production. That means, products and ultimately the social output are the joint result of a highly dynamic process based on human creativity and the transformation of objects without motion, i.e. incorporating former labour of fixed quality. The various types of labour involved accentuate requirements related to capability notions like *personal factors* of production, know-how, appropriate skills characterised by an always evolving nature. Further, various *material factors* of production of a concrete character and static nature support processes of labour taking the form of production techniques.

In the foregoing section we presented an evolutionary picture of processes of labour and introduced the main concepts on which the capability approach rests. With the purpose of explaining distribution of disposable income within a neoclassical framework, we need to conceive factors of productions only as those personal and material factors of production which are explicitly paid for their contribution. The use of the neoclassical mechanisms of distribution introduces serious theoretical inconsistencies which we cannot avoid at the present stage of our investigation. We try however to attenuate some of these by bringing into play various elements representing essential features of the process of social reproduction, which are usually neglected by orthodoxy.

2.3.1   Economic nature of allocation of disposable income

Let us consider the processes $K = (K_t)_{t \in \mathbb{T}}$, $I = (I_t)_{t \in \mathbb{T}}$, $C = (C_t)_{t \in \mathbb{T}}$, $Q = (Q_t)_{t \in \mathbb{T}}$, $Y = (Y_t)_{t \in \mathbb{T}}$, $k = (k_t)_{t \in \mathbb{T}}$ and $q = (q_t)_{t \in \mathbb{T}}$ where the variables K_t, I_t, C_t, Q_t, Y_t, k_t and q_t denote respectively capital stocks, gross investments, (personal) consumption, social output, disposable income, capital-labour and output-labour ratio in the secondary sector at the time $t \in \mathbb{T} = [\tau^j, \tau^{j+1})$. We should write $\mathbb{T}_j$, but we suppress the subindex j for notational simplicity unless any confusion may arise.

Technical conditions and investment

To begin with, we shall assume that *social output* Q can be represented as a function $\widetilde{F}$ of the so-called factors of production K and L which is connected to *disposable income* Y by means of relations linked as follows:

$$Q_t = \widetilde{F}(t, K_t, L_t), \tag{2.7}$$
$$Y_t = Q_t - \delta_t K_t, \tag{2.8}$$

where, for any t, the parameter δ_t stands for the proportionate rate at which existing capital depreciates and satisfies the condition

$$\delta_t > 0, \ \delta_t \in \mathbb{R}_+. \tag{2.9}$$

These relationships reflect the need of maintaining production going at least at the prevailing level. We assume disposable income as determined according to the preceding equation, both at the microscopic and macroeconomic level. This is a fact that common practices of accounting confirm. It is necessary to stress that notions concerning appreciation, depreciation, obsolence, aging and physical deterioration of capital are not clear-cut. However, clarification of this matter would bring us beyond our scope; in this regard see, for instance, [198]. The distribution of disposable income is ruled by the relation

$$Y_t = F(t, K_t, L_t), \tag{2.10}$$

where it is necessary to make clear that the function F has been introduced just for the purpose of allocating disposable income between profits and wages. Further, let us recall in addition to the foregoing well-known relationships of

social output, disposable income or net production, the following expressions for
(net) investments or actual changes in the stocks of capital and consumption:

$$\dot{K}_t = s_K(t)(Y_t - WL_t) + s_L(t)WL_t, \tag{2.11}$$
$$C_t = (1 - s_K(t))(Y_t - WL_t) + (1 - s_L(t))WL_t. \tag{2.12}$$

These equations rely on our assumption economic agents in the secondary or
capitalist sector are either capitalists or workers. That means they make their
living to a great extent from capital and wage income respectively. These
notions are characterised by specifying the parameters s_K and s_L appearing
in the definition of (net) investment and consumption, which are indeed the
functionals $K : t \rightarrow K_t$ and $C : t \rightarrow C_t$. For any $t \in \mathbb{T}_j$, the parameters $s_K(t)$
and $s_L(t)$ satisfy relationships

$$0 < s_L(t) < s_K(t) < 1, \quad \text{with } s_K(t), \ s_L(t) \in \mathbb{R}, \tag{2.13}$$

and the parameter δ_t for any t, $t \in [\tau^j, \tau^{j+1})$, represents the fraction of the
social output that has to be put aside to compensate for (social) capital used
up.

Then *gross investment* is defined as the fraction of social output set aside
to replace depreciated capital and to cover the actual changes in the stocks of
capital, and this is represented by the relationship given by

$$I_t = \dot{K}_t + \delta_t K_t. \tag{2.14}$$

Let us note the initial conditions corresponding to any time interval $\mathbb{T}_j$ are
always adjusted to the special characteristics of the j-th period of time. Moreover, the initial level of capital stock K_0 and labour services L_0 shall meet the
following conditions

$$K_0 > 0, \quad L_0 > 0, \quad \text{with } K_0, \ L_0 \in \mathbb{R}_+. \tag{2.15}$$

Finally, from the assumption that in the short-term (net) investment and savings coincide follows that disposable income can be consumed either for productive or personal purposes, i.e. for any $t \in \mathbb{T}_j$ the well-known *income identity*

$$Y_t = \dot{K}_t + C_t, \tag{2.16}$$

holds and states that the national income can either be directed to the actual
expansion of capital or be consumed. It is in this sense that we distinguish
among net investments, personal and productive consumption, as one easily
recognises looking at (2.16) and (2.8).

In (2.11) and (2.12) above, $s_K(t)$ and $s_L(t)$ denote the fractions of *capital income* $(Y_t - WL_t)$ and *wage income* WL_t saved, building part of an equilibrium
mechanism, which can be properly understood on the basis of the principle of
effective demand and not just representing the assumption that in the short-run
(aggregate) savings equate with (aggregate) investments. This simple distinction between capital income (firms) and wage income (households) enables us
to make savings depend entirely upon the type of income concerned which provides room for the introduction of the mechanisms of multiplicator analysis
explaining thereby how investments come to determine savings, see [23].

Individual preferences play no role in this regard but enter the picture in matters concerning the composition of the social output. We assume at the microeconomic level, relationships of the type (2.7) and (2.8) hold, so that firms keep intact in the average their pre-existing capital. Investments in productive capital are governed, within certain limits, by decisions of firms according to their level of profits and are aimed at increasing productivity, technical knowledge as well as reducing costs.

Let us stress that we shall distinguish between (aggregate variables like) national income Y and social output Q, both of which will represent *potential purchasing power* and include newly generated (aggregate) *social value*. However, only the latter contains the social product necessary to conserve constant capital.

At the macroeconomic level, it is not easy to make precise the meaning of depreciation allowances. The notion of capital is complex and the methods of ascertaining its value are somehow arbitrary. At this stage, it suffices to say constant capital comprises machinery and instruments of labour as well as the materials of labour. However, from the perspective of its durability and other related factors, capital can be seen as made up of fixed capital relative to a certain interval of time, say one year, in the sense that it does not yield up its total value in this time, and circulating capital comprising those items that are fully consumed and paid for in the course of the unit of time considered.

This hints at a characteristic feature of capital, namely, its movement bringing about a continuous change in its form, which can be visualised as a systematic circuit of transformation bringing capital back to the particular form in which it started. This property is known as the turnover of capital and gives rise to the concept of *period of turnover*, which refers to the time it takes to complete the circuit.

Obviously, the period of turnover has a deep effect upon the rate of profit and offers analytical tools to investigate productive capital, its rates of accumulation and profit according to the different rates at which it circulates. We shall have the opportunity to come back to this issue later.

The following representation of social investment and consumption as given by (2.11) and (2.12) is very useful:

$$\dot{K}_t = s_t Y_t, \qquad (2.17)$$
$$C_t = (1 - s_t) Y_t, \qquad (2.18)$$

where for any t the parametric variable s_t stands for

$$s_t = s_K(t)\pi_K + s_L(t)\pi_L, \qquad (2.19)$$

with π_K and π_L denoting the relative *income shares of capital and labour* respectively. Obviously, $0 < s_t \leq 1$ and $\pi_K + \pi_L = 1$. That means, $\pi_K = (\pi_K(t))_{t\in\mathbb{T}}$ and $\pi_L = (\pi_L(t))_{t\in\mathbb{T}}$ with $\pi_K(t)$ and $\pi_L(t)$ given by

$$\pi_K(t) = \frac{Y_t - WL_t}{Y_t} \quad \text{and} \quad \pi_L(t) = \frac{WL_t}{Y_t}. \qquad (2.20)$$

Sometimes, it is convenient to work in the time interval $[0, T]$ with T any arbitrary positive scalar. In that case, we consider $[0, T]$ as the time horizon.

However, to convert any statement restricted to $[0, T]$ into one in $\mathbb{T}_j$, one needs to multiply the functions involved by the characteristic function $\mathbf{1}_{\mathbb{T}_j}$ associated with $\mathbb{T}_j$ and eventually change the integral limits to ones of the type $T \wedge \tau^j$ and $T \wedge \tau^{j+1}$ respectively.

In this sense, we shall also write $\delta = (\delta_t)_{t \geq 0}$ and $s = (s_t)_{t \geq 0}$. Similarly, we shall write $K = K(t)_{t \geq 0}$, $I = (I_t)_{t \geq 0}$, $C = (C_t)_{t \geq 0}$ and so on, stressing the character of the process of economic activities behind these variables. For some accounting purposes natural relationships like

$$Q \;=\; \dot{K} + C + \delta K, \tag{2.21}$$
$$Q \;=\; (Y - WL) + WL + \delta K, \tag{2.22}$$

are helpful, since they illustrate from different perspectives principles concerning allocation of social output, conservation of value and the like. For instance, the latter relationship stands roughly for various activities behind generations of value and reminds us of the accounting identity

$$(u) \qquad\qquad (s) \qquad\qquad (v) \qquad\qquad (c)$$
aggregate value $=$ surplus value $+$ variable capital $+$ constant capital,

an identity often used in the literature to investigate qualitatively rates of profit and surplus value based on differences in the period of turnover characterising various sorts of capital involved.

The information condensed in the period of turnover is of great significance assessing the rate at which firms and/or branches of production can expand and together with the rate of profit it provides a measure of feasible and possible rates of accumulation. See [23, 71, 120, 130].

The technology as a set of alternative techniques

For the purpose of the present section it will suffice to think of techniques as the capital-labour relationship, while technology will mean the capability to create alternative techniques for the production of the bundle of commodities at hand. In a more restricted sense, technology stands for the capability to make actual possibilities from the available technological feasibilities. Technological progress shall mean accordingly an improvement in the capability of creating alternative techniques.

For any time $t \in [\tau^j, \tau^{j+1})$, the technology is specified by the relationship (2.7), i.e. this is given by the functional $Q : t \longrightarrow Q_t$. Moreover, for a fixed but arbitrary time t, let $F(t, K_t, L_t)$ represent a distribution function which acting as an intermediate step in the process of reproduction provides for means of consumption; necessary consumption includes both personal and productive expenditure. The function $F(t, K_t, L_t)$ fulfils the following properties, which are usually attributed to production functions of the neoclassical type.

A 2.1 *Assumptions on the distribution function* $F(t, K_t, L_t)$

- For any $K_t, L_t \geq 0$, it holds that $F(t, K_t, L_t) \geq 0$ and that $F(t, 0, 0) = 0$. This reminds us, means of production and labour power are essential recipients of disposable income, since production has no aim on its own. The argument t allows the function itself to vary over time in the form of time shifts viewed by orthodoxy as technical progress; time shifts are of two kinds. First of all, disembodied in the short run, indeed of the Harrod-neutral type, and secondly of embodied type in the long run, which shall be carried out mainly through capital impulses.

- Y_t is for any time t a smooth function of class $\mathbb{C}^{1,2,2}(\mathbb{R}^3_+ \longrightarrow \mathbb{R}_+)$ of flows of capital K_t and labour services L_t, the evolution of which are given by (2.4) and (2.11) through (2.15). Harrod-neutral progress holds. Further, we assume the capital endowment available at the time of starting the j-th time period, which we denote $K_{\tau j}$, is fully utilised.

- Further, Y_t is homogeneous of degree 1 with respect to K_t and L_t, i.e. for any scalar $\lambda > 0$, we have $F(t, \lambda K_t, \lambda L_t) = \lambda F(t, K_t, L_t)$. This assumption is crucial and entails various other properties, as we shall see. Moreover, accompanying the assumption of technical progress of the Harrod-neutral type we will assume constant elasticity of substitution and thus that σ is always strictly less than unity.

- The marginal rewards of physical and human capital for their contributions to social output are non-negative but diminishing with successive increments of capital and labour services, i.e. $\partial_K Y_t, \partial_L Y_t > 0$, and $\partial_K^2 Y_t, \partial_L^2 Y_t \leq 0$. Let us point out that both marginal rewards start at infinity and diminish to zero.

- Strict quasi-concavity, i.e. for any scalar λ^1 and λ^2, so that $0 < \lambda^1 < 1$, $\lambda^2 > 0$, the relationship $F(t, \lambda^1 Z_t + (1 - \lambda^2)Z_t^*) \geq \lambda^2$ holds, whenever $F(t, Z_t) \geq \lambda^2$ and $F(t, Z^*) \geq \lambda^2$. The variable Z represents an arbitrary admissible combination of capital and labour services, that is $Z = (K, L)$, and Z^* another such combination.

It is necessary to call attention to the fact that to render a feasible technology a possible one, certain additional economic and institutional conditions have to be fulfilled. These are related, for instance, to the availability of the factors of production, the prices and the commodity market but also to the institutional framework within which the process of production in question has to operate. Among these, those concerning the underlying financial problem posed by the eventual mobilisation of the factors, adaptation, training and the like are prominent.

Most of the assumptions listed above are fairly standard and do not require further explanation, see [35, 34, 181, 183, 232]. Sometimes it may result convenient to work with

$$Y_t = F(K_t, L_t), \tag{2.23}$$

which obviously satisfies the above mentioned properties. This function may also reflect the effect of time shifts based on learning.

Relationships in intensive form

It is very useful to rewrite (2.10) in terms of the capital-labour relationship, i.e. k_t and the income-labour ratio, i.e. y_t. Then we obtain the following per worker version of (the neoclassical) distribution function

$$y_t = \mathfrak{f}(t, k_t), \tag{2.24}$$

where the relationship

$$\mathfrak{f}(t, k_t) = F(t, \frac{K_t}{L_t}, 1)$$

holds. The distribution function (2.24) is commonly referred to as the *per capita or intensive form* of (2.10). However, we shall emphasise it is indeed a per-worker form instead of a per-capita one and, for any t, it expresses income per worker as a function of capital per worker. Let us point out further the distribution function in its intensive form satisfies an adequate version of all the assumptions just listed, i.e. $\mathfrak{f}(t, k_t)$ is strictly concave, continuously differentiable on $[0, \infty]$, $\mathfrak{f}(t, 0) = 0$, etc. Denoting per-worker income, capital and consumption by the corresponding small cases, we have also the identity

$$q_t = s_t y_t + \delta_t k_t + c_t. \tag{2.25}$$

For expository purposes we will assume, as we recall, sometimes that the economy lives the j-th time period given by the real interval $[\tau^j, \tau^{j+1})$, where τ^{j+1} is determined by means of the random function $\tau : \theta^j \longrightarrow \tau(\theta^j)$ whose characterisation is part of an optimisation problem associated with the impulsive control. Thus the initial conditions in the j-th time period are given as in (2.31).

By analogy with (2.30), (2.10) and (2.24) shall write $Y_t = F(\tau^j, K_t, L_t)$ and $y_t = \mathfrak{f}(\tau^j, k_t)$ for any $t \in [\tau^j, \tau^{j+1})$. This reflects the fact that within the j-th time period the structural parameters will remain constant. The restrictive feature of this assumption is not as serious as it may appear, since that allows a great deal of operationality, linearity and other interesting properties, and the system regains all its generality when one considers it on the long run.

Concerning the technology, we shall say that the set of alternative techniques it describes remains the same within the j-th time period, but changes qualitatively and quantitatively from period to period. In order to keep the notation simple we shall suppress, without further comment, any indexation with respect to j, once things are well under way, or we just assume that $j = 1$ unless any confusion may arise. That means, we shall be dealing mostly with the generic time period $[\tau^j, \tau^{j+1})$; the long-term problems are treated in [94].

Let $\overline{L}_{\tau^{j+1}}$ denote the upper bound of the level of employment which indicate the amount of labour the system, or the secondary sector, can at most absorb during the j-th time period, if it operates at its normal (productive) capacity, e.g. the (aggregate) level resulting from utilisation by firms of their available productive capacity. Further, for any $t \in [\tau^j, \tau^{j+1})$, let $\overline{L}_t$ denote the employment level at which the (aggregate) wage bill, i.e. WL_t, exhausts income Y_t disposable at time t, and set $L_{\tau^{j+1}} = \min\{\overline{L}_{\tau^{j+1}}, \overline{L}_t : t \in \mathbb{T}_j\}$. Then $L_{\tau^{j+1}}$

corresponds in a certain sense to the upper bound on the quantity of labour firms can employ as a whole. This is a limit imposed by the capital (plant) capacity of the firms, which we can formally compute based on (2.4) provided we anticipate the random time τ^{j+1} and know the initial value $L_{\tau^j} = L_{\tau^j} + N_e$. Here, N_e denotes the fraction of the reserve army N_u which moves to more highly productive activities during the j-th period. However, we take N_e and τ^{j+1} in the short-term as given, and in the long-term we obtain as part of the impulsive control the couple $(\theta^{j+1}, \tau^{j+1})$ which determines straight N_e.

Wages and propensities to invest

Concerning investment and savings, it is necessary to stress that our model is designed to adjust to the principle according to which any autonomous increase of investment brings about the corresponding increase of savings. This is a consequence of the observation that in a commodity producing economy, it is the market that determines the volume of production.

On the other hand, under appropriate conditions an economy with labour-surplus of the type under consideration can be seen as having a reserve of productive capacity that can be released when more capital becomes available, allowing an expansion of its labour force. These issues are usually explained in the light of the principle of effective demand or relying on mechanisms connected with the realisation of profits.

We shall make some additional assumptions about the parameters and arguments of (2.11) and (2.12).

A 2.2 *Assumptions on s_K, s_L and W*

- Let us assume the parameters s_K and s_L, introduced in (2.11) and (2.12) denoting respectively the fractions of capital and wage income saved, satisfy (2.13) and the inequality $s_K \geq s_L$, i.e. workers save less than capitalists.

- Once full employment is reached, the control board is able to require investors to save any fraction s_K of capital income as long as it remains below the upper bound $\overline{s}_K$.

- When unemployment prevails, the capitalists' control of the disposition of their income is limited only to the need to negotiate with the control board the upper bound $\overline{s}_K$ on the saving rate.

- Moreover, there exists a minimum wage rate $\overline{W}$, exogeneously fixed, which remains constant over the time period under consideration but adjusts from period to period according to the prevailing standards of living and technical progress.

Let us condense the information about the parameter $s_K(t)$ and the level of employment L_t resulting from the constraints imposed above in the inequalities

$$0 < L_t \leq L_{\tau^{j+1}}, \tag{2.26}$$
$$0 < s_K(t) \leq \overline{s}_K, \tag{2.27}$$

which will be further explained in the next section.

We shall not specify mechanisms of wage bargaining. However, we want to recall a minimum wage rate is set at the beginning of any time period with the aim of allocating idle labour to more highly productive activities in a way the system can afford, and opening the way for the principle of effective demand to operate. We shall come to this issue in connection with the expansion of productive capacity and choice of technique.

2.3.2   Institutional aspects of social production

Taking into account institutional constraints answers the need of economic theorising to incorporate in an essential way functions and roles played by institutions as organisational structures and as guides in normative issues. That means, normative patterns concerning, say, the legitimate and proper character of economic, social and political actions, the sense of morality, self-interest and individual attitudes, which are socially formed and usually inculcated in early childhood.

In this connection, it is extremely important to understand, that although information and knowledge bear distinguishing individual and subjective features, the mechanisms through which we perceive and acquire them are inevitably social. Thus perception and interpretation of social reality cannot be seen as the result of an isolated individual but as a reflection of norms and practices of social life, which in turn are outgrowths of conventions and routines of culture and tradition.

Cognitive processes and human actions

There is the need to get rid of the "rational man" of orthodox economics in order to make room for issues raised by cognitive theory and psychology. Because economic theorising needs to account for the recognition that our actions are not all the result of rational calculations or conscious deliberation. This fact directs attention to the need of considering the role played by (social) institutions in shaping habits and routines on which both knowledge and consciousness at the practical level rely.

Borrowing some ideas from Hodgson [110], we shall mention that observations and studies about the detection by the human brain of something moving in the sky through the highly complex and efficient procedure of visual perception of an object, strongly suggests that acts of seeing, hearing and remembering are all constructive processes. By means of such processes the mind responds to the vast quantity of sense data received by the eyes, the ears, etc., working to reduce uncertainty at the conceptual level and giving the "chaotic jumble" of stimuli some sense and meaning. In a split second without conscious deliberation, the brain manages to discern an object in a situation characterised by too much data for the mind to process and too little of the sort of information most important for decision-making.

On the other hand, various investigations in psychology on multiple levels of consciousness and a hierarchical organisation of the mind compellingly suggest that human agents are both rational and sub-rational and make their plans for future actions on the basis of a limited number of rational calculations at a higher level of consciousness and deliberations. In this way, there is place for fundamental notions like habits, customs, cultural features, intentions, etc., to play a role in economic processes of the learning type, for human calculations about future economic actions and the evaluation of economic policies. For a complete study of these issues and institutional economics we referred to the highly interesting and stimulating book [110] by Hodgson.

After this short incursion into some aspects of institutional economics, it might appear more natural to rely on notions like social awareness and solidarity, learning and motivation. This is in particular true studying developmental problems from the perspective of (the needs resulting from) the process of social reproduction. These include development of productive forces and increasing economic surplus for the purpose of expanding productive capacity and incorporating idle labour into production.

The relevant range of employment

One of the major problems of the labour-surplus economy is, as a look at eqs. (2.11), (2.12) and (2.14) easily shows, that employment and wages conflict with investment and with growth as well, a point we shall elaborate later. To begin with, let us recall an income distribution favouring capitalists does not necessarily promote investment and growth, since the presence of demonstration effects is likely to limit the allocation of the economic surplus to productive purposes. Furthermore, low levels of employment and/or wages is tantamount to low purchasing power which prevents the effective demand and the home market to expand, besides the usual effects on productivity, motivation and expectations. Fiscal and monetary measures are controversial and they are more likely to burden mainly those which do not have economical and/or political power.

Thus this struggle for shares in the economic surplus ends frequently in conflicts between processes underlying the democratisation of economic and political life which belong to daily life in states nations of the periphery. To seek a way out of such a deadlock, we shall examine the question of freeing human creative potentials blocked by unemployment so as to recover self-sustaining features of the principle of social reproduction. To this end, we shall briefly examine the rationale behind the determination of the range of employment.

The relevant range of employment is given by the real interval $(\underline{L}_t, L_{\tau,j+1}]$, where $\underline{L}_t$ stands for the investment-maximising level of employment. Here $\underline{L}_t$ and $L_{\tau,j+1}$ represent respectively an economic and a physical barrier to increases in employment and in wages. That means, an increase in the level of employment or in wages remains, at least theoretically, possible as long as the associate wage income remains less than that corresponding to net social income, i.e. as long as $L_t \leq L_{\tau,j+1}$. Further, employment beyond the level $\underline{L}_t$ would entail a fall in the rate of return and even in the rate of surplus value. The reason is that at full employment wages would increase or just tend to rise due to greater bargaining power of labour and poor incentives resulting from decreasing marginal rewards.

Besides the customary responses of an economy based on profit, i.e. mon-

etary and fiscal policies, there are measures that appear as the more natural
vehicle to avoid the fall of the rate of profit below a certain level. We mean here
the so-called *rationalisation* and *automation*, i.e. replacement of human labour
by machines, aimed at restoring unemployment and replenishing the reserve
army to the end of weakening bargaining power. The latter in turn acts as
a check on any actual or potential rise of employment and real wages. These
reflections reveal the existence of a sort of institutional barrier, i.e. the power
of capital directed to stop investing and increasing labour, whenever the rate
of profit falls too low. Under a capitalist régime this barrier can be reached
even well before $\underline{L}_t$.

Let us point out that in the conventional world of neoclassical economics
the economy would settle at $\underline{L}_t$, at least under perfect competition, as we
argued above. That means the range of employment splits actually into the
two intervals $[0, \underline{L}_t]$ and $(\underline{L}_t, L_{\tau j+1}]$, representing respectively the possibilities
and feasibilities of employment. For the purpose of the economic policy we are
looking for, it is the second interval that attracts our attention. Thus, we shall
require further that the potential level of employment L_t fulfils the inequality

$$\underline{L}_t \leq L_t \leq L_{\tau j+1}, \qquad (2.28)$$

defining thereby the *relevant range of employment*. Therefore, in order to increase employment beyond $\underline{L}_t$ one needs governmental intervention which is
brought about by means of a policy consisting of an impulsive and an institutional component, the latter acting indirectly through a system of social
preferences and value-judgments, as we already mentioned in Section 2.3.3.
Whenever no confusion arises we shall set $L^{j+1} = L_{\tau j+1}$.

Individual self-interest, social preferences and value-judgements

Economic policy decisions aim at providing directly or indirectly for present and
future needs of society. Present needs are related in one way or another to the
consumption of resources presently available to society, while future needs are
associated with capital accumulation or the investment fund of society, and its
application to actual expansion of resources to become available for consumption in the future. Roughly speaking, economic policy decisions are choices
between alternative pairs of present and potential future states of consumption
and their associated political and social well-being consequences. Many choices
are possible, any of them implying an alternative time-path for consumption,
capital, employment, output, etc., along which society would evolve. In order
to choose among those paths we shall assess their social desirability.

Reality tells us, the diverse social and economic conditions under which
individuals grow, mould them so differently that they come to endorse and
even to invent alternative social and economic theories. Therefore, judging
on the social preferability of an economic policy, we have to face a variety of
conflicting issues, i.e. ranking social against individual values and priorities,
weighing social preferability of policies as viewed by those deriving benefits
from the prevailing state of affairs, with the social costs mostly carried by the
uprooted, the unemployed, etc., all those who are not even able to articulate
their needs, sorrows and the like.

It is our intention to shift evaluative concerns of development achievements from the purely subjective sphere to one that places emphasis on the capability of society to function, in the sense of being able to cope with the challenge posed by the fundamental principle of social reproduction. The intention of this shift is to facilitate moving away from the ill-conceived valuation framework of orthodoxy, which views (production of) commodities and incomes as the ultimate concern of economic progress confusing thereby means and ends.

By viewing human living as a set of functionings, i.e. combinations of "states of doing and being" an individual may achieve, the capability approach makes possible an evaluation of development achievements based upon notions, that reflect more closely *enrichment of human life* and *freedom to actually achieve* different ways of living. These notions include ability to escape unemployment, undernourishment and poverty.

To close these notes, let us underline the need of considering as valuable only achieved states, if we want seriously to get rid of the rhetoric nature of the (subjective) utility-based framework of valuation. We shall assume the existence of a planning authority or control board that can give binding directives concerning production, investment and consumption to all capitalists and workers in the economy, provided one observes certain political and ideological rules and socioeconomic arrangements, that we shall call the *institutional constraints*. To put it in a less authoritarian context, we shall assume a government authorised by one of the usual election procedures to give binding directives relying on the decision of individuals. These decisions are assumed to be made in the light of their own self-interest, this interest being appropriately directed by some system of incentives and rewards, social preferences and value-judgements. Think of this system as a set of statements, on issues of this sort, promoted by the ruling party in the program which brought it to power.

However, we prefer to talk of a control board because at this stage we want to abstract from taxes and fiscal policies that the explicit assumption about the existence of a government brings in. The notion of a control board should help to remind us that behind individuals there is an institutional reality which decisively affects cognitive processes, moulds individual preferences and actions in many ways.

A 2.3 *Assumptions on the functionals f and φ*

- Let the strict concave functional $f : t \longrightarrow f(t, \cdot)$ with $f(t, \cdot)$ being a measurable function of class $\mathbb{C}_b^{1,2}(\mathbb{R}_+ \times \mathbb{R}^n \longrightarrow \mathbb{R}_-)$ be given. The functional f will measure preferences either at a social level or at a per-worker (per-capita) basis. In particular it will reflect the reward per-worker or per available labour in terms of characteristics. Expressed by means of units of functionings achieved per unit of time, the functional f contributes to describe a performance index. It indicates in a sense the system ability to translate consumption characteristics, which attained states of accumulation can afford, into functionings.

- Further, we assume as given a strict concave functional $\phi : t \longrightarrow \phi(t, \cdot)$ where $\phi(t, \cdot)$ is of class $\mathbb{C}_b^{1,2}(\mathbb{R}_+ \times \mathbb{R}^n \longrightarrow \mathbb{R}_-)$. The functional ϕ is a terminal pay-off which measures the value in terms of consumption characteristics of the capital bequeathed to the $(j + 1)$-th time period. The function ϕ contributes also to describe a performance index.

- As a first approximation, let us assume that f and ϕ are essentially related to functionings associated with bundles of characteristics determined by the level of necessary consumption prevailing in the $(j+1)$-th time period. Moreover, f and ϕ fulfil some measurability, Lipschitz and growth conditions that we shall specify later.

The foregoing characterisation of the central ingredients of the performance criterion is, needless to say, far from satisfactory. This is owing mainly to the complexity involved in the task of describing functionings and the wealth of information and studies, that a better understanding of these issues requires. Nevertheless, our approach attempts to capture and direct attention to questions deeply related to quality and enrichment of human life, determinants of human capabilities, basic needs and others. This is the result from shifting human beings to the centre of our inquiry both as means and ends of development achievements. Concerning ambiguities, precision and relevance A. Sen argues as follows:

> There are many ambiguities in the conceptual framework of the capability approach. Indeed, the nature of human life and the content of human freedom are themselves far from unproblematic concepts. It is not my purpose to brush these difficult questions under the carpet. In so far as there are genuine ambiguities in the underlying objects of value, these will be reflected in corresponding ambiguities in the characterization of capabilities. The need for this relates to a methodological point, which I have tried to defend elsewhere, that if an underlying idea has an essential ambiguity, a precise formulation of that idea must try to capture that ambiguity rather than attempt to loose it. Even when precisely capturing an ambiguity proves to be a difficult exercise, that is not an argument for forgetting the complex nature of the concept and seeking a spuriously narrow exactness. In a social investigation and measurement, it is undoubtedly more important to be vaguely right than to be precisely wrong.

Amartya Sen
(see [204] pp.45)

In particular, we should make sure that f consistently evaluates the reward in terms of functionings the representative individual worker derived from consumption, e.g. C_t and/or c_t at any time t, and takes into account essential probabilistic elements typically attached to a development undertaking of the present kind, see in this regard Chapter 4. Further, it should reflect, according to the representative working individual, the evaluation of the performance of the economic policy issued by the government, i.e. what the policy-maker or control board believes one can realistically attach to alternative trajectories of the state process.

Thus, we shall understand f and ϕ as a set of socioeconomic alternatives and social values expressing also the prospective economic and social welfare policy of the ruling party, to the fulfilment of which it should have committed itself. It is in the interest of those in power that their anticipations of the future and reality come as close as possible.

2.3.3   Preliminary characterisation of the control strategy

Next we will describe briefly the main steps and elements upon which the economic policy builds, explaining thereby basic features associated with goals setting. It is usual to call certain configurations of policy issues a control strategy. Formally equivalent terms like control function and control process will be introduced later.

Intuitive outline of policy features

Let us explore requirements an economic policy needs to fulfil in order to put the system upon which it acts in a goal-conducive motion. In particular, we want to inquire into the conditions the fulfilment of which may lead to an increase in the productive capacity of the system and to the development of its productive forces. Moreover, we have to learn about the manner of adjusting socioeconomic reality to a framework tailored to the purposes of goal-adequate development policies. According to the methodology we have chosen, we may take as given and/or as data the *initial* and the *terminal* state of the economic system. The latter establishes the *macro-goal* towards which the system is expected to move. As unknowns may act variables representing concepts underlying:

- The *development path* along which the system can attain the macro-goal. This path is fundamentally related to operational and logical issues that may set up required processes in order to obtain the desired objectives.

- The *behavioural* and *motivational patterns* able to bring the system to such a path, to promote an appropriate pace and keep it there, properties we shall refer to as *attainability* and *irreversibility*.

- An *institutional framework* able to recognise and to elicit a consistent *socioeconomic* and *political environment* which allows enough room for coherent changes.

The *unknowns* obtained in that way are essential ingredients of goal-adequate control strategy aimed at increasing *productive capacity*, i.e. skills, know-how and means of production, and at enhancing the level of *efficient performance*. Together with the patterns of efficiency, the fulfilling of productive capacity goals aims at the *full utilisation of productive resources*. For the macro-goal is not furthered whenever labour remains involuntarily idle, if means of production are still available and the need for output is still unsatisfied.

As with many notions in economics, that corresponding to productive capacity is very vague. The most practical way of getting a more tangible notion

may be suggesting to shift attention to the limit related to the rate of output set by capacity. It it necessary to recall that, at a microeconomic level,
the capacity of a plant is related to the length of the working day, alternative
numbers of shifts per week, and possible overtime among other factors. For
the time being we may assume, to any time there corresponds a certain rate of
output for each type of commodity which represents normal working capacity,
and we assume also, in the short run a higher rate can only be achieved by
raising sharply marginal costs.

Wage differentials derive from differentials in skill and productivity, which
in turn can be traced back to those in capital-labour ratio, capital endowment,
growth potential, etc. These differentials may be thought of as determining the
degree of attractiveness of productive activities, i.e. the appeal and activity or
a sector may have for uprooted labourers or potential migrants. This appeal
may release potential energies, a dynamic whole of activities which may affect
positively and/or negatively the sectors in question.

The *market*, like rationing and planning, is one of the major methods of
coordination of resource use within, between and beyond the modes and covers those processes of *exchange* and *distribution* which arise whenever ultimate
consumption and *accumulation* does not occur within the producing unit. Foresighted perspectives accounting for essential elements of the process of social
reproduction, the principle of maximum entropy and the interaction between
society and nature are in the main out of reach of market forces.

At any rate the market, under the prevailing sense of social justice, fails to
promote a system of incentives and rewards able to sustain productive effort at
a level adequate to generate the socioeconomic progress needed to absorb the
existing idle labour into production. To help the system to overcome the resulting impasses and to take advantage of favourable opportunities as well, one
needs governmental interventions directed to weaken bottle-necks or strengthen
productive power and functioning differentials by a policy aimed at enhancing
the system's growth and development potentials.

This may be achieved first by injecting a certain amount of capital, which
we shall call *capital impulse* or *impulsion*, at a certain time, to be called the
impulsion time. Second, by undertaking any effort to build social awareness,
to shape preference and value-judgements in a way able to release blocked
potential creativities and gain active support so as to promote development.
Third, by providing financial means and appropriate incentives so as to increase investments considerably with the purpose of accelerating the expansion
of (productive) capacity and generating effective demand. The interplay of the
second and third point called forth from forces which are expected to avoid
the emergence of *perverse growth*. That is, a sort of growth resulting from
allocating, under the pressure of demonstration effects and distributional dis-
parity, scare resources to production of consumption goods, which may be seen
as luxuries given a certain stage of development.

The second point is a delicate one. It requires a deep knowledge on available
human resources and is thus based on an extensive interdisciplinary research,
a point we have already referred to. Roughly speaking, our *economic policy*
may be seen to consist of the following two phases:

1. First, one increases the productive capacity of (the three sectors of) the system by means of impulsive controls. These amount to capital injections, which in turn make possible a new choice of techniques at a higher level of employment and bring about a structural change. We shall refer to this phase as the *long-term strategy* or *development policy*.

2. Second, one lets the system evolve according to a policy aimed at unfolding the productive forces of the economy, e.g. absorption of a certain amount of idle labour, the size of which depends upon that of the capital injection. A further instance may be just shifting resources to more highly productive activities. This phase shall be referred to as the *short-term strategy* or *accumulation policy*.

When one starts the second phase either the capital-labour ratio or the fraction of labour force actually employed have to remain fixed until the next impulsion time. Thus any increase in the level of employment will have to rely upon the own capital accumulation, i.e. formation of new capital out of the prevailing income shares of capital and labour.

The time elapsed between two consecutive impulsion times determines a time period, the duration of which depends again on the corresponding capital impulse, i.e. the duration of the time period may vary from period to period. One repeats these two phases again and again until the system attains a sustainable state of balanced growth which moves towards a dynamic equilibrium at which entropy increases at the least possible pace. This basic idea is depicted in Fig. 2.6.

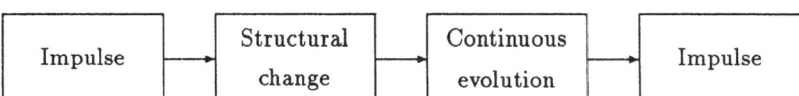

Figure 2.6: Outline of the impulsive control

In regard to the distinction between short-term and long-term economic features of the strategy, let us say, that we shall understand under short-term conditions for the time being, those specifying according to Marshal, the determining characteristics of the supply of commodities. That means these economic conditions that keep the largest possible size of commodity supply at certain niveau. This is more or less given by time dependent considerations concerning construction, installation and starting of operation of additional plant capacity. In this connection, we assume that within the short-term, the level of the stock of capital installed is given.

As an approximation, let us denote by $\overline{K}^\tau$ a certain number of machines that determines the available productive capacity. This level of capital stock determines the highest potential level of commodity supply. The operation of the process of production at its highest potential level would pose some requirements on the supply of labour, skills, availability of resources and the like. On the other hand, let us denote the (maximal) level of available labour by $\overline{L}^\tau$. Hence, we may set the upper bound $\overline{Q}$ at the highest potential level of output in the short-term, with $\overline{Q}$ given by

$$\overline{Q} = \min\{a_l\overline{L}^\tau, a_k\overline{K}^\tau\}, \tag{2.29}$$

where the minimum is taken among the techniques available in the given period represented by the currently feasible technical coefficients a_l and a_k introduced above. The question concerning with the actual level Q of social output will be dealt with in[94].

Further, let us assume that in the long-term we may change the productive capacity. This assumption is stronger as it may appear at first sight, because the productive capacity is in various ways linked to the major potentials of development: labour skills, accessibility and size of the market, diversity of the social output, feasible and possible sets of techniques, self-intensifying mixes of necessary consumption and capabilities are among the most important.

Impulsive and continuous control components

The theory of impulsive control provides techniques by means of which to determine a sequence of capital impulses and impulsion times that over the long-run will advance the economy to a certain state of development. The sequence just mentioned will be called an *impulsive control strategy* or simply an *impulsive control*, the role of which is to enable the economy to make a suitable and flexible choice of technique, one that is not binding for long periods of time but adjusts as capital becomes available and the system matures to more adequate production techniques. These capital impulses generate structural changes which are assumed, for the sake of simplicity, to occur instantaneously. See Fig. 2.6 above.

Furthermore, through a purposeful combination of capital-deepening and capital-widening decisions, one seeks to enhance both human and physical capital efficiency and to achieve a more appropriate allocation of labour, in a sense to become clear later. Consequently, these capital injections will bring about, besides an increase in productivity and in the rate of technical progress, a strengthening of the capability of the economy to absorb the labour-surplus, to smooth out sectorial dissimilarities and to keep in check deterioration of crafts, migration etc.

In order to maintain a certain transparency in the conception and to assure operationality of the *control strategy*, we shall distinguish carefully between *short-term* and *long-term* components of the strategy as follows:

- *Impulsive control* consisting of those decisions linked to capital impulses and impulsion times which are related *inter alia* to choice of techniques, structural change and technical progress. These are the long-term aspects of the process of social reproduction and they act as a data for the short-term component.

- The *continuous control* consisting of those decisions linked to the consumption-investment mix, and the level of employment which are related to the capital endowment per worker, negotiation of an upper bound to the rate of investment, accounting price of investment, intergenerational bequest, etc.

However, we shall point out that this separation of roles between impulsive and continuous control is a modelling device rather than a sharp analytical disentanglement of the socioeconomic processes behind it. There are further various institutional elements of policy-making deeply interwoven with short-term and long-term issues to which we shall turn later. Let us mention at this stage, that in the present essays we are mainly concerned with those aspects of the development policy associated with continuous control. However, we shall briefly touch on a few issues of impulsive control.

Formal outline of the system dynamics

In order to put the foregoing ideas in a mathematical setting, let us consider an abstract probability space $(\Omega, \mathcal{F}, \mathbf{P})$ endowed with a filtration $(\mathcal{F}_t)_{t \geq 0}$ of sub$-\sigma-$fields of $\mathcal{F}$ with the usual properties. Further, let us consider on this probability space a $\mathbb{R}^m-$valued Brownian motion denoted by $B = (B_t, \mathcal{F}_t)_{t \geq 0}$ and a $\mathbb{R}^n-$valued stochastic process $X = (X_t, \mathcal{F}_t)_{t \geq 0}$ representing the state of the economy, where $n, \ m \in \mathbb{Z}_+$. That means that at the time t the economy is in some state which is described by X_t and its associated stochastic characteristics, as we shall see later. Let τ^j and t be in $\mathbb{R}_+$ with $j = 1, 2, 3, ..$ and let $\alpha^j : t \longrightarrow \alpha_t^j$ be an $A-$valued random process on $(\Omega, \mathcal{F}, \mathbf{P})$ with $A \subseteq \mathbb{R}^n$. Then the dynamics of the state of the system is governed by the equations

$$
\begin{aligned}
dX_t^j &= \sigma(\tau^j, X_t^j, \alpha_t^j)dB_t + b(\tau^j, X_t^j, \alpha_t^j)dt, \quad t \in [\tau^j, \tau^{j+1}), \quad (2.30) \\
X_{\tau^j}^j &= X_{\tau_-^j}^{j-1} + \theta^j \quad \text{and} \quad X_\tau^j = X_\tau^{j-1}, \quad t \in [\tau^{j-1}, \tau^j) \quad (2.31)
\end{aligned}
$$

$$
X_0 = x \quad \text{and} \quad X_t = \lim_{j \to \infty} X_t^j, \quad t \geq 0 \quad (2.32)
$$

where θ^j, τ^j are random variables defined on the probability space introduced above, and will be determined relying on the information available. They denote respectively the capital impulses and the impulsion times, with $\theta^j \in \mathbb{R}^n$ and $\tau^j \in \mathbb{R}_+$. The drift vector b and diffusion matrix σ are defined as usual. Finally, $X_{\tau_-^j}^j$ is given by

$$
X_{\tau^j}^j = X_{\tau^{j-1}}^{j-1} + \int_{\tau^{j-1}}^{\tau^j} \sigma(\tau^{j-1}, X_t^{j-1}, \alpha_t^{j-1})dB_t + \int_{\tau^{j-1}}^{\tau^j} f(\tau^{j-1}, X_t^{j-1}, \alpha_t^{j-1})dt.
$$

$$(2.33)$$

Let the pair (θ, τ) denote the sequence of pairs of random variables $\{(\theta^1, \tau^1), (\theta^2, \tau^2), \ldots, (\theta^j, \tau^j), \ldots\}$, called impulsive control strategy, which is a formal representation of the (process of) impulsive control.

The *impulsive control problem* will consist in determining the impulsive control (θ, τ) in a way that the system governed by the eqs. (2.30), (2.31) and (2.32) attains a desired state according to a certain performance criterion and under additional constraints that we do not need to specify here.

While the *continuous control problem* will consist in characterising, for a fixed but arbitrary j, the continuous control process α^j, so that the economy governed by eq. (2.30) attains under certain constraints a desired state according to a performance criterion to be introduced later. See (1.18) and (1.19)–(1.20). As we mentioned already, it is the latter problem to which we shall dedicate most of these essays. The stochastic process $X = (X_t, \mathcal{F}_t)_{t \geq 0}$ is usually called the *basic* or *controlled process*.

Chapter 3

Change and time in development and growth processes

The utopia which has informed industrial societies for the last two hundred years is collapsing. And I use the term utopia in its contemporary philosophical sense here, as the vision of the future on which a civilization bases its projects, establishes its ideal goals and builds its hopes. When a utopia collapses in this way, it indicates that the entire circulation of values which regulates the social dynamic and the meaning of our activities is in crisis. This is the crisis we are faced with today. The industrialist utopia promised us that the development of the forces of production and the expansion of the economic sphere would liberate humanity from scarcity, injustice and misery; that these developments would bestow on humanity the sovereign power to dominate Nature, and with this the sovereign power of self-determination; and that they would turn work into a demiurgic and *auto-poietic* activity in which the incomparably individual fulfilment of each was recognized — as both right and duty — as serving the emancipation of all.

Nothing remains of this utopia. This does not mean that all is lost and that we have no other option but to let events take their course. It means we must find a new utopia, for as long as we are the prisoners of the utopia collapsing around us, we will remain incapable of perceiving the potential for liberation offered by the changes happening now, or of turning them to our advantage by giving meaning to them.

André Gorz
(see [95], pp.8)

3.1   Time and change in economics

As a first approximation we may conceive a *dynamic system* as an object undergoing a process of change and think of time as a fictive device by means of which we attempt to keep trace of change and measure the degree to which it occurs. In order to understand the notion of change one has to distinguish between observations at the microscopic and macroscopic level which in turn requires the ability to distinguish between fast and slow moving processes of change. In this sense systems come to exhibit specific orders and patterns of structure we try to capture and storage in the form of models. These models are representations of systems and are built, storaged and updated as our perception and assessment of processes of change evolve. Thus, models, understood as qualitative and quantitative pictures of perceived reality, constitute the cornerstones of processes of human learning.

The most essential link between change and time is the one connected with the direction along which the system potential for change diminishes, its patterns of structure decay and its states of disorder increase. To understand the nature of time in this sense and its implication for the notion of change in society, we start examining the meaning of logical time, as it is used in classical dynamics, and sketching the main features of the most relevant concept of historical time.

3.1.1   Time arrows and historical nature of the world

At a fairly intuitive level let us say the mechanistic conception of classical physics sees a dynamic system as an object characterised by phenomena describable as a process, the main features of which can be formally separated into *laws of motions* and *initial conditions*. However, knowledge of initial conditions requires the kinetics of the system under consideration has already been defined, so that its states are conceptually given as well as operational rules based on which all its potential states can become unambiguously linked by way of a unique mapping. This notion is associated essentially with the *principle of causality* according to which any event must possess a cause though not necessarily an effect and with the *mechanistic notion of time* which views it just as a geometric parameter in a way that puts past and future on the same footing. The following features resulting from the classical picture of change and time need to be emphasised.

- Time is completely separated from and independent of space according to the belief one could unambiguously measure the interval of time between two events. This corresponds to the notion of absolute time.

- The separation of laws of motion and initial conditions underlies the idea that the laws of classical mechanics are identical with those of pure reason. This is based on a strong deterministic conception of the world and on an oversimplification of the notion of initial conditions.

- The gravitational origin of classical dynamics brought about an extremely simple description of microscopic phenomena. This fact together with the assumption of time symmetry hinders an adequate understanding of the current state of dynamic systems which is a condition for successfully predicting the future evolution of less simple dynamic systems.

- The irreversibility exhibited by certain processes in classical dynamics is explained by the inherent imperfection in perceiving, observing and measuring motion, instead of relating it basically to the dynamic nature of the underlying systems. In this sense it will make a difference shifting certain properties from the macroscopic to the microscopic level and vice versa as, for instance, heterogeneity, viscosity and decay require.

As a natural consequence of thermodynamic considerations it turns out that without taking into account the direction of time, as the presence of irreversible processes entails, one cannot grasp the fundamentals of evolutionary dynamics characterising essentially the unfolding of living, social and other complex systems. The second law of thermodynamics gives rise to a distinction, among the states such systems may attain, according to which certain states may act as attractors of others. Thereby, the notions of equilibrium and systems far from equilibrium acquire a more natural framework in which laws of motion and initial condition bear a probabilistic meaning.

On the impact of classical dynamics upon economic thinking

It is not at all surprising that the paradigm of classical dynamics came to influence economic theorising. However, what is a kind of puzzle is the fact that various shifts of paradigm along the centuries have not brought fundamental changes in thinking of economic orthodoxy. The mechanistic view of the world, in particular, Newton's theory of gravity, holds independent of what the bodies are made of and requires the notion of absolute time as well as an intermittent intervention by God to support the maintenance of order and regularity.

This appealing, simple and successful model built on the mechanistic view of Bacon, Descartes and Newton was taken without major changes as a description of a social system where individuals are reduced to the role of social atoms and nature to that of a frame without life of its own. The interaction of social atoms among themselves and with nature have to reflect in essence the functions of physical phenomena as envisioned in the mechanistic paradigm. Actions and ideas rooted in tradition and culture did not find any place, instead rationality enters as the principle according to which social atoms, i.e. individuals acting solely on its self-interest, ruthlessly pursue maximisation of their own wealth. The state has here just to guarantee the freedom of social atoms to do so and to protect their property rights. Freedom extends to the full subjugation of nature or better the "negation of nature" and full subjugation of non-European people or better the "negation of non-European people".

Four centuries after the emergence of the mechanistic world paradigm the operation of mechanistic rationality guided by the "invisible hand" acts upon two thirds of mankind and nature as a nightmare with an almost unchallenged dominance. However, this dominance of orthodoxy does not rely on scientific superiority but rather on the lack of social perspective and ignorance of the essential principles upon which the order of society and nature depend. See [178, 170, 104] concerning critical views of classical dynamics and [223, 178, 153, 107] regarding the mechanistic picture of society in particular as represented by John Locke and Adam Smith.

In a purely dynamical sense the laws of motion describe the time dependence of system states, e.g. state equations in differential form and logic or rational time. Nevertheless, the concept of time introduced thereby involves certain circularity due to the fact that motion itself is defined as change with a functional dependence on time. Further, a basic feature of causality is based on the notion of nonevent associated with the case of an absence of causes. The mechanistic description of causality, which has played such a central role in economics, considers the concept of force as cause of motion, and it is in this way related to determinism. However, the main effect of a force acting upon a particle is the change of its motion. Moreover, this above characterisation of causality enables to combine generalised (position) coordinates and momenta by introducing the Hamiltonian concept of state, represented by dynamics governed by differential equations of first order in time, with the phase space of the system consisting of the configuration space and the conjugate momentum space.

Since knowledge of the Hamiltonian determines the (laws of) motion of the system, it is necessary to call attention to the relevance of the initial conditions and the assumption of a highly simplified microscopic behaviour. In its standard form this notion of causality determines uniquely a state at any time from an arbitrary state at another earlier or later time. That means, it singles out neither a direction of time nor a point corresponding to the present.

Motion as a socioeconomic description of change

By contrast, dealing with social systems one needs to consider a historical time so that a causal structure gets added in which the past is fixed and carried in an informational structure available as memory in the present, while the future is open and in principle cannot be unambiguously determined from the present. A comparison of classical dynamics with thermodynamics will shed light on these issues and improve our ability of describing social reality, a point to which we shall come back later in this chapter.

Evolution and growth are the most impressive phenomena in the world and constitute the basis of various processes responsible for the reproduction of systems of whatever nature. These essential ingredients are therefore present in a variety of ways in virtually all socioeconomic processes and are responsible for the always changing character and increasing complexity of our social, economic and cultural environment. This fact renders old methods of policy making and management of society inappropriate and challenges social scientists to understand the dynamics of society, to learn to cope with *time* and to manage *change*, since behind any features related to development and growth we find always the basic concepts of change and time.

Therefore, understanding these two concepts will necessarily help us to grasp socioeconomic phenomena and to discern how states of society and activities are interrelated in time, how changes occur over time and how we can improve our ability to influence the chance of occurrence of what might be called more desirable outcomes.

3.1.2 Equilibrium and forward moving systems

The mechanistic origin of orthodox economics finds its vivid expression in the dominance of equilibrium analysis as the most used working tool. A pendulum tending toward a vertical equilibrium position has been applied, as a metaphor according to which the dealers in a market bear in their minds a clear concept as to what the equilibrium price is, even if it never settles there. As Joan Robinson argued, this concept of equilibrium is incompatible with history. It applies a metaphor based on movements in space to an adjustment process taking place in time. It confuses historical with logical time, since it translates the space analogy of remeding misdirections by going to and fro into logical time distinguished by symmetry of past and future, while in historical time the past is irrevocably gone and the future unknown.

Curious features of the orthodox notion of equilibrium

The conception of an equilibrium position of state which is unlikely to be ever attained at any particular moment of time can hardly be explained from experience, since this depends on how others behave and on the assumption all parties concerned know clearly at each moment of time, what the equilibrium position is, see [181]. This sort of equilibrium models bases on an awkward knowledge of the rules and motives governing human behaviour, and exhibit various curious features. As a way of example:

(1) an equilibrium position which can never be attained but exists in virtue of the fact that it is believed to exist;

(2) expectations, formed on the basis of an unattainable equilibrium which is believed to exist, are not revised in order to keep the system going;

(3) an equilibrium position lying in the future towards which the system moves owing to the belief those influences in the past that frustrated the attainment of equilibrium will be prevented to occur in the future.

Joan Robinson's sharp distinction between logical and historical time derives from the recognition of two different types of economic arguments. The first proceeds constructing models by specifying a sufficient number of equations so as to determine the involved unknowns and then finding values for them that are compatible with each other. These equations provide a path through time which is not confined to stationary equilibrium relations. Such a model is following a path in *logical time* approaching in one direction a "future" state within some given compatibility requirements and in the other a "past" state also fully determined. In this sense a model depicting an equilibrium position consists of a close circle of simultaneous equations and exhibits only a causation in the mechanistic sense. Indeed this is only a pseudo-causation due to the presence of time reversal. Think of supply and demand curves which determine a price compatible with a quantity to be traded.

Robinson's notion of historical time follows from the second type of argument according to which constructing a model one specifies a particular set of values obtaining at any moment of time and indicates how their interactions may be expected to play themselves out. Values in this particular set hence do not need to be in equilibrium with each other. Thus, in a historical model, what happens in the future results from the interactions of the behaviour of agents described in it and, as in actual history, it has to be capable of being and/or getting out of equilibrium. The initial conditions contain, beside physical data, the state of expectations of the agents concerned, be these based on past experience or on traditional beliefs, and these are liable to be disappointed. Thus, initial conditions may be, but do not have to be, causae efficientes and/or causae finales pointing out an eventual divergence with the classical mechanistic picture.

The distinguishing feature of *historical time* may be seen in the fact that in a historical model causal relations have to be specified and initial conditions have to be set in a way rendering change a forward movement, making the present a break in time between an unknown future and an irrevocable past, see [181]. This is the spirit prevailing in thermodynamics where representative ensembles are introduced corresponding to the idea of considering a large number of copies of the system distinguished only by their generalised (position) coordinates and momenta. As we will see later this idea is highly useful whenever initial conditions cannot be given with precision. In this sense the notion of historical causality although still fairly restrictive comes closer to that of overdetermination inherent in social systems and reflects the idea that any event must possess (at least) a cause in the past, i.e. an antecedent, and not necessarily an effect in the future.

Before we pass to examine some particular issues on time, let us remark that although *a priori* comparisons of equilibrium positions may illustrate some actual situations, it is necessary to keep them in their logical place and start our analysis rather from the rules and motives governing actual human behaviour. That means, we have to include in an essential way processes of learning and forgetting as a manner of counteracting the usual simplicity of microscopic descriptions.

Role of knowledge and historical causality

For the purpose of our inquiry, we shall understand *time* as an analytical instrument by way of which we make records of our perceptions of change occurring in reality. This view stresses the fact that our notion of reality is linked to our own capacity to observe and memorise it. Therefore awareness and memory build in our mind a decisive link to an enormous storehouse of information codified in models and images of systems, i.e. those representations of characteristic features of order and structure by means of which our mind identifies systems building our reality. Being itself a dynamic system the human brain is extremely active; it stores, creates and updates incessantly models of systems, i.e. pictures of order and patterns of structure. The role of historical time in this connection is to fix the present and separate the past, which is in principle recognisable backwards, to form the future which can at best be probabilistically determined from the present state.

It is important to bring the impact of the passing of time in its historical guise upon memory and know-how, as an outcome of learning, in connection with their qualitative and quantitative changes. Because these changes are linked with the capability of individuals to function and with their potential skills to produce. Knowledge and information cannot evolve by themselves and without proper assistance. In this sense, the historical nature of the world becomes the basis of any fundamental concept of becoming and it is already entailed in the time-directed concept of causality as introduced above.

As we shall see later, based on concepts like learning, forgetting, information and entropy, one can interpret the behaviour of human beings to some extent as a result of social evolution on the light of historical models of causal relations. While learning increases our information and potential to function, forgetting increases our lack of potential to function and entropy suggesting in a sense a direction in time. On the other hand learning is an evolutionary process consisting in a large family of processes all accomplishing a manifold of testing and selection activities by means of which they create and/or destroy, reinforce and/or update models of systems.

Hence we shall focus on those linked essentially with production and activities which through human memory-like and other physiological processes help us to visualise a direction of time, i.e. arrows of time.

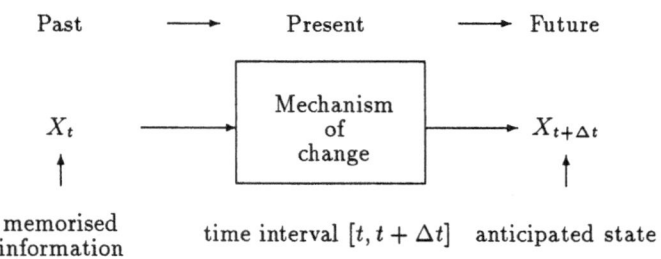

Figure 3.1: The process of time-change

The increased pace at which time-changes occur gives rise to the need of constructing dynamic models as simple and transparent representation of the processes of change taking place in a system. On the basis of such models we expect to increase our ability to understand and manage change. Intuitively, this idea seems at first simple and we depict it in Fig. 3.1 above. It reflects the capacity of mankind to accumulate knowledge and to change the prevailing patterns of behaviour in the light of past experiences with the aim of adapting in various ways to the challenge emerging from attempts of anticipating states of the future. Although the weight of the accumulated knowledge in the form of lived experiences, traditional beliefs or the like does constitute a powerful constraint, it is necessary to stress one cannot establish unambiguously relationships between the past, the present and the future from the point of view of historical causality as it is possible along logical time following the logic of determinism.

3.1.3 Probabilistic view of causality and time structures

Thinking in the framework of systems theory let time be that variable based on which one records the evolution of a process taking place in a given system. Let us denote this process by X, so that at time t, in the past, its state is represented by X_t and let us assume, it contains all the information on the system available at that time.

Processes without and with after-effect

Using the notion of representative ensembles introduced above, we shall suppose the state of the system at time t may be characterised by the probabilistic relationship $\mathbf{P}(X_t \in B) = m(t, B)$, which is a natural consequence of the fact that the initial conditions cannot be given with precision. For the sake of simplicity we will write also m_t instead of $m(t, B)$ and set $m = m_0$. Here we have assumed that the (attainable or) possible states of the system form a set $\mathfrak{X}$ which can be equipped with a σ-algebra $\mathcal{B}$ of (observable) subsets $B \subset \mathfrak{X}$, so that $m(t, B)$ measures the probability of the event $\{X_t \in B\}$, i.e. the probability that at time t the system attains a state situated in the set B. The measurable space $(\mathfrak{X}, \mathcal{B})$ is called the *phase space* of the system.

A dynamic characterisation of the state the system attains at time $t + \Delta t$ is based on the assumption for any set $B \in \mathcal{B}$, there exists a family of functions $\{P_{t,x}(X_{t+\Delta t} \in B) = P(t, x, t + \Delta t, B), \ t \in [0, T]\}$. Any function of this family indicates the probability with which a copy of the system that starts at time t from the state $X_t = x$ reaches at time $t + \Delta t$ a state $X_{t+\Delta t}$ situated in B. An essential feature of this description is deeply connected with the observation that as the system changes in time so does the probability measure $m(t, B)$, which describes probabilistically the present state X_t.

Hence, for any arbitrary but fixed time t and any positive Δt, we get at most an average functional $m(t + \Delta t, B) = \mathbf{P}(X_{t+\Delta t} \in B)$ determined by

$$m(t + \Delta t, B) = F_{t+\Delta t}(t, m_t) = \int P_{t,x}(X_{t+\Delta t} \in B) \, m(t, dx), \qquad (3.1)$$

as an adequate probabilistic description of the state the system attains at time $t + \Delta t$. The average is taken with respect to the measure $m(t, B)$ prevailing at time t. For $t > s$, the mapping $F_t(s, \cdot) : m_s \longrightarrow m_t$ acts on $\mathbb{P} = \mathbb{P}(\mathfrak{X}, \mathcal{B})$, the space of all probability measures on $\mathcal{B}$.

This average relationship relies on the assumption the time interval Δt is small compared with the time of observation and large enough to guarantee independence of the changes in the system attributable to two consecutive intervals, i.e. it assumes at time $t + \Delta t$ there is no after-effect attributable to the movement of the system up to time t.

Let $P_{t,x}(X_{t+\Delta t} \in B)$ be the probability that at time $t + \Delta t$ the system attains a state $X_{t+\Delta t}$ situated in B knowing that at time t it started from the state x, under the condition that the movement of the system up to time t is completely known. A system is termed *without an after-effect* if the probability $P_{t,x}(X_{t+\Delta t} \in B)$ does depend solely on the state $X_t = x$ the system occupied at time t.

This property is the underlying characteristic of Markov processes. It is easy to check using (3.1) that for any $s < u < t$ the relationship

$$m_t = F_t(u, m_u) = F_t(u, F_u(s, m_s))$$

holds, which is a sort of composition rule. Further, let us assume that the family of probabilities defined as $\{P_{s,x}(y \in B) = P(s,x,t,B), \; s,t \in [0,T]\}$ is a family of stochastic kernels. This family is usually known as the *transition probability* of (the Markov process governing the dynamics of) the system. Then using the Chapman-Kolmogorov equation, one gets easily, that

$$m_t = F_t(u, m_u) = F_t(u, F_u(s, m_s)) = F_t(s, m_s).$$

That means that defining arbitrarily the probability measure m_s and setting as above $m_s(B) = P(X_s \in B)$, one obtains from (3.1) the probabilistic characterisation of the system at any succeeding instant of time t, $t > s$.

This is precisely the probabilistic version of causality in logical time, which suggests that the process building the representative ensembles exhibits a motion following probabilistic laws but the probability itself propagates according to the principle of causality. For $s = 0$ the distribution $m = m_0$ is called the *input (probability) distribution*.

An infinitesimal characterisation of the mechanisms of change driving the probabilistic structures associated with the foregoing notion of causality, results from the relationship concerning the right-hand derivative $\frac{\partial m_t}{\partial t}$

$$m(t + \Delta t, B) - m(t, B) = \frac{\partial m(t, B)}{\partial t} \cdot \Delta t + o(\Delta t),$$

which leads to *Kolmogorov's forward equations*

$$\frac{\partial m(t, B)}{\partial t} = A_t^* m(t, B) \quad \text{and} \quad \lim_{t \downarrow s} m(t, B) = m(s, B) . \qquad (3.2)$$

Here A_t^* is a linear operator defined on the linear space $\mathcal{D}^*$ which depends parametrically on $t \in [s, t^*) \subset [0, T]$. Considering the family of operators $\{T_{ts}^*, \; t \geq s \geq 0\}$ associated with the transition probability of the system which acts on the space $\mathbb{V}$ of all finite charges on $\mathcal{B}$, i.e. the set of all finite countable additive set functions $q(B)$ with $B \in \mathcal{B}$, and setting $q(t, B) = T_{ts}^* q(B)$ with $q(B) = q(s, B)$ and $t \geq s$, the space $\mathcal{D}^*$ is the subset of $\mathbb{V}$ consisting of those charges $q(B)$ for which the limits

$$\lim_{h \downarrow 0} \frac{T_{t+h,t}^* q(B) - q(B)}{h} = A_t^* q(B),$$

$$\lim_{h \downarrow 0} T_{s+h,s}^* q(B) = q(B),$$

exist for all $(t, B) \in [s, t^*) \times \mathcal{B}_0$, where $\mathcal{B}_0 \subset \mathcal{B}$ and s is fixed. The operator T^*_{ts} maps $q(s, B)$ into $q(t, B)$ in a similar way as $F_t(s, \cdot)$ maps $m(s, B)$ into $m(t, B)$ by means of (3.1). The fact that $q(t, B) = \mathsf{T}^*_{ts} q(B)$ for $t \geq s$ points out the forward direction of the causal relationship mentioned above.

The *equation of motion* or mechanism of change underlying the evolution of process X, sometimes called the state equation, may be given by means of SDE's, stochastic PDE's or stochastic difference equations. However, any such formulation requires a deep understanding of the process of change and the sort of causal time-ordering of the underlying activities. Indeed, the notion of causality takes in many cases the form of a Markov property, a heavily irrealistic assumption; an alternative is the more general concept arising in stochastic analysis known as the nonanticipation principle, see [5].

Reflections on the interaction of determining forces

Let us consider, following Robinson [181], some issues related to the construction of a model applicable to actual history. As a first approximation let us say one has to specify the technical conditions to prevail in the economy at hand and the behavioural pattern of its agents, households, firms or whatsoever. Then by specifying the initial conditions one dumps down the economy, represented by this model, in a particular situation at a particular date in historical time and works out what will happen next. The path the model follows will be that which corresponds to the level of fulfilment or disappointment of the expectations and behavioural pattern of its agents on the one hand and on the capability of the system as a whole to adjust to the resulting state of affairs. That means, such a model has to recognise to some extent various complications attached to concrete reality.

Let us mention as an illustration that an overall rise or fall in prices, employment, interest among others, may be followed by relative changes in particular markets and regions and this in turn may give rise to changes in the current and expected values of relevant variables. Then, a historical model will be able to follow a smooth path, provided interactions and processes interplay – which build the major basic events of changes incorporated in the system dynamics – bring under utmost discrepancies between current and expected values. This world enables the system to maintain confidence in the state of expectation of its agents.

On the other hand whenever there exists a propensity for the current situation to overweigh past experience, confident expectations cannot be held and a discontinuous path may result from a switch of the system to more appropriate technical conditions or the like. Such a break of continuity becomes the most natural consequence of a variety of quantitative and qualitative processes of change underpinning the continuous struggle of mankind to adapt to and to handle better the always changing reality.

Nevertheless, we cannot unambiguously think of a certain socioeconomic configuration causing another one. First owing to the fact that in reality the state of affairs prevailing today may vary heavily from those featuring the situation, when the relevant decisions concerning today were made, in the past.

Second, because processes of change have as a rule after-effects which may not have been anticipated at the time such decisions were made. Needless to say, anything related to overdetermination, the principle inherent in social systems, derives from the complexity underlining a great degree of interaction and a not always transparent interplay of manifold processes building our reality.

A key role on the way to understand time and change is played by the principle of social reproduction, a point that becomes apparent whenever we focus on the structures of organisation of social systems and on their potentials of renewal. Since available potentials and patterns of organisation are used up and decay, owing to processes forwarding the system and to aging without recreating and renewing properly their basic elements, the system structures may rigidify and eventually loose their identifying features.

To approximate the intricate nature of events against the background of time and change, one resorts to probabilistic analysis which switches from the causal explanatory ground to assessment of characteristics reflecting occurrence of such events. Alternatively, one may seek insight on the basis of qualitative analysis. Following issues are essential:

(1) only those processes emerge and/or go on for which a potential exists, a feature that determines a sort of causal relationship going from the existence of potential to the emergence, the ongoing or continuation of processes;

(2) using up available potentials is a process akin to entropy and determines therefore the direction or arrows of time, since any event is a product of its own virtue in motion from the potential to the actual;

(3) learning as a process consisting of coding, transforming and transferring information and know-how becomes basically an anti-entropic process; and

(4) failure of off-setting the course of entropy may result in a break of structure which emphasises the unity of continuity and discontinuity.

Hence the picture of time and change in the logical sense resulting from the need of recording our observations of evolutionary states of a dynamic system remains inadequate without adding to it the arrows of time, since they indicate the direction on which available potentials are used up. It is in this sense that we shall understand the notions of historical time and change introduced above, which suggests a static level of analysis without changes in the system potentials and a dynamic level reflecting the course of entropy.

Time structure of processes intervening in economic intercourse

Descriptive pictures of structure as a moment-to-moment relationship predominate at the static level while rates of change, in general varying in the course of time, constitute the essence of pictures of structure at the dynamic level. The application of mathematical techniques of systems theory is aimed at sorting out certain elements of the real-world structure, and at arranging them in a model capable of reflecting those system features, we attempt to understand, which are most important to us. Constructing such a model we intend to translate complex structural features into simplified (structural) parameters, in a way that gives rise to analogous processes still exhibiting the system properties we want to investigate but in a manner amenable to mathematical tools of analysis.

In order to appreciate the degree of difficulty of the problem at hand let us outline, following to some extent [217], some issues associated with the structures of time and change distinguishing various processes to which one attributes an enormous explanatory power. For the time being we shall however dispense with a detailed presentation of the elements behind them.

- The *time structure of social and individual needs* and/or, the commonly used surrogate, *preferences* and *timing decisions* including patterns determining mechanisms by means of which to appraise the relative desirability of outcomes occurring over time. Directing attention to the potential resulting from functionings, as a general form of states of being and doing, we avoid the conventional lack of unambiguous assessment criteria of future against current outcomes and provide flexibility in decision making associated with uncertainty.

- The *time structure of transformation processes* related to the technology available at an arbitrary time for productive activities. Thinking of system potentials helps us to visualise the notion of technology as means of transforming a set of inputs into a set of outputs at any specific instant of time under certain hypotheses about the process of change.

- The *evolution of temporal processes* is reflected in the time-path of certain variables and in the presence of feedbacks of several kinds. To determining such time-paths there are numerical and descriptive difficulties, i.e. time units to be adopted, length of the time interval, frequency and accuracy of temporal records.

- The *temporal character of decisions and information*, as well as their coupling point out the acting of various learning mechanisms assessing the effect of decisions on information and know-how, and for the purpose of providing decision makers with an appropriate coupling of information. It also requires that learning mechanisms be devised to create and to code knowledge properly as well as to reduce uncertainty as information unfolds over time.

The fact that the process of social reproduction comprises a manifold of learning mechanisms and feedbacks renders it eminently evolutionary. Thus an everlasting sequence of reinforcing, rejecting and updating of system models may be the most natural response to our attempt to describing basic processes of consumption and production which reflect exhaustion and decay of system potentials. However, these basic processes of production do not solely require the availability of potentials to be used up in the course of fulfilling production aims. They may also contribute to increasing knowledge, information and complexity in terms of differentiation and structure, provided these newly acquired qualities are properly coded into something, e.g. means of labour, working skills of individuals and production techniques.

The role of the principle of social reproduction here is precisely to point to the necessity of restoring used up potentials and/or increasing self-sustaining power so as to maintain the system capability of meeting its needs for production. Turning attention to the characteristics underlying commodities making up the social output, puts us in a better position to understand basic features of various circular flows of social reproduction. Because system potentials available at any moment of time are deeply related to the capability set of individual agents, and this set consists of alternative functionings, which in turn depend on the states of being and doing deriving from the bundle of characteristics to which an individual agent is entitled. See in this regard Subsections 1.2.2 and 2.2.1.

In order to capture the time structure of the basic processes supporting social production we have thus to examine in addition to the static description: the flux of change in the structure, patterns of organisation and order. It is evident, that if a significant number of changes occur in one time interval the quality of information and accuracy of records will depend critically on the number and quality of the factors on which we base our attempt to capture and measure changes. More or better information concerning the impact of an action upon future states render relevant choices more manageable, while less information or lack of it make them less efficient.

However, additional information may entail more observables entering our picture of the system and thus more complexity in the process of change relating the past to the future. For the sake of maintaining operationality one may have to give up potential sources of information which in turn adds to uncertainty. These reflections on the time structure of fundamental magnitudes on which we base the description of economic systems stress the disarming simplicity of orthodox wisdom. In a world which is changing all the time, it does not make any sense to give changes the character of exceptionality and undesirability, as the word "shock" suggests. We know a lot about reality and can learn a lot more about time and change as needed to understand issues behind structural dynamics.

Before closing these notes on the time structure of processes, let us stress the uncertainty attributed to the fact that certain processes are basically random is an essential and thus an irreducible feature of the system. That means, at any moment of time the state of such a process is a random variable which links our knowledge of values of its state to our information on the time structure of its distribution. In this sense attempts to ascertaining closely the values of the system state are offset by the behaviour of its transition probability over time, regardless how accurate and efficient may be the feedback and learning mechanisms incorporated in the state equation. Recall in this connection Heisenberg's principle of uncertainty.

3.1.4   Planning and risk management

Next, we shall deal with two common procedures concerning the attempt to anticipate uncertainty and offsetting some of its effects. These procedures rely basically on information and learning mechanisms which may evolve in the course of time. The first one focuses on the organisation of activities over time and is based on a clear understanding of causes of change. It is aimed at delineating temporarily linked events, activities and actions in a way capable of transforming them into a coherent and eventually lasting whole. However, structures of organisation rigidify and decay, as we have already mentioned, a fact that may itself give rise to uncertainty.

The second one is directed to reducing the uncertainty associated with the process of change. It combines a variety of operations like inquiring, estimating, updating and predicting with the purpose of anticipating future outcomes so as to replace a set of strings of unknown events in time by one or more sets of strings of presumed events. A natural consequence of anticipating the future may be postponement of actions usually described by discounting operations and/or taking a risk against the prospect of larger but uncertain returns, see [217].

The need of means of counteracting uncertainty

Stochastic control theory provides adequate tools for model design and analysis of processes of development and growth in dynamic systems. Stochastic control techniques combine purposely the two procedures introduced above so as to increase our ability to ensure that what is intended will likely occur and underpin in this sense the most basic planning objectives. Namely, these techniques attempt to replace a set of possible and relatively uncontrolled events by a set of possible and more desirable ones, i.e. to organise events in time and to control the uncertainty unfolding over time. Roughly speaking, planning objectives are a response to a natural demand for enhancing the likelihood of future occurrence of a set of desired states on the basis of an adequate organisation of relevant activities and an accurate assessment of uncertainty which build the essence of planning strategies and risk management.

Let us make more precise what we shall understand under the notion of likelihood of temporarily linked events and the like. For example, let us think of the equation of motion of a system described by a process X. Let $X = (X_t)_{t \geq 0}$ be driven by a SDE defined on the probability space $(\Omega, \mathcal{F}, \mathbf{P})$ endowed with a filtration $(\mathcal{F}_t)_{t \geq 0}$, or current of σ-algebras $\mathcal{F}_t$ of $\mathcal{F}$ holding the usual hypotheses. In the sequel a letter in lower cases shall denote values which may be assumed by variables denoted by the corresponding upper cases. Drawing on the characterisation of the process X in terms of probability and letting x_t stand for the values of the state X_t at time t, we know that to each one of these values corresponds a probability $\mathbf{P}(x_t | \mathcal{F}_t)$, see (3.1) and (3.2). Let us recall for any time t, the σ-algebra $\mathcal{F}_t$ describes the history of the process X up to time t or represents a system of information resulting from our records on the states of the process X up to the time t.

Then following Tapiero [218] we shall associate, as a first approximation, *knowledge* with our ability to specify properly P and $\mathcal{F}_t$, i.e. a state of knowledge becomes a function of the information we have about the process X. On the other hand, *information* is a quantitative and qualitative structure based on which we may reduce our uncertainty. Thus, if planning in a changing environment is to be understood as the method by means of which we increase our ability to ensure that what is intended will occur in the prospect of uncertain future, it turns out uncertainty is an element one has to avoid and/or reduce. At a fairly intuitive level we may recognise the direction along which uncertainty reduces is precisely that along which knowledge increases. This is an observation that agrees with the relation in statistical mechanics between probability and entropy.

Modelling future and risk management are at the bottom of any planning endeavour for the fulfilment of which stochastic analysis and control theory offer a rich and flexible toolkit of powerful devices, which may become helpful understanding the unfolding of time and managing emerging changes. With respect to future modelling let us say it involves the elaboration of goals, which comprises an outline of desired states and alternative courses of goal-conducive actions as well as a conceptualisation of necessary structures of organisation and feasible timing schemes for the allocation of available means to achieve desired states.

Distinguishing between two extreme states of knowledge, certainty on one hand and complete ignorance on the other, all intermediate states can in principle be identified with a special type of information or lack of it, from which the prevailing states of knowledge as visualised by $P(X_t|\mathcal{F}_t)$ results. Although the aim of planning, namely altering in a desirable manner the states a system may reach and their probabilities of occurrence, becomes in this sense self-explanatory and possesses indisputably great relevance, we are far from getting ready-made solutions and/or clear-cut strategies with practical significance. This is not just because mathematicians, as K. L. Chung put it, are more inclined to build fire stations than to put out fires, but owing to the fact we know about the past and/or about certain mythical states of equilibrium or optimality a lot more than about the paths toward it. Even our knowledge of the present is rougher and more cursory than what we may have to approximately characterise $P(x_t|\mathcal{F}_t)$.

Aleatory contracts and other heuristic notions

Needless to say, the future cannot be entirely determined by actions and strategies derived from mathematical considerations. However, we expect the mathematical techniques may offer us alternatives to learn how to modify through the equation of motion those mechanisms describing the way the past decisively affects future states of the system. This is certainly not an easy task but one which due to the relevance may be highly rewarding. As a way of motivating probability researchers let us recall in this regard following considerations from Aldous' stimulating book on *Probability Approximations via Poisson Clumping Heuristic*.

A mathematical area develops best when it faces hard concrete problems which are not in the "domain of attraction" of existing proof techniques. An area develops worst along the "lines of least resistance" in which results are slightly generalised or abstracted. I hope this book will discourage theoreticians from the pursuit of minor variations of the known and the formalisation of the heuristically obvious, and encourage instead the pursuit of the unknown and the unobvious.

D. Aldous
(see [7], p. 252)

The recognition of an inherent irreducibility of uncertainty in the future course of certain phenomena underlying the evolution of dynamic systems, leads naturally to the problem of quantifying (this kind of) uncertainty and managing the accompanying risk. It is necessary to mention at this stage that risk is usually associated with deviations in the future from a sort of natural extrapolation of the known state of affairs, which is amenable to prediction and is in principle insurable.

In general one conceives *risk* as an exchange of here-and-present certain goods for uncertain goods according to the terms stipulated in an aleatory contract, a question which has attracted a great deal of attention. An *aleatory contract* is a binding arrangement involving an element of chance. Although the idea that those who share risks deserve larger shares of the resulting rewards is intuitively simple and appealing, quantifying uncertainty and estimating risk shares depend to a major extent on value-judgements of particular circumstances which render prevailing risk management practices susceptible to manifold charges. The *instruments for risk management* are classified, broadly speaking, in *passive* ones which are those based only on exchange mechanisms substituting payments today for potential damages in the future and in *active* instruments which are mainly related to loss prevention and technological innovation.

Risk and the framework of perfect knowledge

To begin with let us mention, risk is the characterisation of a situation with a specific informational structure. Namely, a state from which risk may obtain underlies a state of knowledge lacking only selected information items. These missing information items correspond to events the occurrence of which is dominated by homogeneity and replicability. In this sense a situation of risk exhibits a substancial body of knowledge and contextual information which requires an analytical framework suitable to highly homogeneous events. This might be the reason why orthodoxy, which assumes a state of perfect knowledge and rationality, frequently alludes to situations of risk. In this connection, it is necessary to say that the notions of risk and decision-making cannot be accomodated in the framework of perfect knowledge and economic rationality, since as Loasby argued: "If knowledge is perfect and the logic of choice is complete and compelling the choice disappears; nothing is left but stimulus and response. If choice is real, the future cannot be certain; if future is certain, there can be no choice." Quoted in Hodgson [110], p. 10.

In the framework of national economics in underdeveloped areas the use of passive instruments for the management of risk is not widely propagated and mainly limited to certain kinds of crop, medical and unemployment insurance. The main reason might be the difficulty assessing the probability of damage, lack of knowledge and imperfect distribution of information. Loss prevention may also fail for the same reason, i.e. inadequate knowledge concerning a machine maintainance policy as means of altering the effects of costly machine breakdowns, lack of expertise selecting among various sorts of seed corn for a given type of soil, inefficient organisational skill in regard to the proper timing of harvesting and distribution, etc.

However, the implications of risk management on the effectiveness of planning are crucial due to facts already mentioned, making it a powerful instrument by means of which structural processes are conveniently altered. This is of great relevance since structural processes are responsible for various qualitative and quantitative features of the transformation of a given set of input into output states with varying probabilities of occurrences. What risk management ought to do is avoiding neither capitalists nor workers get caught in the extremes of certainty and full risk. Certainty is linked in different ways to inhibiting outcomes known in the literature as moral hazards, while full risk may yield resignation, bringing down the level of aspirations and counteracting motivation and innovations as well as the effects of incentives expected to follow from planning activities.

Therefore, future modelling and risk management need to rely on a deep examination of attitudes toward preference and risk over time involving cultural elements, tradition and beliefs. It necessitates also to separate clearly risk management from suspicious business practices, usury and speculation. For instance, think of the links between avarice and foresight reflected in the proportion between risk bearing and gain, or the links between social responsibility and wisdom, as involved in calculations of the relation between risk taking and provision for genuine contingencies inherent in the process of social reproduction.

Because as long as these links are not sufficiently clarified, the notion of rationality based on self-interest and the logic of "the more, the better" and "more of the same" as represented by orthodoxy will continue to obscure the need of a new rationality. Namely, a rationality based on culture, social values and solidarity in a way that let self-interest and responsibility to one's family transcend the fairly limited sphere of oneself, encouraging thereby environmental and intergenerational thinking at large, as social reproduction entails.

3.2   The search for evolutionary power

A short examination of the process of social reproduction quickly reveals the manifold roles played by capital: As a *social relation* which allows to distinguish between those who owning capital can hire the labour of a certain number of individuals and organise their work directly or through hired managers and those who do not have it and as a consequence have to lend their labour power to the former; as *means of finance* or purchasing power by means of which one can provide a place to work, some equipment and some materials to work on as well as a wage fund from which to pay salaries before the cycle of production and sales is completed; as *factor of production* which may take the form of plants, equipment and other means of labour; as a *source of income* which gives its owner command over purchasing power and this can either be consumed or assumes the form of means of finance; as a *value* in the sense or means of storing purchasing power which in principle can assume at any time the form of purchasing power at certain transaction costs; and many other forms.

3.2.1   The nature of capital in the course of time

A characteristic feature of capital is movement, in the sense of a continuous change in its form. In general, capital assumes at any time only one of the first three forms the value of which varies in the course of time according to the specific nature of its dynamics. Capital in any of its different forms is in one way or another related to the process of production, a fact that makes its nature highly dynamic and random.

Capital and investment as sources of change

Any examination of capital hence necessarily extends economic analysis into the domain of time and change. This view is an answer to the fact that capital and any process in which it participates cannot be understood without considering the effect of the unfolding of time. Because along the unfolding of time investment – viewed as a first approximation as using means of finance with the purpose of creating capital goods or adding to the existing level of plant, equipment and other means of labour – starts various processes of capital formation, the materialisation of which takes place in the course of time and depends on the fulfilment of several uncertain operations. Without the fulfilment of expectations and realisation of basic operations, capital can never return to its original form. The idea that along the process of social reproduction capital has intrinsically the property of going back to its original form gives rise to the notion of the circuit of capital. This reflects the property of capital initiating in the form of means of finance to go through the process of production in the form of labour power and means of production toward the market as commodities and purchasing power to return systematically to the form of means of finance. This movement of capital from the first pole of the circuit to the last – here from means of finance to means of production to commodities to means of finance – is called its *turnover*, and the time it takes to complete a curcuit the *period of turnover*. The fact that capital may assume a variety of forms, e.g. fixed and circulating capital as well as finance and industrial capital, and that they all have a specific turnover and period of turnover makes the quantification of the process of capital accumulation extremely difficult.

Regardless of the form capital may assume, it is always committed to the process of production with the purpose of adding to its value at an expected rate of return. However, since there is no guarantee all the operations involved will be implemented successfully, say because the relevant market for inputs and/or outputs may unexpectedly change unfavourably, elements of chance and risk are always present. In addition to the capital differential in their periods of turnover one needs to take into account the differentials in their mobility among the various lines of production, a factor which reduces considerably the investors' capacity to adapt to new information on the prevailing rates of return. The first factor affects crucially measurements of value, while the second brings about a sort of sluggishness in the drift of capital accumulation. On the other hand financial capital is associated with relatively high speeds of adaptation and its value undergoes highly irregular fluctuations.

Several questions emerge concerning the causal mechanism underlying the evolution of capital and the state of fulfilment of expectations. In a model applicable to historical reality however, it is not easy to specify the basic connections between current and expected level of rates of return which give rise to the relevant flows of capital, owing to complexity and eventual interactions in the markets for the kind of production concerned. Furthermore, the notions of realised and expected rate of return to capital are vague and complex, mainly due to available conventions to assess realised returns to capital and the type of uncertainty behind the state of expectations. Complexity responds to the variegation in the past experience, accumulated knowledge and informations available to the large number of economic agents taking part in the process; similarly it reflects manifold configurations and irregular ground on which the constituent operations of this process take place.

A detailed examination of the background of the process of capital accumulation will be presented elsewhere. For the time being let us briefly mention this would require to distinguish among a great variety of means of finance like bonds and shares as well as differentiating among owners of capital like rentiers, shareholders and entrepreneurs. At a fairly intuitive level one may differentiate between capital entering the spheres of production and circulation of commodities. This enables us to differentiate between productive and circulation capital pointing to the type of activities each supports. Looking closer these spheres of activities suggest further differentiations like industrial and finance capital as well as commercial and fictitious capital among the most relevant. The closer one looks at these activities the sharper the differentiation of them and hence of the capitals underpinning them.

A general notion of capital

However, it is necessary to keep in mind the most essential feature of capital is its ceaseless movement, i.e. its inherent quality of changing its form incessantly. This extremely dynamic feature of capital will be helpful in understanding the process of production as the acts of transforming, rearranging and/or transporting natural objects and/or value matters which entail the availability of potentials to be used up in the course of production and distribution. Natural objects or value matters mean anything that may be coded in any form of capital or become means of value.

Full realisation of potentials characterises states which deprive capital of its most essential feature, i.e. movement, defining a state of equilibrium. It is necessary to stress, all states of equilibrium are in a deep sense temporary and hence the result of limited perception. Because changes in the quality of our perception, for instance by means of an appropriated rescaling of time and/or space, may add to our knowledge of the flux of change thereby unravelling (and/or revealing still available) potentials.

At this stage we know about capital as a social relation and as means of production. Further we are acquainted with the notion of finance capital which bears some features of money and industrial capital, since it exhibits links, to some extent fictitiously, between capital outside the production process and capital embodied in productive plants. A fairly general definition may be this according to which *capital* is any value that can be increased by means of surplus-value. This comprises the categories of wealth and money as a source of future income both involving time and risk.

Furthermore, if we see the material production process as a scheme by way of which a flow of inputs is converted over time into a flow of outputs, one easily recognises our definition also applies to physical capital or capital goods as well. This makes apparent a basic objective and function of capital is to provide potential for and to steer change. However, in order for capital to materialise its property of generating surplus-value, as means of increasing its current value, it has to be committed by its owner to a productive, financial, or exchange activity or the like. In a broad sense this commitment is the consequence of an *investment decision* reflecting the view that any *investment* can be visualised as transferring an asset by its owner to someone else or to nature with the prospect of future returns.

In general investments involve liabilities held by someone who is supposed to provide the returns. On the other hand, investments *per se* are operations of capital allocation bearing liability held by nature and acting as a rule through the sphere of production. The type of capital we are dealing with determines the kind of investment, i.e. (purely) financial investments associated with money or finance capital, (industrial) investments associated with the creation of physical capital and the like. The foregoing reflections call attention to the delicate and intricate nature of matters concerning capital, a topic which has attracted a great deal of attention and has caused long lasting controversies and still awaits for a serious and profound scientific scrutiny.

3.2.2   Notes in the vein of Robinson's historical time

At this stage we are aware of the fact the notions with which we are forced to work are far from being unambiguous and the need of elaborating a theoretical framework which provides our accumulation models with a sound and flexible footing. For this purpose we will pick some concepts and reflections from the highly illuminating contributions of Joan Robinson and her followers which we believe bring out the essence of forces driving change and time intrinsic to processes shaping development and growth. "In a historical model", said Joan Robinson [181], "the stock of capital goods at some based date is taken to be whatever it happens to be. It can be valued at historic cost or at a current reproduction cost, or in terms of its prospective earning power discounted at whatever is considered to be the appropriate rate of interest. Each measure (unless, by a strange fluke, perfect equilibrium obtains)", she stresses, "is vague and complex, and each gives a different result." Then, she concludes, "This is certainly a very tiresome state of affairs for both private and social accountants, but it cannot be amended by pretending that it is not so."

Purchasing power and available potentials

In a capitalist system anyone who has enough command over means of finance and knows how to combine and organise work and property in production may in principle become an entrepreneur. The fact that modern institutions separate in general ownership and control of production introduces in the concept of entrepreneur a certain ambiguity. Shareholders and managers alike may legally and formally be seen as owners of production and as entrepreneurs according to the degree of ownership concentration and production control. As long as actual entrepreneurs see opportunities for profitable investments and owners of means of finance seek opportunities to turn finance into earning power there exist potentials for future production. On the other hand, those who have no means to become entrepreneurs need to seek opportunities of acquiring and maintaining earning power by schooling or training and by lending their labour power.

Thus command over means of finance and/or working skills or knowledge may entitle individuals to an income or certain measure of purchasing power, which can be differentated as income from property and income from work. Since the significance of production lies in the consumption which it makes possible, it is fairly natural to ask about eventual imbalances at the macroscopic and microscopic level between the consuming power and purchasing power that production provides for. The relevance of this issue lies in its connection to the availability of potentials for future production as sources of means of finance, means of acquiring and maintaining earning power and finally as means of providing opportunities for profitable investments.

Thinking of commodities in terms of characteristics may help to clarify the main points behind this issue, since we may then focus on the sets of functionings – as states of being and doing – which one can attain given certain purchasing power and get a clearer picture of the availability of potentials. Identifying consuming with purchasing power distorts essential features of economic theorising and astutely hides greedness and cynismus nourishing widely used practices in production, distribution and trading. Let us call attention to the following issues:

- Consumption is an involved process unfolding in the course of time with the purpose of restoring used up potentials. Purchasing commodities just initiates it and its fulfilment depends crucially on characteristics of commodities and personal utilisation features. Markedly slow moving elements like culture and habits are among the most influential factors in the process of moulding consumption patterns even beyond the category of essentials, i.e. food, shelter and schooling.

- The power of restoring used up potentials depends more deeply on consumption rather than on purchasing power. This is particularly true in

economies where marketing commodities has become more troublesome
than producing them, and a large part of their prices is directed to cover
costs arising from the need of persuading consumers to buy them and/or
gaining access to more favorable market shares.

- A close examination of the command (i) certain groups of individuals
  may have over money; (ii) a unit of money has over goods and services;
  and (iii) the price of a commodity bundle has over their characteristics,
  at different moments of time and points in space, shows how confusing
  and distorting may be identifying consuming and purchasing power.

- Under the prevailing rules of economic intercourse in large areas of the
  world the motives sustaining the efforts of individuals are in general increasingly less related to acquiring command over flows of goods and
  services suitable to meet human needs than to getting command over
  money. As a byproduct consumers get a lot less in terms of commodity
  characteristics than what they pay for, a fact leading to a heavy distortion
  in terms of internal and international terms of trade.

- A consequence of the interplay of foregoing features is the emergence
  of mechanisms of income drain and sucking available potentials which
  adversely affect manifold branches of production at a national and international level. Siphoning off economic surplus from state nations of
  the periphery through a variety of mechanisms of this sort has been institutionalised in the International Order initiated by Bretton Woods
  agreements, apparently aimed at counteracting multiple irregularities in
  large-scale international flows of finance accompanying the collapse of the
  open colonial system. See for instance [137, 144, 174].

The meaning and delicacy of these issues derive from vital links between notions like wealth and money, expenditure and consumption, earning power and
property to mention only a few. The common and distinguishing features of
these notions are essential and conflictive, difficult to grasp and measure, since
they represent processes undergoing change in space and time. The principle
of social reproduction sets underlying processes to work together and lets them
get their true essence by going through time cycles, along which their working
potentials are either maintained, renewed, used up, stored, created or expanded
on the basis of the dynamic properties of capital and the need of restoring the
earning power of labour.

Ability of maintaining productive power

The availability of potentials, as the main source of social dynamics, does not
depend essentially on thinking of economic wealth either as a quantity of value
coming into commodities by acts of labour or as a quantity of utility belonging naturally to commodities which can be got out of them by consumption.
The availability, storing and creation of potentials depend critically on characteristics of commodities and on personal utilisation features, as presented in
various paragraphs on the capability approach, from which arises an enormous

who's who of miscellaneous functionings. Potentials are in a sense a measure of available functionings and reflect the ability of the system of maintaining its productive power for the future.

The principle of social reproduction embodies a rationality that prevents the system of preying upon available potentials in a manner which impairs its future potentials and jeopardises its long-term viability either at the microscopic or macroscopic level. Since the economic surplus of society is seen as representing its future potentials, the value the level of (actual) economic surplus may possibly attain, exerts an influence upon distribution and effective demand in terms of means of finance and purchasing power which in turn determine attainable levels of means of acquiring, maintaining and increasing earning and/or productive power. The realisation of the latter leads to expenditure, the levels of which affect the degree of fulfilment of profit expectation and provide confidence on the profitability of future investment opportunities. Profit expectations and investment plans are more liable to erratic movements and uncertain outcomes, the more capital leaves the sphere of production towards finance markets, where it may assume properties more adequate to increasing than to storing value.

Under given technical conditions, we may associate with any given but arbitrary rate of profit per unit of time-period a set of prices at which social output pays the wage bill and claims on the (value of the) inputs entering production at the beginning of the period. In this sense, reversing the situation a little bit we could derive a causal relation going from a given wage rate to a rate of profit per unit of time-period, provided we know how to value the stocks of capital entering at the beginning of the period. For the time being, let us just mention we shall to a large extent follow the alternative worked out by Robinson in [184] to close our models of accumulation. This shall briefly consist in applying the principle of effective demand, introducing a ceiling to profits by fixing a minimum wage rate and arguing along the lines of a modified version of Robinson's model of the "golden age".

3.2.3   Peering into the value of capital

The property, according to which capital continuously changes its form going through specific circuits of production, finance and/or trading as given by its nature, makes extremely difficult to grasp the meaning and to assess the value of capital. One can convince oneself thinking of capital, for instance, as a factor of production and in particular as long-lived plant.

Practical issues resulting from the motion of capital

One may be inclined to understand the value of a long-lived plant in terms of the income a firm derives from it, i.e. the profit viewed as the excess of receipts over costs associated with its operations of production over its life time. As receipts one may consider proceeds accruing to the firm from sales of the corresponding output at prices prevailing at the market. Costs may be related to purchases of materials, wage bill corresponding to labour needed to operate the plant, maintenance and repairs among the most important. Needless to say that calculating receipts and costs is not a simple thing, since they depend on the expected earning life of the plant and changes in the input and output markets and many other factors. Even more, the value of the plant over the course of its life is likely to change owing to wear and tear, physical aging due to the mere passage of time, obsolence and loss of profitability of relevant markets and so on. None of these factors is easy to calculate. One usually avoids some difficulties by using backward techniques of assessment based on certain expected values and experience.

Thinking of capital as means of finance does not put us in a better position to assessing the value of capital. Heterogeneity and yield, mobility, liquidity and risk differentials act as veils hindering us to ascertain value properly. It is in this sense that our perception of reality depends crucially on the sharpness of focus and on the scaling of time and space, since it is complexity rather than size which represents the more basic feature of a system.

Changes in our perception attributed to rescaling time and/or space may either bring out or blur fundamental differences and/or similarities in the notions underpinning the conceptual framework of production and capital. As an example think of the drawing line between amortisation funds reflecting costs of production due to wearing out physical capital of different ages and returns to capital as known from accounting practices of firms. This entails fixing amortisation quotas according to a certain scheme of expectations and establishing flows of finance valued at historical units linking the sales value of output and costs of work in progress, maintenance, replacement and eventual expansion of the plant. It is well known that the border lines between these notions are elusive and may vary heavily going from short-period to long-period analysis.

The forward view of capital value

If acts of production are viewed as consisting of operations for whose accomplishment available potentials are used up, then it becomes natural to rank acts of production in accordance to the lack of potentials they generate. Lack of potential amounts to lack of motive power or latent change capacity, since only those events happen for which potentials exist. Therefore, starting our considerations about the process of social reproduction with social output as given, puts us in the situation of rationally deciding the fractions of it to be directed to purposes of maintaining producing or earning power of available labour and capital as well as to purposes of improving or increasing it.

In terms of characteristics of commodities this amounts to the allocation of social output to the aim of maintaining the capability of individuals to functioning and to the purpose of keeping in the productive power of means of production; on the other hand, allocation of social output to the extent of increasing future capabilities and productive power as the principle of social reproduction requires. That means, consuming a fraction of social output with the aim of maintaining the current state of being and doing, and putting aside, i.e. saving from current consumption, with the aim of quantitatively and qualitatively improving the future states of being and doing which characterises the system.

This rearrangement of the current state of affairs with the aim of changing the future and improving the coming state of affairs constitutes in a general sense a *decision of investment*, i.e. setting aside a fraction of available social output with view of getting a larger fraction of future (uncertain) social output. If we do not make any distinction between investments and savings, we may also interpret an investment of social output as the sacrifice of fraction of current consumption for a larger fraction of (uncertain) future consumption. Saving shall be understood as foregone current consumption and represents an asset for the society as a whole. This social asset may be seen as held by the government or control board, which acting as liability holder has to provide for the returns in the form of claims on future consumption. That means, the control board acts as a financial intermediary to whom foregone consumption generates a liability and the owner of the asset can theoretically exercise control over the operation through periodic votes. On the other hand, returns are supposed at least to make up for the social value given up together with foregone consumption. This kind of reasoning brings us back in a broad sense to the prevailing definition of the notion of investment; namely, investment is the sacrifice of certain current value for (possibly uncertain) future value.

How we can measure this value is a matter that will be dealt with later. For the time being, we shall just state that the problem at stake is of precisely the same nature and size of the one associated with the measurement of a stock of capital. As a first approximation we may think of the value of capital as accumulated investment or equivalently accumulated foregone current consumption. The problem of measuring foregone consumption at historical units and a proper rate of discount will be discussed later. Let us mention that our considerations are dominated by an emphasis of macroscopic issues so that we shall speak rather of social value of capital. The value of capital as emerging from finance markets will be treated in a forthcoming work.

The value of capital and viability

Underlying the problem of measuring *capital value* as the total sacrifice of consumption made in the past is that of characterising properly the process of material production; this includes investigating distinctive properties of the time-profile of processes having an input-output relationship regarding the sphere of production. Measuring inputs in terms of available potential given up and outputs in terms of future potentials allows to evaluate the efficiency of acts of production according to the rate of change in the lack of potential or in a thermodynamic sense in the currency of entropy. Without forgetting the characteristics of commodities are at the centre of our reflections, this amounts to searching for arrangements which shift inputs to the prospect of future outputs in a way that diminishes available deficits. Here inputs are measured in terms of sacrifice of current consumption, outputs in terms of increased consumption, and deficits in terms of renounced earning or producing power, i.e. consumption in terms of characteristics.

To understand what this means, let us argue on the basis of means of finance, an analogy which although veiled by the presence of money has the virtue of being an issue of daily life. According to this whenever net inputs occur, i.e. inputs exhibiting value in excess of the value of corresponding outputs, funds coming from outside of the process have to be drawn in either by borrowing or by any other transaction for which interest will have to be paid. On the other hand, the occurence of net outputs, which shall be understood similarly, may give rise to drawing a fraction of the excess value out of the process for the purpose of repaying past debts or investments outside of the process. Thus examining investment opportunities entails checking the viability of the process at stake, i.e. making sure this process yields at least the market rate of interest. Then, we shall call a process *viable*, whenever at the end all liabilities and borrowings can be repaid.

The problem of time valuation of an asset committed to a productive process is that of ascertaining the value of streams of the future income to which it gives rise; this fact stresses a causality order running from an available potential – or an item of capital, which may represent an asset – to future streams of income attributed to the creation of earning and/or productive power derived from the asset acting as an input of a productive process. Consequently, the value of capital can also be ascertained from the value of the future income that it generates. However, such a notion needs to be supplemental by entropy considerations in order to account for processes flourishing without maintaining the productive power, since this amounts to preying upon the potentials of future generations and/or exploiting external potentials in a way which impairs the future viability of underlying processes.

3.3   Understanding socioeconomic change

The description of processes of change may be accompanied by uncertain outcomes and varying expectations, which have occupied us at different stages. Next we shall elucidate the sense in which structural changes will play a role in examining questions regarding growth and development.

3.3.1   Horizontally and vertically integrated structures

The concept of structure is related in one way or another to a set of relationships among individual agents and/or processes describing essential features of the system, where the role of time either in its logical or historical guise is essential regarding continuity and rupture. Thus thinking of structure as a set of qualitative and quantitative relationships among certain magnitudes and arrangements ruling the behaviour and interplay of individual agents, constitutes a fairly adequate starting point. The notion of structure conceived in this way provides a natural framework against which most essential elements evolve, as needed for the maintenance and renewal of objective conditions, underlying thus the coherence and continuity of a system and compatibility of its events as well.

Logical and historical causality

At the present stage we are interested in certain relationships building the core of processes associated with causality at the logical and historical level. In this connection we shall make the distinction between logical and historical causality as a way of approximating the more essential and less tractable principle of overdetermination which overrides any causal ordering of social events. Thus on the basis of vertically and horizontally integrated structures supplemented with purposely constructed feedbacks, we attempt to reflect the fact that in social systems an overwhelming majority of events may act both as cause and effect of processes shaping patterns of organisation and order inherent to underlying structures.

Logical causality acts through vertically integrated structures which are largely independent of historical time. Logical causality reflects mainly causal implications among essential aspects of a system and attempts to explain the substance of causal relations distinguished by the presence of persistent and deeply rooted forces, which remain largely unalterable in the course of time. While searching for causes behind events, logical causality points to the less visible layers of reality and focuses rather on content than on the form of the considered phenomena.

On the other hand, *historical causality* acts through horizontal integrated structures and feedback mechanisms depending essentially on historical time and on virtual changes in logical time. Historical causality points to the more visible layers of reality and focuses hence on forces shaping the form of the phenomena under consideration as (historical) time goes by. Needless to say that such a distinction between logical and historical causality is merely a fictitious device to improve our limited perception, since in reality processes of change go simultaneously through vertically and horizontally integrated structures building a complex interplay of forces. The forces at work give rise to a flux of change consisting in

(1) a relatively stable and largely invisible component representing coherence and continuity features of the system;

(2) an indirectly observable component embodying cyclic effects attributable to the efficiency of the system to renewing used-up potentials and eventually to increasing them relatively and/or creating new ones; and

(3) a rapidly changing and more visible component reflecting the effect of using up available potentials on the system ability to shape emerging strings of events.

Within the perspective of social reproduction we may break down processes of production into sequentially recurring phases aimed at replenishing, using up, renewing and/or accumulating the productive power of society and its environment. A careful examination of these always recurring phases gives rise to distinction between aspects related to space and time. The spatial features concerning these phases are essentially linked to the social character of the process of production, in particular in regard to social division of labour and cooperation. In this connection, the structures of society are largely dominated by issues of functionality resulting from patterns of organisation and order.

On the other hand, from considerations of temporal features of these phases arises the need of concentrating on the material content of productive activities or alternatively on their form. This depends on the sharpness of focus and scaling of time used while perceiving reality. Regarding scaling of time, focusing on material content suggests slow motion and hence a negligible account of the effects of time passing by. This is a situation we shall approach from the perspective of logical causality and thus in terms of logical time. By way of example, we may consider the set of activities related to production, all along the way from the state of raw materials toward the (final) commodity, provided these activities and their intermediate outcomes are casted in the world of logical time as in a Walrasian exchange economy; this is a visualisation of vertically integrated features of production processes. This amounts to say that logical causality focuses on essential properties of commodities and intermediate outputs, viewing them as substance. Therefore, (historical) time does not essentially exert any effect upon them, so that one may direct attention solely to causal relations largely independent of (historical) time.

As soon as historical time enters the analysis, commodities need to be imbedded in a field so as to reflect the effect of the passing of time, with regard to contingency deriving from realisation of profits or effective demand, obsolence and technological innovation, spoilage, vagaries of the market and the like; this corresponds to visualising horizontally integrated features of production processes. That means on the way from the factory towards final consumption, social output is persistently subject to changes which are more visible as random fluctuations of less essential characteristics of items of social output but fundamentally affecting at a less visible level structural features of social reproduction. These manifold changes acting at various speeds and levels of reality are mainly attributed to changes in the system environment. Historical causality draws attention to causal relations less directly linked to substantial characteristics of items building and to operations underpinning the social output, and to manifestations of changes attributable to the impact upon structures and institutions of the passing of time; see in this regard [160].

3.3.2   From worker/producer to worker/consumer

To the aim of understanding the impact of change and time upon growth and development, we have to direct our attention to the working of the principle underpinning social reproduction in regard to gradual increases in the rate of utilisation of available potentials and/or freeing latent potentials from eventual constraints. In this sense phenomena related mainly to certain measures of qualitative and quantitative utilisation of available potentials will be considered as constituting basic processes of growth with the purpose of serving social reproduction. On the other hand, phenomena not primarily linked to enlarging utilisation rates of currently available potentials will be viewed as defining processes of development. Development forces involve mechanisms allowing and causing growth as well as those aimed at uncovering and unfolding potentialities in a way essentially related to self-renewal, while forces of growth go along with exhaustion and advance toward maturity and rupture.

The logic of fragmenting processes of labour

Central to the process of social reproduction are nature and human resources, since they embody any potential and capability of utilisation. Further, working out latent potentialities and replenishing renewable natural and human resources are at the heart of processes of labour and (material) production. Essential elements are hereto processes of knowledge, learning, management and planning, which are also space and time dependent and sensitive to the notion of rationality to prevail.

In regard to production, management and planning necessitate that all spheres of production, society and life of individuals are amenable to fragmentation so as to rendering calculable cost of factors, making predictable demand for commodities and enhancing feasibility of schedules of production, investment and finance. In regard to labour, management and planning went to the extreme of breaking down knowledge embodied in processes underlying production into tiny units as means of obtaining a relatively high homogenisation of labour skills and establishing standard patterns of efficiency. In this manner a process once aimed at dominating nature and emancipating labour ends sweeping away harmonious interactions between nature and humankind and reducing labourers in the main to mere appendages of machines.

Therefore, after separating human labour from means of production and its produce, divorcing the spheres of living and working, stripping labour processes of its humanity and culture, divesting work of its creative and vital links, and depriving productive activities of meaning and motivation, it becomes necessary to invent scientific methods to bring in labour processes organisation, motivation and efficiency now discovered to be lacking; see Fig. 2.3 and accompanying explanatory paragraphs. Management and planning then guided by the principle of economic rationality come to investigate into human nature, now assumed to be averse to work, with the purpose of modernising and rationalising labour. Under modernising one should understand organising and stimulating processes of labour, so as to render them more methodological and better adaptable to current objectives of production. While rationalising should mean freeing processes of labour from cultural roots and the sort of know-how primarily linked to a harmonious interaction with nature and work as a way of life.

The fact that the modern notion of labour needs to rely on the worker/consumer relation rather than on the more natural, time-honoured and self-regulating worker/producer relation introduces a manifold of genuine irrationalities. Due to the relevance and delicacy of these issues we shall briefly mention a few points and refer to [84, 95, 147] for a detailed study.

Regulatory instruments and the labour process

To begin with, let us recall that in regard to the notion of rationality, we have to distinguish among a variety of manifestations according to the domain upon which it applies. Hence, there is a system rationality, an individual rationality, enterprise rationality, worker rationality, economic rationality and so on.

In general rationality can be visualised as a goal-conducive quality of outcomes resulting from human intentionality, motivation, disposition, function, responsibility, capacity, knowledge and many other factors interacting in a relatively complex manner, and on which the accomplishment of tasks rely. Needless to say, the qualities involved are individual, agency, group and system specific. In this sense the notion of rationality depends on the prevailing patterns of organisation and structure as well as on the manner individuals, groups, agencies and the various spheres of activities are socially and functionally integrated in the society and the system at stake. Taking into account logical and historical causality emphasises furthermore the degree to which rationality may be of the external, autonomous, heteronomous, self-regulated and/or hetero-regulated variety. See [100, 192].

Having in mind the essentials on rationality introduced above, let us recall the ideas sketched in Fig.2.5 concerning interactions set in motion once evolutionary processes, as assumed there, take place in society. A natural point to initiate our brief analysis is the box with label conception/execution, which represents processes of labour identified by given personal futures b_i and functionings f_i as well as by available production techniques codified in the coefficients a_l and a_k, and the consumption characteristics $c(x_i)$. The main distinguishing features of the worker/producer relation are its harmonious interaction with nature and knowledge guided by intuition and culture, strong reliance upon social integration on the basis of consensual agreements and/or communicative actions. Moreover, since the relation worker/producer involves manifold acts of self-regulation and self-organisation dominated by its coherence with laws of social reproduction, it develops wide networks of human creativity and social cohesion rooted in a rationale of minimum waste of renewable motive power.

By contrast the rationality associated with the relation worker/consumer already described involves a minimum of self-regulated and self-organised cooperation and solidarity between those engaged in the same task, and as a result a minimum of social integration. Workers are attached like a cog to a highly complex machinery and assume functions dictated from outside which they have to perform unquestioningly. The great disproportion between individuals and gigantic technical installations as well as mechanistic organisation functioning by hetero-regulation aggravate the individual's loss of sight and motivation, and intensify their lack of capacity to adhere to alien goals.

To the end of counteracting these features and with the purpose of bringing in the system functionality as a type of rationality beyond individual comprehension and intentionality, one introduces, in addition to the usual kind of compensatory consumption, regulatory instruments, i.e. incentive and prescriptive regulators.

To the first type belongs money, security, prestige and/or power made available to individuals through a carefully and hierarchically conceived field of functions, tasks and responsibilities aimed at a functional rather than social integration so as to prevent cohesion and solidarity. This gives rise to a dualistic fragmentation of labour consisting of a stable core of employees highly skilled and functionally flexible and a numerically flexible labour periphery, which needs to adjust to the vagaries of the market. The periphery splits itself into a stable subcore of not highly skilled employees directed to administrative jobs, services and the like. And a subperiphery of external labour consisting again of a core of highly specialised professionals and a periphery of labourers with no particular skills, see [95].

The second type of regulatory instruments consists of sophisticated prescriptions aimed at forcing individuals to adapt to functional forms of conduct which result from heteronomous imperatives of the system.

Social integration versus economic rationality

After having described the main features associated with the basic relations worker/producer and worker/consumer, we are better prepared to give a quick look to irrationalities attached to the worker/consumer relation resulting from viewing society as consisting of individuals persuing their own goals independent of each other and without meaningful links to the ultimate objectives of the organisation in which they work. This brings out an increasing isolation of humankind in a huge social machine, where workers to a large extent are deprived of coordinating, self-regulating and self-managing activities.

Humankind is thus let alone with compensatory consumption, a sort of reward which the system cannot afford in the long run, since it runs against the principle of maximum entropy. The always recurring claims of saving time associated with modernising, rationalising and other dishumanising transvestities of the social division of labour can be better examined in the light of thermodynamics. As a matter of fact, with the help of the notion of entropy we may learn that transition to more complex and robotised technologies can never save time as long as this transition goes at higher speeds of (eventually increasing) inflows of energy. On the contrary, since these technologies generate more entropy they consume more time. This undeniable issue has so far remained hidden from orthodoxy, a doctrine based on an analogy of the theory of energy, owing to a failure of recognising the central role played by the second Law of Thermodynamics. Let us stress that the production of entropy accompanying a hasty movement towards the next technological energy watershed goes only on expenses of future generations, state nations and people in the periphery and nature. A treatment of these crucial points is beyond the scope of the present work, thus we refer to [178, 5].

Setting aside reflections from the perspective of the principle of maximum entropy, let us focus on the centre of conflicts between objectives of system functionality and individual rationality within the framework surrounding the worker/consumer relation characterised by social disintegration. In this sense we have conflicts between the system rationality defining functional integration and the individual rationality, which give rise to social disintegration. A first point one needs to observe is that functional integration of individuals, in the sense of prescribing them functionally fragmented tasks, prevents social integration. The role of social integration would be achieving social objectives on the basis of cohesion, cooperation, solidarity *inter alia*.

The reason is the belief of orthodoxy that by eliminating the play of human factors, in the sense of replacing living labour and free workers by labour processes coded in complex machinery (dead labour) and fairly constraint labourers, economic rationality and functionality render operations of the social machine calculable and predictable. The aim of compensatory consumption is then to move workers to freely accede (*sic*) entering a fully degraded labour process. Thus, the rationale is to make workers seeing constraints, frustrations and suffering inherent in functionally programmed labour processes, as compensated outside work, where freedom is recovered by acts of consumption. Giving up fulfilment inside work – inherent in the worker/producer relation – for fulfilment outside work – resulting from the worker/consumer relation – becomes associated with a double way of living, a schizophrenic situation with devastating implications for the great majority of those for which no alternative remains. See in this regard [31, 233, 149] and the works cited there, in particular Max Weber, Durkheim and Habermas.

At this stage, it should be clear that an intrinsic aim of compensatory consumption is to render compatible system and individual rationality, an issue on which the success of the system depends on manifold ways. Let us draw attention on a few of them.

- First, it necessitates an ideology of consumption. It sets consumption as a magic, reifying force through which human beings attain an identifying individuality and happiness by setting them apart from the crowd.

- Second, it necessitates an alien institution – commercial advertising and propaganda – aimed at shaping tastes and consuming drives without manipulating (*sic*) human freedom and rising complicity in wrong doings.

- Third, it necessitates a reserve army of unemployed labour to ensure the efficiency of regulatory instruments. This gives also an ethic and moral sense to the pains resulting from refusal of accepting alienation in work.

- Fourth, it necessitates insatiable consumption and profit drives in order to provide the market with adequate clearing power. This inherent unstability is linked to the failure of an ever-increasing financial compensation to reconcile human refusal to alienation and the concentration in the core of higher wage shares.

- Fifth, it necessitates sufficient availability of potential for the accumulation of wealth and finance so as to afford the challenge resulting from compensatory monetary rewards and sufficiently developed production capacity to fulfil the released consumerist drive.

These five points bear decisively explanatory power in regard to recurring crises in the world capitalist system and are at the heart of Keynes' principle of effective demand, Marx's theory of profit realisation and the French characterisation of regulation known as Fordism, see [1, 2]. To conclude, let us say state nations of the periphery provide the required potentials both as source of negative entropy as well as flexible crisis outlets and sink of entropy flows.

3.3.3   Evolutionary aspects of change processes

The capability set describes, at any moment of time, alternative states of being
and doing available to individual agents. It depends, as we have seen, on the
individual's entitlements and on commodity characteristics on the one hand and
on personal features and utilisation functions on the other. Since the capability
set reflects the abilities to act an individual agent may achieve in a given system,
it determines the agent's freedom to function given her/his entitlements. In
this sense capability sets emphasise the essential labour quality of embodying
know-how and the quality of means of labour, in particular physical capital, of
being dead labour or codified processes of labour.

   These remarks have the aim of clarifying our interpretation of economic
development as a learning process, i.e. a process primarily intended to increase
human knowledge. This perspective sets economic theorising in an evolutionary
framework and renders the description of change an information act. In this
regard it is extremely important to note that information like energy is an
abstract entity and possesses a physical reality. Thus, it can be transformed
from one form to another, i.e. coded and decoded, it can be transferred from one
system to another through interaction, cooperation and the like. See Chapter 7.

Information content of basic economic concepts

Let us begin observing that from a system theoretical point of view, the state
of information of a system may reflect the state of organisation exhibited in
terms of patterns of structure and order. Further, the information content
of a system describes its organisation capacity and its quality of arranging
its parts according to specific patterns of order. Thus, lack of information,
which is obviously linked to entropy, becomes a function of system organisation.
With these simple notions of information theory we may attempt to see basic
economic concepts from the perspective of knowledge and learning.

   Essential to the understanding of processes of labour, as sketched in Fig.
2.3, are first the interdependence between consuming of means of production
and bringing out anew means of production, and second specific interactions of
the social, political, technological and intellectual spheres of society in a way
conditioned by the stage of development of productive forces and the prevailing
relations of production. These relevant features are a consequence of viewing
society as a reproduction system, where any human being is a biological and
social artefact as reflected in the corresponding bundle of characteristics.

   On the basis of the information contained in the characterisation of individuals, as biological and social artefacts, and means of production, as human
artefacts reflecting know-how, we shall attempt to visualise production as consisting of evolutionary processes centring on learning and knowledge. Considering the three traditional factors of production as varying combinations of
know-how, energy and matter, let us consider processes of production as social
activities, involving space and time, by means of which structures of know-
how acting upon energy give rise to operations of selection, transportation,
arrangement and transformation of materials with the purpose of yielding human artefacts and organisation. Learning as a basic feature of mankind makes

sure qualitative and quantitative changes in the structure of know-how result in
evolutionary changes in processes of production. The basic properties of learning, from the point of view of cognitive sciences, suggest in a natural manner the
most appropriate counterparts of evolutionary notions in biology like mutation,
replication, cooperation, competition, coevolution, selection among others. See
[110, 121, 27, 28].

From the point of view of information theory, information is an inherent
property of the universe, i.e. it is a part of its internal structure as matter and
energy. Thus, information exists independent of our capacity of perceiving,
processing and understanding it. Human information is just one of the forms
that it can take, and can be understood as the information created, decoded,
interpreted, organised, stored and transmitted by human beings. Therefore,
the historical experience of mankind is reflected in our knowledge of the world
and the history of humankind is similarly reflected in the history of the evolution of human artefacts. Both historical experience and artefacts of human
beings contain the essence of human knowledge and represent information on
the efforts mankind directed to attaining specific ends.

It is necessary to mention, human information may involve the perception
of the internal structure of the universe. This is so because it is not only
necessary to get access into the information content of a certain entity, but one
has to be able to put the information conveyed into an adequate context, i.e.
organise bits of information into coherent patterns comprehensible to human
beings. Let us point out in this regard following vital issues:

- Any system is susceptible to be organised so as to exhibit patterns of
  structure and order.

- Order should be understood as a purposive arrangement of the parts of
  the system, which is usually thought as reflected in its structure and
  organisation.

- Disorder, as opposed to order, destroys information conducive to an end
  and can provide thereby the mechanisms needed for mutations to appear.

- In this connection, order provides the mechanisms for replication and
  continuity, growth those for duplication of information, and coherence
  those mechanisms accentuating replication.

It is in this sense that the quest of humankind for knowledge lies at the heart
of any process of labour and that human information manifests mankind's
capacity to perform work, to purposely interact with matter and energy to
the end of establishing coherent patterns of organisation and maintaining them
in an organised and functional state. Therefore, since evolutionary processes
involve descriptions about change in the charactistics of the system and its
elements, they reflect acts of organisation and information going on in the
system. Structural parameters in turn reflect relatively stable properties of
organisation, and motion condenses change in the information status of the
system as reorganisation procedes.

It is from this general perspective that we have to look at processes of production, where technology is a surrogate for know-how or structure of information. Labour characteristics comprise features of human beings resulting from their nature as biological and social artefacts. Thus the characteristics of individual agents (persons) derive from processes of interaction, communication and learning, and acquire meaning within the own information environment and existing organisations, e.g. learned images, knowledge, skills, behavioural traits acquired in the course of life like language. Characteristics of physical capital reflects those features of human artefacts, from simple working tools to automatic machinery. Space and time determine qualitatively and quantitatively the resulting bundle of characteristics, as state of being and doing, as an enhancing and/or inhibiting factor of efficiency and/or functionality. A detailed analysis of production processes from an evolutionary point of view will be given in [94].

Entropic processes and information

A system theoretical description of processes of interaction, cooperation and competition underpinning social production necessarily has to rely on human observation of reality, knowledge, cognition, culture and institutions. Acquisition of knowledge, storaging and processing information, learning and forgetting are essential features of any act of production as presented above, first introducing processes of labour and second reviewing essential notions from system theory; see in this connection Chapter 7. Next, we shall briefly recall some links between information and production in the context of entropic processes.

Let us begin calling attention to a hidden feature linking disorder, uncertainty and lack of information on the one hand and organisation, certainty and information on the other. This hidden feature refers to a common property, namely that of having high entropy for the first group and low entropy for the second. Since at equilibrium the entropy of a system is highest, it follows that to drive the system from equilibrium one has to add (free) energy and information into the system with the purpose of increasing its organisation. At this point, it is important to mention again that any act of production uses up available potentials, which contributes to exhaustion of energy and decay of structures. Thus, in order that a system maintains its low level of entropy, it necessitates to suck negative entropy from its environment. This conslusion can be written quantitatively rearranging Boltzmann's equation in the form

$$ -S = k \ln(I/I_0), \qquad (3.3) $$

where S stands for entropy, k for Boltzmann's constant, I and I_0 for a measure of information corresponding to the current and initial state of entropy. The relation (I/I_0) measures quantitatively the state of order attained by the system in question; see in this regard [5, 195].

Processes in the course of which available potentials are used up, are present in almost all systems and their entropic effect may be offset by anti-entropic processes, as they act upon their environment. Water exhausts, for instance, its gravitational potential running downhill to the sea. In an open system like the earth however, its gravitational potential is recreated by the throughput of energy from the sun that acting upon water in the sea gives rise first to evaporation and then to precipitation of water on the land as rain or snow.

In social systems the principle of social reproduction attempts to ensure that the continuous exhaustion of potentials are offset by their continuous recreation. Since acts of production provide the means of consumption as starting point of all social existence, it becomes indispensable to replenish the potentials from which production originated, so as to allow the basic circuit of reproduction to go on. The degree to which the recreation of potentials occurs, gives rise to critical states of disorder from which structural changes (mutations) may result, order associated with continuity (replicative fidelity), growth liked to duplication of information, and coherence deepening processes of replication. The recreation of potential or equivalently the creation of order in a social system always involves the creation of a greater disorder elsewhere, a topic which has received a great deal of attention in the literature on imperialism. A treatment of this topic in the light of the Second Law of Thermodynamics is given in [5].

Switching to the information content (energy) of social phenomena and viewing socioeconomic development primarily as a process in the increase of human know-how, adds realism and relevance to economic theorising although at a higher level of complexity. In Chapter 7 of Part III we have stated and explained eleven theses which are fundamental to the understanding of the role of modelling, as a bridge between reality and its mathematical appraisal. Theses 1, 2 and 3 stress lack of perfect information, Theses 3, 5 and 11 warn about the fragility of emerging knowledge, while Theses 4, 8 and 9 relate to essential problems studying the whole and its parts. Thesis 6 has relevance with respect to change in the probability of events as suggested above, and Thesis 10 hints to the need of analysis from the point of view of statistical thermodynamics. However, it is important to say that these Theses have themselves a great relevance.

Characterising states of knowledge

Having identified the role of human information and various sources of continuity and rupture, let us sketch some features characterising the knowledge state of a system, moving according to the Second Law of Thermodynamics from states with low entropy toward states exhibiting higher entropy. Doing this, one has to keep in mind that under the rationale of social reproduction, mechanisms capable of recreating used-up potentials are expected to drive the system away from equilibrium. Mechanisms of this sort are, for instance, those behind the principle of effective demand and profit realisation. Recall first that equilibrium is a state distinguished by full realisation of available potentials and second that freeing energy, for instance by removing constraints acting on the systems and increasing disposable knowledge and technology, by acting on latent information, need to procede the release of any anti-entropic process.

From this perspective one recognises the presence of strange loops, a particular form of feedback mechanism of the type depicted in Fig. 2.5, exhibiting features peculiar to processes on the basis of which manifold systems maintain low levels of entropy, as self-reproduction, self-organisation and self-regulation necessitate, or give rise to more complex systems, as bifurcation and catastrophe theories attempt to describe.

Since states of disorder, uncertainty and lack of information are definitional features of a state of entropy, a few words on these and related notions may help us to understand the meaning attached to the action responding to the need of driving a system away from a state at which the entropy is highest. To this end we shall learn to distinguish qualitatively and quantitatively between various states of knowledge.

- *Ignorance* is that state from which the acquisition of knowledge initiates. A process of knowledge is generally realised by passing from a state of lack of relations with objective reality to one less lacking of such relations. These emerging relations are in principle partial, approximative and temporal, and they are based on ideas, views, judgements and propositions founded in turn on active observations about objective reality.

- *Uncertainty* is a state of knowledge resulting from lack of appropriate information, while a wealth of less relevant and unstructured data to process. Situations of this sort are often supplemented with considerations resorting to judgement, guesswork, confidence, belief, inspiration and intuition.

- *Practical and tacit knowledge* result from living in a particular world and the demand, created by practical intercourse with nature and with one another, to make our ideas, opinions and judgements correspond to the reality from which they arose. Apart from practical dealings in the material world in which we exist, we have no way of testing the reliability of our perceptions.

- *Routines and habits* represent a relatively sluggish state of knowledge by means of which human beings functionally respond to the impossibility of maintaining a full conscious deliberation over all aspects of behaviour due to the amount and complexity of information involved. Routines and habits built by individual agents and institutions may lead however to conscious actions of others.

This brief description of likely states of knowledge may give us an idea of the quality and quantity of information and energy, one may have to put in a system with the purpose of offsetting the effect of entropic processes. It is extremely important to emphasise that within the present framework the alternatives to choice are real, the outcomes are uncertain and costly, and entropy and thus time play a fundamental role.

At the edges between disorder/order and exhaustion/renewal emerge alternatives to change taking the form of mutations and rupture which exhibit a great deal of innovation, novelty and surprise. These types of events pose challenging questions in regard to the characterisation of the resulting information structure. Behind these problems there exists a wealth of issues resulting from dualism in science associated with the separation in the parts and the whole, strange attractors and strange loops and the neglect of their corresponding information content. These will be treated in [94].

3.4   Essentials on political economy

It has been said, every child knows that a social formation which does not re-produce the conditions of production, i.e. the productive forces and the existing relations of production, at the same time as it produces would not last a year. This might be true with regard to the means of production but to the extent that the reproduction of labour power gets into consideration things are more involved and deeper. We attempt to grasp briefly the main issues involved.

3.4.1   Ideological roots of economic theorising

Political economy is the study of the general principles governing the social processes of production and distribution aimed at satisfying the human needs resulting from the historical stage of development attached to society. This succinct and widely accepted definition stresses the following crucial issues.

- *Renewal of used-up potentials versus rupture:* Whenever perceptions on social reproduction and renewal of nature fail to reflect objective reality, the perspective of change and rupture emerges at a continuously changing speed dictated by the exhaustion pace of available potentials.

- *The motive power of change as source of development:* The object of study is of undeniable highly dynamic nature, since processes underpinning society are characterised by a delicate and deep dependency on his-torical time. Thus, the major challenge derives from the necessity of anticipating change and launching actions conducive to development of productive forces.

- *Unity and separation of organiser from organised system:* The interaction of humankind and social reality is primarily reflected on the role stages of social development play determining human needs and culture, and on recognising human activities involved in processes of production and distribution, in turn shape decisively this social reality.

- *Coherence and agreement as a way toward social integration:* The social character of the processes at stake arises and evolves from man's basic association in production. In the course of history manifold relations are won and consolidated step by step on the basis of cooperation and processes of labour.

- *Communicative action and consciousness:* The increasing complexity of modern society may throw a veil upon the intricate structure of inter-actions preventing communication and consciousness to unfold proper-ly. From the limited perception of a new emerging reality may result problems concerning coherence and agreement between competing and conflicting interests.

These observations on the definition of political economy suggest many different ways of perceiving the reality around us according to our own standing in this social world and explain in a sense the diversity of priorities attached to social problems and endorsement or refusal of theories and methods to be used in solving them. Next, we shall elaborate a little on these matters.

Impact of ideology upon social reality and vice versa

Alongside the emergence of social production, humankind started forming ideas about causal processes in nature and about things. In the course of development of society, views on the conditions of material life, on modes of production, social relations and the like penetrated human thinking.

A more or less systematic body of ideas, beliefs and attitudes intended to act as a characteristic outlook of a whole society, but built on the basis of a given social structure and with the purpose to serve the interests of definite social groups in definite stages of social development is called an *ideology*. Ideologies are intellectual instruments corresponding to the material position and requirements of certain groups or classes. They must satisfy the logic of common sense and reflect a particular part of reality without becoming inconsistent with the principal facts which emerge in the experience of society at a given stage of development; see [42].

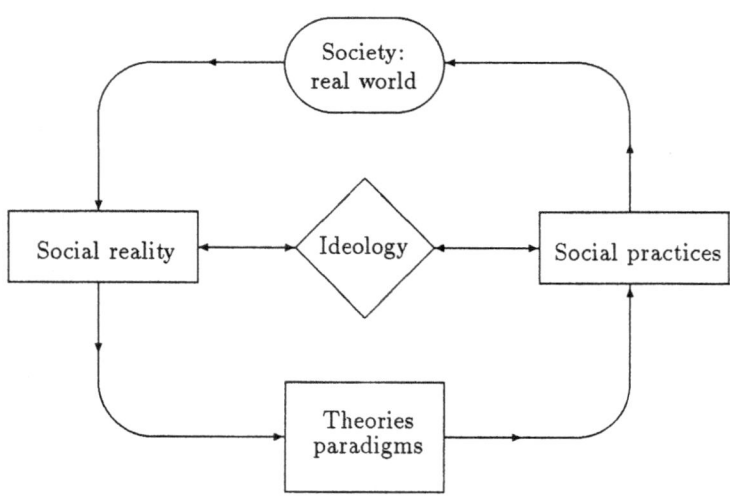

Figure 3.2: Human theorising of social reality

As we have already observed, different social environments shape individuals in such a way they come to form judgements and theorise about causes and cures for poverty, unemployment, inflation, and recession. This sort of engagement leads individuals to further reflections out of which theories are built. On the other hand, different economic theories appeal to personal feelings and beliefs in such a way that they come, quite often independently of their scientific validity, to endorse them. These feelings and beliefs are themselves rooted in a certain sense in a sort of common knowledge or popular ideology consisting of socially acceptable values and attitudes, social and religious beliefs which are in part inherited and in part learned and enforced by the ideological state apparatuses; see [8, 136, 230].

The knowledge people have of social reality is a human product inherited from the past, it is further a cumulative social creation resulting from man's own active intervention in the world. We like to stress the orientation of causality emerging from this observation, first running from social reality as represented by people's knowledge toward man and on the other hand from man in a context of human beings reproducing and altering this knowledge in their own way toward social reality.

At a higher level, we have theories shaping society and society in turn shaping theory. Figure 3.2 summarises these observations and depicts the orientation of causality. Due to the dynamic character of society, it turns out that social reality, ideology and social practices are dynamic as well, in spite of the fact that ideology in its various forms, i.e. religious, ethical, legal, political, etc. plays a major role screening out heretical perspectives and values, and through a sophisticated set of rewards, penalties and value-judgements, aims at maintaining the status quo. For instance, the higher level belief that man is greedy by nature and acquisitive is highly predestinated to call forth the lower level economic belief in competition and in the need of material rewards, as an incentive for engagement and dedication in activities associated with the social process of production.

Further instances of higher level issues of philosophical character concerning destiny, existence, human nature, morality *inter alia*, suggest additional causal links to the emergence at the lower level of beliefs and value-judgements, nourishing concrete representation of authority, democracy, individualism and private property among others, which are particularly suitable for the preservation of the status quo. A variety of issues of this sort are spread out in the immense literature on the foundations of society, see [8, 84, 99, 140, 196, 226].

At a preanalytical, precognitive stage, we are usually faced with a world view and vision of reality that acts as a raw sketch of guiding principles, directions and purposes based on which any further intellectual activity and search for understanding rely. In this sense a *vision* is a general perspective of reality aimed at pointing to phenomena and problems of society, preventing us from seeing others and suggesting even a ranking for eventual research. Thus the causality goes from ideology to vision as a conceptual "raw material" for theorising and analysis. See [136].

Theories and paradigms

Having called attention upon the ideological bias appearing in (wo)man's representation of her/his social environment and her/his theorising about social reality, it goes without saying that any theory focuses on particular things, i.e. the objects of the theory, and defines them in a specific manner. The selection of objects and definitional specification coming about reflect merely events, and at a certain stage of history, problems have provoked human beings to theorise about them – the chosen objects and the visualised links – in particular ways attaching thereby characteristic meanings to them.

Therefore, the knowledge people have of the world is then shaped by the theories with the help of which they attempt to understand it and by the very world in which they live. Hence, the real challenge of social life and advanced civilisations consists in enriching the own vision of the world through communication with those which visualise it differently and in the eagerness to learn from all the cultural diversity this world is made up. This reflection shows a way of reaching a better understanding of the problems of society and getting better prepared to solve them satisfactorily.

Broadly speaking, we shall conceive a theory as composed of a selection of specific objects, definitional characterisations of these objects and their various links; besides a set of propositions concerning the behaviour of these objects shall be logically derived from their characteristics and a set of assumptions as well as principles with ideological background. We pass on to distinguish between a partial and a general theory according to the body of phenomena the theory is aimed at describing and explaining.

To the extent the underlying ideology prevails, so do its associated theories and as a consequence the core of the dominating theories begins to be seen as a *paradigm*, i.e. a body of theories accepted by its adherents as containing basic principles and elements based on which one can apply and advance their knowledge with purpose of understanding and explaining reality.

The challenge of the facts of life

Reformulating the main conclusions from the contribution of processes of labour to the development of humankind as a biological and social artefact, we come to the recognition of mankind as a *being of praxis*. The idea of human(kind) reflecting the essence of man(kind), as a being of praxis, focuses on human actions upon nature aimed at humanising the natural environment and creating in turn human nature, by developing human capacities to interact with nature, fulfilling and generating needs from which to start anew. Autonomous, self-reflective and creative ways of acting are among the major features that have brought out human(kind) in a long biological evolution and specific forms of self-development.

Notwithstanding the enormous anthropological ground for the emergence of human(kind) after a long range evolutionary process as a biological and social artefact, orthodoxy sticks to a model of human nature detached from any social environment and in the most radical version even detached from nature. Human nature bears then a metaphysical character, it becomes an image of (wo)man emanating from a sort of transcendental agency akin to the notion of abstract individual. In this way it escapes any scientific scrutiny, critical analysis and evaluation, becoming therefore an appropriate ideological entity.

From this perspective we should not be surprised by the overwhelming agreement among the ideologies of laissez-faire capitalism viewing human nature as characterised by features like inertia, sluggishness, aversion from labour, egoismus, greediness, aggressiveness, cruelty and other misrepresentations. Less surprising is the fact that future-oriented projects committed to removal of unfavourable conditions giving preponderance to negative human characteristics, are fiercely opposed by ideologies of the *status-quo* on the ground of animal instincts in human beings which must not be unlashed. That Locke, Hobbes, Fichte and Smith never used the same line of argumentation to critically analyse slavery, colonialisation and other manifestations of European barbarism becomes in this way self-evident.

Needless to say, paradigms cannot answer satisfactorily all the questions they raise, so that no standard approach or explanation develop. A consequence of this fact is a continuous struggle among competing paradigms to provide answers to the largest number of problems and to adapt their cores according to the reality. As the latter changes over time, paradigms like ideologies are often forced to adjust to the economic, political and technological processes of society or to give way for more acceptable ones.

Every social formation tends to reproduce its conditions of production at the same time that it produces with the purpose of maintaining its productive capacity. Since the productive capacity of society is not obviously linked to the need to reproduce its productive forces and the existing relations of production, this gives rise to questions on the role of paradigms in the formation and transformation of human subjectivity and theorising about society. This is linked to issues concerned with the operation of ideology at an individual and institutional level, in the organisation, maintenance and transformation of power in society, including everyday notions and experiences as well as the elaboration of intellectual doctrines.

At this stage we shall recall inconsistencies and conflicts resulting from the necessity of continuous renewal of used-up potentials and the failure of doing so. Some are related to the inconsistency between rationality prevailing at the working place and outside of work, or between individual rationality and functional integration. As we already mentioned, this gives rise to system irrationality in the form of consumerism, which aggravates the problem of entropy production in terms of waste, pollution and moral deterioration. Let us point, following Gorz [95], that when getting paid becomes the primary objective in the working activity of an individual, the compensatory consumption, which acts as incentive to withdraw into the private sphere and give priority to the pursuit of self-interest, deprives individuals of networks of solidarity, mutual assistance, social and familiar cohesion; these networks break down as functional integration advances. The following quotation illustrates another fallacy resulting from the rationale of the more the better.

> The paradoxical and perfectly predictable consequence of this *asocial socialization* will be that the state will find it necessary to reinforce its prescriptive powers. Indeed, the pursuit of personal advantage could only produce a collective optimum in an environment in which there were

no shortages, in which sufficient resources existed for there to be unlimited increase in overall wealth and in which, in the absence of all forms of rigidity and inertia in the material sphere, the advantages reaped by one sector of society were never gained at the long-term expense of another. Glimmerings of this world of 'unlimited opportunities', which forms the basis of the liberal utopia, have been seen on a number of occasions throughout history, particularly during the colonization of North America. Otherwise, we know that in an environment where there is scarcity, the actions of individuals each pursuing their own immediate advantage produce in the material sphere an overall result which frustrates the aims of each and, in the words of Marx, 'thwarts their calculations and destroys their hopes'. Sartre calls the way in which the collective effect runs counter to the individual strivings which have produced it 'counter-finality': for example, the actions of individual farmers each clearing trees in order to increase the amount of cultivable land they have, leads to soil erosion and catastrophic flooding.

André Gorz
(see [95], pp. 47)

A further point concerning functional integration, to which we shall call attention, is the attempt to merge functional and social integration with purely ideological purposes. At a fairly intuitive level, we may say degradation of labour results from transforming the worker/producer into the worker/consumer – abolishing thereby the satisfaction associated with producing works and the pleasure of doing tasks – and bringing in compensatory consumption instead. As we have seen this put the worker/consumer in a purely capitalist factory environment to perform jobs in a fully fragmented process of labour. These jobs fail to offer individuals any self-identifying value, work fulfilment, or employment stability, they fail to motivate, keep interest or willingness to cooperate. Therefore, earning enough money becomes the only incentive to accepting that type of functional work, but this requires a relatively high level of development so as to provide sufficient finance and means of consumption.

At this stage we shall recall, ideologists from various origins postulated the reconciliation between public and private interest, surrogates of functional integration and the resulting social disintegration, resorting to the notion of socialist consciousness and to exaltation of national pride and patriotism. Socialist consciousness came to mean for socialist parties of soviet type a body of moral and intellectual virtues on the basis of which social and functional integration conflate into one. On the other extreme for fascist parties, the exaltation of patriotism embodied various elements of charismatic and omniscient guide akin to religious cult, aimed at promoting faith in traditional and nationalist values and at attaining social cohesion by reinforcing chauvinism and racism; see [95].

Socialism of soviet type has capitulated and fascism seems to disappear at least in its open and uncompromising guise, however the problems remain and challenge society with increasing threat. Current orthodoxy is incapable of addressing these questions, since it has become increasingly separated from practice and still maintains in its core the most obstinate ideological machinery. This results from the need of hiding behind the laws of the market the successful and well tested policies of colonialisation, an imperative function the assumption of which elevated orthodoxy to dominance but paralysed it as a science. It remains to hope that the post-cold-war period experiences a decolonialisation of mind from which a true dedication to the pressing questions of socioeconomic reality may follow and that Eurocentric driving forces seriously begins questioning its way of viewing non-European cultures as a step toward abandoning the universal imposition of a model asphyxiating by now almost two thirds of mankind.

3.4.2   Brief characterisation of the major schools

Basically, an economic theory consists of a set of *objects of study* or categories, a set of *assumptions* delineating what shall be seen as important and unimportant and *hypotheses* about economic behaviour which shall be derived from assumptions of the theory.

Concerning the modern neoclassical, Marxian and post-Keynesian theories, we shall first say that they all may include in a certain way demand and supply functions, markets for capital, both human and physical, and commodities. The three are compatible with continuous factors substitution and marginal productivity analysis.

As a first approximation, we can get the following pictures susceptible to a comparative study, see [146]. *Neoclassicists* relate the objects of the theory in a way so as to determine an equilibrium system, where the consumer plays a crucial role in accordance with the behaviour pattern derived from an equilibrium-tailored set of assumptions on household preferences and value-judgements. Full-employment is a decisive although ambiguous notion. The causality runs from growth towards distribution.

Forcing the *Marxian picture* into the framework of equilibrium, one can get a description according to which equilibrium obtains starting from a conventional wage rate and differing saving habits attributed to capitalists and workers. Class struggle is crucial and the causality runs from distribution towards growth.

Finally, in the *post-Keynesian alternative*, one goes from the investment demand of capitalists, their saving propensities and combines the remaining elements in order to determine equilibrium. Investors play here the decisive role, while growth and distribution are endogenous and determined simultaneously.

However, in order to understand how the essential forces at stake interact, one needs to know more about the objects and assumptions characterising the corresponding theories, a point to which we turn.

Neoclassical paradigm: mechanisms of pure exchange

Let us begin our sketch of the neoclassical paradigm presenting briefly the main ingredients of its ideological base. Neoclassical theory focuses on presumptions about human nature of the individual and three economic acts attributed to all individuals: owning, buying and selling. At the centre lies the notion that the innate human nature of individuals determines economic outcomes. This essential feature reveals the dualist origin of neoclassical thinking, i.e. the detachment of the neoclassical observer (rational man) from society and even more from her/his human nature. Accordingly, people behave rationally and act in their own self-interest and under conditions of maximum freedom, their own productive abilities are such as to generate the maximum wealth possible in a society. People's material wants are unlimited and hence can never be satisfied.

It is extremely important to note that in neoclassical thinking human needs are not objects of the theory; human needs gave way to wants as a requirement to elevating the rationale of "the more the better" to the driving principle of economic rationality. The limited nature of human needs constituted a formidable obstacle for the rise of the capitalist spirit, see for instance [227, 209, 30]. Society is then assumed to be acquisitive and growth-oriented so that more material wealth is always preferred to less. Competition, as opposed to cooperation, is a necessary and desirable fact of life, ownership is a natural right. Material incentives are essential for human motivation and inequality in material well-being is also a fact of life.

The *neoclassical vision* emerging from this ideological base gravitates towards the market as the place where individuals, with diverse property holds, voluntarily come to sell and/or buy with the purpose of maximising their satisfaction and hence obtaining maximal wealth for the society as a whole, since any transaction is mutually beneficial or else it will not occur. Since unlimited wants render resources scarce, the efficiently allocation of resources becomes the question of utmost interest. Lichtenstein [136] suggests that this vision can best be described as *scarcity-choice-harmony* and from this becomes apparent all efforts shall be put in the establishment of the required societal institutions that make sure this inner human essence works itself out to the greatest happiness of the greatest number.

Concerning the objects of neoclassical theorising, let us stress individuals, markets, commodities, technologies and prices; they are followed by money, income, saving and investment. One can, in the way to understand these basic objects, write however a long list of further objects including for instance, individual preferences, utility, supply and demand, production, distribution, labour and capital among others. According to the object at stake, one obtains a full body of economic theories such as the supply and demand theory, the theory of value and price, the general equilibrium theory, which belong to the heart of the neoclassical paradigm.

A distinguishing feature of neoclassical theories is its notion of causality, i.e. a conceptual relationship of the sort cause-effect with which to connect the particular set of objects. This has been called *essentialism* or *determinism* which is based on the presumption any event can be shown to have certain causes or determinants that are essential to its occurrence. At the most general levels, the individual, pursuing her/his self-interest according to the pattern of rationality attributed to human nature, is seen as an ultimately determining cause for economic development. At a lower level, supply and demand are elevated to the category of (market) forces by way of which market prices are determined; they have an ultimately determining cause in the mind of individuals concentrated mainly on innate notions of pleasure (utility) and pain (disutility), too.

Since neoclassicals deny any causal relationship going from economic change to rational preferences and postulate that natural and cultural issues are outside the main body of the theory, given wants and productive ability of human beings the curious construction of steady-state equilibrium emerges. Joan Robinson and associates point the wealth of ambiguities, puzzles and inconsistencies inherent with the notion of equilibrium in its different guises. As a response to the claim that the concept of equilibrium is incompatible with history, neoclassicists forwarded the view that equilibrium lies in the future and that provided the forces are removed that prevented the economy, in the past, to reach it, the economy will tend towards equilibrium and get there in due course, see [181] for precise references on these questions.

The neoclassical concept of equilibrium is in fact scientifically wrong and constitutes a striking misrepresentation of the physics on which it relies. Namely, it is based on the Newtonian model of gravitation in the field of astronomy, which consists in regular, continuous, repeated motions of celestial bodies that never collide; yet collisions provide the mechanisms of pure exchange in the neoclassical model. More recent efforts to let mechanisms of pure exchange operate in a framework incorporating utility as a surrogate of energy still misrepresent the physical model on which they are inspired, since they neglect the most important concept of the paradigm of energy, namely entropy; see [5, 150].

Let us mention in passing that the Second Law of Thermodynamics, in particular the notion of entropy, cannot be introduced in the neoclassical framework without shaking its fundaments, as a quick analysis of the assumptions on perfect knowledge (information), unlimited wants and maximising behaviour in the light of Boltzmann's equation (3.3) shows.

The political consequences of the orthodox view continue to have dramatic impacts upon the lives of millions of people all around the world and in particular in the so-called Third World, since powerful nations of the centre supported by illegitimately deduced conclusions of ideological nature continue leading the way toward the irreversible annihilation of the natural and living environment of societies in powerless state nations of the periphery. They are vested with the power conferred by their missionary commitment and dedication to the progress of mankind (sic), a task glamourously visualised as the worldwide struggle for free markets.

Contrasting sophisticated rhetoric with the truths of real life, let us mention that Josué de Castro in [49], for instance, shows with appalling evidence, how in the way to "economic development and progress" prosperous regions and societies in Latin America, Africa and Asia were transformed into areas and conglomerates of hunger and depression by forcing them to monoculture and large-scale production, be it agricultural, livestock or whatever, directly or on behalf of Eurocentric companies. It is very hard to see how these or associated operations are mutually beneficial to societies in the centre and the periphery of the Free World.

Marxian picture: principle of social reproduction

Before we pass to sketch the ideological core of Marxian theories, let us make two preliminary comments. First, Marx's theory is a critique of capitalism, i.e. an attempt to understand the essence of capital accumulation. It is extremely important to realise at the outset that Marx's critique was not aimed at exposing the abuses of capitalism but rather its inherent dynamics. Marx clearly considered capitalism to be a relatively progressive system that driven by intrinsic imperatives of expansion and continuous quests for markets, raw materials and other means of offsetting always recurring crises, would ultimately pave the way to the dissolution of pre-capitalist structures and generate world-wide vigorous dynamics of capital accumulation and growth. This Marx believed to be an indispensable condition for the coming of socialism. Therefore, while Marx was well acquainted with its destructive and exploitative character, he was aware of the need to save and enhance positive contributions of capitalism. The main point was thus to analyse capitalism with the aim of transforming it by removing its oppressive components, liberate its potential with the purpose of bringing also the fruits of technology and cultural creativity to the vast majority of mankind. That this was an illusion and that in this sense Marx himself surrendered to technology fetishism is witnessed by today's concentration under the sway of the international division of labour of poverty, immiseration and starvation in state nations of the periphery. This point remains true independent of the fact that condemning almost two thirds of humankind to unjustifiable denial and suffering is made less visible through the subtle veils of racism and apologetical theories, which builds an essential part of orthodox ideological core, see in this regard [173, 174].

Let us conclude this first comment, recalling that Marx himself did not spend time or effort analysing communism, he rather concentrated on capitalism, see [230]. Thus, Marxian theory is essentialy an analytical machinery purposely tailored to the investigation of the working of capitalism.

The second preliminary comment refers to the notion of human nature and the individual in Marx's theory, where the individual is always placed within a social framework which is historically understood. People make their history themselves, however they do so in a given environment. Therefore, the search for a coherent image of human nature seems to be limited by the need to realise the effect of the interplay of manifold forces shaping society and determining its stage of historical development, as well as the eventual priority of social life, which concerns the way human beings actually are in society, over social organisation, which describes social reality, identify the power and relevance of structures. From this perspective results an approximative, partial and temporary notion of human nature as a projection in each historical epoch of a more comprehensive conception of human nature. This notion delineates characteristics fundamental to existence constituting the essence of humankind as a human animal. For a detailed analysis of these topics we refer to [72, 26, 228, 81].

Marxian theory focuses on the presumption that social relationships inherent to the social arrangement into which individuals are understood to be born shape and change what human beings are, think and do. The capitalist system is viewed as consisting of a complex set of relationships in a society divided in classes. Those who own means of production obtain an income derived from this ownership and build the *capitalist class*, while those who do not own means of production get their income by working for those who do and thus they build the *working class*.

Means of production acquire the quality of capital, if they are made available by their owners to members of the working class, based on their ownership relation, with the purpose of accomplishing the process of material production and generating a surplus. Marxian theory starts from the presumption of class exploitation and proceeds to explore what and how other aspects of the society interact with that specific class structure. It claims that individualism and free markets serve to hide and perpetuate class injustice.

A crucial feature in Marxian theory is the direction of causality, which goes from social relationships to the resulting pattern of individual behaviour. Associated with the process of material production one considers highly interwoven natural, cultural and political processes, so that a further distinguishing feature connecting causes and effects comes out, attached to the rejection of essentialism. Namely the presumption that no single event can ever be conceived to occur by itself, independent of the existence of the others, i.e. all events, not only some essential few, add their unique effectivity or influence to bring into being anyone occurring.

This kind of reasoning, associated with dialectics or overdeterminism in a modern guise, is a major tool of understanding how events exist, and stresses that any aspect of society is always a cause and an effect, playing thereby an active role in shaping and bringing about a state of tension or change. This calls forth the view that in the face of a highly complex world, human beings cannot more fully appreciate and understand social reality than they can fully observe, capture and explain somebody else's perception of her/his individual reality regarding familiar, social or environmental events. Because what constitutes anyone's reality depends upon anyone's individually available apparatus—observations, capacity, objects of interest, qualification and notions of causality—to bring forward anyone's individual explanation of events. Thus, theories and explanations are partial, approximative and temporary, Marxian and non-Marxian alike.

These presumptions give rise to a *Marxian vision* based on the recognition of the eminent social character of the process of material production and the need of identifying the always evolving forces on which the principle of social reproduction gravitates. This vision is largely dominated by issues concerning the volatile process of capital accumulation, a complex variety of interacting processes, not rarely involving conflicting configurations resulting from the existence of social classes, in a setting usually not limited to market institutions. This complex variety of interacting processes can best be described as a system consisting of ever changing processes of economic, natural, cultural and political nature underlying a complex structure of conflicting social relationships. These processes, as a whole, build the process of social reproduction and each process is shaped and comes into existence by all the others, since they emanate basically from the process of material production and contain within themselves its conflicting determinants.

The objects of the theory are class, surplus, capital, labour, labour power, commodities, values, production, distribution, accumulation, crises and imperialism. These objects of study as well as any aspect of society are viewed as processes in order to stress the endless change happening in them.

Post-Keynesian alternative: reconstructing political economy

The alternative that we shall sketch next may be understood as the post-Keynesian reconstruction of classical political economy. The core of ideas building the post-Keynesian alternative incorporates thoughts and efforts of a variety of investigators working outside dominating orthodoxy. Thus, it integrates a wealth of ideas, perspectives and developments which may as well characterise a post-classical, post-Marxian or institutional alternative.

Post-Keynesian objects of study are production, household demand, growth and distribution, prices and pricing mechanisms, money, credit and finance, wage determination and inflation. However, these objects are studied from the perspective of historical time rather that logical time, so that post-Keynesian theory deals with processes evolving in a forward direction. In particular, the post-Keynesian approach distinguishes between long-term and short-term analysis, as directed to the study of secular trends and cyclical fluctuations, or deviations around those trend lines. Since price is not simply viewed as a marketplace phenomenon post-Keynesians, as classicals alike, have three types of prices: mark-ups, prices of production or natural prices, and market prices. Post-Keynesian price theory is, however, mainly linked to the work of Piero Sraffa.

The classical roots of post-Keynesian and Marxian paradigms lead them to share a vision of capitalism based on the principle of social reproduction, social classes, circular flows, surplus and others. Four institutional dimensions are considered: normative, political, economic and anthropogenic. A complete introduction of post-Keynesianism is given in [58, 136].

3.4.3   Some political issues of employment

Development planning demands a lot of efforts characterising potentials of economic changes called forth by an economic policy and assessing the impact that these changes may have on vital aspects of society. Before we go on to outline an employment policy as a vital element of a learning process focusing on the acquisition of skills and competences, let us give a look at various issues which result from examining the prospect of application of such a policy.

On so-called noneconomic benefits and external costs

In a broad sense theories of economic growth and development are concerned with matters related to the evolution of factors of production, their markets and technical progress. Planning and decision making are in particular activities resulting from the presence of partial information and uncertainty, and the need to anticipate certain features associated with the evolution of various socioeconomic factors, their relationships and underlying institutions.

As we have already pointed out, the purpose of planning is to increase our ability to cope with change and encourage the occurrence of more desirable future states. However, the evaluation of the achievement of planning goals leads to a variety of problems that seem to require distinguishing between economic and noneconomic benefits or external costs. Nevertheless, a careful analysis may lead to the conclusion that what we have is just certain kinds of benefits and costs, the economic impact of which may be less direct than others.

For instance, under the concept of standard of living people understand, besides unambiguous economic factors, something like availability of educational and health services, environmental conditions, various kinds of communication and information and the like. Very few people will seriously doubt on the fact that the greater the consum of these services the greater would be the productivity in the corresponding society. Further instances may be national self-reliance, different kinds of infrastructure, the presence of a well-understood system of value-judgements which actively encourages social responsibility, solidarity and cooperation, egalitarian and redistributive views, and others.

Hence, efforts to distinguish between economic and noneconomic benefits appear to be rather unsatisfactory unless they have a tentative and approximative character as a means of simplifying reality and easing measurements. Behind the tendency of discriminating between benefits of such type appear the intention of hiding the main recipients of economic benefits deriving from them, in much the same way as similarly ambiguous terms like external and secondary casts attempt to dilute origin and responsibility of damaging effects associated with technology.

The most important of these types of benefits and costs are in fact related to anti-entropic and entropic processes and deserve utmost attention. Only a serious analysis of development and growth processes within a socioeconomic system in the light of the Second Law of Thermodynamics will make apparent how the continuous neglect of the course of entropy embodied in energy dissipation, pollution, health hazard, unemployment, malnutrition, moral and social disintegration and others, can explain why socioeconomic problems world-wide spread out faster than the required problem solving capacity, why the more complex are the technologies available the more disordered becomes the world, why the more sophisticated are the means of communication and information the higher the fragmentation of society and its danger of malfunctioning, and a lot more.

"The truth is", says Rifkin [178], "every technology creates temporarily islands of order at the expense of greater disorder to the surrounding"; "true that secondary disorders caused by a particular technology can be temporarily solved by the application of additional technology. But the solution will inevitably result in even greater disorders than the one it solved". Thus, since pseudo-solutions just potentiate the residual problems, the alternatives remaining are learning to act according to the principles of social reproduction and entropy or under the guide of the "invisible hand" precipitate humankind and nature to the final disaster.

The appeal of employment

It is necessary to emphasise employment needs not to be an objective in itself, so that the desirability of additional employment has to remain subordinated to its potential generation of output. Let us list some reasons, drawn to some extent from [44], that attract our attention concerning when employment policies may lead to creating employment opportunities.

- Labour-power as an expression or materialisation of manpower is an important economic resource that is wasted in the case of unemployment. Therefore, an economic policy that directly or indirectly contributes to employment creation shall be viewed as pertaining the goal of fuller exploitation of productive potential.

- Integrating idle labour or maintaining active labour in the production process is an efficient way of improving the quality of the labour force which otherwise would deteriorate physically and morally.

- Associated with the fact of being employed or being idle are learning and forgetting processes, i.e. acquisition, improvement and deterioration of skills, that in a way or another affect present and future labour productivity, costs of production and the attached rewards and these in turn influence purchasing and consuming power.

- With the participation of labour in production there are associated a variety of social outgrowths which derive from its character of being a social process that may impact in different ways the fundaments of society. Let us mention a few: Qualities like mobility, cooperation, solidarity, awareness and others may be enforced or weakened through the experiences resulting from it.

Thus, employment policies are also efficient mechanisms for the fulfilment of short- and long-term social consumption objectives. The fact that the impact of such policies comes out only on the long run and that discounting rates may be needed for its evaluation is another matter; namely, one of methodology of measurements. In this regard, this is a matter concerning the need of appropriate method of evaluating the links between social goods like schooling, training, health, nutrition, discipline, organisation and production and the corresponding social reward.

Similarly think of the advantage of getting a transparent understanding of notions of human rights, freedom of action *inter alia*, and the impact this will have upon the conceptual elaboration of labour markets based on the human features of labour, going from a theoretical perspective of production processes to the concretisation of actual ranges of meaningful functionings open to individual agents. To close this subsection let us recall unemployment acts upon society bringing out damaging effects ranging from psychological and social distress, poverty, undernourishment, disease, to lawlessness, vagrancy, crime and others. One might be inclined to ignore the human costs involved but one cannot define away the really immense social costs that unemployment entails in the short and in the long run.

The labour market and human nature

Besides the analytical consequences deriving from alternative theories, we need to consider practical consequences and political dilemmas associated with policies attached to contending theories. Let us begin remarking that different theories bring about different actions and these impact lives of individuals, shape their attitudes, expectations, confidence and come ultimately to determine their behaviour. Therefore, policy issues related to employment need to be studied from the economic, social, political and psychological point of view. It is extremely important to emphasise as an essential and distinguishing feature between the labour market and other markets the human nature involved: first of all labour processes are in the first place processes of learning and second labour power as an essential carrier of creativity is critically linked to biological requirements of a labourer as human animal.

The inability of orthodoxy to differentiate between the labour and commodity market is best explained by its fundamental inconsistency with historical time, entropy and information problems inherent in processes of learning and social reality. In this sense the continuous renewal of labour power depends critically on the existence of anti-entropic processes and on the acquisition of skills and competences along learning processes which develop in the course of time. Consciousness, creativity, motivation, aspirations and the like exert a profound influence on the acquisition of skills and competences and since these features are linked to the standard of living prevailing in the given social environment, they are crucial determinants of labour heterogeneity. As a first approximation, we shall consider a relatively stable core and dynamic periphery of skills and competences which are both critically linked to the stable core and dynamic periphery of technology. The notion of individual freedom needs to be essentially coupled with available potentials in regard to mobility, information and flexibility, to adequately adapt to the unfolding of technology, which is at the heart of the concept of capability to function presented above. See [203, 58, 57].

Part II

Models with Persistent Structures

The division of labor among nations is that some specialize in winning and others in losing. Our part of the world, known today as Latin America, was precocious: it has specialized in losing ever since those remote times when Renaissance Europeans ventured across the ocean and buried their teeth in the throats of the Indian civilizations. Centuries passed, and Latin America perfected its role.

... The more freedom is extended to business, the more prisons have to be built for those who suffer from that business. Our inquisitor-hangman systems function not only for the dominating external markets; they also provide gushers of profit from foreign loans and investments in the dominated internal markets. Back in 1913, President Woodrow Wilson observed: "You hear of 'concessions' to foreign capitalists in Latin America. You do not hear of concessions to foreign capitalists in the United States. They are not granted concessions." He was confident: " States that are obliged ... to grant concessions are in this condition, that foreign interests are apt to dominate their domestic affairs ... ," he said, and he was right.

... Latin America is the region of open veins. Everything, from the discovery until our times, has always been transmuted into European— or later United States—capital, and as such has accumulated in distant centers of power. Everything: the soil, its fruits and its mineral-rich depths, the people and their capacity to work and to consume, natural resources and human resources. Production methods and class structure have been successively determined from outside for each area by meshing it into the universal gearbox of capitalism. To each area has been assigned a function, always for the benefit of the foreign metropolis of the moment, and the endless chain of dependency has been endlessly extended.

... *Our defeat was always implicit in the victory of others; our wealth has always generated our poverty by nourishing the prosperity of others— the empires and their native overseers. In the colonial and neocolonial alchemy, gold changes into scrap metal and food into poison.*

Eduardo Galeano
(see [78], pp. 11–12)

Chapter 4

Stopping times and other probabilistic issues

A world will be considered where nothing is absolutely fixed. At any given point of time, there is a stock of means of production and of technical knowledge which is inherited from the past, but as time goes on this stock is continuously changing. Any of its components is susceptible to being increased and improved, if not directly, through substitutes; provided that enough time is allowed for the members of the community to learn (through investigation and experiment) how to do it. It is this learning activity that represents the spring moving the whole system, by its constant application to improving the production processes and to starting new productions. Thus the commodities which are exchanged and consumed are not goods that may be found in nature. They are goods produced by men, practically in whatever quantity men like, provided that they think it worth devoting to them the amount of effort they technologically require, this effort being itself changing as an effect of technical progress.

In this scheme, therefore, Man and not Nature represents the central focus. Man is the mover of the whole system, in a double role: by providing with his likings and preferences the criterion for deciding on the quantities and types of commodities to produce, and by inventing and operating the process of production. Accordingly, the gravitational centre of the whole analysis that follows will reside in the learning power of the members of the community and not – as it has been in traditional economics – in the limited amount of natural resources. It will reside, in other words, not in the caprice and scarcity of Nature, but in the progress and ingenuity of Man.

Luigi L. Pasinetti
(see [160], pp. 22–23)

4.1   Choice of technique and accumulation

In this section we shall start spelling out the logic of allocation of national income on which most of our work rests. To this end, we shall call attention to various issues related to learning as a key notion underpinning the evolutionary conception of social reproduction and to production as the crystallisation of know-how and information purposely directed to the transformation of energy and materials into the form of human artefacts and commodities.

4.1.1   A glance over the choice of technique

Learning and information play the most important contribution to understanding the nature of production:

- First of all, learning and information are essential factors inherent in labour processes which, as we have seen, originate as a plan for the future in human minds;

- Second, human artefacts and commodities, which result from processes of labour, are a kind of information structure. Energy and materials, which provide means of coding information, are limiting rather than determinant factors of evolution;

- Third, economic development, as a process along which the factors of production unfold, is primarily a process in the increase of knowledge and information. Thus, crucial aims of development are removing constraints hindering the realisation of available potentials and preventing their irreversible exhaustion;

- Fourth, the extremely important role energy plays in the process of evolution is manifested by the fact that human artefacts are subject to aging. This entropic process may be counteracted on the basis of knowledge and information through acts of nutrition, repair and renewal.

After having mentioned some basic aspects of the nature of production, let us recall that organisation, as a specific category of human artefacts, sets to coordinate rich and manifold sorts of knowledge and information naturally disseminated among various segments of society. In this sense, organisation leads to relatively complex processes of production based on interaction, cooperation, flexibility and diversification.

Exhaustion versus replenishing

At this stage we know about the highly evolutionary character of learning and that acts of production start off with potentials which are gradually used up. In this sense, it is natural to see social reproduction as a sequence of pairwise coupled processes going in opposite directions: consumption and production. Behind the first process we have exhaustion of available stocks of energy and materials with the purpose of recreating the used-up potentials. The second is aimed at replenishing stocks of energy and materials in the course of which available potentials are used up. Needless to say, this sequence of pairwise coupled processes changes the environment and with it the availability of energy and materials, thus knowledge changes and with it any act of production. Since human evolution depends critically on the availability of appropriate forms of energy and means of storaging information, we shall understand a technology as a set of techniques exhibiting common energetic and informational features.

The warning eye of entropy

Corresponding to any technology human society gives rise to a tendency of expanding the essential features of its techniques into all aspects of social life. However, since the availability of specific energy forms acts as a constraint upon the development of productive forces, any further expansion needs to go hand in hand with the development of underlying forms of energy. At this point the principle of entropy reminds us that technical progress cannot be taken as automatically leading to social progress. On the contrary, a failure of understanding the nature and implications of technical progress may jeopardise society's ability to recreate potentials.

In foregoing chapters we have called attention to the fact that generation of economic surplus or surplus product opens the way for improving the resources available to society as well as its efficiency in the use and allocation of them. Since labour and resources, in the specific form of means of production, set a bound to productive capacity, it becomes apparent that economic surplus affects critically the set of (available) production techniques and further technological progress. This set of production techniques indicates the alternative combinations of capital and labour against the resulting social output, given specific features of the endowment of capital and labour available at any time.

Before we introduce in a precise manner the notion of choice of technique, let us point out that as long as techniques are chosen from a given technology the principle of entropy acts as a law of diminishing potential. Out of the struggle between exhaustion and recreation, begins to shape a potential for a new technology as exhaustion takes the overhand. However, the actual emergence of the technology depends crucially on the availability of appropriate energy forms and information media.

4.1.2   Choice of technique and economic rationality

The following paragraphs are dedicated to present a concise picture of productive structures, a topic of utmost importance to be treated in detail in [94] including technological change on the basis of evolutionary watersheds, innovation and creativity.

A deterministic description of choice

Our conception, which is partially based on [160, 121, 73], attempts first to capture in a static description the most outstanding structural patterns and then to visualise the manner how this static description or flashlight picture of our dynamical system evolves over time.

Therefore, in the first step the aim is to capture the patterns prevailing in the vertically integrated structure of production. The notation is that of foregoing chapters in particular in regard to the model of social reproduction and the process of production, except for additional sub- or superindices which account for specific features of the structure we shall shortly bring to the forefront. To save notation we shall abstain from making particular references to commodities building the output bundle Q, however considering explicitly $Q^{(1)}, Q^{(2)}, \ldots, Q^{(n)}$ will not affect our reflections. Neither shall we write the time subindex in variables like Q_t, K_t and L_t, unless it becomes necessary.

Let us assume that at any time t, $t \in [0, T]$, the technology $\mathfrak{T}$ prevailing is known to consist of m production techniques $\widetilde{F}_1, \widetilde{F}_2, \ldots, \widetilde{F}_m$ all of which bring about the same quantity of output $\bar{Q}$, i.e. $\widetilde{F}(t, K_{(j)}, L_{(j)}) = \bar{Q}$ holds for any $j = 1, 2, \ldots, m$. Here each $K_{(j)}$ (respectively $L_{(j)}$) stands for the vector of characteristics representing inputs of physical machines and intermediate commodities (respectively physical quantity of labour) required in the vertically integrated production of $\bar{Q}$. Further, let the vectors $P_{(j)}$, $\Pi_{(j)}$ and W represent respectively the output price and rate of profit associated with technique j and the wage rate about which we may suppose they are given. This may be seen as a natural consequence of considering time arbitrarily fixed at the point t and placing ourselves within the vertically integrated structure level of a single production process from which $\bar{Q}$ results.

At this point we may resort to economic rationality and select a technique with an optimising cost of production. That means, we may select a technique $\widetilde{F}(t, K, L)$ with a production cost denoted by J,

$$J(t) = \inf_{j=1,2,\ldots,m} \{J(t, j) : \widetilde{F}_{(j)}(t, K_{(j)}, L_{(j)})\},$$

where $J(t, j)$ stands for the production cost $P_{(j)} K_{(j)} + W L_{(j)}$ associated with the technique j. Further, K and L above represent the optimal level of capital and labour. We have purposely used for the chosen technique the notation $\widetilde{F}(t, K, L)$ assigned already to the production function in Section 2.3.

The stochastic picture of choice

Nevertheless, the foregoing characterisation of $\widetilde{F}$ is fully inappropriate within a dynamical framework, since as soon as we pass to examine how this static description evolves over time a wealth of inconsistencies emerge. One is linked with the need of physical capital to get committed to a specific production process over a relatively long period of time. This gives rise to various questions concerning the presence of uncertainty with respect to the rate of profit, physical deterioration and obsolence of capital. A further inconsistency is linked to the random nature inherent in the output demand and in the level of employment, in particular when the point under consideration moves far in the future, the impact of which comes to affect future rates of profit through the mechanisms of effective demand.

To make a long story short, about the rate of profit we can realistically assume at most, the vector $\Pi_{(j)}$ is known only probabilistically. Thus, at any time t, $t \in [0, T]$, we are solely in a position to define the conditional optimal cost of production J by means of the formula

$$J|_{\mathcal{F}_t} = \inf_{j=1,2,\ldots,m} \left\{ \mathsf{E}[J(t,j)|\mathcal{F}_t] : \widetilde{F}_{(j)}(t, K_{(j)}, L_{(j)}) \right\}, \tag{4.1}$$

where the expectation is computed on the basis of the probability measure characterising $\Pi_{(j)}$ and $\mathcal{F}_t$ represents the information available at time t. The conditional optimal production cost J can further be characterised by a SDE using control methods like those introduced in this and the following chapters

From the formula (4.1) and the recognition that the rate of profit Π can at best be assumed to be known probabilistically it turns out that the chosen technique bears a marked stochastic nature, too. The difficulty intrinsic to the question of specifying satisfactorily initial conditions leads furthermore to interpreting the technique $\widetilde{F}$ as a representative ensemble, where any of its elements is characterised by a rate of profit the occurrence of which is determined by the underlying probability measure. Following the same line of argumentation we consider the function F, representing prevailing mechanisms of income distribution, as being characterised by a stochastic nature of similar origin.

To conclude, let us say that the above characterisation of $\widetilde{F}$ and $\mathfrak{T}$ comes close to considering prevailing techniques as being grouped in a stable core and a periphery of less stable techniques.

4.1.3   A basic model of accumulation

At this point we are in a position to expound a relatively simple model of stochastic accumulation, which is susceptible of extension to a general multi-sectorial dynamical framework of the Pasinetti type. It is extremely important to keep in mind that in our framework, flows of labour and capital services are defined in terms of characteristics, which in the light of the principle of entropy are inexorably subject to exhaustion, aging and break down. Therefore, from the point of view of social reproduction, models of accumulation need to incorporate mechanisms aimed at recreating potentials.

At a fairly elementary level these mechanisms take the form of processes of nutrition, repair and population growth. However, as human artefacts increase in complexity, processes of innovation, cooperation and solidarity come into being, which are attributable to human creativity, interplay of individual generations and the fact that living beings need a period of development until they reach a certain level of maturity. This is a feature contrasting sharply with capital and commodities, which start loosing structure as soon as they come into being.

Fundamental relation of growth

The following derivation of a model of capital accumulation relies on assumptions and notions presented in Chapter 2, Subsection 2.1.4 and Section 2.3, and goes as follows. Considering (2.2) and (2.3), we get after simple operations

$$\frac{1}{a_l} = c_l + (\delta_t + g_t)\frac{a_k}{a_l}$$

$$h(k_t) = c_l + \delta_t k_t + g_t k_t. \tag{4.2}$$

The obtained expression, as it stays at the moment, says that any unit of output per worker is allocated to meet the requirements per worker, to make up for the deterioration of capital as represented by c_l and $\delta_t k_t$ respectively. The first two terms account for allocation of social output with the purpose of maintaining the productive capacity of labour and capital, while the third and last term account for allocation of output responding to the decision of expanding the level of output to become available in the future.

Thus the coefficient g in the expression $g_t k_t$ needs to rely on experience, tradition or cultural factors. To the aim of getting insight into the mechanisms determining the size of intended increases in the stock of capital per worker, let us set $g_t k_t$ equal to change with respect to time of the stock of capital per worker, i.e.

$$\frac{\dot{K_t}}{L_t} = g_t k_t \quad \text{or} \quad dK_t = g_t K_t dt \tag{4.3}$$

Then, taking $k_t = K_t / L_t$ and applying logarithmic derivation with respect to time, we obtain

$$\dot{k_t} = \left(\frac{\dot{K_t}}{K_t} - \frac{\dot{L_t}}{L_t} \right) k_t,$$

which after setting $L_t = L_0 e^{a_t t}$ and a few arrangements gives

$$\dot{k_t} = g_t k_t - a_t k_t \quad \text{or} \quad g_t k_t = a_t k_t + \dot{k_t}.$$

Provided the relative rate of growth of the labour is given by a_t, the last relation above says that from the fraction of capital per worker $g_t k_t$, set aside with the purpose of expanding the productive capacity, the fraction $a_t k_t$ is directed to endow workers entering the production process so as to maintain capital equipment at the prevailing standard as described by k_t, and the rest amounting to $\dot{k_t}$ represents actual improvements in productive capacity.

Coming back to (4.2) and substituting $g_t k_t$, we get a picture of capital accumulation which exhibits essential features of social reproduction, i.e.

$$\mathfrak{h}(k_t) = c_t + \delta_t k_t + a_t k_t + \dot{k_t}. \tag{4.4}$$

The first three terms at the left-hand side of this accumulation model are directed to maintain or replenish potentials, while the last term is directed to removing constraints with the purpose of enhancing the potential power of society, think for instance of potentials unreleased due to lack of working tools.

Next, we shall derive formally a stochastic version of the model of accumulation (4.4), based on a simple application of Itô's rule and postpone for a later point making a few comments on mathematical properties of the relation $\mathfrak{h}(k_t)$ at the left-hand side of (4.4).

A preliminary model of stochastic accumulation

To obtain a model of stochastic accumulation, we just go back to (4.2) and try to get from it a stochastic version of (4.4). As before we seek to compute the differential dk_t for which one needs a differential expression for dK_t and dL_t. For simplicity we may assume, dK_t is given deterministically by (4.3) and dL_t, viewed as the demand for labour services, is the only stochastically determined factor. Let dL_t be given by (2.4) and assume further that both K_t and L_t are scalars. A simple application of Itô's rule gives

$$
\begin{aligned}
dk_t &= d(K_t/L_t) \\
&= \frac{\partial}{\partial K_t}\left(\frac{K_t}{L_t}\right) dK_t + \frac{\partial}{\partial L_t}\left(\frac{K_t}{L_t}\right) dL_t + \frac{1}{2}\frac{\partial^2}{\partial L_t^2}\left(\frac{K_t}{L_t}\right)(dL_t)^2,
\end{aligned}
$$

which after substituting dK_t and dL_t by (4.3) and (2.4), and a few operations leads to the expression

$$
dk_t = (g_t - a_t + \zeta_t^2)k_t dt = \zeta_t k_t dB_t.
$$

Next, we get from (4.2) an expression for $g_t k_t$ and substitute it in the foregoing equation. We then obtain

$$
dk_t = \left[\mathfrak{h}(k_t) - c_t - (a_t + \delta_t - \zeta_t^2)k_t\right] dt - \zeta_t k_t dB_t, \tag{4.5}
$$

which is the stochastic version of (4.4). This can be easily checked by setting $\zeta_t = 0$, which means there is no uncertainty in (2.4). The interpretation is as before, except for the stochastic term $\zeta_t k_t dB_t$ which reflects a dispersion around the drift term. The diffusion coefficient $-\zeta_t k_t$ makes the Brownian motion act upon the drift term so that the latter appears fuzzy. That means, the presence of the stochastic terms prevents us to observe the drift term sharply. A helpful alternative is that of working with the mean drift, a point to which we will come back later.

To conclude, let us make a few comments concerning mathematical properties of the current form of the drift term. The mathematical characterisation of the drift term depends on how we define the technical coefficients in (4.2) and/or on the distinguishing features of available techniques as well as those properties carried over to the chosen technique after the inf operation in (4.1). That means, the drift term may be characterised as a set-valued function or after the choice of technique, it may be a function which does not necessarily exhibit the properties of the available techniques. Recall for instance that the sup and inf of continuous functions are not necessarily continuous. Therefore, unless we introduce appropriate restrictions, the expression (4.5) leads in general to the study of differential inclusions, a case we will treat in [94]. For the purpose of the present monograph we shall substitute $\mathfrak{h}(k_t)$ in (4.5) by the distribution function of the neoclassical type as already explained in Section 2.3.

4.1.4 The stochastic process of capital accumulation

Understanding basic characteristics of evolutionary economic processes in society, like time structure of change in needs, preferences, attitudes, decisions among many others, is a basic task of intertemporal analysis. One expects thereby to enhance our ability to construct dynamic models which reflect adequately social reality and to strengthen our ability to learn, to gather information and to adapt to predictable and nonanticipative events. From the perspective of political economy the ultimate aim of this analysis should be the design of actions so as to replace possible but unwanted future events by more probable and desirable ones.

Let us assume that the process of capital accumulation per worker, denoted by $k = (k_t)_{t \geq 0}$, is governed by the stochastic differential equation

$$dk_t = \left[s_t y_t - (a_t - \zeta_t^2) k_t \right] dt - \zeta_t k_t \, dB_t, \qquad (4.6)$$

where the term in brackets, known as the *drift vector*, represents the average economic surplus per worker at time t and the second term, known as the diffusion coefficient, describes fluctuations about it driven by the Brownian motion process $B = (B_t)_{t \geq 0}$ defined as usual. Setting $\mathfrak{h}(k_t) = q_t$ and using (2.25), we are able to rewrite the drift term of (4.5) as above, where $s_t y_t$ stands for the net investment per worker, see (2.17). In the sequel we shall distinguish between the level of disposable income socially available for purposes of net investment and the level of disposable income required to meet changes in the demand while keeping capital per worker at the prevailing standard. We shall call the first type social investment and the second one, required investment. In the stochastic model, required investment includes disposable income set aside to meet uncertainty, i.e. $\zeta_t k_t$. Thus, in eq. (4.6) at the right-hand side the first term of the expression in brackets represents *social investments per worker* while the second *required investments per worker*, so that the difference of them indicates the average rate of change of capital stock per worker. The coefficient of the Brownian motion reflects the deviations from this average due to uncertainty. Eq. (4.6) shall be referred to as the *stochastic equation of the accumulation of capital per worker* and has been used under different frameworks to investigate a variety of problems in economic dynamics.

The fundamental equation of growth

For the purpose of the present study, we should interprete the drift vector as the mean conditional forward derivative, see [24, 154]. Thus, we shall denote the drift of (4.6) by $D_+ k_t$ and assume that the relationship

$$D_+ k_t = \lim_{h \downarrow 0} \mathsf{E} \left[\frac{k_{t+h}(\omega) - k_t(\omega)}{h} \Big| \mathcal{F}_t \right],$$

holds; in certain situations it may appear convenient to use the operator $\mathsf{E}_{s,x}$ which stands for the expectation operator conditioned at the level associated with the initial point (s, x), i.e. $k_s = x$. To save notation, we shall often use the more suggestive expression $\dot{k}_t = D_+ k_t$, unless any confusion arises. Recalling the income identity (2.16), the definition (2.8) and setting $\eta_t = a_t - \zeta_t^2$ for any $t \in [0, T]$, one easily understands on the basis of the new interpretation of the drift the averaging accumulation relationship

$$y_t = c_t + \eta_t k_t + \dot{k}_t, \qquad (4.7)$$

which states that in the average the disposable income per worker y_t shall be
allocated to the purpose of maintaining the level of consumption per worker c_t
and the level of capital per worker, i.e. $\eta_t k_t$, as well as to yield net increases
in the level of capital per worker k_t. Eq. (4.7) is a stochastic version of the
well-known *fundamental equation of economic growth* and holds only within the
ranges determined by the corresponding diffusion coefficient.

The choice of a control action shall yield the highest possible net increase
in the pace of accumulation. However, higher accumulation paths may require
heavy social sacrificies which may harm the process of development in various
ways.

Relevant range of capital and choice of initial conditions

To capture some of the issues associated with eventual social damages to development in a way amenable to economic and mathematical scrutiny, let us
rewrite (4.7) as

$$\dot{k}_t + \underline{c}_t = y_t - \eta_t k_t,$$

where $\underline{c}_t$ stands for the minimum level of personal necessary consumption per
worker resulting from the assumption on the existence of a floor on the admissible wage rate W, i.e. from the assumption that the prevailing wage rate W has
to satisfy the inequality $W \geq \overline{W}$ with $\overline{W}$ given. In the sequel we shall refer to
$\underline{c}_t$ as the level of *necessary personal consumption*, which has not to correspond
to a minimum level of subsistence but rather adjust to the prevailing niveau of
socioeconomic progress of the society.

Due to the fact that our approach focuses on (the principle of) social reproduction, it shall suffice to mention that the characteristics of the commodities
represented by $\underline{c}_t$ shall include housing, education, skills, medical care, food,
leisure, sanitary convenience among others at a qualitative and quantitative level capable of ensuring, at least, the maintenance of the existing labour power.
However, the notion of necessary personal assumption can be put in a precise
setting resorting to concepts like functionings and capabilities as in [203, 202].

At the right-hand side of the foregoing equation, we have disposable income remaining after allowances associated with required investment and this
remaining income can be allocated either to consumption or to the purpose
of expanding productive capacity. The left-hand side hints to the potential
allocation of economic surplus per worker to consumption beyond the level $\underline{c}_t$
and to the expansion of the level of capital accumulation. Any increase of the
level of personal consumption above $\underline{c}_t$ expands the range of states of being and
doing of individuals, but slows down directly the pace of capital accumulation.
On the other hand, keeping personal consumption at the level $\underline{c}_t$ enables a
higher pace of accumulation at which the economy moves to higher and more
stable levels of capital. Some conflicts and dilemmas associated with this type
of decision will be studied later in this essay.

The levels of capital $\underline{k}_t$ and $\overline{k}_t$ attached to $\underline{c}_t$ exhibit interesting behavioural
features. First of all, let us mention that these two levels of capital $\underline{k}_t$ and $\overline{k}_t$ are
equilibrium states, the first unstable while the second stable. The qualitative
behaviour exhibited by the equation above places some restrictions of economic

policy nature, since it limits the range of the potential admissible levels of capital per available worker. For instance, accumulation paths starting from levels of capital at the left of $\underline{k}_t$ are driven by the system dynamics to the zero level of capital. Moreover, not any path starting from initial levels of capital at the right of $\underline{k}_t$ moves towards the upper level $\overline{k}_t$, since the Brownian motion acting on the accumulation path may still drive it out of the stable region.

Let us call attention to the fact that causation runs here from capital to consumption, so that for purposes of economic planning one has first to look at the levels of capital falling within the domain of technological possibilities at any arbitrary time t. Taking into account various political and social considerations on the prevailing economic policy, one works out feasible levels of consumption c_t which are then open to choice. Hence by setting the level of personal necessary consumption $\underline{c}_t$ according to the priorities dictated by the economic policy, one gets the relevant range of capital per worker. Any feasible path of capital accumulation has therefore to start from an initial level of capital within the technologically relevant range.

With regard to developing economies, one assumes in general low levels of available capital per worker which is only true whenever one ignores the phenomenon of capital flight. From this perspective one may opt to set the level of personal necessary consumption $\underline{c}_t$ as low as to ensure bigger initial levels of capital become feasible, or other policy measures directed to the enlargement of the relevant range of capital. However, by doing this one makes also the social burdens of development heavier and jeopardises the future realisation of expected profit as the principle of effective demand tells us. See [90, 93, 91].

At different points we have called attention to two major consequences of keeping personal necessary consumption at a low level: the first has to do with blocking creative potentials of individuals by inhibiting the unfolding of their "states of being and doing", and the second has to do with narrowing the domestic market by maintaining purchasing power at low levels which in turn frustrates expectations of capitalists for profit realisation. Therefore, the next section is aimed at examining the impact on the socioeconomic behaviour of individuals of economic policies arising from configurations we just have sketched.

4.2   General outline of stopping rules

An interesting control strategy of special structure is that characterising *stopping time problems*, i.e. those problems related to the question of stopping optimally the evolution of a system described by means of a stochastic process. At any admissible stopping time τ, one has in this type of stochastic control problems the alternative options of stopping or continuing the evolution of the system at hand. One associates with these options accordingly a *stopping reward* ϕ (or penalty) and/or a *continuation cost* f (or gain).

In a certain sense the decision to stop at τ entails that observations of the system after this time are neither available nor necessary. Intuitively, this means the decision of stopping depends only on the history of the process up to the time τ and not upon what happens afterwards. This interpretation accentuates the feature that the action of stopping shall be regarded as irreversible and as the only option available. However, cases where there are grounds for a restarting are also considered in the literature as alternating, interventions and impulsive controls. See [89, 91] and the references given there.

Stopping techniques have been widely applied in a large variety of fields, i.e. mathematics, engineering, biology, medicine and economics among others. The problems of devising strategies for drilling in oil exploration, for identifying some parameters of a process, for quickly detecting the appearance of a disorder in the functioning of a machine that deteriorates with age and for the appropriate timing of maintenance and machine replacement policies fall into this category.

Let us assume experiments are made in order to identify some parameters of a process. The accuracy of the identification will improve as the number of experiments increases. However, the greater the number of experiments the larger the associated costs, so it becomes natural to ask when it is optimal to stop. Slightly different is the situation in the problem of quickest detection of appearance of a disorder in a machine deteriorating with age, where one shall avoid the production of faulty items and a premature replacement of the machine, since both involve costs. Coupling a replacement with a maintenance policy gives an example of a control problem considering continuous and stopping controls, i.e. maintenance and replacement.

Increasing the number of drillings, experiments, inspections etc., one improves the accuracy of parameter estimation and adequacy of the strategy. However, the greater the number of them, the larger the associated costs, so that it becomes natural to ask when it is optimal to stop. In medicine one finds applications of stopping policies of similar nature, too. Think of a patient subject to a therapeutical treatment and consider the effects of the therapy on the disease growth as well as the harmful influence of the therapy on the health state of the patient. Stopping time techniques may shed some light on questions related to when to stop the therapy and on the timing of the corresponding doses.

An economic application of these techniques that have become highly fashionable is the one related to the investigation of the effects of uncertainty upon the value of a capital asset with stochastically varying returns which can be marketed at any time under conditions partially determined by its owner but limited on the basis of available information. The key questions concern how timing and valuation issues can be altered so as to get the maximum expected benefits or alternatively the minimum average risk. In the present essay we shall be concerned with the application of this interesting and intuitively simple control technique to problems of development planning, to stop optimally the evolution of a system governed by a SDE.

4.2.1   Common features of stopping problems

Let us stress some of the common features of the just mentioned problems. They involve a time parameter t and a state variable measuring the state of affairs that we summarise assuming that the evolution of the system is described by a stochastic process $X = (X_t)_{t \geq 0}$, with $t \in [0, T]$ and X_t taking values on the phase space $\mathbb{R}^n$. This process is assumed to be defined on a probability basis $(\Omega, \mathcal{F}, \mathbf{F}, \mathbf{P})$. The scalar T is arbitrary but fixed and denotes the end of the time horizon. In this chapter G denotes a bounded open set in $\mathbb{R}^n$ with boundary ∂G to be specified in short; further we have $Q = (0, T) \times G$, where ∂Q stands for its boundary. The variable τ_G shall denote the first exit time from the set G. For the sake of simplicity, we shall set $n = 1$ unless it is explicitly specified otherwise. At this point we like to warn the reader about the unusual sense some inequalities will have in the sequel, a feature which has to do with the fact that several of our functionals will assume only negative values.

Basic characterisation of a stopping problem

The further specification of the problem requires fixing a terminal *reward function* $\phi = \phi(t, y)$ defined on Q and assumed to be Borelian and negative-valued, or alternatively according to the case of a cost function. Hence at time $s \in [0, T]$, when the state of affairs is given by $X_s = x$ or in a more general setting by the history of the system up to the time s which is represented by $\mathcal{F}_s$, the decision maker faces the following choice:

- stopping the evolution of the system precisely at the time s, an action with which one associates the reward $\phi(s, x)$, or

- continuing for a possibly random period of time τ obtaining then an average terminal reward given roughly by $\mathbf{E}\{\phi(s + \tau, x + X_\tau)|\mathcal{F}_s\}$.

Let us point out, the decision maker is assumed not to be able to foresee the future, a requirement embodied in the assumption that τ represents a nonnegative random variable depending, without anticipation, on X which renders the variable τ a stopping time. That means, the time τ is such that for any $t \in [0, T]$, it satisfies the condition $\{\tau \leq t\} \in \mathcal{F}_t$ with $\mathcal{F}_t$ describing the σ−algebra generated by the observations of the process X up to the time t. Intuitively, this says the decision of stopping depends only on the history of the process up to the time t and not upon what happens afterwards. This interpretation accentuates the feature that the action of stopping shall be regarded as irreversible and as the only option available. Cases where there are grounds for a restarting shall be considered elsewhere.

Optimising problem 4.1 (Stopping time problem) *Consider a process X defined on the probability basis $(\Omega, \mathcal{F}, \mathbf{F}, \mathbf{P})$ and the reward function ϕ defined as above. Let $\mathcal{T}$ denote the family of admissible stopping times with respect to $\mathbf{F}$ for which the variable $\mathbf{E}\{\phi(s + \tau, x + X_\tau)|\mathcal{F}_s\}$ is well defined and that $\tau \leq T$. Then, find a stopping time $\hat{\tau} \in \mathcal{T}$, so that the average reward attains a maximum and there is a payoff function $u = u(s, x)$ satisfying the relationship*

$$u(s, x) = \operatorname*{ess\,sup}_{\tau \in \mathcal{T}} \mathbf{E}\{\phi(s + \tau, x + X_\tau)|\mathcal{F}_s\} \qquad (4.8)$$

$$= \mathbf{E}\{\phi(s + \hat{\tau}, x + X_{\hat{\tau}})|\mathcal{F}_s\}.$$

It is necessary to point out that an existence result does not suffice to solve
this problem. One needs furthermore an explicit constructive procedure for
the optimal stopping time as the hitting time of a particular Borelian set.
Roughly speaking, under certain regularity conditions an *optimal rule* can be
described as "Stop as soon as $u(s, x) = \phi(s + \tau, x + X_\tau)$" and this can be
characterised by the *continuation set* defined as $C = \{u(s, x) > \phi(s+\tau, x+X_\tau)\}$,
where proceeding for a little while may open the possibility of improving ones's
reward provided that the payoff-rate per unit of time is positive. Observe that
the characterisation of the continuation set C amounts to the specification of
the set G. It is interesting to note moreover that the characterisation of the
continuation set C, or its complement, the *stopping set* S, does not depend
on the initial point (s, x), thus leading to solution for all initial points (s, x)
simultaneously, see [12, 38]. On the other hand, the evolution of the process X
is governed by the probability law attached to the initial conditions. To close
this preliminary description of the optimal stopping problem, we shall stress
that specifying measurability and other properties of the function $u(s, x)$ lies
at the heart of the theory of optimal stopping.

A sequential construction of the stopping rule

Under suitable regularity conditions on $\phi = \phi(t, y)$, Chernoff's sequential construction of an optimal stopping rule is based on the idea of approximating
the stopping problem (4.8) by discrete-time versions of it, giving rise thereby
to a finite sequence of stopping times $\tau^{(n)} = (\tau_1, \tau_2, ..., \tau_i, .., \tau_n)$ with $\tau_i \in \mathcal{I}$
for $i = 1, 2, .., n$ amd $\tau_i \in [0, T]$. Due to the fact that $\tau^{(n)}$ permits less choice,
the resulting optimal payoff function $u|_{\tau^{(n)}}$ associate with $\tau^{(n)}$ attains lower
values than the solution u of (4.8) and the associate continuation set $C^{\tau^{(n)}}$ in-
tersects the events $\{\tau = \tau_i\}$, with $i = 1, 2, ..., n$, on larger sets than does C.
As more elements are added to the sequence $\tau^{(n)}$ of admissible stopping times,
the corresponding $u|_{\tau^{(n)}}$ increases and the set where $C^{\tau^{(n)}}$ intersects the events
$\{\tau = \tau_i\}$ decreases. In this way u and C can be derived as limits of monotone
sequences.

Now let $g^{(n)}$ denote the expected reward function associated with Chernoff's
sequential approximating procedure and $u^{(n)}$ the corresponding optimal payoff
function. If we consider the family of functions satisfying the inequality (4.10),
it turns out that $u^{(n)}$ is characterised by the fact that, among those which
are also bounded below by $\phi(s, x)$, it is uniformly minimal restricted to finite
sequences $\tau^{(n)}$ of admissible stopping times. This is a special case of Snell's
envelope.

By using this characterisation, the payoff $u(s, x)$ can be constructed as
the limit of an increasing sequence of functions $\{u^{(n)}(s, x)\}$, where $u^0(s, x) = \phi(s, x)$ and

$$u^{(n+1)}(s, x) = \operatorname*{ess\,sup}_{\tau^{(n)} \in \mathcal{T}^{(n)}} \mathsf{E}\{u^{(n)}(\tau, X_\tau)|\mathcal{F}_{\tau^{(n-1)}}\}, \qquad \tau_i > s, \; x \in \mathbb{R}^n,$$

where $\mathcal{I}^{(n)} \subseteq \mathcal{I}^n$ and $n > 0$. It is obvious that the greater becomes n the less
will deviate the sequential optimum from the continuous optimum. However,
accuracy has a cost in terms of the additional sampling this entails.

At this point, we know that from a Brownian motion B we may get a solution of the heat equation $\partial_t u = \frac{1}{2} \sum_i D^i D^i u$ in $(0, T) \times G$, where G is a bounded domain in $\mathbb{R}^n$ and τ its first exit time. Setting $M_s = u(t - s, B_s)$ and assuming u solves the heat equation above, it follows from the Itô formula that M is a local martingale in $[0, \tau \wedge t)$ and that it is a bounded martingale if u is also bounded. This, together with the appropriate boundary condition, provides an alternative way of characterising the continuation set as above. Knowing that, the stopping set $\mathcal{S} = \{u = \phi\}$ determines the first boundary condition. In this sense we assume u is continuous in $[0, \infty) \times \bar{G}$, $u|_{\partial G} = \phi$ and $u(t, x) = 0$ for $t > 0$ and $x \in \partial G$. About the function ϕ we assume it is bounded and continuous. Thus, the martingale convergence theorem implies $M_s = \mathsf{E}\{\mathbf{1}_{\{t \leq \tau\}}\phi(B_t)|\mathcal{F}_s\}$, since M is uniformly integrable.

However, since the position of the boundary is unknown, the solution cannot be determined. Thus, adding to the characterisation of the continuation set for $i = 1, 2, \ldots, n$ the relationships

$$D^i u = D^i \phi \quad \text{and} \quad \partial_t u = \partial_t \phi, \tag{4.9}$$

leads to the free boundary problem. Having this characterisation of the continuation set, we have opportunity of introducing a specifying condition with the help of which one may distinguish between minimisation and maximisation. Hence, we add the constraint $u \geq \phi$ which completes our rough description of the stopping problem using the heat equation. There are many details that remain to be done but our outline has merely heuristic and expository purposes. We shall just mention the smoothness assumptions on the boundary are both unjustified and unnecessary; see in this regard [12]. Let us add at this stage that from the point of view of applications the characterisation of the boundary is to some extent a matter of modelling. In passing we should mention, situations allowing for violations of boundary conditions are taken into account with the help of penalisation methods or randomised stoppings, see [231, 128].

For any $t \in [0, T]$, the function u satisfies the general property known as the "supermartingale property" given by the inequality

$$u(s, x) \geq \mathsf{E}\{u(t, X_t)|\mathcal{F}_s\}, \qquad s \leq t, \quad x_t \in \mathbb{R}^n,$$

which means, roughly speaking, that no advantage can be derived by first waiting for a fixed time $t - s$ and then using the optimal policy. On the contrary, failure to switch at time t to the optimal policy entails a reward loss given by

$$u(s, x) - \mathsf{E}\{u(t, X_t)|\mathcal{F}_s\}.$$

Moreover, it has been shown in special cases, that for any well-defined stopping procedure, the expected reward function $g(s, x) = \mathsf{E}\{\phi(t, X_t)|\mathcal{F}_s\}$ is uniformly maximal if and only if

$$g(s, x) \geq \mathsf{E}\{g(t, X_t)|\mathcal{F}_s\}, \qquad s \leq t, \quad x \in \mathbb{R}^n, \tag{4.10}$$

and $g(t, X_t) \geq \phi(t, X_t)$ always. Considering the class of functions g with the property (4.10) which are also bounded from below by $\phi(t, y)$, one may characterise the solution of the optimising problem (4.1) as the function u in this class that is uniformly minimal. This idea is the starting point of a more probabilistic treatment of this problem according to which the solution to the optimisation problem (4.8) is characterised as the smallest regular supermartingale greater than or equal to ϕ.

4.2.2 Free boundary and Stefan's problem

For the sake of convenience, let us assume the process X introduced above satisfies the following stochastic differential equation (SDE)

$$
\begin{aligned}
dX_t^x &= \sigma(t, X_t^x)dB_t + b(t, X_t^x)dt, \qquad t > s, \qquad (4.11) \\
X_s^x &= x \in \mathbb{R}^n,
\end{aligned}
$$

where $B = (B_t, \mathcal{F}_t)_{t \geq 0}$ is an m-dimensional Brownian motion on the probability basis $(\Omega, \mathcal{F}, \mathbf{F}, \mathbf{P})$. Furthermore, we shall assume that for any $t \in [0, T]$ the functions $\sigma(t, y) : [0, T] \times \mathbb{R}_+^n \longrightarrow \mathbb{R}^n \times \mathbb{R}^m$ and $b(t, y) : [0, T] \times \mathbb{R}_+^n \longrightarrow \mathbb{R}^n$ are Borel measurable with $\sigma(t, y) = (\sigma^{(ij)}(t, y))$ and $b(t, y) = (b^{(i)}(t, y))$, where $i = 1, 2, .., n$ and $j = 1, 2, .., m$ and satisfy the conditions associated with (8.13). In such a case the equation (4.11) is said to be (or more precisely to define a diffusion process) of the Markovian type, i.e. it has continuous trajectories which satisfy the Markov property. Note that an equation of the Markovian type reduces to a system of differential equations (a dynamical system) $\dot{X}_t = b(t, X_t^x)$ when $\sigma \equiv 0$. Thus, the notion of SDE above generalises the notion of an ordinary differential equation by adding the effect of random fluctuations.

The generalised Stefan's problem

Let $\mathbf{A}_t$ represent the characteristic operator associated with (4.11) which for the case of a diffusion process of the type under consideration coincides with the differential generator of X; this fact directs attention to the set of functions in the domain of the aforementioned operators. A more general stopping time problem considering explicitly running or observation costs, may then be obtained substituting the relationship

$$
u(s, x) = \sup_{\tau_s^x \in \mathcal{I}^*} \mathbf{E}\left[\phi(\tau_s^x, X_{\tau_s^x}^x) + \int_s^{\tau_s^x} f(t, X_t^x)dt \bigg| \mathcal{F}_s \right], \qquad (4.12)
$$

for (4.8), where the supremum is taken with respect to all admissible stopping times τ_s^x corresponding to the process $X = (X_t^x)_{t \geq s}$, and the integrand is a (negative) measurable right-continuous mapping $f(t, y) : [0, T] \times \mathbb{R}_+^n \longrightarrow \mathbb{R}_-$. At a first approximation, we may think of the system starting at time s from the state x with $\mathbf{P}_x\{X_s = x\} = 1$, and with a reward and cost structure consisting of two components: (1) an integral cost evaluated at the continuation rate $f(t, y)$ as long as the control decision prescribes not to stop and (2) a terminal reward determined by $\phi(t, y)$ evaluated at the moment a trajectory first exits the continuation set.

The characterisation of the continuation set leads to formulation of a problem in which part of the boundary is not given. This part of the boundary which we have specified by the additional boundary conditions (4.9) is then called a free boundary and has to be determined with the solution of the differential system. Stefan's problem has indeed to do with the phenomenon of melting a thin ice block occupying an interval $a \leq x < \infty$, with a possible heat input or heat output, vanishing internal temperature, with heat flux proportional to the rate of melting ($\mathbf{A}_t u \leq 0$) and variable heat capacity. For the Brownian motion $\mathbf{A} = \frac{1}{2}\Delta$, where Δ stands for the Laplace operator, a fact that justifies the term generalised Stefan's problem for the problem to follow.

Under appropriate smoothness conditions on ϕ and f, as well as adequate options to extend continuously the boundary functions in the continuation set and beyond the boundary which guarantees at least two continuous partial derivatives in x and one in t, one can demonstrate the corresponding optimal payoff u solves the generalised Stefan's problem given by

$$\mathbf{A}_t u \;=\; -f, \quad \text{in } G, \qquad \mathbf{P}_x \otimes dt \quad \text{a.s.} \tag{4.13}$$
$$u \;=\; \phi, \quad \text{on } \partial G, \qquad \mathbf{P}_x \otimes dt \quad \text{a.s.,} \tag{4.14}$$

where $G = \{y | \phi(t, y) < u(t, y) < \infty\}$ and $\partial G = \{y | \phi = u\}$. Formally, the eq. (4.13) follows by expanding $u(t+\delta, y)$ according to Taylor's formula, considering the state of the process X associated with a stop between $t + \delta$ and t, and a limit argument. The boundary condition (4.14) is less obvious. The set G is the continuation set as before.

A glance at (4.16) quickly suggests two points. First, staying in the continuation set G decreases the expected terminal reward since any additional unit of time in G increases the sojourn expenses in the amount $f(t, y)$. Hence, there is a *prima facie* of a need to look after the most favourable time of first departure from G. Second, since as long as the process lives in G the prevailing régime of capital accumulation may go ahead without major changes, it is necessary to make sure the continuation rate $f(t, y)$ does not frustrate expectations of a favourable payoff. This suggests the expected value of the infimum of $\phi(t, y)$ and $f(t, y)$ along any trajectory of X need to be greater than $-\infty$.

A quick reminder on excessive functions

Provided $\mathbf{E}\left\{\int_0^{\tau_G} |\nabla_x u(s+t, x+X_t)|^2 dt\right\} < \infty$, the relationship (4.13), and a simple application of Itô's formula give

$$\mathbf{E}\{u(s+\tau, x+X_\tau)|\mathcal{F}_s\} - u(s, x) = \mathbf{E}\left\{\int_0^\tau \mathbf{A}_s u(s+t, x+X_t) dt \Big| \mathcal{F}_s\right\},$$

from which we can conclude $-u$ is an excessive function and it is characterised as the greatest excessive minorant of $-\phi$. Dynkin's approach to stopping time problems rests on the study of *excessive functionals* which are for the Brownian motion precisely those nonnegative functions v for which $\Delta v \leq 0$.

In this connection, let us recall Doob's analysis reveals that a natural way to approach the boundary is along Brownian trajectories. In fact, almost every trajectory starting from the continuation set hits its complement – the stopping set – at a regular point of the (ramified) boundary. Observe that the first

time of departures from G coincides with the first time the process X hits
the set ∂G. A point $x \in \partial G$ is said to be a *regular boundary point* of G
with respect to X if almost all paths starting from x leave G immediately, i.e.
$\mathbf{P}_x\{\tau_G > 0\} = 0$. Otherwise a point $x \in \partial G$ is said to be an *irregular* boundary
point, i.e. $\mathbf{P}_x\{\tau_G > 0\} = 1$. The expression $\mathbf{P}_x\{\tau_G = 0\}$ is called the *probability
of immediate escape*. Considering regular boundary points is aimed at avoiding
pathological behaviour of trajectories so as to make sure that if a trajectory
hits the boundary at a regular point or initiates from such a point then it
leaves it immediately. On the other hand, irregular points provide a reserve of
nontrivial exit times and thus a reserve of reward functions for which one can
consider a nontrivial stopping time.

For the process X above (or for its semigroup $\mathbf{T}$), a positive universally
measurable function g is said to be *excessive* if (1) $\mathbf{E}\{g(X_t)|\mathcal{F}_s\} \leq g(X_s)$ for
any $t \geq s$ and (2) $\lim_{t\downarrow s} \mathbf{E}\{g(X_t)|\mathcal{F}_s\} = g(X_s)$. That means, the information
resulting from the infinitesimal behaviour of the process X in the continua-
tion set G and its limiting behaviour as reflected by the function g is essential looking for a characterisation of the stopping rule, since this information
gives the necessary link with the supermartingale property and provides moreover a right-continuous version of the supermartingale $g(X)$. This observation
accentuates the relevance of excessive functions for stopping problems, a feature which becomes crucial for excessive functions associated with a standard
Markov process.

4.2.3   Supermartingale characterisation of stopping rules

To see the role excessive functions play studying the structure of the optimal
payoff u, we shall consider briefly two particular situations. The first, when the
underlying process X is a general supermartingale and the second when it is a
Markov process. To begin with, let us recall a positive universally measurable
function h which is excessive for the process X is said to be and *excessive
majorant* of the Borel measurable function g if $h(t, x) \geq g(t, x)$ for any $t \in \mathbb{T}$
and $x \in \mathbb{R}$. Furthermore, an excessive majorant of g is said to be the *smallest
excessive majorant* of g if h is less or equal to any excessive majorant of the
function g.

A general supermartingale setting

Let us consider a probability basis $(\Omega, \mathcal{F}, \mathbf{F}, \mathbf{P})$. Further, let us consider a real-
valued function $\tau : \Omega \longrightarrow [0, \infty]$, which viewed as a Markov time with respect
to $\mathbf{F}$ describes a nonanticipative stopping rule. The first hitting time τ_G of
∂G defined before is a Markov time. We shall denote by $\bar{\mathcal{I}}_m$ (respectively $\bar{\mathcal{I}}_m^s$)
all Markov times (respectively all Markov times τ for which $\tau + s \geq s$ holds),
while $\mathcal{I}_m$ (respectively $\mathcal{I}_m^s$) denotes all finite Markov times (respectively all
finite Markov times τ for which $\tau + s \geq s$ holds).

Let us consider a real-valued process $Z = (Z_t)_{t\geq0}$, where $Z_t = \phi(s+t, x+X_t)$
for $t > 0$ and $X_s = s$. We shall assume Z is right-continuous and adapted with
respect to $\mathbf{F}$ and such that $\mathbf{E}\{\sup_{t>0}|Z_t|\} < \infty$. To the end of questions

concerning optimal stopping, we draw attention to Markov times τ in $\mathcal{I}_m$ (respectively $\mathcal{I}_m^s$) representing the class of Markov times at which the process Z eventually attains its maximal value and for which the variable $\mathsf{E}\{Z_\tau\}$ (respectively $\mathsf{E}\{Z_\tau|\mathcal{F}_s\}$) is well-defined. We introduce the following notation,

$$\mathfrak{s}_s = \operatorname*{ess\ sup}_{\tau \in \mathcal{I}_m^s} \mathsf{E}\{Z_\tau|\mathcal{F}_s\}, \quad \mathfrak{m}_s = \sup_{\tau \in \mathcal{I}_m} \mathsf{E}\{Z_\tau\},$$

and $\bar{\mathfrak{s}}$ and $\bar{\mathfrak{m}}$ represent the notion corresponding to Markov times in $\bar{\mathcal{I}}_m^s$ and $\bar{\mathcal{I}}_m$. It is intuitively clear the formulated optimisation problems have trivial solutions for processes Z which are supermartingales.

Theorem 4.1 *The process $\mathfrak{s} = (\mathfrak{s}_t)_{t \geq 0}$ is the smallest supermartingale majorising $Z = (Z_t)_{t \geq 0}$ or equivalently that aggregates the conditional maximal reward. Moreover, for each $\tau \geq 0$ in $\mathcal{I}_m$, respectively $\mathcal{I}_m^s$ we have $\mathfrak{m}_t = \bar{\mathfrak{m}}_t$ and*

$$\mathfrak{s}_s = \bar{\mathfrak{s}}_s \qquad \mathsf{P} \quad a.e.$$

and one can construct an optimal or ϵ-optimal $\hat{\tau}$ at which the relationship

$$\mathfrak{s}_s = \mathsf{E}\{Z_{\hat{\tau}}|\mathcal{F}_s\} \quad or \quad \mathfrak{s}_s \leq \mathsf{E}\{Z_{\hat{\tau}}|\mathcal{F}_s\} + \epsilon$$

holds for any $\epsilon \geq 0$.

Proof: See Fakeev [64] pp. 324–331. ∎

The process $\mathfrak{s}$ is known as *Snell's envelope* of Z. There are a variety of results characterising such processes, some of them working with weaker conditions. Let us look next at the Markovian case.

Some features of the Markovian case

Let P_x be the probability measure associated with a process X starting from the initial point (s, x) and let $X = (X_t, \mathcal{F}_t, \mathsf{P}_x)$ be a standard Markov process. Then, the reward obtained at the stopping time τ can be expressed by a random variable of the type

$$Z_\tau = e^{-\alpha\tau}\Phi_\tau^{s,x}, \tag{4.15}$$

where $\Phi_\tau^{s,x} = e^{-\alpha s}\Phi(x + X_\tau)$ and Φ is a positive measurable function on $\mathbb{R}_+$ and $\alpha \geq 0$. Due to the Markov character of the process Z it shall suffice to concentrate on those functions Φ that render the reward process Z a supermartingale in order to deal with the stopping problem following roughly the philosophy of the general setting sketched above. From potential theory, one knows that for this to be the case Φ ought to be an α-excessive function. Let us make precise under which condition a function is called α-excessive. Let h be a (nearly) Borel function such that $0 \leq h \leq \infty$, then h is said to be α-*superaveraging* iff

$$e^{-\alpha t}\mathsf{T}_t\, h \leq h \qquad \text{for all} \quad t,$$

and it is said to be α-*excessive* iff in addition we have

$$\lim_{t \downarrow 0} \mathsf{T}_t\, h = h,$$

where T_t denotes the semigroup (or transition function) associated with X, i.e.

$$\mathsf{T}_t\, h(x) = \mathsf{E}\{h(X_t^x)|\mathcal{F}_s\} = \mathsf{E}_{s,x}\{h(X_t)\}$$

provided that $X_s = x$. The following lemma ensures that the process Z given by (4.15) builds a supermartingale, whenever g is α-superaveraging.

Lemma 4.2 *If Φ is α-superaveraging and $\Phi(X_t)$ is intergrable for each t, then the process Z given by (4.15) is a supermartingale.*

Proof: See [40, 54, 205]. ∎

In the Markovian setting characterising Snell's envelope attached to the process $(e^{-\alpha t}\Phi(X_t))_{t\geq 0}$ aims at the following complementary issues:

- To establish that Snell's envelope of a reward process taking the form $(e^{-\alpha t}\Phi_t^{s,x})_{t\geq 0}$ is of the form $(e^{-\alpha t}\widehat{\Phi}_t^{s,x})_{t\geq 0}$, where $\widehat{\Phi}^{s,x} = e^{-\alpha s}\widehat{\Phi}$ and the function $\widehat{\Phi}$ is the so-called *reduced of order α* associated with Φ.

- To prove that the function $\widehat{\Phi}$ can be chosen independently of the probability measure P_x associated with the initial point (s, x), i.e. the existence of a version of the Snell's envelope is independent on the initial law.

The following theorem due to Dynkin summarises the corresponding results.

Theorem 4.3 *Let us assume the standard Markov process $X = (X_t, \mathcal{F}_t, P_x)$ starting from an arbitrary initial point $x \in \mathbb{R}$ with $X_s = x$ has continuous or lower semi-continuous trajectories. Then the Snell's envelope of the reward process $Z = (Z_t)_{t\geq 0}$ given by (4.15) takes the form*

$$\mathfrak{s}_t = (e^{-\alpha t}\widehat{\Phi}_t^{s,x})_{t\geq 0}, \qquad t \geq 0,$$

where $\widehat{\Phi}^{s,x} = e^{-\alpha s}\widehat{\Phi}$ and $\widehat{\Phi}$ is the smallest α-excessive function majorising Φ. Moreover, if for all $x \in \mathbb{R}$, we have $\mathsf{E}_{s,x}\{\sup_{t\geq 0} e^{-\alpha t}\Phi_t^{s,x}\} < \infty$, then the hitting time τ^ϵ given by

$$\tau^\epsilon = \inf\{t \geq 0 | e^{-\alpha t}\widehat{\Phi}_t^{s,x} \leq e^{-\alpha t}\Phi_t^{s,x} + \epsilon\}$$

is ϵ-optimal for any $\epsilon \geq 0$. Under further assumptions the first hitting time τ^0 defined as

$$\tau^0 = \inf\{t \geq 0 | x_t \in \partial G\},$$

is the optimal Markov time and $\partial G = \{x_t \in \mathbb{R} \mid \widehat{\Phi}_t^{s,x} = \Phi_t^{s,x}\}$ the stopping set.

Proof : See [231], pp. 232–236 and [64]. ∎

Let us recall the family of excessive functions for a standard Markov process which are well-defined on the continuation set G belong to the domain of the characteristic operator A_t, i.e. the set of all continuous solutions of (4.13) for $f = 0$. Using Bellman's principle of optimality, one shows also that the α-excessive functions belong to the domain of the characteristic operator. One obtains then the inequalitites $\widehat{\Phi} \geq \Phi$ and $\mathsf{A}_t\widehat{\Phi} \leq \alpha\widehat{\Phi}$, and the relationships condensed in the equation $(\mathsf{A}_t\widehat{\Phi} - \alpha\widehat{\Phi})(\widehat{\Phi} - \Phi) = 0$ which together give as detailed view of the stopping time problem; see [20, 59]. In the next subsection we shall attempt to see these issues from a perspective we can apply to our economic problem.

4.2.4 Boundary processes and random time change

In this subsection we shall present a quick description of the mechanisms by means of which a trajectory starting in the (continuation) set G speeds up (respectively slows down) its movement toward the boundary ∂G. To this end we consider an additive functional φ on the interval $[s,t]$ of the process X, see Subsection 4.3.4. Studying $\varphi_t^s = \int_s^t \varrho_r \, dr$, one directs attention first to the differential generator of the process transformed on the basis of acceleration or deceleration of its motion, and second to the characteristics of φ_t^s. The first gives insight into the qualitative behaviour of the underlying motion and the second reflects the impact upon the performance of the system driven by X.

Moreover, exponentials of the additive functional φ of X build multiplicative functionals which may expand or contract the probability structure underlying the process X. Precisely, to these contractions or expansions one attributes the resulting acceleration or deceleration of motion of the process. In particular, we shall consider the multiplicative functional $\exp\{-\varphi_t^s\}$ which is of the contracting type and in its socioeconomic guise plays a central role in the sequel.

Therefore, we collect in this subsection only those notions which are central to grasp the impact of this functional upon the trajectories of the process. Dynkin [54] gives a rigorous and complete presentation of the topics included here.

Some features of additive functionals

Let $(\Omega, \mathcal{F}, \mathbf{F}, \zeta, \mathcal{O}, \mathcal{T}, \mathbf{P}_\mu; \tau \in \bar{I}_m)$ be a controlled system governed by a standard process X of the Markov type; note the control actions are Markov stopping times. We shall concentrate on trajectories starting at time s from the state $X_s = x$ known only probabilistically, i.e. on the basis of the probability measure μ. These trajectories live in the continuation set G and terminate at its boundary ∂G at the first hitting time τ_G which as defined above belongs to $\bar{I}_m$.

For any $t \in [0,\zeta)$ the function φ_t^s takes values in $(\mathbb{R}_+, \mathcal{B}(\mathbb{R}_+))$ and depends only on the evolution of the process X during the time $[s,t]$, and for arbitrary $s \le t \le u$ it satisfies the condition $\varphi_t^s + \varphi_u^t = \varphi_u^s$. Then φ is said to be an *additive functional* of the process X, if for any $B \in \mathcal{B}(\mathbb{R}_+)$ one has $\{\omega : \varphi_t^s(\omega) \in B\} \in \mathcal{F}^0$ and $\{\omega : \varphi_t^s(\omega) \in B, \zeta(\omega) > t\} \in \mathcal{O}_t$.

Let φ be an arbitrary additive functional which is homogeneous and right-continuous up to $\mathbf{P}_\mu$-null sets. If for any initial distribution μ and all $t \ge 0$ one has first $\mathbf{E}_\mu\{\varphi_t^s\} < \infty$ and second $\sup_{x \in \mathbb{R}_+} \mathbf{E}_x\{\varphi_t^s\} < \infty$, then φ is said to be a *W-functional*. Recall $\mathbf{E}_\mu\{\varphi_t^s\} = \int_{\mathbb{R}_+} \mathbf{E}_x\{\varphi_t^s \mid X_s = x_s\}\mu(dx)$. Then the function $f_t(x) = \mathbf{E}\{\varphi_t^s|\mathcal{F}_s\}$ is said to be the *characteristic* of φ.

Furthermore if $f_t(x)$ is finite, it is called a *W-function*, which means it satisfies: (1) For each $t \ge 0$, the function $f_t(w)$ is measurable with respect to $\mathcal{B}(\mathbb{R}_+)$; (2) for $s,t \ge 0$ and $x \in \mathbb{R}_+$ one has $f_s(x) + \mathbf{T}_s f_t(x) = f_{s+t}(x)$, where $\mathbf{T}$ is the semi-group associate with X; (3) $\lim_{t\downarrow 0} f_t(x) = 0$. Note the function $f_t(x)$ is nondecreasing, since for $s,t \ge 0$ it holds $f_{s+t} - f_s(x) = \mathbf{T}_s f_t(x) \ge 0$. Moreover the function $f(x) = \lim_{t\uparrow\infty} f_t(x)$ is well defined and if $f(x) < \infty$ the relationship $f_s(x) = f(x) - \mathbf{T}_s f(x)$ holds. It is not hard to see that $f(x)$

is nonnegative and that (1) $\mathbf{T}_t f(x) \leq f(x)$ for $t \geq 0$ and $x \in \mathbb{R}_+$ and (2) $\lim_{t \downarrow 0} \mathbf{T}_t f(x) = f(x)$ for $x \in \mathbb{R}_+$, which means $f(x)$ is excessive. Let φ be some W-functional and define φ_∞ by $\varphi_\infty^s = \lim_{t \uparrow \infty} \varphi_t^s$, then by the relationship $f(x) = \mathbf{E}\{\varphi_\infty^s | \mathcal{F}_s\}$ one defines an excessive function which is called the *rough characteristic* of φ.

Various types of operations with functionals of this sort and their characteristics have been studied; these operations lead to important classes of transformations of processes. Let us mention in this connection, the functional φ given by $\varphi_t^s = \int_s^t \varrho(X_r) \, dr$ enables the construction of subprocesses of X; they correspond to certain conditional probability distributions. To see this, let $\mathbf{T}$ and $\widetilde{\mathbf{T}}$ denote the semi-group corresponding to the process X and the subprocess $\widetilde{X}$. Further, we define for the Borel set Γ the characteristic $f_t(x)$ by

$$f_t(x) = \widetilde{\mathbf{T}}_t \mathbf{1}_\Gamma(x) = \mathbf{E}\{\mathbf{1}_\Gamma(X_t)e^{-\varphi_t^s} \,|\, X_s = x\}$$

which defines clearly the transition function of the new process. Observe that the multiplicative functional $\exp\{-\varphi_t^s\}$ which appears in the expectation above is of the contracting type; indeed for any $0 \leq s \leq t$ and $x \in \mathbb{R}$, it is almost surely contracting with respect to $\mathbf{P}_x$ and it satisfies the inequality $\mathbf{E}_{s,x}[\exp\{-\varphi_t^s\}] \leq 1$. Thus the characteristic $f_t(x)$ defines uniquely the transition function $\widetilde{P}(t, x, \Gamma)$ of the subprocess $\widetilde{X}$ corresponding to the multiplicative functional $\exp\{-\varphi_s^t\}$. It is easy to see that for any $x \in \mathbb{R}$, $\Gamma \in \mathcal{B}(\mathbb{R})$ and all $t \geq 0$ the inequality $\widetilde{P}(t, x, \Gamma) \leq P(t, x, \Gamma)$ holds, which reflects the action of the multiplicative functional upon the process X.

Another way of examining the effect of the operation of a multiplicative functional on trajectories of the process X, is to see how the infinitesimal operator $\widetilde{\mathbf{A}}$ of the process $\widetilde{X}$ corresponding to $\exp\{-\varphi_s^t\}$ looks like; in passing let us recall that for any standard Markov process X in the framework we are considering its infinitesimal operator, its differential generator and its characteristic operator are equivalent and their domains of definition coincide.

Therefore, it is interesting to know what happens to these operators under the action of a multiplication function. It is not difficult to see the domain of the infinitesimal operators $\mathbf{A}$ and $\widetilde{\mathbf{A}}$ associated with the semigroups $\mathbf{T}$ and $\widetilde{\mathbf{T}}$ coincide and that $\widetilde{\mathbf{A}} = \mathbf{A} - \varrho$. Note this situation is a more general compared with the one treated in the foregoing subsection, since ϱ depends on the state while α was assumed to be constant. Since ϱ is positive-valued, we may conclude that for any u in the domain of definition $\widetilde{\mathbf{A}}u \leq \mathbf{A}u$ holds. Therefore, if the function represents the system performance as before, we can say the effect of the functional $\exp\{-\varphi_s^t\}$ on the process X, is slowing down the rate of change of the payoff. Furthermore, the expansion

$$e^{-\int_s^t \varrho(X_r) dr} = 1 - \varrho(X_s)(t - s) + o(t - s),$$

suggests by analogy to exponential decay that a trajectory starting at time s from the state $X_s = x$ does not terminate during the interval $[s, t]$ with a probability given by $\varrho(X_s)(t - s) - o(t - s)$ independent of the nature of the motion up to the time s. This remark will play an important role in the sequel.

Considering the first exit time τ_G and fixing the random variable ϱ at $X_s = x$, we may set

$$\int_s^{\tau_G} \varrho(x)dr = t,$$

from which one obtains $d\tau_G = \varrho(x)^{-1}dt$, a relation giving the rate at which the motion is accelerated or decelerated due to the action at time t of $e^{-\varphi_t^s}$ on the trajectory of the process starting from s at the state x. Setting $t = \tilde{\tau}_G$ in the foregoing relationship gives a way of comparing the first exit time of X and $\tilde{X}$ as well as their rate of change.

Characteristics of transformed processes

To close this short incursion into the role of additive functionals of processes, we shall mention that under the action of the functional φ the motion of the system in the continuation set becomes to be governed by the equations

$$\begin{aligned}
\mathbf{A}_t u - \varrho u &= -f & \mathbf{P}_x \otimes dt \quad \text{a.s.} \\
\mathbf{A}_t u - \varrho u &= 0 & \mathbf{P}_x \otimes dt \quad \text{a.s.}
\end{aligned}$$

for the inhomogeneous and homogeneous case respectively. On the other hand, as customary all the solutions of the inhomogeneous equation are obtained by adding any solution to all solutions of the corresponding homogeneous equation. Following theorem describes the structure and properties of the (excessive) functions which solve the dynamics governed the evolution of a continuous standard Markov process in the continuation set.

A 4.1 *Assumption on escaping trajectories.*

- For every $x \in G$ there is a neighbourhood $U \subset G$ such that the quantity $\mathbf{E}\{\tau(U) \mid X_s = x\} < \infty$, where $\tau(U)$ is the first exit time from U.

- For every $x \in G$, $\mathbf{P}_x\{\tau_G = \infty\} = 0$, where τ_G is the first exit time from G.

- Let ϕ be an arbitrary continuous function defined on ∂G and define the function $g(s,x) = \mathbf{E}\{\phi(\tau_G, X_{\tau_G}) \mid X_s = x\}$, then for $x \in G$ we assume $\lim_{x \to y} g(s,x) = \phi(s,y)$ with $y \in \partial G$ and s fixed.

Theorem 4.4 *Let X be a continuous standard Markov process and G be a bounded open set for which the assumptions in A 4.1 hold. Further let ϕ and f be defined by (4.24) and satisfy A 4.3 and A 4.5. Then the continuous function g and the bounded continuous function F given by*

$$\begin{aligned}
g(s,x) &= \mathbf{E}\left\{e^{-\varphi_\tau^s}\Phi(X_\tau) \mid X_s = x\right\}, & x \in G \\
F(s,x) &= \mathbf{E}\left\{\int_0^\tau e^{-\varphi_t^s}\psi(X_t)dt \mid X_s = x\right\}, & x \in G
\end{aligned}$$

solve respectively the homogeneous and inhomogeneous equation.

Proof: It is an adaptation of a result by Dynkin, see [54], Vol. 2, pp. 46–48. ∎

Finally, adding solutions of the inhomogeneous case to one of the homogeneous equation gives the characteristic function $u(s,x)$ which represents a more general solution of the inhomogeneous case, i.e. $u(s,x)$ is given by

$$u(s,x) = \mathsf{E}\left\{e^{-\varphi_i^s}\Phi(X_{\tau_G})|X_s = x\right\} + \mathsf{E}\left\{\int_0^\tau e^{-\varphi_i^s}\psi(X_t)dt \mid X_s = x\right\}$$

where ϕ is as before a function defined on the boundary ∂G, and Φ is defined on the continuation set G as above. We should mention here that using excessive functions, additive functionals and the like to construct characteristics is the stochastic counterpart of the method using characteristics to solve PDE's. We refer to Dynkin [54] for a detailed study of this matter.

4.3 Stopping times associate with planning

A thorough analysis of benefits deriving from development and costs it entails necessitates to resort to a careful study of mechanisms underpinning incessant processes of exhaustion of available productive power, its renewal, recreation and expansion. To understand the most basic features of processes along which one uses up potentials available to society and releases acts aimed at replenishing and recreating used up potentials, we shall recur to the model of capital accumulation introduced in Subsection 4.1.4 To this end one makes use of the notion of social supply price of investment and social demand price of capital. The first measures the amount of currrent consumption one has to give up in order to forward one additional unit of investment and the second evaluates social costs associated with that investment policy in terms of the resulting lack of potentials to function, e.g. earning power, consuming power and many other features determining states of being and doing.

Based on concepts related to accounting prices and the adjoint state, we have established various interesting relations among them and derived various equations which provide a powerful framework suitable to the purpose of studying essential links of capital accumulation and social reproduction, see [48, 89]. Moreover, we have obtained a representation for the underlying process that facilitates addressing relevant questions associated with benefits and costs emanating from the development process.

4.3.1 Socioeconomic underpinning of system performance

To the end of casting the central issues of conceptualising benefits and costs deriving from the process of development in a manner amenable to scientific scrutiny, one needs to free these concepts from a wealth of ambiguities susceptible to orthodox rhetoric. Central to our approach is the idea of focusing on the command over characteristics attached to commodities rather than on command over commodities. This is related to the need of distinguishing clearly between purchasing and consuming power, freedom of will and freedom of action, and many other sources of ambiguity we have referred to in Part I.

Benefits versus costs

The process of "given up current consumption" has an own dynamics related to various socioeconomic factors going from production and technological aspects over social relations of production to a variety of issues like human, social maturity, awareness and motivation. On the other hand, the process of "benefits to be derived" from the policy under consideration has too an own dynamics, the effects of which are expected to maintain and/or reestablish productive potential, earning and consuming power, and in this sense to mitigate the social burdens resulting from development policies. It is essential to keep in mind that benefits accruing to individuals release a wealth of interactions, give rise to linkage effects, externalities and the like, e.g. realisation of profits necessitates purchasing power which through acts of consumption generates earning power and so on. Thus social benefits may stimulate *inter alia* interaction, cohesion, social solidarity and higher performance standards, ultimately bringing about the fulfilment of development requirements provided that the benefits overcome the costs.

Grasping the basic mechanisms ruling the logic between flows of benefits and costs, is essential to capture the presumed conflict between current consumption and employment on the one hand, and investment, which means future consumption, and employment on the other. Actually, what we expect to understand is the associate conflict taking place at a higher and more subtle level generated by various responses of individuals to their socioeconomic environment. We shall condense this conflict into the race between the rates of increase of the benefits, emerging as the development policy at stake materialises, and changes in the social price that this policy entails. Stopping techniques may be helpful exploring the appropriate timing and dosification of development policies of the sort mentioned in Subsection 4.1.4 and figuring out their impact on the rate of flows of social benefits and price. In that way, we expect to gain insight into various aspects on which social aspects like motivation and impatience rely.

Impatience

In general impatience relates to the premium on the exchange between present and future states of being and doing and as such is often seen as relying on subjective assessments of the corresponding desirability differential. For the present purpose *impatience* should be interpreted roughly as the percentage excess of present states of functioning getting out of date over anticipation of desirable states of functioning. Obsolescence is a notion reflecting states of being and doing per unit of characteristics prevailing currently in contrast to states of becoming and anticipated doings per unit of future consumption characteristics.

To begin with, let us mention the notion of impatience carries the presumption that present commodity characteristics are preferred and depends heavily on the socioeconomic features of individuals, in particular on entitlements. Let us summarise, following [67], some crucial features of impatience dependence.

- *The size of expected income streams.* In general one may say the smaller the level of income the greater the impatience to reach a higher level of income as early as possible. The pressure of present needs blinds in a rational sense an individual to the needs of the future. To the extend that this pressure follows from the man's desire of keeping alive and maintaining the ability to cope with the future, this influence of impatience can be viewed as rational.

- *The time shape of income streams.* The effects of the time shape of an income stream upon impatience vary greatly according to the size of the income. For instance, at lower income levels slightly increasing income streams may suffice to diminish impatience for present income.

- *Its composition.* That means, to what extent it consists of nourishment, shelter, medical care, education, amusement and so on. Any decrease, that in a way or another decreases the characteristics attached to c influences the capability of the individual to function and acts upon impatience having a similar effect to the one of a diminution of the total income. To the extent that this decrease generates uncertainty, it will increase or decrease impatience, according to the effect obtained in the income.

- *The degree of risk or uncertainty.* Future incomes are always subject to some uncertainty and are for that reason equivalent in a sense to lower incomes. Therefore, whether the impact of uncertainty upon impatience is weakening or enforcing it, it turns out the main effect resulting from the presence of uncertainty in a particular period of time is enhancing the value of a unit of certain income in that particular period.

- *The individual factor.* Impatience varies with individual's own characteristics like foresight, self-control, habits, uncertainty of life, and with the motivation on which she/he is acting like agency achievement, personal well-being or standard of living.

The information available to individuals, as well as their speed of adaptation to changes in their socioeconomic environment, shape impatience in a way that brings about bounds upon their capacity to behave perfectly according to the principle of rationality and may inhibit the process of development. On the other side, motivation and solidarity among other attitudes emanating from a successful implementation of economic policies, including those of capital accumulation and unemployment elimination, may contribute greatly to outweigh the disadvantages of the strategy at issue and enhance the development process.

4.3.2   Economic development as a learning process

At different points in the course of this work learning and evolution emerge as major sources of dynamics and progress. Learning and evolution are indeed processes with deep similarity and constitute therefore the ultimate resource on which human and social development rest. Human intelligence and creativity represent a virtually inexhaustible source of evolutionary power.

Development like learning is a very complex process and remains like learning largely ungrasped. Learning reflects states of interaction with our environment through irreversible links with the brain. Our states of being and doing are manifestations of interaction of individuals with society and nature irreversibly linked to development. Both notions are associated with processes of change, time, memory, adaptation, flexibility, motivation and many others. Thus, we will not try to make the idea precise but just embark on it and attempt to learn by doing. To point out how these factors are taken into account, we shall interpret economic development as a *learning process*. The topic to be learnt is the economic development as a structure with sociological, economic, cultural, technical and political components. The learner is a representative individual enjoying an average level of know-how, experience, training, motivations, aspirations, skills, *inter alia*. In fact representative means here average or better normal in a sense akin to the notion of normal prices or prices of production. However, the ambiguity involved may diminish as one shifts analysis to the framework of many interacting agents; see [6].

The learning process

Then the process goes as follows: The learner chooses a study intensity which depends on his level of ignorance and the degree of difficulty of the topic. The topic is as already suggested the process of economic development, i.e. we suppose that the individual can engage in a developmental activity in varying amounts. The learning assumption entails that the higher the intensity of dedication to the activity, the sooner the degree of difficulty will start decreasing. Nevertheless, to every degree of difficulty there corresponds a study intensity beyond which the study becomes unpleasant. To the extent that the activity remains pleasant the engagement can be intensified; but to the extent that the individual reaches this critical intensity, he will start lowering intensity of the activity in the future. Therefore, we shall face a race between the rate at which the individual learns and the rate at which he changes the study intensity.

To put these ideas in economic terms it would suffice, as a first approximation, to think of the study intensity as the personal necessary consumption $\underline{c}_t$. A high study intensity amounts then to a low level of necessary personal consumption $\underline{c}_t$. This interpretation helps to visualise the advantages, from the point of view of accumulation of capital, as well as the disadvantages of a low $\underline{c}_t$. Coming back to the learning process let us sketch the various alternative that we may get according to how high or low is the chosen level of $\underline{c}_t$.

- Let us think first that the learner chooses a high learning intensity, i.e. a low level of $\underline{c}_t$ because he knows that this fosters the pace of accumulation. The individual learner may give up before the learning process has been completed or may learn so much that a point comes, where study intensities previously considered unpleasant become pleasant. He may even increase his study intensity and hence accelerate the learning process. Let us mention saying that an activity becomes unpleasant simply means there is an alternative that may be regarded as more preferable. Lack of such an activity may cause frustration.

- It may also happen that the individual starts with low study intensities, i.e. high levels of c_t, because the learner knows that the high study intensities are unpleasant at the beginning. Hence, the learning process remains pleasant throughout and the learner may slowly increase the study intensity until he masters the topic.

- As in daily life, it may very well happen that, in a touch of enthusiasm, the individual may commit himself to the development process without realising the difficulties associated with it. When these become evident, he associates with the development process a price measured in terms of individual limitations, social strains and forced loss of traditional behaviour and values. Then, even low learning intensities become unpleasant generating frustration and damaging economic and political decisions. The development process may slow down or even stop.

- But if the development process reaches a point where it is evident that the benefits dominate the disadvantages, then even high intensities, considered as unpleasant before, become pleasant and the economic development process could be intensified enabling the economy to approach its goals at a faster pace.

The foregoing sketch and a careful analysis of the economic implications of the various expenditure alternatives of the social economic surplus point out crucial links between the material, social and individual aspects of the process of development. In the next subsection we shall briefly refer to some of them from a rather macroeconomic perspective, while in Subsection 4.3.4 we shall discuss them from a microeconomic perspective.

Employment and wages versus investments and growth

Let us focus on a policy aimed at stimulating capital accumulation and eliminating unemployment of the sort analysed above that we illustrate diagrammatically in Fig. 4.1. That employment and wages conflict with investment and growth can be easily recognised examining the corresponding effects in the fundamental equation of growth (4.7). Tracing the social costs and benefits associated with the trade-off's between current and future consumption and employment is more difficult. However, having concentrated on the problem of accumulation and allocation of labour a little reflection on the policy issues and the social impact behind the decision of setting the minimum wage rate $\overline{W}$ and determining thereby the level of necessary personal consumption c_t, together with a glance at Fig. 5.1, may help to visualise the situation. We refer to [90, 93, 91] for a detailed presentation of these issues.

Individual contributions to the developmental goals of society are articulated in terms of amounts of current consumption and traditional working habits that are given up. The success in balancing up these sacrifices is intimately connected to the extent that society manages to persuade or induce individuals to actively support policies conducive to full employment and to the extent that these policies materialise. Thus, it is safe to say that pleasantness in-

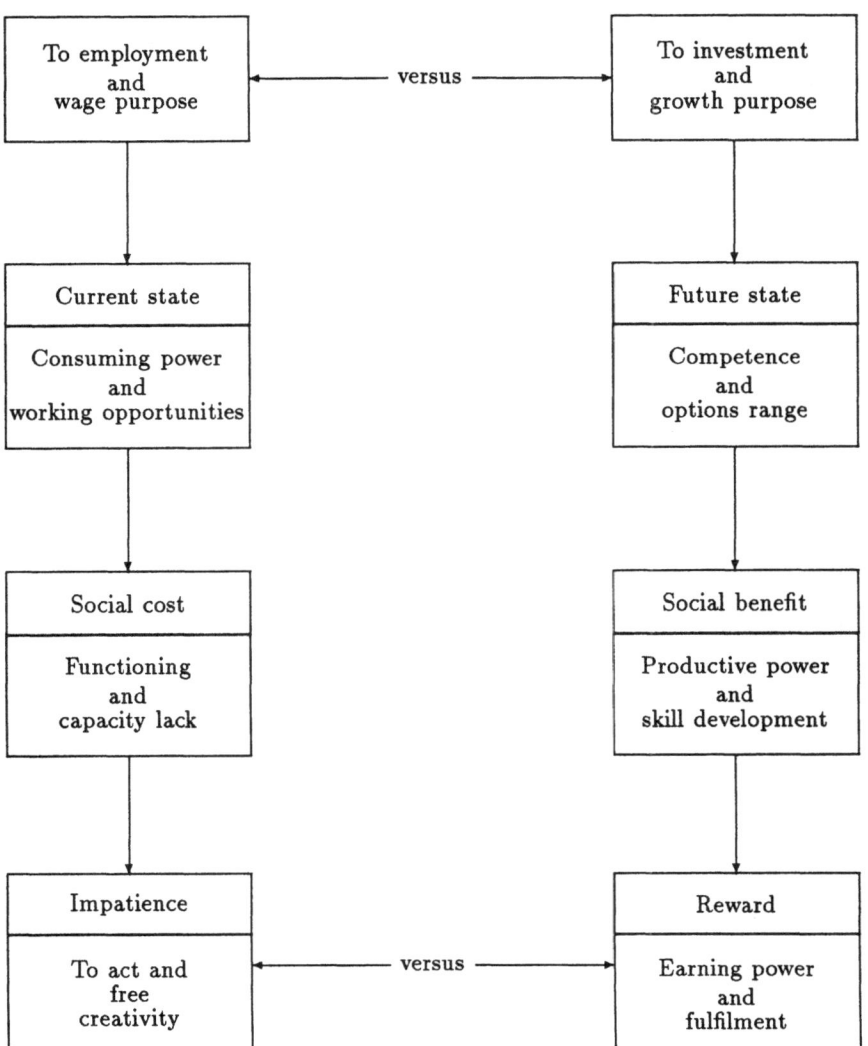

Figure 4.1: Conflict associated with resource allocation

creases with the degree of inducement and decreases with the amount of the contribution.

However, to really understand the essence of the presumed conflict associated with resource allocation one needs to see society at large. One has to consider society interacting with nature, available potentials and acts of depletion and renewal of resources. The most simple way of approaching states of being and doing is to relate them to productive potentials and to understand any act of production is inevitably an act of consumption. This crucial reflection is the heart of social reproduction. Once so much is understood, we should attempt to describe the characteristics of individuals and means of production in connection to productive power. Here again any act leading to exhaustion of productive power has to be accompanied by its corresponding renewal or replenishing of productive power.

Putting these notions in terms of characteristics directs attention to states of functioning which are related to training, working skill, know-how and many other characteristics underpinning human learning. In this sense decisions of investment direct resources to purposes of maintaining productive power and to expanding it. At the right-hand side of Fig. 4.1 one sees those features of investment which one may visualise as stimulating productive power and giving rise to motivation. At the left-hand side of Fig. 4.1 one recognises various issues related with current consumption akin to maintaining working power. In this connection it is crucial to recall the fact that humankind's greatest source of social change is the process of human learning as the major component of creativity, flexibility and adaptability.

Thus, decisions of employment are directed to freeing creativity and blocked productive potentials. These are related to the search for knowledge, know-how and training opportunities. The conflict may consist in blocking human creativity and frustrating the human drive to learn. To ensure social and individual potentials are continuously renewed, it is necessary to realise self-sustaining mechanisms are critically linked to humankind's drive for freedom of action as opposed to barely freedom of will.

4.3.3 Free boundary and stopping rules

The purpose of the present section is to give a heuristic description of the mathematical model corresponding to the learning process outlined above. We shall use the same notation of Section 4.2.1 according to which the evolution of the system is described by a stochastic process $X = (X_t^x)_{t \geq 0}$. However, identifying the process X with the process $k = (k_t)_{t \geq 0}$ introduced in Section 4.1.4 may help us to visualise the issues we shall deal with next.

A 4.2 *Assumptions on f and ϕ*

- Let the function $f : Q \longrightarrow \mathbb{R}_-$ be nonnegative measurable and the function $\phi : Q \longrightarrow \mathbb{R}_-$ be nonpositive measurable and strict concave.

- The functions f and ϕ are sufficiently smooth, i.e. they have two continuous partial derivatives in x and one in t.

- $Q = (0,T) \times G$ is a bounded domain and G a bounded open set in $\mathbb{R}^n$, with regular boundaries denoted respectively by ∂Q and ∂G. The function ϕ is continuous on ∂G and on G the function f satisfies a Hölder condition.

- The variable τ_s^x represents the first exit time from the interior of G. As admissible are viewed those stopping times τ_s^x for which the quantity (4.16) is well-defined, i.e. $\phi(\tau^x, X_{\tau^x}^x)$ is integrable with respect to $\mathbf{P}_x$ and $\mathbf{E}_{s,x}\{\tau_s^x\} < \infty$ for any trajectory starting at time s from the state $x \in G$.

The foregoing assumptions allow for application of Itô's rule; see [54] Chs. 5 and 13 as well as [20, 75] Ch. 3 and Ch. 16 respectively. These assumptions enable also an intuitive discussion of the issues under consideration, however one can do with weaker assumptions; see [59].

Furthermore, let us assume that $\phi(t, X_t^x)$ measures the flow of benefits deriving from the development process as perceived by the representative individual at time t, when the state of affairs is given by X_t^x. Moreover, let $f(t, X_t^x)$ represent the rate of flow of social costs that at time s one anticipates to hold at time t. As a first approximation, one may think of the rate of flow of consumption reward to be foregone, reflecting in a sense the dissatisfaction associated with the lack of current consumption, that furthering the development undertaking entails; see in this connection Subsection 4.1.4. Thus the performance functional can be defined by

$$\mathbf{E}\left\{\phi(\tau_s^x, X_{\tau_s^x}^x) + \int_s^{\tau_s^x} f(t, X_t^x)dt \Big| \mathcal{F}_s\right\}, \qquad (4.16)$$

where the dynamics is as in(4.11). The second term in (4.16) is the flow of social costs and interacts somehow with inducement and impatience, while the first shall reflect current motivation and rewards; in the next section we will come closer to the way of understanding impatience and motivation. Summing up, the performance functional reveals the capacity of the system to actively contribute to the unfolding of its productive forces.

Then, we are interested to find the stopping time τ_s^x among those that allow the expectation operator in (4.16) to be well defined, and such that the relationship (4.17) holds.

$$u(s,x) = \operatorname*{ess\,sup}_{\tau_s^x} \mathbf{E}\left[\phi(\tau_s^x, X_{\tau_s^x}^x) + \int_s^{\tau_s^x} f(t, X_t^x)dt | \mathcal{F}_s\right]. \qquad (4.17)$$

Theorem 4.5 *The assumptions in A 4.2 hold. Let X be a standard Markov process defined on $(\mathbb{R}^n, \mathcal{B}(\mathbb{R}^n))$ and let $\mathbf{A}$ be its characteristic operator. Assume moreover the behaviour of X is governed by the probability $\mathbf{P}_\mu$ for any arbitrary but fixed initial probability law μ. Then the stopping time $\hat{\tau}_s^x$ at which the performance functional (4.16) attains an optimal value as defined by (4.17) is uniquely determined up to $\mathbf{P}_\mu$-null sets by the relationships (4.20) and (4.21).*

Sketch of the proof: Consider an arbitrary but fixed $x \in G$. Then, consider the functional $u(t, X_t)$ with $t \in [s, \tau_s^x)$ where τ_s^x is any arbitrary admissible stopping time. The next step is to show Dynkin's formula holds, i.e. that we have

$$\mathsf{E}_{s,x}\{u(\tau_s^x, X_{\tau_s^x})\} - u(s,x) = \mathsf{E}_{x,x}\left\{\int_s^{\tau_s^x} \mathsf{A}_t u(t, X_t)\, dt\right\}.$$

This step can be visualised applying Itô's formula to the functional $u(t, X_t)$ from $t = s$ to $t = \tau_s^x$, a step we do under the assumption u satisfies the same smoothness requirements as f and ϕ. Then one can easily rewrite (4.17) in a form equivalent to the well-known Bellman equation which is the starting point of various analytical methods to solve the stopping problem. That means, we get then the equation

$$0 = \sup_{\tau_s^x} \mathsf{E}\left\{\phi(\tau_s^x, X_{\tau_s^x}) - u(\tau_s^x, X_{\tau_s^x}) + \int_s^{\tau_s^x} \mathsf{A}_t\, u(t, X_t^x) + f(t, X_t^x)]dt\Big|\mathcal{F}_s\right\},$$

$$(4.18)$$

where A_t is given by the relationship

$$\mathsf{A}_t = \frac{\partial}{\partial t} + \mathsf{L}_t, \tag{4.19}$$

and L_t stands for the differential operator associated with eq. (4.11). Taking into account the boundary conditions associated with (4.17), we get from (4.18)

$$\mathsf{A}_t u \;=\; -f, \quad \text{in } G, \qquad \mathsf{P}_x \otimes dt \quad \text{a.s.} \tag{4.20}$$
$$u \;=\; \phi, \quad \text{on } \partial G, \qquad \mathsf{P}_x \otimes dt \quad \text{a.s.} \tag{4.21}$$

where $G = \{y|\phi(t,y) < u(t,y) < \infty\}$ and $\partial G = \{y|\phi = u\}$. It is necessary to point out that the more difficult part of the proof is obtaining the properties of measurability and smoothness of the payoff u; in this connection we refer to [59]. ∎

The foregoing theorem justifies our reflections on the stopping problem in Subsection 4.2.1. Thus our next objective is to identify the most adequate interpretation of the various ingredients of the stopping problem and to transfer the structure underlying the performance functional and infinitesimal behaviour of the process X into sensible socioeconomic features.

Summing up, the value function u can be legitimately conceived as the solution of the generalised Stefan's problem, where G is the *continuation set* as introduced by Chernoff. It says, that as long as the process lives in G, the best strategy is not to stop. The reason is that, since the terminal reward function ϕ is in that set always less than the payoff u, stopping in G cannot be optimal. On the other hand, as soon as the process hits the boundary ∂G, it is optimal to stop. This strategy is thus called a *stopping rule*.

For a full treatment of this problem and further bibliography we refer to [205]. [20, 21] deal with stopping time problems using variational inequality methods. From a mathematical point of view, we have to characterise f and ϕ, so that u belongs to the domain of the operator A_t and solves (4.20). From a computational and practical point of view, it is of great importance to characterise G and ∂G, so that they depend only upon the drift and diffusion coefficients of (4.11), and the initial conditions. For more details we refer to [86, 91].

4.3.4 A few aspects of microeconomic nature

The most crucial issues in planning problems are related to investment decisions and thus entail postponement of current consumption. These may have consequences, say benefits and costs, extended over a relatively long period of time and have to be evaluated in a way reflecting the relative advantage or disadvantage of the moment at which they materialise.

General assumptions

In this section we shall attempt to capture the most fundamental features underlining the evaluation of benefits and costs deriving from such an act. Let us concentrate for the time being in the following points:

- Features concerning the time horizon. Let us say, we consider an infinite time horizon, which splits in a finite period of time $[0, T]$ and the rest.

- Concerning the reward function to be used, let us state that its basic properties are separability and homogeneity which implies the constant-elasticity form. See [89, 91].

- On time preferences, we shall assume that a sort of positive time preference prevails. Mainly due to the expected level of income, uncertainty in the length of life and personal maturity.

The evaluation of an action, the consequence of which should come to bear in the future, turns out to be sensitive to the subjective rate of discount being used. Conventionally, the social (subjective) rate of discount in dynamic economics is understood as the path of the relative rate of change of the weights assigned by society, i.e. by its representatives, or thought to be assigned by it, to the expected flow level of social benefits.

In [89], we have derived the consumption rate of discount and proved relying on an intertemporal consistency requirement, resulting from the application of the Stochastic Maximum Principle, that the former rate equates the consumption rate of return; see also [87, 88]. Furthermore, the optimal policy calls forth the elimination over the long run of the labour-surplus and the premium attached to investments furthering directly or indirectly employment. These facts entail future rewards of consumption higher than present rewards, diminishing accounting prices of capital and that the consumption rate of return be less than the own rate of return.

The basic idea underlying the corresponding working framework is that of accompanying any increase in the level of employment with shifts of claims on present to claims on future consumption precisely in the amount that wages exceed the contribution to the social output attributable to the additional labourers. The basic idea mentioned above motivates us to shift to thinking in terms of consumption characteristics or better deficit of them. That is why we assume nonpositiveness of the functions describing social benefits and costs which makes a criterion of the form "the less negative the better" the prevailing rationale. However, this is not at all the point, where we would like to stop our search for a meaningful performance criterion. Let us mention shortly that behind the performance functional (4.16), we expect to have processes reflecting various changes around the standard of living in a sense that low levels of it are associated *inter alia* with deprivation of decent housing, adequate food, basic medical care and schooling and as a result of it undermine the unfolding of the functionings and capabilities of individuals and in turn of society, see [202].

Survival motivation and personal maturity

Using the notation of Section 4.1, let us consider the process $k = (k_t)_{t \geq 0}$ of capital accumulation given by (4.6) and the associated process of consumption per available worker $c = (c_t)_{t \geq 0}$ related to the former by (4.7). Let ϑ_t^s denote the conditional probability of survival until the time t of the representative individual under consideration, given that she (or he) was alive at the time s. Further, let the random variable ζ denote the terminal life time of the (representative) individual considered. Thus, the variable ϑ_t^s describes the conditional probability of the event $\{\zeta > t\}$ and this is connected via conditional distributions with additive functionals of the basic process k.

On the other hand, the probability of being alive at time t depends on the probability of dying at each point in time between the reference time s and the future time t. But the latter is the product of survival until the time t, which has been denoted by ϑ_t^s, and the probability of dying if alive. Therefore, it is not difficult to see that the relationships

$$\dot{\vartheta}_t^s = -\vartheta_t^s (d_t + \xi_t) \quad \text{and} \quad \vartheta_t^s = \exp\left\{ -\int_s^t (d_r + \xi_r) dr \right\} \qquad (4.22)$$

hold, where d_r and ξ_r reflect respectively the probability of dying at any time r independent of the level of consumption and dependent on the individual's level of consumption $c = (c_r)_{s \leq r \leq T}$. More precisely, we should have $\vartheta_t^s = \vartheta^s(k_t)$ and $\xi_r = \xi(k_r)$. However, recall that c_t itself depends on k_t, see (4.7). Further, we shall assume that the parameter functions $d(\cdot)$ and $\xi(\cdot)$ are known and that ξ_r increases as c_r decreases which integrates statistical information on various nutritional aspects and concerning demand for food, housing, clothing as well as medical and educational services providing thereby a powerful theoretical framework that reflects differential issues of intertemporal microeconomic behaviour, like the survival motivation at low income levels and like the profit motivation at high income levels.

4.4   Stochastic waves and inherent issues

In the course of the present chapter we have become acquainted with the main tools of the theory of stopping control. We know for instance, the process X approaches the boundary at speed rate determined by its infinitesimal operator

A; this is in fact linked at a basic level with the fundamental theorem of calculus which in the framework of Markov processes takes the guise of Dynkin's formula. Similarly, we have learnt the effect of an additive functional φ acting on the process X is a change in the infinitesimal operator which amounts to speeding up or slowing down the pace of the process toward the boundary. From a slightly different perspective we learnt about other operations – random change of time is an example – which lead to a sort of expansion of the domain where the process lives up to its exit. The objective of this section is to use this knowledge to put our intuitive picture of development as a learning process on a firm footing.

Under certain circumstances a subprocess $\widetilde{X}$ generated by the action of a measurable functional ϕ upon the strong Markov process $X = (X_t, \mathcal{F}_t, \theta_t, \mathbf{P}_x)_{t \in \mathbb{T}}$ can be termed a stochastic wave. Indeed, for the subprocess $\widetilde{X}$ to be a stochastic wave corresponding to X and ϕ, it suffices first that $\widetilde{\mathcal{F}}_t = \mathcal{F}_{\tau_t}$, $\widetilde{X}_t = X_{\tau_t}$ and $\widetilde{\theta}_t = \theta_{\tau_t}$ for a finite random change of time represented by τ_t for all $t \in \mathbb{T}$ hold with $\tau_0 = s$ $\mathbf{P}_x$ a.s. for all x, and second that $\phi(X_t)$ is continuous in t. Here θ denotes the shift operator.

As we already mentioned a consequence of our structural setting is that when considering the n-dimensional state X, we may assume the components $X^{(i)}$, $i = 1, 2, \ldots, n$ are independent Markov processes which do not need to be observed at the same time; this feature is crucial regarding our interpretation of X as a representation of a development process. Two remarks are appropriate: (1) in this connection we centre attention on the system performance – as a functional of the state – rather than on the states themselves; (2) we examine on the basis of multiplicative functionals the impact upon the boundary associated with changes in stochastic waves. However, our analysis remains at an intuitive level leaving for a later point a detailed study of these issues.

According to our general assumption, the representative individual shall value higher a given level of (necessary) personal consumption today than the same level sometime in the future. This behavioural assumption is justified on the grounds of uncertain longevity, which is given by (4.22), and on the level of individual immaturity. That means, there are two main reasons why the representative individual may value future benefits and costs differently than current ones leading thereby to discounting due to uncertain longevity and to discounting due to the level of immaturity.

The discount factor corresponding to the level of immaturity would be given by $\exp\{-\int_s^t m_r \, dr\}$, where m_r stands for the rate of discount due to personal immaturity at time r of the representative individual. It might be preferable to think of maturity, for the sake of positive thinking, what would entail reversing the direction on which the weights apply, i.e. the more mature the larger the weights by means of which to discount future consumption. It goes without saying that this is a very difficult concept intimately related, at the micro-level, to the prevailing stage of socioeconomic development. In this sense it changes like development rather slowly. However, since we have decided to concentrate on lack or deficit of income and necessary personal consumption rather than on the availability of them, it becomes obvious our weights, in a general sense, ought to be growing rather than discounting factors. Further, due to the fact that our prevailing rationale is of the type "the less negative the better", it follows that future deficits corresponding to discounted future income, which are shrunken by means of the discounting factor or the positive preference effect, are to be viewed today as more negative or larger deficits.

4.4.1   Towards an adequate rate of discount

Drawing on ideas forwarded in [232], we shall in the present section derive or better model the subjective rate of discount, a notion fundamental to the understanding of the formation of expectations concerning the flows of future social benefits and costs resulting from the process of development. From the outset we shall make clear that calling our rate of discount subjective refers to the fact that it reflects individual perceptions and anticipations formed at time s expected to materialise at time t. However, perceptions and anticipations formed at time s are outgrowths of the state of affairs represented by $X_s = x$, and this is condensed in the corresponding values of the processes d, ξ and m which we have already introduced; see (4.22) in the paragraphs about it. Thus, our subjective rate of discount rests on a solid ground and has no resemblance at all with the obscure neoclassical notion which bears the same name.

   For the time being, we shall consider individual benefits rather than benefits accruing to the society as a whole. More precisely, we shall consider the standard concept of the representative individual and refer to [5] for a more general treatment based on Gibbs potentials defined on a system consisting of many interacting agents.

Probabilistic meaning of discounting weights

The factors defined in the foregoing section correspond to expectations formed by the individual at the time s about the state of affairs to prevail at the time t. Putting both factors together results in a global discounting factor of the type

$$\exp\left\{-\int_s^t [(d_r + \xi_r) + m_r]dr\right\}$$

depending on the process k as stated before. To save notation, let ϱ_r denote the integrand in the expression above, i.e. $\varrho_t = d_t + \xi_t + m_t$ for any t, and $\varphi_t^s = \int_s^t \varrho_r\, dr$. Here again ϱ_t and φ_t^s are assumed to be parametric functions given by the relationships $\varrho_r = \varrho^s(k_r)$ and $\varphi_t^s = \varphi^s(k_t)$, where the upper index s points that the underlying expectations are formed at the time s.

   With these notational simplifications we define the factors Λ_t^s and D_t^s by means of the following expressions

$$\Lambda_t^s = \exp\{-\varphi_t^s\} \quad \text{and} \quad D_t^s = 1 - \exp\{-\varphi_t^s\} \tag{4.23}$$

with the usual convention that $\Lambda^s = \Lambda^s(k_t)$ and $D_t^s = D^s(k_t)$ for any t and any s fixed but arbitrary. We shall call Λ_t^s and D_t^s respectively the *subjective discounting factors*, the first is due to *motivation* and the second is due to *impatience*. The process $\varrho = (\varrho_r)_{s \le r \le T}$ shall be termed the *subjective rate of discount*.

Loosely speaking, we may expect motivation decreases and impatience increases with uncertainty; uncertainty should be seen in relation with the probability density or rate of change in the probability measure of an event. High values of Λ_t^s are associated with low levels of motivation and vice versa, a statement that will become clearer once we give a way of interpreting the multiplicative functional $\exp\{-\varphi_t^s\}$.

Alternatively, one can as well set $\Lambda_t^s = {}_*P_K(t)\exp\{-\varphi_t^s\}$, where ${}_*P_K(t)$ stands for the accounting price of capital at time t and measures the amount of consumption to be given up in order to forward one more unit of investment. This definition is linked to questions of choice of techniques and causes Λ_t^s to decrease with ${}_*P_K(t)$. Let us recall that ${}_*P_K(t) > 1$, at any time $t \in [0, T]$, as long as unemployment prevails and approaches the nominal price of unity either as unemployment becomes technological rather than institutional or full-employment results and labour becomes as capital scarce.

A 4.3 *Assumptions on the discounting weights*

- The process $\varrho = (\varrho_r)_{s \leq r \leq T}$ is characterised for any time r by the function ϱ_r which is nonnegative, bounded, right-continuous in the continuation set.

- The processes $d = (d_r)_{s \leq r \leq T}$, $\xi = (\xi_r)_{s \leq r \leq T}$ and $m = (m_r)_{s \leq r \leq T}$ are assumed to be given by the state variables which are for any r nonnegative, bounded, right-continuous in the continuation set.

A glance at the discounting factor Λ_t^s in (4.23) suggests that the motivation, solidarity and enthusiasm attached to the development process dissipates at the rate $\varrho = (\varrho_r)_{s \leq r \leq t}$, since the probability that the representative individual survives until the time t, given that the individual was alive at the time s, is described by the expression ϑ_t^s and since the factor $\exp\{-\int_s^t m_r \, dr\}$ due to the personal immaturity induces positive time-preferences. On the other hand, the factor D_t^s in (4.23) suggests also that impatience grows at the rate ρ, due to a great extent to the fact that the uncertainty implicit in ρ operates, in the eyes of the representative individual, as a virtual empoverishment and as a shortening of life; both effects are known to increase impatience.

Relevance of discounting factors

Before closing this section on subjective discountings we shall call attention to the microscopic nature of the issues on which the subjective rate of discount $\rho = (\rho_t)_{t \geq 0}$ relies. A central outgrowth of the capability approach is providing a quantitative setting for the assessment of the ability of the representative individual and its socioeconomic environment to transform various characteristics of the necessary personal consumption bundle into their corresponding functioning, in particular those responsible for longevity, health and maturity. We shall call this property the factor of characteristic conversion.

Thus taking into account that living conditions are states of existence and functionings the reflections of various aspects of them, we like to think of the parameters involved in ρ as the quantitative indicators of the individual factor of characteristic conversion.

In this vein one may interpret the multiplicative functional $\exp\{-\varphi_s^t\}$ in a manner which is fairly natural. To this end, we need to transfer the probabilistic reflections on the survival of a representative individual behind (4.22) to probabilistic reflections on the survival of her/his characteristics. Survival of characteristics should be understood as the event according to which the characteristics under consideration remain updated at the time t, given that at the time s they were updated. Saying that certain characteristics are updated means, the states of functioning to which they give rise guarantee their owner an employment opportunity and with it the corresponding command over commodities. From this point of view we obtain following interpretations which we summarise as assumptions.

A 4.4 *Probabilistic interpretations underlying weights*

- φ_t^s: updating potential and $-\varphi_t^s$: lack of updating potential.

- $e^{-\varphi_t^s}$: probability of outdating states of functioning; that means, the probability that states of functioning which at time s are up to date, turn out to be also up to date, according to the standards anticipated to prevail by time t. Note this is a decreasing quantity less than or equal to unity.

- $1 - e^{-\varphi_t^s}$: probability of updating states of functioning; that means, the probability that states of functioning which at time s are out of date, become up to date according to the standards anticipated to prevail by time t. Note this is an increasing quantity less than or equal to unity.

- ϱ_s: the probability of getting an employment at time s, given that underlying states of functioning are out of date; that means, the rate of change in the probability of updating states of functioning.

Thus, the factors D_t^s and Λ_t^s, besides providing some information at time s concerning the probability of survival on various grounds and the readiness to postpone rewards to the future time t, offer us a link of interaction between policy making and its impacts upon individuals. That is to say, the parameters involved in ρ together with our understanding on impatience and motivation, as introduced above, allow us to assess the state of impatience resulting from the economic policy at stake.

As a way of example, let us mention that a high rate of discount m associated with a low level of personal maturity of the representative individual brings about a strong need for present gratification, i.e. high degree of impatience due to the individual's weakness in the ability to anticipate the future. Furthermore, a high uncertainty of life of the representative individual, due for instance to undernourishment and a precarious medical service, brings about a virtual impoverishment by shortening the expected length of life and tends to increase impatience.

A further aspect that we shall emphasise is that Λ_t^s and D_t^s, as just explained, bear various personal characteristics concerning the (representative) individual at stake which are carried over to the performance functional (4.16). This is a property that contrasts greatly with standard utility functions used in orthodox economics that hold likewise for any individual and at any time. However, the factors Λ_t^s and D_t^s describe mainly subjective features of individual's functionings. Therefore, in the next section, we shall briefly deal with some of the objective aspects attached to these issues. At a fairly intuitive level, we may interpret Λ_t^s as the probability of keeping up to the prevailing standards and D_t^s as the probability of catching up. These intuitive interpretations reveal more closely the relation of Λ_t^s with motivation and D_t^s with impatience.

4.4.2 Learning benefits against burdens

The purpose of this section is to put in a more precise setting the learning process we outlined above and to point out some quantitative and qualitative questions underlying the interplay of development and learning.

Characterising social benefits and costs

As a first approximation, one may think of Φ and ψ as a function of the current value of the process $c = (c_t)_{t\geq0}$ representing (necessary) personal consumption. However, we are not interested in assessing social benefits and costs according to the commodity-command of the (representative) individual but rather on the ability of the individual, its socioeconomic and institutional environment to transform commodity-characteristics into personal achievements of functionings ("states of doing and being").

For that reason, in the notion of (necessary) personal consumption we shall see a bundle of consumption goods and consider that the characteristics of these goods are what the process c indeed represents. Hence, the process c shall convey among others information on individual patterns of nourishment, shelter, housing, schooling, medical care, amusements. Furthermore, since we stress social reproduction, the level and composition of these characteristics should be in a way to ensure, in a broad sense, the renewal of the labour power in a way that the individual keeps on enjoying these.

A 4.5 *Assumption on benefits and costs*

- Let the measurable functions $\Phi : \mathbb{R}_+^n \longrightarrow \mathbb{R}_-$ and $\psi : \mathbb{R}_+^n \longrightarrow \mathbb{R}_-$ be nonnegative and strictly increasing in any of the components of the current value of the state process. They are explicitly defined by (4.24), where Φ is defined on the boundary ∂G and ψ on the continuation set G.

Now, we are prepared to put the performance functional (4.16) in a more precise setting. Thus, let us assume that ϕ and f can be represented as the product of a function of time and a function of the state. Namely, we shall assume that for any $t \in [0, T]$ and any $y \in \mathbb{R}_+^n$, the relationships

$$\phi(t, y) = \Lambda_t^s \cdot \Phi(y) \quad \text{and} \quad f(t, y) = D_t^s \cdot \psi(y) \tag{4.24}$$

hold, by means of which we shall represent any kind of utility-like index deriving from the development process, i.e. representing social benefits and costs as perceived by the representative individual. In (4.24), we view the subjective discounting factor due to impatience D_t^s and the factor Λ_t^s due to motivation as functions of time t.

Hence, taking into account (4.24) the performance function (4.16) can now be given the following more enlightening expression

$$\mathsf{E}\left[\Lambda_{\tau_s^x}^s \cdot \Phi(X_{\tau_s^x}^x) + \int_s^{\tau_s^x} D_t^s \cdot \psi(X_t^x)dt \,\Big|\, \mathcal{F}_s\right], \qquad (4.25)$$

where the function ψ measures deficiency of functionings attributed to a low level of $\underline{c}$ resulting from policy making purposes, while the function Φ measures on the other hand the improvement in the choice set of alternative functionings. The factors Λ_t^s and D_t^s bring in the subjective aspects of performance in the sense that they reflect the impact of development upon individuals, and the way they perceive and measure observed reality. Thus D_t^s may reinforce the existing deficiencies of functionings, while Λ_t^s may strengthen the capability improvement.

Qualitative features of the impact of uncertainty upon impatience

Finally, we shall analyse the impact upon benefits and costs as well as various interactions, call forth by an economic policy based on the choice of a learning intensity, or in other terms by setting personal consumption at a low level, as required for policy making purposes of accumulation.

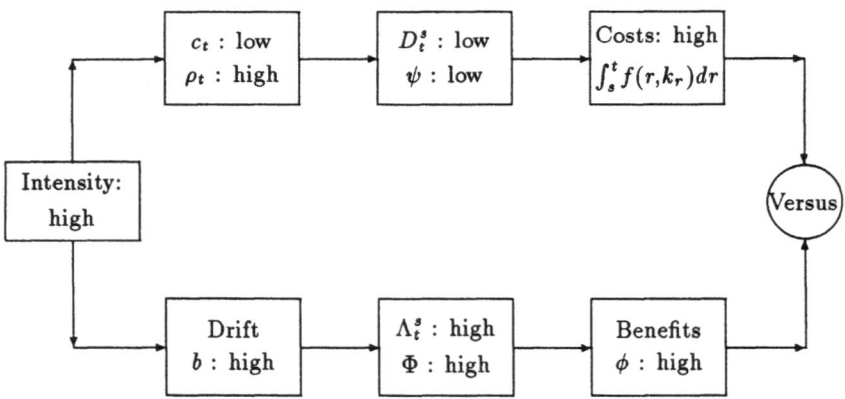

Figure 4.2: Interplay between benefits and costs

To be able to do that, in a transparent fashion, we need some simplifying assumptions and specification of the socioeconomic framework. To begin with, let us assume our economic system is of the type referred to in Section 4.1.4, for more details see [47, 89, 91]. Further, we assume that there is no delay between actions and the appearance of corresponding effects. That means, lowering the level of $\underline{c}$ comes to bear immediately upon the state of existence of individuals. Similarly, the increasing of investments attached to the foregoing action is assumed to materialise at once in a higher pace of accumulation. However, it is necessary to account for the potential anomalies mentioned in Section 4.1.4.

Last, we assume that the changes involved in φ_t^s do not exhibit any pathologies and that φ_t^s as well as ϱ_t^s are always strictly positive. In Fig. 4.2, we depict the main features of the flows of benefits and costs that we shall explain next.

For explanatory purposes let us assume there exists a government that acts relying on the decisions of individuals made in the light of their own self-interest; this interest is in turn appropriately directed by some system of incentives, social preferences and value judgements. Then, in accordance to the commitments resulting from the electoral contest which rised the ruling party to power, we assume a planning board, acting on behalf of the authority just mentioned, sets for policy making purposes various institutional constraints aimed at furthering capital accumulation, that means, shifting of claims on consumption from the present to the future, and gives binding directives concerning investments out of profits. That corresponds typically to the case of a high learning intensity or low level of consumption $\underline{c}$. Further, let us recall that the system at hand starts from low levels of capital k.

Then, it is fair to conclude, under the circumstances described above and taking into consideration our definition of the subjective rate of discount ϱ, that as the uncertainty behind $1-\exp\{-\varphi_t^s\}$ increases a higher degree of impatience comes about, as suggested by the low values of D_t^s. On the other hand, due to a low level of $\underline{c}$, it follows a higher degree of deficiences in the state of existence of individuals bringing about higher deficits in their functionings that correspond to low levels of ψ. Therefore, the flow of social costs becomes more negative as the flow at the top of the diagram in Fig. 4.2 indicates.

A precise specification and evaluation of the social costs attached to low levels of necessary personal consumption is a very difficult matter and need a thorough study of the interaction of the components of ϱ, a deep understanding of impatience, uncertain longevity and a more structured view of functionings of individuals and adequate assessing methods.

Let us have a look at the flow of benefits resulting from the terminal reward, going along the flow at the bottom of the diagram in Fig. 4.2. A high intensity, as we already know, amounts to low levels of consumption and thus to a higher commitment of the social output to aims of investments. As a consequence, a larger pace of accumulation comes about, as pointed out by $\dot{k}$ in (4.7), resulting in higher levels of future social output and a bigger capability set for individuals and a high flow of benefits. On the other hand, associated with a positive drift, i.e. $\dot{k} > 0$, there are shifts of claims on present to future consumption which is embodied in the factor Λ_t^s. However, due to the presence of uncertainty at low levels of consumption, the factor Λ_t^s moves to states with higher values which reveals a high level of motivation, the effect of which is rising up social benefits.

4.4.3   Stochastic waves and a capital asset analogy

As an alternative one might appeal to a social assessment of investments which parallels the standard techniques of evaluating assets, which is based on the present value of all future rewards. One believes this measure would yield the best estimate of the value of the increments in social output and in turn the best estimate of the value of the increments in employment and consumption attributable to the present additions to capital. We will present briefly the main lines of argumentation.

Renounced consumption as a capital asset

In order to put the corresponding learning process in the setting outline in Section 4.3, one may consider the representative individual has at any time t the alternative to allocate her/his share of the social economic surplus, right-hand side of eq. (4.7), to meet the requirements associated with her/his personal necessary consumption and to the commitment deriving from the acquisition of a special capital asset and servicing the loans linked with the asset purchase. Concerning the asset one may assume that it is a kind of natural resource, i.e., planted stands of trees or herds of livestock, characterised by a highly random age-dependent evolution of its size. On the other hand, the market value of the asset is also subject to uncertainty and evolves as a geometric Brownian motion. One may assume further that an appropriate size evolution of the asset demands a minimum level of capital endowment embodied in working tools and labour services below which the size of the asset may decrease, say due to pestilence, fire, wind etc. The stopping time problem consists here in finding optimal harvesting rules that take into account timing and valuation issues like size and market value of the asset as well as the comsumption requirements of the asset owner from the point of view of her/his capabilities, i.e. states of being and doing.

However, we prefer to resort to a direct assessment based on the flow of instantaneous social reward in terms of consumption characteristics per available labour generated by the chosen investment-consumption mix and on the social value of the terminal state measured in terms of the deficit in consumption characteristics compared with the level resulting from the golden age. This approach will be used in Chapter 5, where we obtain a formula for the evaluation of social benefits which echoes the procedure used for asset valuation in capital markets, without given *a priori* the rate of discount.

Two simple learning strategies

Going back to the learning process let us sketch two alternatives that we may get according to how high or low is the chosen level of $\underline{c}_t$. Thus, in this case anything depends on the available information on the qualitative behaviour of ϕ and u in G. For the sake of simplicity, we assume that $\mathbf{A}_t\phi = \mathbf{A}_t u$ and $\nabla_x \phi = \nabla_x u$. Then, we may have the strategy: (1) Choose a low learning intensity, i.e. the relative values of $\dot{k}_t$ and $\underline{c}_t$ are low and large respectively. Since with this choice one associates a relatively low development pace, it may take longer to reach the boundary ∂G, where $\phi = u$. One advantage is that the development process is likely to remain pleasant throughout and the learner may slowly increase the study intensity, i.e. increase $\dot{k}_t$ by decreasing $\underline{c}_t$, till the harvest time comes. However, the longer it takes until the harvest the greater are the chances for uncertainty to come into operation; for instance changes in the market value of the asset. Let us see the second strategy: (2) Choose a high learning intensity, i.e. the relative values of $\dot{k}_t$ and $\underline{c}_t$ are large and low respectively. Under this strategy a higher growth pace of the asset obtains so that the development process may hit the boundary sooner. Due to a relatively low level of $\underline{c}_t$ the learner may either give up before the learning process has been completed or may learn that much that a point comes, where study intensities previously considered unpleasant become pleasant, for instance, reflecting the prospect of an abundant harvest. Then, a further increase in the study intensity may become feasible and even accelerate the learning process. However, relatively low levels of $\underline{c}_t$ are associated with uncertain longevity and impatience as suggested in the foregoing section. Let us mention that saying an activity becomes unpleasant simply means that there is an alternative that may be regarded as more preferable. Lack of such an activity may cause frustration.

A front of stochastic waves

To close this chapter we like to mention that since in a certain sense the effect of an additive functional on the process X is shifting away the boundary, then a purposely chosen sequence of interventions in the process may be seen as building a front of stochastic waves. The idea is connected with Chernoff's sequential construction of a stopping rule we presented in Subsection 4.2.1 and Dynkin's idea of stochastic waves.

Given an measurable functional ϕ defined as before, we construct for $n = 1, 2, \ldots$ a sequence of stopping times τ_n such that $\tau_n = \inf\{t + \tau_{n-1} > \tau_{n-1} : \phi(t + \tau_{n-1}, X_{t+\tau_{n-1}}) > \phi(\tau_{n-1}, X_{\tau_{n-1}}) + t\}$, $\tau_0 > 0$ and $t > 0$; see (4.8) and (4.10). Intuitively, the stopping times τ_n indicate the first time the process X hits successively higher level sets of ϕ. The subprocess corresponding to ϕ is given for $n = 1, 2, \ldots$ by the relation $\widetilde{X}_t^n = X_{\tau_n}$ and $\widetilde{X}^n$ is said to be the *front of stochastic waves* corresponding to ϕ and $\{\tau_n, n = 1, 2, \ldots\}$.

From the point of view of applications a central feature of this notion is the derivation of the adjoint and the adjoint state equation on the basis of perturbations of functionals associated with stochastic waves; for instance perturbations of the function g defined in connection with (4.10) in a manner similar as in Chapter 6 and Chapter 7. This idea will be persued further in [5, 94] to which we refer; in connection with stochastic waves see [55, 56, 159].

At the end, we would like to emphasise that most of the ideas presented here are of an eminently modelling character and since modelling itself is a learning process it represents just a stage of it, hence a temporary and relative projection of reality, susceptible of extension and improvement. See [90].

Chapter 5

The martingale tuning of driving forces

In trying to find an explanation for these phenomena in those years, I put special emphasis on the fact that the countries of Latin America formed part of a system of international economic relations which I named the "center-periphery" system. In reality, this concept had been turning over in my mind for some time. At first I gave it a cyclical character, considering that it reflected the active role of the industrial centers and the passivity of the periphery, where the consequences of the economic fluctuations of the centers were intensified. There was in effect an "economic constellation," at the center of which were the industrialized countries. Favored by this position and by their early technical progress, the industrialized countries organized the system as a whole to serve their own interests. The countries producing and exporting raw materials were thus linked with the center as a function of their natural resources, thereby forming a vast and heterogenous periphery incorporated in the system in different ways and to different extents.

For each peripheral country, the type and extent of its linkage with the center depended largely on its resources and its economic and political capacity for mobilizing them. In my view, this fact was of the greatest importance, since it conditioned the economic structure and dynamism of each country—that is the rate at which technical progress could penetrate and the economic activities such progress would engender. Similarly, this system of international economic relations exaggerated the degree to which income in the periphery was siphoned off by the centers. Moreover, the penetration and propagation of technical progress and its fruits in economic activities oriented toward exports became characteristic of a heterogeneous social structure in which a large part of the population remained on the sidelines of development.

Raúl Prebisch
(see [169], pp. 176–177)

5.1   Contribution to growth of accumulation

Changes in an economic system reflect a highly complex process, since they involve spatial structures as well as a manifold of time varying attitudes and preferences for potential outcomes. At the level of the individual, changes that imply increased skills and capacity, greater positive freedom, creativity, self-discipline, motivation, responsibility, and material well-being are thought to be conducive to the development of the system. At the level of the society as a whole, changes furthering development are not easy to identify. Broadly speaking, we will characterise these as changes that lead to an increased capacity to regulate internal and external relationships, to interact with the environment in a fashion conducive to harmony with nature and its self-regenerative dynamics, to understand the entropic laws and to accumulate know-how. These features are in different ways reflected in the capacity of a system to generate economic surplus and in the manner in which work is organised.

In this chapter we initiate a deep analysis of various processes around the working activity of people relying on our belief that labour, as the most basic element of human society determines its entire development and character.

The process of change in the working activity of people, its pace and impact are essentially linked to changes in the productive forces including means and techniques of production and to changes in the entire pattern of production relations among the participants in the social process of production.

However, in order to understand the system as a whole one is forced to distinguish between material-technical and socioeconomic processes. The high degree of abstraction that this requires calls for the introduction of interacting processes aimed at recovering the links between changes in the production relations and changes in the productive forces of society.

One of the most simple processes within the category of material-technical is the one of capital accumulation. This is explained usually by means of a model which describes social and economic changes deriving from the time evolution of the social endowment capital per actually employed worker as we shall see next.

5.1.1   The role of control interventions

At various points in the course of the present study we have been confronted with the dilemma resulting from the alternative of allocating currently available disposable income to purposes of increases in the current level of employment and wages against the alternative of favouring rather investment and growth with the prospect of higher increases in the future levels of employment and wages. A glance at (2.11), (2.12) and (2.14) confirms, a higher level of employment and/or wages means a higher wage bill, a lower capital income and thus lower investment. On the other hand, the higher purchasing power deriving from a higher wage bill may as well stimulate demand and gives rise to profitable investment opportunities and higher rates of growth. It is precisely this two-sided effect resulting from employment and wage policies which originates contradictory and ambiguous views. The fact that favouring current investment and growth also exhibit two-sided effects does not clarify the issue, since the *status-quo* has itself a vested interest.

Thus we shall direct our attention to a strategy stimulating the expansion of productive capacities and higher rates of investment which in accordance with the principle of effective demand, needs higher levels of employment and wage income, if profit realisation is expected to follow.

In this regard two points have to be taken into account. The first hints to eventual changes in the prevailing pattern of tastes that an expanded and diversified production and social output requires. The second refers to changes in the bundle of labour characteristics resulting from the need to adapt to changes in the core of prevailing techniques. In this connection, we argued in Chapter 2 on the basis of recent developments in psychology and anthropology for the need of considering tastes and technology as changing along with the transformation process of society. This gives rise to a vital interaction of means and ends in the economic process, and stimulates to a dynamic interplay of judgements of fact and value. These points are necessary in order to understand motivations and expectations behind intertemporal decisions; see also Section 4.3.

Dialectic of change as a basis for policy making

Dynamic systems undergo processes reflecting the presence of dialectic relationships. To the purpose at hand, let us draw attention to the dialectic relationship between motion and rest which requires us to visualise a state of rest as a limiting case of motion and vice versa a state of motion as consisting of a sequence of consecutive states of rest. Translated into the framework of social systems we are faced with a relationship between change and steadiness or invariance. However, the natural interaction between both is obscured by the prevailing distribution of power among those who get advantage from change or steadiness.

The natural questions associated with these issues draw attention to choice of priorities and the degrees of relevance assigned to social objectives, as well as to methods used to free and mobilise potential energies, and to the links among incentives, rewards, ranges of initiatives and responsibilities of economic agents. It is extremely important to realise, margins of choice and associate potential changes are both determinants and consequences of the stage of development of productive forces of society.

In particular, the ability to generate product surplus and to allocate it efficiently are in this sense outgrowths of the prevailing practices determining access to available resources and the distribution of the social output. The product surplus and its allocation determine at once social productivity and affordable measures of economic and cultural freedom. This happens according to the resulting actual and potential levels of employment, the creative energies of the system and to the room the mechanisms of self-perpetuation allows to those of change.

Therefore, it is crucial to recognise that parallel to self-perpetuating social forces emanating from vested interest, be it of economic, financial, political or

social nature, there come up ideologies destined to serve the *status quo*. Under
these circumstances a state of the affair gets into being which increasingly un-
dermines development of structures, constrains social processes and reinforces
established pattern of behaviour, so as to ensure maintaining the flow of benefits at the expense of the weak. Nation states of the periphery are a living
evidence of this successful *modus operandi*. Naturally, there emerge anew social
tensions and desire for change as pressing response to fulfil the most basic needs
and human aspirations usually integrated in the process of social reproduction.

Hence, in the light of the mentioned dialectic perspective, it is necessary to
assess possibilities for change, to evaluate consequences from a given situation
and its alternatives, including organisation, will and ideologies, with the aim
of characterising the sources of dynamics underlying opposing social forces and
structures, their drive to continuity and dragging away or rupture.

The social content of the notion of capital per worker

At this stage we know already capital does not stand simply for means of production but it involves also social relations. From our simplifying assumption
according to which the economic system, along this essay, consists of a relatively advanced sector and a reserve army of unemployed labour, we can safely
concentrate upon commodity production. Consequently, labour takes the social
form of wage labour and capital may represent produced means of production.
Means of production include crystals of dead labour, codified labour processes
and organisation, a matter essential to the role of information and learning in
the evolution of society.

Behind the process of (commodity) production by means of commodities
there exist various circular flows which are better understood from the point of
view of national accounts and build the heart of models of social reproduction.
Roughly speaking, a certain combination of capital and labour, as produced
means of production and labour power, generates social output in the form of
commodities. Wage and profit rates indicate the rewards accruing to labour
power and capital. The first corresponds to production services and is measured
in units of money per unit of time; the second represents utilisation units and
is given barely as a number per unit of time. This obeys the fact that means of
production (fixed capital) maintain to a large extent their natural form whereas
by contrast labour power needs to be renewed at any unit of time (production
period).

The ratio $k = \frac{K}{L}$ is usually seen as representing a technique of production.
It has become customary to talk about the units of social output that come out
under the application of technique k. Putting aside the most essential problems
connected with the measurement and distinction of capital as produced means
of production and as source of income, let us think of k as the capital units,
in the form of the productive endowment, available per unit of time to any
employed labourer committed to production at the prevailing working time
rate.

Let us consider k in the following equivalent ratio:

$$\frac{\text{(capital/output)}}{\text{(labour/output)}} = \frac{\text{capital required to produce 1 output unit}}{\text{labour (power) required to produce 1 output unit}}.$$

It is easy to recognise the wealth of information needed in order to make precise the amount of capital units necessary to produce 1 output unit. However, matters become more embarrassing, when one moves to ascertain the labour power requirements to produce 1 output unit, owing to the fact that this would involve various socioeconomic, anthropological and spatial issues in addition to the questions ignored usually by orthodox economics.

This stresses the need of taking into account institutions, social relations of production as well as some factors constraining social behaviour in order to understand the highly simplified character of the process $k = (k_t)_{\{\tau^j \le t \le \tau^{j+1}\}}$ as a representation of capital accumulation. Our way of capturing the wealth of information hidden by the ratio k is to direct attention to characteristics of capital and labour underlying the production of one unit of output; see Subsection 2.2.1.

5.1.2   Stochastic accumulation of capital per worker

In this subsection we will focus on material-technical aspects of production. However, let us begin stressing some issues on their social form so that we can keep in mind the ultimate goal of our undertaking: the adjustment of production relations to change in the productive forces so as to ensure an adequate development of the social process of reproduction. In a preliminary outline of the state of any economic system, its capital structure may constitute, when appropriately supplemented with social components, a suitable description. Indeed, given the technology and institutions, the characterisation of social output will be the result of determining the time path of capital, both physical and human, and the structure underpinning decisions of its allocation over time among different uses. In turn, with the time path of social output as given, allocation between consumption and capital accumulation determines the capital structure to follow. Therefore, we shall start considering the process of capital per worker $k = (k_t)_{\{\tau^j \le t \le \tau^{j+1}\}}$ as the appropriate state description of the system and focus our efforts in representing as accurate as possible the process of change and the implications of policy decisions over time.

The accumulation process and its properties

At a fairly elementary level, production, services of labour and (units) of capital utilisation are jointly called the outlay of the process of (material) production, while the quantity of products turning out is called the return of the process. It is necessary to keep firmly in mind that outlays and returns are flows of physical units in given time period. Hence, they depend on the time period under consideration.

ODE's in general and SDE's in particular are appropriate means of describing the relation between returns and outlays underlying social reproduction.

Nevertheless, the fact that this relation would need to reflect developmental changes in the conditions of production concerning *inter alia* productiveness of labour, state of science and know-how, social organisation of production, the extent and capabilities of means of production, reminds us of the limitation of such models and calls for caution interpreting them.

The process of capital accumulation in this essay is thought of as describing qualitatively the prevailing tendency in the allocation of social output for the purpose of increasing means of production (per worker). In this essay we shall work on the basis of a dynamics of accumulation derived from a relatively narrow framework; it is instructive to contrast this way with the one followed in Section 4.1.

In this sense, let us recall, the capital-labour ratio $k_t = \frac{K_t}{L_t}$ at time t can be viewed in this context as a function of the capital stock K_t and labour participation L_t; one should be aware of the fact that the most one may realistically expect to have is a set-valued function, so that convex analysis and differential inclusion appear to offer more appropriate working tools. Notwithstanding this observation, we shall assume K_t and L_t satisfy respectively the ODE (2.17) and the SDE (2.4), the latter being of the type (8.16) with $a_t(\omega) = a_t L_t$ and $\sigma_t(\omega) = \zeta_t L_t$. We have the following Lemma.

Lemma 5.1 *Let us assume that A 2.1 holds and that the processes of capital stock per worker $k = (k_t)_{\{\tau^j \leq t \leq \tau^{j+1}\}}$ and actual labour services $L = (L_t)_{\{\tau^j \leq t \leq \tau^{j+1}\}}$ satisfy (2.17) and (2.4). Then, the evolution of process k is governed by the SDE of the diffusion type (5.1).*

$$dk_t = \left[\frac{\dot{K}_t}{K_t} - (a_t - \zeta_t^2) \right] k_t dt - \zeta_t k_t dB_t. \tag{5.1}$$

Proof. Simple application of Lemma 8.32, see Chapter 8, gives

$$dk_t = \frac{\partial}{\partial t}\left(\frac{K_t}{L_t}\right) dt + \frac{\partial}{\partial L_t}\left(\frac{K_t}{L_t}\right) dL_t + \frac{\partial}{\partial K_t}\left(\frac{K_t}{L_t}\right) dK_t +$$
$$+ \frac{1}{2}\frac{\partial^2}{\partial L_t^2}\left(\frac{K_t}{L_t}\right) d\langle L_t, L_t \rangle,$$

which, taking into account (2.4) and (2.17), turns into

$$dk_t = \left[a_t L_t \frac{\partial}{\partial L_t}\left(\frac{K_t}{L_t}\right) + \dot{K}_t \frac{\partial}{\partial K_t}\left(\frac{K_t}{L_t}\right) + \frac{1}{2}(\zeta_t L_t)^2 \frac{\partial^2}{\partial L_t^2}\left(\frac{K_t}{L_t}\right) \right] dt +$$
$$+ \zeta_t L_t \frac{\partial}{\partial L_t}\left(\frac{K_t}{L_t}\right) dB_t.$$

Some operations and arrangements yield the SDE (5.1). ∎

It is of crucial importance to understand the dynamic nature of the economic processes acting behind the drift coefficient of (5.1). With this purpose, let us rewrite using (2.21) the term in brackets at the right-hand side of (5.1) as the relationship

$$\frac{\dot{K}_t - (a_t - \zeta_t^2)K_t}{K_t} = \frac{(Q_t - \delta_t K_t - C_t) - (a_t - \zeta_t^2)K_t}{K_t},$$

indicating the rate of social output per unit of capital available for the expansion of the productive capacity. The drift coefficient of (5.1) can be written as

$$(q_t - \delta_t k_t - c_t) - (a_t - \zeta_t^2)k_t,$$

where the first term stands for economic surplus per worker and the second represents the fraction of capital dedicated to meet employment requirements of production plans of capitalists and uncertainty in per-worker units. The case corresponding to $a_t - \epsilon_t$ taking on positive values suggests a reduction of the existing reserve army of unemployed labour; otherwise, a building up of it results. As we mentioned already, the parameter a_t is deeply connected with issues concerning the realisation of profits and this in turn is linked to questions of size of the market, the principle of the effective demand and national accounting about which we have referred to in foregoing chapters. See [23].

As it stands, the drift term points to potential relative changes in the stocks of capital and planned employment-changes at any time t. In eq. (5.1) at the right-hand side the first term of the expression in brackets represents *savings per worker*, while the second *required investments per worker*. The difference of them indicates thus the actual average rate of change of capital stock per worker. The coefficient of the Brownian motion reflects deviations from this average due to uncertainty underlying the demand for labour and it is thus linked to uncertainty governing profit realisation. Eq. (5.1) shall be referred to as the *stochastic equation of the accumulation of capital per worker* and has been used under different frameworks to investigate a variety of problems in economic dynamics.

Taking into account (2.17) and (2.24), we rewrite eq. (5.1) in the following more compact although less transparent form, that we shall often use for the sake of simplicity,

$$dk_t = b(t, k_t, \alpha_t)dt + \sigma(t, k_t)dB_t, \tag{5.2}$$

where the drift $b^\alpha(t, k_t) = b(t, k_t, \alpha_t) = s_t y_t - (a_t - \zeta_t^2)k_t$ and $\sigma(t, k_t) = -\zeta_t k_t$. Furthermore, q_t and $s_t y_t$ related by means of (2.25) and α^j is the functional $\alpha^j : t \longrightarrow \alpha_t$ with $\alpha_t = \alpha(s_K(t), L_t)$ a function of class $\mathbb{C}_b^{1,1}([0,1] \times \mathbb{R}_+ \longrightarrow \mathbb{R}_+^2)$. The movements of k_t can be gauged based on observation about the level or relative value of the functions $s_t y_t$ and $a_t k_t$ at time t or $q_t - \delta_t k_t - c_t$ and $(a_t - \zeta^2)k_t$, i.e. the actual and required investments per worker respectively. The difference between these two terms at any time t portrays the expected rate of change of the level of capital per worker as a function of k_t; more precisely this is a function depending on the information available at time t about the level of k_t, and provides helpful insights into the prevailing stochastic dynamics of accumulation of capital per worker. Knowing for instance that at time t the inequality $b^\alpha(t, k_t) \geq 0$ holds, means that accumulation takes place and that k_t increases.

However, one word of caution is in order, since these statements hold only almost surely according to the probability measure prevailing at time t; thus a more adequate way of writing the drift of capital accumulation would be $b(t, k_t) = \mathsf{E}\{b(t + \Delta t, k_t + \Delta k_t)|\mathcal{F}_t\}$ which makes apparent the nature of the conditional average characteristic of the drift and the role of the information structure. The information structure is represented by $\mathcal{F}_t$ and determines in a sense the probability measure needed to compute the average. In the next section we shall have more to say on this point.

As suggested already, we are interested by way of appropriate selections of α^j to ensure that a desired process of capital accumulation comes about. This poses two further problems, one related to the requirements on α^j so as to enable it to elicit the desired accumulation, see A 5.1, and the second dealing with existence and uniqueness questions that the following Lemma answers affirmatively.

Lemma 5.2 *Let us make the assumptions A 2.1 and A 2.2. Then, for any fixed but arbitrary control process $\alpha^j = (\alpha_t)_{\{\tau^j \le t \le \tau^{j+1}\}}$, the coefficients $b(t, k_t, \alpha_t)$ and $\sigma(t, k_t)$ of eq. (5.2) satisfy the conditions for uniqueness and existence of a strong solution of the SDE (5.2). Furthermore, it turns out that this solution is a diffusion process with a drift and diffusion coefficient given respectively by $b(t, k_t, \alpha_t)$ and $a(t, k_t)$, where $a(t, k_t) = \sigma(t, k_t)\sigma(t, k_t)^*$.*

Proof: See Chapter 8 and Corollary 1 [82], p. 53.   ∎

The fact that the SDE (5.2) depends on the control α does not need to detain us at the moment, since at any time t α_t is indeed a parameter. Thus we may think of (5.2) as a SDE which depends on a parameter. The fact that k is expected to remain positive should not preoccupate us due to the fact that we will be working with SDE's which are linear, a consequence of our structural setting; for that type of SDE's we show in Section 6.2 under which conditions they remain positive. Concerning the *diffusion property* of (5.2) or equivalently of (5.1), we have the following information. At any time t, random changes in the process of accumulation per worker consist locally first of an average macroeconomic tendency of the rate of change of k_t given by $b(t, k_t, \alpha_t)$ that we shall denote, abusing a little the notation, as $\dot{k}_t = b(t, k_t, \alpha_t)$ and secondly of fluctuations about the average given by $\sigma(t, k_t)$. At this point, let us mention again that uncertainty is essentially linked to the presence of a stable core of characteristics of capital and labour skills in front of a relatively unstable periphery of such characteristics, about which we have only probabilistic knowledge; see [5].

Dealing with deterministic systems knowledge of the state at some time $t = t_0$ and the choice of the policy instruments determines the whole course of the future system evolution. On the other hand, stochastic systems depend critically on the information structure available at any moment of time; this point is emphasised by measurability requirements. Thus saying that the process k is (progressively) measurable means that at any time t the variable k_t depends on the history of the process up to time t; this is written for simplicity as $k_t \in \mathcal{F}_t$ and measurable should be understood as observable. These issues are treated in Chapters 8 and 9 to which we refer.

However, one may as well assume the process k depends at any time t solely on the values the state k_t may attain, a simplification which sets stochastic systems in a position quite similar to deterministic ones. The *Markov property* is a natural generalisation of the assumption that with respect to systems evolving in a random fashion knowledge of the current state and the instruments, completely determines the probability of occupying the various possible states at all future times. As we have already seen, the Markov property has a marked causality provenance and it forces us to think of any state as being described by a set of alternative values with a given likelihood of occurrence, which depends critically on the current state of knowledge; see [5]. Among others [29], [148] study eq. (5.1) and stationary properties of its solution, while [37] deals with existence and uniqueness questions. [140] gives a full account of applications of Itô's calculus to economic problems.

5.1.3 Policy aims and conflicting decisions

Control variables are usually called in economics policy instruments, reflecting the idea that they can be purposely chosen by the policy maker so as to bring about certain desired system behaviour. Policy instruments are typically seen as means of allocating resources to alternative productive uses and to consumption. Fiscal and monetary interventions, designed usually on the basis of taxes and inflationary measures, are further instances of policy instruments. The latters act rather indirectly on the determination of the allocations. We will restrict ourselves to instruments applied directly towards investment and consumption decisions.

Parameter choice underlying investment decisions

Let us point out that to any couple $(s_K(t), L_t)$ corresponds also a couple $(\dot{K}_t, C_t)$. For any t, any couple $(s_K(t), L_t)$ determines a function $\dot{K}_t$, as a look at (2.11) indicates, and C_t results from the income identity (2.16) as a residual, since at any time t the aggregate income Y_t satisfies the income identity given above. The map $\alpha_t \longrightarrow b(t, k_t, \alpha_t)$ will stress the dependence of the drift of (5.2) on the time paths of the process $(s_K, L) = ((s_K(t), L_t))_{\{\tau^j \leq t \leq \tau^{j+1}\}}$, see on the right-hand side of (5.1) at the first term in the expression in brackets and (2.11).

A little cumbersome, although maybe more informative version of the foregoing control map is then given by $(s_K(t), L_t) \longrightarrow b(t, k_t, (\dot{K}_t, C_t))$, which makes also evident that α_t may be allowed to be a piecewise-continuous function of class $\mathbb{C}_b^{1,1}([0,1] \times \mathbb{R}_+ \longrightarrow \mathbb{R}_+^2)$. The function $b(t, k_t, \alpha_t)$ representing the accumulation drift, turns out to be of class $\mathbb{C}^{1,2,1}(\mathbb{R}_+ \times \mathbb{R}_+ \times \mathbb{R}_+ \longrightarrow \mathbb{R}_+)$. Hence, it becomes natural to assign to α_t the role of a policy instrument, and this renders the parameter $s_K(t)$ and the random variable L_t the elements of choice.

Let us close these remarks pointing out that by choosing a time path of the couple (s_K, L) we determine b^α and σ characterising thereby a unique process k. In the language of control theory $\alpha^j = (\alpha_t)_{\{\tau^j \leq t \leq \tau^{j+1}\}}$ is called the *control process* which we shall make precise later. From an economic point of view, the choice of alternative control processes α^j implies choosing a set of alternative drift and diffusion coefficients $b(t, k_t, \alpha_t)$ and $\sigma(t, k_t)$ and these in turn entail picking a set of alternative time paths of accumulation $k^\alpha = (k_t)_{\{\tau^j \leq t \leq \tau^{j+1}\}}$.

Many such accumulation paths are possible and they behave differently according to the properties of the instruments. Before we introduce some additional notions based on which we will design a criterion of social desirability or means of comparing alternative accumulation paths, let us make precise the requirements one expects the policy instruments to fulfil.

A 5.1 *Assumptions on the control process α^j*

- The control process $\alpha^j = (\alpha_t)_{\{\tau^j \leq t \leq \tau^{j+1}\}}$ has to be chosen so that for any $t, t \in [\tau^j, \tau^{j+1}]$, the pair of policy instruments α_t is a piecewise-continuous function of class $\mathbb{C}_b^{1,1}([0,1] \times \mathbb{R}_+ \longrightarrow \mathbb{R}_+^2)$.

- Only control processes α^j associated with time paths of (s_K, L) for which $s_K(t)$ and L_t fulfil the constraints given by (2.13) and (2.15) are under consideration.

- The control process α^j is feedback and Markovian. It satisfies some measurability and Lipschitz conditions aimed at ensuring the existence of a unique solution of (5.1), which we will make precise later.

An *admissible control process* α^j shall be one for which a uniquely defined controlled process k, given by eq. (5.1), exists. The family of admissible control process α^j will be denoted $\mathcal{A}$. As we shall explain in the next sections, the control process α^j is linked in a fundamental manner to a probability measure $\mathbf{P}_\alpha$, which gives sense to saying that the choice of instruments determines a set of alternative states distinguished with a corresponding likelihood of occurrence. In this sense, the Brownian motion in (5.1) and (5.2) is implicitly determined by the choice of the control process. However, for expository reasons we ignore in this section this delicate point.

Brief analysis of conflicting decisions

Recalling that $s_t y_t + \delta_t k_t$ stands for gross investment per worker and the identity $q_t = s_t y_t + \delta_t k_t + c_t$ holds, it turns out that accumulation has to follow in the average the relationship

$$q_t = c_t + (a_t + \delta_t - \zeta_t^2)k_t + \dot{k}_t, \qquad (5.3)$$

which states that in the average the social output per worker q_t will be allocated to the purposes of maintaining the level of consumption per worker c_t, and the level of capital per worker, i.e. $(a_t + \delta_t - \zeta_t^2)k_t$, as well as to yield net increases in the level of capital per worker $\dot{k}_t$. At this stage, we already know that a_t represents the relative rate of change of labour participation, δ_t stands for the rate at which existing capital deteriorates and includes in a certain sense obsolescence. Hence, $(a_t + \delta_t)k_t$ denotes the amount of social output per worker that has to be put aside in order to endow workers entering the production process at least at the existing standards and replace used up working capital equipments. The last term, i.e. $-\zeta_t^2 k_t$ corresponds to that fraction of q_t put aside in order to meet the requirements imposed by unpredictable events; we can see $-\zeta_t^2 k_t$ also as a correction term due to an over or under estimation

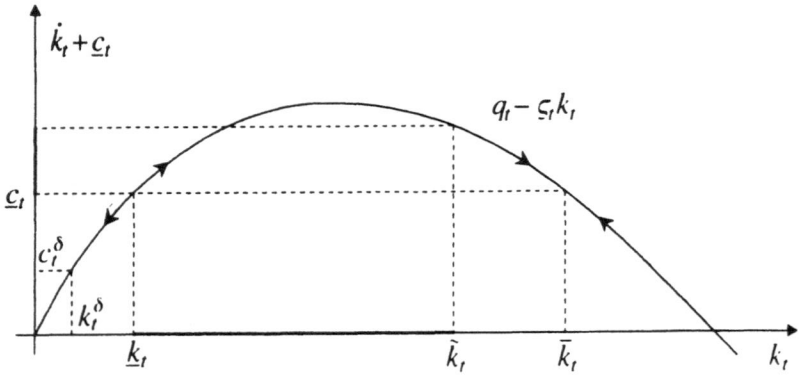

Figure 5.1: Phase diagram of accumulation of capital per worker

on the actual level of y_t brought about by uncertainty. To understand the eventual detrimental or constructive effects of economic policy decision, let us look at the qualitative behaviour of the process of accumulation of capital per worker. With this purpose we will examine the drift of (5.1) in the form given by (5.3) and draw a phase diagram, see Fig. 5.1. Phase diagrams are very helpful in qualitative analysis of differential systems and are standard working tools in the theory of optimal economic growth, see [35, 114]. We will refer to $\underline{c}_t$ as *necessary consumption per worker* which comprises personal consumption necessary to maintain earning power of direct producers – denoted by c_t^p – and productive consumption necessary to replace used up means of production – denoted by c_t^δ so that after setting $\varsigma_t = a_t + \delta_t - \zeta_t^2$ we may rewrite (5.3) as

$$\dot{k}_t + \underline{c}_t = q_t - \varsigma_t k_t.$$

The level of consumption $\underline{c}_t$ represents a threshold below which the productive power of capital and/or labour decay critically. To prevent this situation one sets limits to the values of the capital rate of return and to the rate of wage, i.e. the rate of interest for the first and for the second one assumes the existence of a floor on the admissible wage rate W denoted by $\overline{W}$; see Assumption A 2.2. The left-hand side in the foregoing equation hints to the potential allocation of social output per worker, after provision for planned changes in production level, beyond the level $\underline{c}_t$ and to further expansion of the level of capital accumulation. Any increase of level of consumption above $\underline{c}_t$ slows down directly the pace of capital accumulation. In Fig. 5.1 we depict a phase plane $(\dot{k}_t + \underline{c}_t, k_t)$ for any time t arbitrary but fixed, portraying social output per worker net of planned changes in investments per worker against the level of capital per worker. According to this $\underline{c}_t = c_t^\delta + \underline{c}_t^p$, where $\underline{c}_t^p$ is the personal necessary consumption corresponding to $\overline{W}$.

Associated with $\underline{c}_t$ there are two levels of capital, i.e. $\underline{k}$ and $\overline{k}$, which attract our attention. These two levels of capital $\underline{k}_t$ and $\overline{k}_t$ are also equilibrium points, the first unstable while the latter stable as the arrows in Fig. 5.1 show. The qualitative behaviour displayed in the diagram below places some restrictions on the potential admissible initial levels of capital per worker; because accumulation paths starting from levels of capital at the left of $\underline{k}_t$ are driven by the system dynamics to the zero level of capital. On the other hand, not every path starting from initial levels of capital at the right of $\underline{k}_t$ moves towards the upper level $\overline{k}_t$, since the Brownian motion acting on the accumulation path may still drive it out of the stable region.

Let us point out that the level of capital $\tilde{k}_t$ corresponds to that level of the average rate of return beyond which further investments are not longer worth. This leaves us with a relevant range of capital k_t given by $[\underline{k}_t, \tilde{k}_t]$. All these features have to be taken into account designing a control strategy and point to various difficulties of economic and mathematical nature inherent to the problem at stake.

5.1.4   The stochastic optimising problem

At this point it should be clear that central to the problem of resource allocation is maintaining and developing the productive potentials of society. Thus at a first level one has the decision of allocating social output to purposes of consumption – meaning aimed at renewing used up productive power – and to the end of expanding and improving the capacity to produce which is indeed the investment proper. At the second level one has to cover necessary consumption which has a personal component related to the direct producers and a component connected to deterioration of capital and its eventual depreciation. Fundamental to these two levels is the idea of a circuit of social reproduction; any analysis of investment at a further level is beyond the scope of the present work.

A potential of characteristics to function

Thus, the main ingredients on which to base the decision on the allocation of social output to the end of consumption and investment are already present in(5.3). How the fulfilment of investment evolves in the form of expansion and improvement in the capacity to produce, is reflected by the process of accumulation (5.1). To see how human labour in the form of states to function evolves, one needs an equation independent of (5.1) which when coupled with it describes the motion of the system. With this purpose in mind we introduce a potential which describes social costs attached to the accomplishment of the accumulation objectives.

The social cost of the accumulation policy consists in turn of an integral and a terminal reward. The first is related to deficits in the states of functioning attributed to deprivation of consumption characteristics resulting from shifting additional units of social output to investment purposes. The second describes enlargements in the range of the capability set that the attained terminal state of capital accumulation makes affordable by closing an *a priori* fraction of the

initial gap between the state of capital and a reference state. As a matter
of fact these two components reflect cost since they represent deficits in the
current states of functioning, however we shall term them rewards because of
qualitative improvements embodied.

In the present work we will not deal with the question of empirical assessment of the notions involved. For the time being, we shall assume that they can
be evaluated in a certain sense, i.e. we claim that one can express preferences
between alternative accumulation paths and that we can give them numerical
values. In short, we assume that to any admissible time path α^j, or equivalently to any admissible path $(s_K, L) = ((s_K(t), L_t))_{\{\tau^j \leq t \leq \tau^{j+1}\}}$, corresponds
a process k^α governed by (5.1), the social desirability of which can be assessed
by the mathematical expectation of the negative-valued functional

$$k^\alpha \longrightarrow \int_{\tau^j}^{\tau^{j+1}} f(t, C_t, \alpha_t)dt + \phi(\tau^{j+1}, k_{\tau^{j+1}}), \qquad (5.4)$$

where we write the control variable α_t explicitly to point out the dependence of
f on the control process α^j. In (5.4) we shall consider the integrand $f(t, C_t, \alpha_t)$
instead of the more general functional $f(t, k_t, \alpha_t)$. As we argued before, it
results more convenient to deal with the optimisation problem in terms of
the pair $(s_K(t), L_t)$ rather than in terms of the investment-consumption mix
$(\dot{K}_t, C_t)$.

A 5.2 *Assumptions on the performance structure*

- For any t, the functions $f(t, C_t, \alpha_t)$ and $\phi(t, k_t)$ are bounded from below,
  real-valued and negative. Further, they are strict concave in the second
  variable and have one continuous derivative in all variables.

A precise characterisation of these functions will be given in the next section. A
fairly complete characterisation of f and ϕ from the point of view of economic
theory was given in Chapter 4.

Formulation of the control problem

Substituting Y_t, as given by (2.10), in C_t and $\dot{K}_t$ it is easy to see that the control
variable $\alpha_t = \alpha(s_K(t), L_t)$ emerges in a natural way, so that the objective
functional (5.4) can be modelled properly in terms of the fraction of capital
income invested s_K and the labour participation L_t, as the control α^j suggests.
Finally, recall that the choice of the path $(s_K, L) = ((s_K(t), L_t))_{\{\tau^j \leq t \leq \tau^{j+1}\}}$
amounts to the choice of the path $((\dot{K}_t, C_t))_{\{\tau^j \leq t \leq \tau^{j+1}\}}$, as one easily recognises
looking at eqs. (2.11) to (2.12).

From the outset we should say that focusing on deficits the rationale underlying optimisation implies getting the least possible deficit. As a first step,
we can state our *dynamic economising problem* as that of choosing an admissi-
ble path $\alpha = (\alpha(s_K(t), L_t))_{\{\tau^j \leq t \leq \tau^{j+1}\}}$, so that the social desirability attached
to its associated process of capital accumulation k^α given by (5.1) attains a
supremum. From a mathematical point of view this amounts to the following
control problem with complete information.

Optimising problem 5.1 (Control with complete information)
*The assumptions A 2.1–2.3, A 5.1 and A 5.2 hold and assume complete infor-
mation. Then solve the stochastic control problem*

$$\sup_{\alpha \in \mathcal{A}} \mathsf{E} \left\{ \int_{\tau^j}^{\tau^{j+1}} f(t, C_t, \alpha) dt + \phi(\tau^{j+1}, k_{\tau^{j+1}}) \right\}, \qquad (5.5)$$

subject to (5.1) and (2.7) to (2.16).

For notational simplicity we shall set in the sequel $\tau^j = 0$, $\tau^{j+1} = T$ and
suppress the index α, whenever it is clear that the considered path is the
one associated with a given control. Those readers willing to see merely the
solution of the optimising problem just stated, may go directly to Section 5.4,
where we construct the optimal accumulation policy according to the tools to
be introduced in the forthcoming sections.

Next, we shall present some basic concepts of stochastic control theory from
the perspective of the semimartingale calculus. Since we are mainly interested in the application of control techniques that may help us to construct a
development strategy aimed at eliminating unemployment, we proceed in an
intuitive way and refer those readers interested in a more rigorous treatment
to the received literature.

5.2 Controlled diffusion processes

The aim of this section is to outline from the probabilistic point of view, known
as the martingale approach, the main steps by means of which one controls the
evolution of a diffusion process $X = (X_t, \mathcal{F}_t)_{\{0 \le t \le T\}}$ that can be represented
by a SDE of the Itô type. We shall present formally a few basic concepts
on control and optimality, and wait until the next section for a more precise
formulation.

5.2.1   Control of diffusions with complete information

In this subsection we shall discuss in some detail the problem of controlling
a stochastic process governed by a SDE of the diffusion type. Owing to emphasis in probabilistic methods more care is given to the actual paths and the
probability law of the process rather than to certain average functionals. An
immediate consequence of this, is restricting the control process to act directly
through the drift term. Hence, the diffusion coefficient will not exhibit any
explicit dependency on the control. Two special features of the martingale
approach, presented in Chapter 9, shall be clear at the outset.

- The effect between switchings of control interventions is to change the
  probability measure on the space of trajectories of the state process.

- Failure of using the optimal control during an interval of time entails
  according to the principle of optimality a loss of reward.

These two features are outgrowths of the fact that the martingale approach works with the actual sample path or laws of the controlled process itself and places the information structure in the forefront of analysis.

System dynamics and control

Let Ω be the space of continuous functions defined on $[0,T]$ with values in $\mathbb{R}^n$; it should be appropriate to take $\widetilde{\Omega}$ to distinguish it from the space appearing in the reference space, however we shall use Ω instead of $\widetilde{\Omega}$ unless confusion arises. Further, let $\{B_t\}$ be the family of coordinate functions and $\mathcal{F}_t^0 = \sigma(B_s, s \leq t)$. Let $\mathbf{P}$ be a Wiener measure on $(\Omega, \mathcal{F}_T^0)$ and $\mathcal{F}_t$ the completion of $\mathcal{F}_t^0$ with the $\mathbf{P}$-null sets of $\mathcal{F}_T^0$. Let us make the following assumptions on the drift function $b(t, x, \alpha)$ and diffusion coefficient $\sigma(t, x)$.

A 5.3 *Assumptions on σ*

Suppose $\sigma : [0,T] \times \mathbb{R}^n \longrightarrow \mathbb{R}^n \times \mathbb{R}^n$ is a matrix-valued function with the following properties:

1. The function $\sigma^{(ij)}(t, x)$ is $\mathcal{F}_t$-predictable, where i and $j = 1, 2, \ldots, n$; with $\sigma(t, x)$ one associates an $\mathcal{F}_t$-predictible $n \times n$ symmetric matrix $a(t, x) = \sigma(t, x)\sigma^*(t, x)$, where $\sigma^*(t, x)$ denotes the transpose of $\sigma(t, x)$.

2. The Lipschitz condition $|\sigma^{(ij)}(t, x) - \sigma^{(ij)}(t, y)| \leq \sup_{0 \leq s \leq t} |x_s - y_s|$ holds, where i and j are as above.

3. The matrix $\sigma(t, x)$ is non-singular for each (t, x) and $(\sigma^{-1}(t, x))^{(ij)} \leq \gamma$, where γ is a fixed constant independent of t, i and j.

In the sequel we shall concentrate on the nondegenerate case when the least eigenvalue of $a(t, x)$ is uniformly bounded away from zero. We have the following Lemma.

Lemma 5.3 (Strong solution) *Let $(\Omega, \mathcal{F}, \mathbf{F}, \mathbf{P})$ be a probability basis and $\sigma(t, x)$ be defined as in A 5.3 above. Then, there is a unique strong solution to the SDE*

$$dX_t = \sigma(t, X_t)dB_t, \qquad x_0 \in \mathbb{R}^n \quad given. \tag{5.6}$$

Proof: The proof of this Lemma follows from the definition of stochastic integrals. See Chapter 8. ∎

For an integrable random variable x_0 with a prescribed law, the SDE above describes a system dynamics condensed in the arrangement $(\Omega, \mathcal{F}, \mathcal{F}_t, \mathbf{P})$, where the process B is an n-dimensional standard Brownian motion independent of x_0. However, in general the process B should be m-dimensional with $m \leq n$, since there is no need to consider all components of the process X as random.

A 5.4 *Assumptions on the control α*

- Let $A \subseteq \mathbb{R}^n$ be compact and $\mathcal{A}$ the set of admissible $\mathcal{F}_t$-predictable A-valued processes $\alpha = (\alpha_t)_{t \geq 0}$.

Under the action of the control, there will emerge a drift term characterised by the assumptions to follow in short. For practical reasons, it appears natural to consider control processes whose values at any time t are chosen on the basis of the information on the state process gathered up to time t. These controls are said to be the feedback form or briefly to be feedback controls, since they are expressed as a function of the system state the values of which are thought of as being observed and fed back into the control.

Thus an admissible control $\alpha \in \mathcal{A}$ is said to be a *feedback control* if it is progressively measurable with respect to the natural filtration of X. Since α cannot be specified *a priori*, one cannot in general construct a process X on a prescribed probability space with x_0 and B given. In the nondegenerate case, one proves that given x_0 and B as above there may construct processes whose corresponding probability laws agree. This is the essence of the notion of weak solution to which we turn; see also Chapter 8.

The controlled system and its probability laws

Let us model a system which enables us to examine changes in the trajectories of the process X under consideration. To this purpose we shall work on the space $(\Omega, \mathcal{F}, \mathcal{F}_t)$ endowed with a family of probability measures P_α defined on $\mathcal{F}_T$ for $\alpha \in \mathcal{A}$. We need the following assumptions.

A 5.5 *Assumptions on b relative to α*

Let $b : [0, T] \times \Omega \times \mathcal{A} \longrightarrow \mathbb{R}^n$ be a function satisfying the properties to follow. For simplicity, we shall set $b^\alpha(t, x) = b(t, x, \alpha)$.

1. The function $b(t, x, \cdot)$ is continuous in $\alpha \in \mathcal{A}$, for each (t, x),

2. The vector b^α is $\mathcal{F}_t$-predictable in (t, x) for each $\alpha \in \mathcal{A}$,

3. The inequality $b(t, x, \alpha) \leq \gamma(1 + \sup_{s \leq t} |X_s|)$ holds, where the constant γ is defined as above.

At this point let us stress, the dependence of $b^\alpha(t, x)$ on x does not need to be smooth so that "bang-bang" and other noncontinuous controls fit naturally into this framework. Now, let us define for $\alpha \in \mathcal{A}$ and $0 \leq r \leq t \leq T$,

$$m_{r,t}(\alpha) = \tag{5.7}$$
$$= \exp\left\{ \int_r^t \left(\sigma^{-1}(s, X_s) b^\alpha(s, X_s) \right)^* dB_s - \frac{1}{2} \int_0^t |\sigma^{-1}(s, X_s) b^\alpha(s, X_s)|^2 ds \right\}.$$

The family $\mathcal{A}$ of admissible controls is a subfamily of $\mathcal{F}_t$-predictible processes with values in A for which the stochastic integral in (5.7) is well defined, and the semimartingale $m_{t, t \wedge T}(\alpha)$ is uniformly integrable and strictly positive. The boundedness of $\sigma^{-1}(t, x)$ and the growth condition on $b^\alpha(t, x)$ imply the Novikov condition, i.e. $\mathsf{E}(\frac{1}{2} \int_0^t |\sigma^{-1}(t, x) b^\alpha(t, x)|^2 ds) < \infty$ and this in turn guarantees $\mathsf{E}\{m_{0,T}(\alpha)\} = 1$.

In Chapter 9 we have presented in a brief outline on the semimartingale approach to stochastic control the main steps concerning change of control, trajectory and probability law. There we have surveyed those essential issues which make sure we can define a measure $\mathbf{P}_\alpha$ on $(\Omega, \mathcal{F}_T)$ as in (9.16); see also Chapter 8. We shall set $m_t = m_{0,t}$ for any t. We need the following theorem.

Theorem 5.4 (Weak solution) *The assumptions A 5.3, A 5.4 and A 5.5 hold. Under $\mathbf{P}_\alpha$ the process $X = (X_t)_{0 \leq t \leq T}$ satisfies*

$$dX_t = b^\alpha(t, X_t)dt + \sigma(t, X_t)dB_t^\alpha, \qquad (5.8)$$

with initial condition $X_0 = x_0 \in \mathbb{R}^n$, where $B^\alpha = (B_t^\alpha)$ is the $\mathbf{P}_\alpha$-Brownian motion given by

$$dB^\alpha = dB_t - \sigma^{-1}(t, X_t)b^\alpha(t, X_t)dt. \qquad (5.9)$$

Proof: This theorem follows by showing that B^α is a Brownian motion and observing that stochastic integrals under $\mathbf{P}_\alpha$ or $\mathbf{P}$ give the same process. See Chapter 8 or Example B in [59], pp. 90–91. ∎

The Theorem 5.4 gives us a clear mechanism on the basis of which the probability laws of X, more precisely X^α, described by (5.8), and X, governed by (5.6), are interconvertible via a prescribed absolutely continuous change of measure. The arrangement $(\Omega, \mathcal{F}, \mathcal{F}_t, \mathbb{P}, T, \mathbf{P}_\alpha; \alpha \in \mathcal{A})$ represents thus a controlled system.

5.2.2   Main ingredients of optimality

The aim we pursue next is to present a model of stochastic control on the basis of which optimality can be naturally formulated either in terms of supermartingales or martingales. At the heart of the following development are properties of the control family which enable commutation of operations like taking lattice supremum and conditional expectation. To this end we will introduce some functionals of the process X and examine some of their properties.

The value function and some properties

Let us assume the instantaneous rate of reward, which reflects the stage of running reward, is determined by a nonpositive real-valued, bounded measurable and strict concave function $f^\alpha(t, x)$ satisfying the same conditions as $b^\alpha(t, x)$ and assume further, the terminal reward function or terminal reward is given by a real-valued, bounded measurable and strict concave function $\phi(t, x)$ as before. We shall record these properties in the following assumption

A 5.6 *Assumptions on f^α and ϕ*

- Let the (current rate of) reward functional $f : [0, T] \times \Omega \times \mathcal{A} \longrightarrow \mathbb{R}_-$ and terminal reward $\phi : [0, T] \times \Omega \longrightarrow \mathbb{R}_-$ be nonpositive, real-valued, bounded measurable and strict concave. Further, the functionals f and ϕ satisfy the assumption A 5.5 with the obvious modifications.

Needless to say, the definition of the conditional reward and related functionals have to be modified so as to incorporate a better insight into the set $\mathfrak{P}$ containing the probability measures attached to chosen control interventions. According to the above remark we shall take expectation with respect to the measure P_α. Thus, if control $\alpha \in \mathcal{A}$ is used, the expected total reward R^α is now given by

$$R^\alpha = \mathsf{E}^\alpha \left[\int_0^T f(t, X, \alpha_t) dt + \phi(T, X_T) \right], \qquad (5.10)$$

where the notation E^α suggests the expectation is assessed with respect to the measure P_α. By analogy, for fixed $t \in [0, T]$ and $X_t = x$, the conditional remaining reward $v^\alpha(t, x)$ is given by

$$v^\alpha(t, x) = \mathsf{E}^\alpha \left[\int_t^T f^\alpha(s, X) ds + \phi(T, X_T) \Big| \mathcal{F}_t \right], \qquad (5.11)$$

As a conditional expectation, the function v^α is defined only almost surely. For a fixed $\alpha \in \mathcal{A}$ there exists for any (t, x) an equivalence class of random variables $v^\alpha(t, x)$ and we will adhere in this connection to the customary abuse of using interchangeably equivalence classes and representative elements of them; see [59, 179]. Thus, statements of the value function and related notions should be understood to mean there is at least one member in the equivalence class at stake for which the statement holds. From (5.11) follows that $v^\alpha(t, x)$ only depends on a control process α restricted to $[t, T]$ and since all the measures in $\mathfrak{P} = \{ \mathsf{P}_\alpha, \alpha \in \mathcal{A} \}$ are equivalent, as we have already seen, the null sets up to which the function $v^\alpha(t, x)$ is defined are also control-independent.

Since f and ϕ are bounded, the function $v^\alpha(t, x)$ is also bounded and hence a well-defined element of $L^1(\Omega, \mathcal{F}_t, \mathsf{P})$ for each $\alpha \in \mathcal{A}$. Therefore, the supremum $v(t, x)$ given by

$$v^\beta(t, x) = v(t, x) = \operatorname*{ess\,sup}_{\alpha \in \mathcal{A}} v^\alpha(t, x) \qquad (5.12)$$

exists and is $\mathcal{F}_t$-measurable. This result is due to the fact that $L^1(\Omega, \mathcal{F}_t, \mathsf{P})$ is a complete lattice and that the value function $v(t, X_t)$ evaluated along any trajectory corresponding to a control feasible for its initial state is a nonincreasing function of time.

To save notation, we shall write v_t for $v(t, x_t)$. Further, let $\mathcal{A}_s^t$ stand for the set of controls α, $\alpha \in \mathcal{A}$, restricted to the interval $[s, t] \subseteq [0, T]$. Let us recall the following lemma due to Rishel [179].

Lemma 5.5 (Principle of optimality) *For each fixed $\alpha \in \mathcal{A}$ and $0 \le r \le t \le T$, the value function v satisfies the inequality*

$$v_r \ge \mathsf{E}^\alpha \left[\int_r^t f^\alpha(s, X_s) ds + v_t \Big| \mathcal{F}_r \right] \qquad \mathsf{P}_\alpha \quad a.s. \qquad (5.13)$$

Sketch of the proof: One has to apply the characterisation of the value function v given by eq. (5.12) restricted to controls $\alpha \in \mathcal{A}_T^r$ and consider the subset $\hat{\mathcal{A}}_t^r, \hat{\mathcal{A}}_t^r \subseteq \mathcal{A}_t^r$, of admissible controls $\hat{\alpha}$ which are equal to α when restricted to

$(r, t]$. Let α be a fixed but arbitrary control with $\alpha \in \mathcal{A}_T^r$ and $\hat{\mathcal{A}}_T^r \subset \mathcal{A}_T^r$. The inequality

$$\bigvee_{\alpha \in \mathcal{A}_T^r} v_r^{\alpha} \geq \bigvee_{\hat{\alpha} \in \hat{\mathcal{A}}_T^r} v_r^{\hat{\alpha}}$$

holds. Using the formula for iterated expectations the right-hand side of the foregoing inequality gives

$$\bigvee_{\hat{\alpha} \in \hat{\mathcal{A}}_T^r} v_r^{\hat{\alpha}} = \mathbf{E}^{\alpha} \left[\int_r^t f^{\alpha}(s, X_s) ds \big| \mathcal{F}_r \right] + \bigvee_{\hat{\alpha} \in \hat{\mathcal{A}}_T^r} \mathbf{E}^{\alpha}[v_t^{\hat{\alpha}} | \mathcal{F}_r].$$

Now the lemma follows, since the family of random variables $\{v_t^{\alpha}, \alpha \in \mathcal{A}_T^r\} \subset L^1(\Omega, \mathcal{F}, \mathbf{P})$ has the ϵ-lattice property and hence the supremum lattice and conditional expectation operations commute.   ∎

The intuitive meaning behind the principle of optimality is that the system performs better whenever the optimal control is used right from the beginning rather than when one applies some other control for a short while and then switches to the optimal control.

Lemma 5.6 *Show that for the family $\{v_t^{\alpha}, \alpha \in \mathcal{A}\} \subset L^1(\Omega, \mathcal{F}, \mathbf{P})$ the ϵ-lattice property holds.*

Proof: Let us consider $\alpha^1, \alpha^2 \in \mathcal{A}$ and $A = \{\omega \mid v_t^{\alpha^2} \leq v_t^{\alpha^1}\} \in \mathcal{F}_t$. Then define, for $s \in (t, T]$, the control α^3 by finite mixing as

$$\alpha_s^3(\omega) = \left\{ \begin{array}{ll} \alpha_s^1(\omega), & \omega \in A, \\ \alpha_s^2(\omega), & \omega \in A^{\complement}. \end{array} \right.$$

For $\epsilon = 0$ and thus for every $\epsilon > 0$, it follows

$$v_t^{\alpha^3} \geq v_t^{\alpha^1} \vee v_t^{\alpha^2} \qquad \mathbf{P} \quad \text{a.s.}$$ ∎

5.2.3   Martingale characterisation of optimality

The principle of optimality obtained by Rishel can be phrased in terms of martingale properties underlying the same function. To this end we resort to the notion of conditional optimality on the basis of which we introduce the process of conditional optimal reward.

The process of conditional optimal reward

From the stochastic analogue of Bellman's principle of optimality we gain insight into basic properties of the value function. Let us consider (9.10) and (9.11). Using the simple notation introduced above, let us rewrite (9.11) as

$$M_t^{\alpha} = \int_0^t f^{\alpha}(s, X_s) ds + v_t \tag{5.14}$$

and note that for $t = 0$ and $\alpha \in \mathcal{A}$, we have

$$M_0^\alpha = v_0 = \bigvee_{\alpha \in \mathcal{A}_T^0} v^\alpha(0, x_0) \tag{5.15}$$

since x_0 is assumed to be a fixed constant, see 5.6 and (5.10). A glance at (5.11) and (5.10) makes clear that the function v_0 represents the "expected total reward" or "maximum expected reward". Further, for any $\alpha \in \mathcal{A}$ and $t = T$ we get

$$M_T^\alpha = \int_0^T f^\alpha(s, X_s)ds + \phi(T, X_T) = R^\alpha, \tag{5.16}$$

which represents on the other hand the total reward or "sample reward" associated with the control α, as a glance at (9.1) confirms. The principle of optimality given by Theorem 9.4 can now be restated as follows.

Theorem 5.7 *The process $M = (M_t^\alpha)_{0 \le t \le T}$ is a P_α-supermartingale for any admissible $\alpha \in \mathcal{A}$. The control $\alpha \in \mathcal{A}$ is optimal if and only if the process M is a P_α-martingale, i.e. if and only if α gives the maximum expected reward.*

Proof: Since the term $\int_0^r f^\alpha(s, X_s)ds$ is $\mathcal{F}_r$-measurable, adding this expression to both sides of (5.13), one gets

$$M_r^\alpha \ge \mathsf{E}^\alpha\left[M_t^\alpha | \mathcal{F}_r\right]$$

for $0 \le r \le t \le T$. That means, the process M is a P_α-supermartingale. Now, if M is a P_α-martingale then from (5.14) and (5.15) follows

$$\mathsf{E}^\alpha\left[M_T^\alpha\right] = \mathsf{E}^\alpha\left[M_0^\alpha\right] = v_0$$

and α is optimal.

Conversely, if α is optimal then for any t we obtain

$$v_0 = \mathsf{E}^\alpha\left[\int_0^t f^\alpha(s, X_s)ds + v_t^\alpha\right] \tag{5.17}$$

and from (5.13) follows

$$v_0 \ge \mathsf{E}^\alpha\left[\int_0^t f^\alpha(s, X_s)ds + v_t\right] \tag{5.18}$$

Hence combining (5.17) and (5.18) one has

$$0 \ge \mathsf{E}^\alpha\left[v_t - v_t^\alpha\right]$$

which together with (5.12) gives

$$v_t = v_t^\alpha \qquad \mathsf{P}_\alpha \quad \text{a.s.} \tag{5.19}$$

Now, adding $\int_0^t f^\alpha(s, X_s)ds$ to both sides of (5.19) and using (5.11), (5.14) and (5.16), follows

$$M_t^\alpha = \mathsf{E}^\alpha\left[M_T^\alpha | \mathcal{F}_t\right]$$

and the theorem is proved.   ∎

The martingale version of the principle of optimality we have obtained is based
on the process of conditional optimal reward and this has been constructed
with the help of the value function which we have introduced without any direct
connection with the solution of PDE's linked to the (HJB) problem. Therefore,
in order to get the adjoint state process we should follow also another way; this
will be done in the next section. We close this section examining the right-
continuity of the value function.

Some reflections on right-continuity

The objective is then to prove that under the prevailing conditions the func-
tion $t \longrightarrow \mathsf{E}\left[M_t^\alpha\right]$, for a fixed $\alpha \in \mathcal{A}$, is right-continuous. Hence, the P_α-
supermartingale $M = (M_t^\alpha)_{t \geq 0}$ would have a right-continuous modification
which is càdlàg, i.e. continuous on the right and has limits on the left; see
[50], [60] gives complete proofs of this result. One concludes also that the value function v has a càdlàg version. Finally, since f^α and ϕ are bounded the
P_α-martingale M is of class $\mathcal{D}$ and has therefore a Doob-Meyer decomposition.

To obtain the desired right-continuity, one uses the fact the state process
has a continuous version, the filtration it generates is right-continuous and the
underlying exponential martingale $m(\alpha)$ for fixed $\alpha \in \mathcal{A}$ is right-continuous and
uniformly integrable. These features deliver the right-continuity of $\mathsf{E}^\alpha[M^\alpha]$,
from which follows M^α has a right-continuity modification.

Let us observe that behind the function M_t^α we have a control process
consisting of α up to time t and the optimal control β from t on. Let us
consider further a function $M_{t+h}^{\widehat\alpha}$ with $\widehat\alpha \in \hat{\mathcal{A}}_{t+h}$, where $\hat{\mathcal{A}}_{t+h}$ stands for the
family of controls coinciding with α up to time $t + h$ and $h > 0$. Here α
represents a fixed but arbitrary control. Further, we assume α^2 is given by

$$\alpha_s^2 = \begin{cases} \alpha_s, & s \in [0, t+h] \\ \alpha_s^1, & s \in [t+h, T] \end{cases}$$

This suggests the controls α^1 and α^2 will differ on $(t, t+h]$. Therefore, to
examine $M_t^{\alpha^1} - M_{t+h}^{\alpha^2}$ and $v_{t+h} - v_t$ we will need to concentrate only on this
interval. Concerning the underlying probability laws, one needs also to study
the incremental difference on $(t, t+h]$, i.e. $m_{t,t+h}(\alpha^1) - m_{t,t+h}(\alpha^2)$, as (9.16)
and (9.17) suggest, provided one takes care of working with truncated control
processes. Thus, we have

$$\widehat{m}_T = m_t(\alpha) \left[m_{t,t+h}(\alpha^1) - m_{t,t+h}(\alpha^2) \right] m_{t+h,T}(\alpha^1),$$

on the basis of which one computes the average of the mentioned incremental
differences.

Theorem 5.8 *The value function v and the conditional optimal reward M^α
have a continuous version which are càdlàg.*

Sketch of the proof: Let us consider $\mathbf{E}^\alpha[M_t^\alpha]$. Then we have

$$
\mathbf{E}^\alpha[M_t^\alpha] = \int_\Omega \left\{ \int_0^t f^\alpha(s, X_s) ds + v_t \right\} d\mathbf{P}_\alpha
$$

$$
= \int_\Omega m_t(\alpha) \left\{ \int_0^t f^\alpha(s, X_s) ds + v_t \right\} d\mathbf{P}.
$$

From which we obtain that for any $\epsilon > 0$ and $\alpha^1 \in \mathcal{A}$ with $\mathcal{A}(\alpha^1, t) = \mathcal{A}(\alpha, t)$, the inequality

$$
\int_\Omega \int_0^t f^\alpha(s, X_s) d\mathbf{P}_{\alpha|_t} + \int_\Omega v_a^{\alpha^1} d\mathbf{P}_{\alpha|_t} \geq \mathbf{E}^\alpha[M_t^\alpha] - \epsilon,
$$

holds, where $\mathbf{P}_{\alpha|_t} = m_t(\alpha) d\mathbf{P}$. Knowing the cost function and terminal reward are bounded, we may choose K so that

$$
\left| \int_\Omega \int_t^{t+h} f^\alpha(s, X_s) ds\, d\mathbf{P}_{\alpha|_t} \right| \leq Kh.
$$

Further, we have

$$
\Delta_{t,t+h} = |v_{t+h} - v_t| \leq |v_{t+h} - v_t^{\hat\alpha}|
$$

$$
\leq Kh + \left| \int_\Omega \widehat{m}_T \left\{ \int_{t+h}^T f^\beta(s, X_s) ds + \phi(T, X_T) \right\} d\mathbf{P} \right|,
$$

from which follows that for $h \to \infty$, $\Delta_{t,t+h} \to 0$. Moreover, since M^α is a $\mathbf{P}_\alpha$-supermartingale, we have

$$
\int_\Omega M_t^\alpha d\mathbf{P}_\alpha \geq \int_\Omega M_{t+h}^\alpha d\mathbf{P}_a.
$$

However, for $\widehat\alpha \in \hat{\mathcal{A}}_{t+h}$, we have

$$
\int_\Omega M_{t+h}^\alpha d\mathbf{P}_\alpha \geq \int_\Omega \left\{ \int_0^T f^{\alpha^2}(s, X_s) ds + \phi(T, X_T) \right\} d\mathbf{P}_{\alpha^2}
$$

and

$$
\int_\Omega \left\{ \int_0^T f^{\alpha^1}(s, X_s) ds + \phi(T, X_T) \right\} d\mathbf{P}_{\alpha^1} + \epsilon \geq \int_\Omega M_t^\alpha d\mathbf{P}_\alpha.
$$

Therefore, as h goes to zero,

$$
\int_\Omega [M_t^\alpha - M_{t+h}^\alpha] d\mathbf{P} = \left| \int_\Omega [M_t^\alpha - M_{t+h}^\alpha] \, d\mathbf{P} \right|
$$

goes to zero, which shows $\mathbf{E}^\alpha[M^\alpha]$ is right-continuous. Then, M^α has a right-continuous modification and v as well. ∎

5.3 Semimartingale principle of optimality

At this point we know how the theory of semimartingales facilitates a treatment
of the problem of steering a dynamic system in a relatively general framework
without loosing transparency and intuitive appealing. In this sense a control
intervention causes changes in the informational structures of the system and
these are transferred to its probabilistic patterns of evolution, as reflected in
an absolutely continuous change of probability measures. These in turn are
accomplished through a clearly defined switching mechanism resorting to operations among admissible control actions. The resulting control techniques
make possible to deal with questions of existence of a unique value function,
see (9.8), and about its properties, which are essential for the establishment of
an appropriate principle of optimality.

In this connection the particular form of the process of conditional reward
based on the functional R_s^α, the characterisation of the value function $v(s, x)$
and the equivalence of the probability measures P_α to the same reference measure P play a crucial role. To see this we shall in the present section focus on
the following features of the martingale approach:

- The Doob-Meyer decomposition as a bridge between the basic controlled
  processes and the adjoint process.

- The characterisation of an optimal control chosen among those optimising
  a certain Hamiltonian.

- Study of the adjoint state based on the stochastic integral representation
  of martingales.

However, we shall restrict ourselves to the case of continuous processes of the
diffusion type under relatively standard regularity assumptions on the coefficients.

5.3.1   Doob-Meyer decomposition and optimality

A nice feature of semimartingale techniques solving a control problem like (5.5)
is that of breaking the old tradition of resorting to the construction of an
optimal control β by maximising the Hamiltonian defined by (9.13), a definition
usually forwarded on the basis of a circular argument. Namely, setting $p = \nabla_x v$, a fact which entails at least implicitly, the Bellman equations of dynamic
programming, characterising the value function $v(s, x)$, have been solved.

Existence of the adjoint and the Hamiltonian

As our heuristic reflections around (5.14) suggest, an alternative emerges elaborating on the Doob-Meyer decomposition of the process of conditional optimal
reward $M^\alpha = (M_t^\alpha)_{t \geq 0}$ and looking for a stochastic integral representation of
its martingale term; see [47]. In this way one gets a process $p = (p_t)_{t \geq 0}$, the
"adjoint process" in the terminology of the control theory, which is defined
independently of the existence of any optimal control. This process shall play
the role of the gradient of the value function, i.e. $p = \nabla_x v$, as customary.

For any $\alpha \in \mathcal{A}$ the Doob-Meyer decomposition guarantees, under the conditions discussed above, the existence and uniqueness of a predictable decreasing process $A^\alpha = (A^\alpha)_{t \geq 0}$, with $A_0^\alpha = 0$, and a uniformly integrable $\mathbf{P}_\alpha$-martingale $N^\alpha = (N_t^\alpha)_{t \geq 0}$, with $N_0^\alpha = 0$, such that

$$M_t^\alpha = v_0 + A_t^\alpha + N_t^\alpha. \tag{5.20}$$

Let us note, a representation of the martingale N^α as stochastic integral with respect to the $\mathbf{P}_\alpha$-Brownian motion B^α would follow from standard results, if the filtration $\mathcal{F}_t$ was generated by B^α and N^α was square integrable, see [139, 129]. However, one can prove that all square integrable $\mathcal{F}_t$-martingales are representable as stochastic integrals of B^α, see [76, 48, 139].

It is worth noting that from (5.6) and the Lipschitz property of $\sigma(t, x)$ follows that the filtration $\mathcal{F}_t$ is generated by $\{X_s : s \leq t\}$ or, equivalently, by $\{B_s : s \leq t\}$, i.e. $\mathcal{F}_t^B = \mathcal{F}_t^x = \mathcal{F}_t$. Since the process X given by (5.8) is only a weak solution, the Brownian motion B^α given by (5.9) does not necessarily generate $\mathcal{F}_t$, i.e. $\mathcal{F}_t \subseteq \mathcal{F}_t^B$ and the inverse inclusion may not be valid. We have the following results.

Lemma 5.9 *Let $N^\alpha = (N_t^\alpha)_{t \geq 0}$, with $N_0^\alpha = 0$, be the uniformly integrable $\mathbf{P}_\alpha$-martingale which occurs in the Doob-Meyer decomposition, (5.20). Then, there is an $\mathcal{F}_t$-predictable process $p = (p_t)_{t \geq 0}$, for which there is an increasing sequence of stopping times τ^n with $\lim \tau^n = \tau$ a.s. and $\mathbf{E}\left[\int_0^\tau |p_s| ds\right] < \infty$, such that*

$$N_t^\alpha = \int_0^t p_s \, \sigma(s, X_s) dB_s^\alpha \quad a.s. \tag{5.21}$$

Proof: See [60], Corollary 16.23. ∎

Lemma 5.10 *Let $\bar{\alpha} \in \mathcal{A}$ be an arbitrary but fixed admissible control process and p be the adjoint process obtained in Lemma 5.9. Then, for any arbitrary admissible control α we have*

$$A_t^\alpha = A_t^{\bar{\alpha}} + \int_0^t [H_s(\alpha) - H_s(\bar{\alpha})] \, ds, \tag{5.22}$$

$$N_t^\alpha = \int_0^t p_s \, \sigma(s, X_s) dB_s^\alpha, \tag{5.23}$$

where the functional $H_s(\alpha)$ is defined by

$$H_s(\alpha) = p_s \, b^\alpha(s, X_s) + f^\alpha(s, X_s). \tag{5.24}$$

Proof: Let $\bar{\alpha} \in \mathcal{A}$ be an arbitrary but fixed control process. Then, taking into account the definition (5.14) of the conditional optimal reward M^α, the Doob-Meyer decomposition (5.20) and the stochastic integral representation (5.21) we have for $\bar{\alpha} \in \mathcal{A}$

$$v_t = v_0 + A_t^{\bar{\alpha}} + \int_0^t p_s \, \sigma(s, X_s) dB_s^{\bar{\alpha}} - \int_0^t f^{\bar{\alpha}}(s, X_s) ds. \tag{5.25}$$

Further, for any other arbitrary admissible control function $\alpha \in \mathcal{A}$, the process $M^\alpha = (M_t^\alpha)_{t \geq 0}$ can be written, according to (5.14) and (5.20) as

$$M_t^\alpha = \int_0^t f^\alpha(s, X_s) ds + v_t, \tag{5.26}$$

$$M_t^\alpha = v_0 + A_t^\alpha + N_t^\alpha. \tag{5.27}$$

Substituting v_t in (5.26) by (5.25) and making a few arrangements give

$$M_t^\alpha = v_0 + \left\{ A_t^{\bar{\alpha}} + \int_0^t f^\alpha(s, X_s) ds - \int_0^t f^{\bar{\alpha}}(s, X_s) ds \right\} +$$
$$+ \int_0^t p_s\, \sigma(s, X_s) dB_s^{\bar{\alpha}},$$

$$M_t^\alpha = v_0 + \left\{ A_t^{\bar{\alpha}} + \int_0^t [p_s\, b^\alpha(s, X_s) + f^\alpha(s, X_s)]\, ds - \right.$$
$$\left. - \int_0^t [p_s\, b^{\bar{\alpha}}(s, X_s) + f^{\bar{\alpha}}(s, X_s)]\, ds \right\} +$$
$$+ \int_0^t p_s\, [\sigma(s, X_s) dB_s^{\bar{\alpha}} + b^{\bar{\alpha}}(s, X_s) ds] -$$
$$- \int_0^t p_s\, b^\alpha(s, X_s) ds,$$

which taking into account (5.24) and (5.9) twice, once for $\bar{\alpha}$ and then for α, leads to

$$M_t^\alpha = v_0 + \left\{ A_t^{\bar{\alpha}} + \int_0^t [H_s(\alpha) - H_s(\bar{\alpha})]\, ds \right\} + \int_0^t p_s\, \sigma(s, X_s) dB_s^\alpha. \tag{5.28}$$

Since for any admissible control the Doob-Meyer decomposition is unique, comparing (5.27) and (5.28) gives for $\alpha \in \mathcal{A}$ the desired representations for A^α and N^α.   ∎

The Doob-Meyer decomposition (5.28) or, equivalently, (5.22) and (5.23) are very helpful examining efficiency of alternative controls. For instance, keeping the control $\bar{\alpha}$ fix and letting α vary in $\mathcal{A}$, these relationships shed light on the impact of both controls upon X as reflected, say, in the Hamiltonian $H_s(\bar{\alpha})$ for s fixed. At this point we like to stress the fact, the Brownian motion B and the adjoint process p as well are control independent.

The stochastic maximum principle

The probability structure and assumptions characterising the system driven by the controlled diffusion studied in the foregoing section, put us in a position to formulate a relatively simple criterion for optimality. This is the object of the following theorem, which states the necessary and sufficient conditions for any admissible control $\alpha \in \mathcal{A}$ to be optimal.

Theorem 5.11   *a) (A necessary condition for optimality) If the admissible control process $\beta \in \mathcal{A}$ is optimal, then it is also $(\mathsf{P}_\beta(d\omega) \otimes ds$ a.s.) optimal for the Hamiltonian H_s defined by (5.24).*

b) (A sufficient condition for optimality) For an admissible control $\beta \in \mathcal{A}$ consider the P_β-martingale $R^\beta = (R_t^\beta)_{t \geq 0}$ defined by

$$R_t^\beta = \mathsf{E}^\beta \left[R^\beta(\omega) | \mathcal{F}_t \right] = \mathsf{E}^\beta \left[M_T^\beta \middle| \mathcal{F}_t \right] = \int_0^t f^\beta(s, X_s) ds + v_t^\beta. \quad (5.29)$$

If for any other admissible control $\alpha \in \mathcal{A}$, the process $I^\alpha = (I_t^\alpha)_{t \geq 0}$ with the function I_t^α given by

$$I_t^\alpha = \int_0^t f^\alpha(s, X_s) ds + v_t^\beta, \quad (5.30)$$

is a P_α-supermartingale, then the admissible control process $\beta \in \mathcal{A}$ is optimal.

Proof: a) If the admissible control β is optimal, then it follows from Theorem 5.7 that $M^\beta = (M_t^\beta)_{\{0 \leq t \leq T\}}$ is a P_β-martingale. Thus, using our comparison formula (5.22), after setting $\bar{\alpha} = \beta$ and the fact that for the optimal control process the conditional optimal reward becomes a martingale which entails $A_t^{\bar{\alpha}} = A_t^\beta = 0$, we conclude the integrand in (5.22) cannot be positive. More precisely, we have for any $\bar{\alpha} \in \mathcal{A}$

$$H_s(\alpha) - H_s(\beta) \leq 0 \qquad \mathsf{P}_\beta(d\omega) \otimes ds \quad \text{a.s.}$$

so that for any admissible $\alpha \in \mathcal{A}(s, \beta)$ the inequality

$$H_s(\alpha) \leq H_s(\beta) \qquad \mathsf{P}_\beta(d\omega) \otimes ds \quad \text{a.s.} \quad (5.31)$$

holds. This proves part a) of the theorem.

b) Suppose now that for any admissible control process $\alpha \in \mathcal{A}$ other than β the process I^α is a P_α-supermartingale. Then, adding and subtracting $\int_0^t f^\beta(s, X_s) ds$ in (5.30) and using (5.29) one gets

$$I_t^\alpha = R_t^\beta + \int_0^t \left(f^\alpha(s, X_s) - f^\beta(s, X_s) \right) ds.$$

Then, setting $t = 0$ in (5.29) and (5.30) one obtains

$$v_0^\beta = R_0^\beta = I_0^\alpha = \mathsf{E}^\alpha \{I_0^\alpha\} \geq \mathsf{E}^\alpha \{I_T^\alpha\} = v_0^\alpha.$$

That is, the process β maximises the total expected reward and is hence optimal. ∎

5.3.2   Integral representation of the adjoint process

So far we have made some progess which puts us in the way of characterising the unique adjoint process $p = (p_t)_{t \geq 0}$, obtained in Lemma 5.9 in the stochastic integral representation (5.21) of the martingale N^α, with the help of which the Hamiltonian function $H(\alpha) = (H_t(\alpha))_{\{0 \leq t \leq T\}}$ has been defined; regarding the integral representation of a martingale see Chapter 8. The Hamiltonian is helpful comparing alternative control processes and builds the starting point, together with the value function, the conditional remaining reward and other functionals, to the elaboration of a nice criterion of optimality. Nevertheless, our conditions of optimality are bound to hold only for fixed although arbitrary points of time. Thus, in order to be able to learn more about the dynamic features of our statement of optimality, it is necessary to derive the adjoint state equation. As a first step, one needs to explore some questions on the differentiability of functionals on which this optimality rests.

Fréchet differentiability

At a fairly general level, one needs a stochastic integral representation of Fréchet differentiable functionals of paths of the underlying diffusion process. In its more simple form it is known, the functional of a Brownian motion can be represented as the sum of a constant and stochastic integral of the Itô type. Similar representations are also known for homogeneous additive functionals of a Brownian motion with finite expectations. Various known results associated with Clark's representation attempt to characterise an adapted process ϕ with

$$\int_0^T \mathsf{E}|\phi_t|^2 dt < \infty$$

such that

$$M_t = \mathsf{E}\{M_0\} + \int_0^t \phi_s dB_s, \qquad 0 \leq t \leq T,$$

for a square integrable $\mathcal{F}_t$-adapted process M with continuous sample paths. The most basic result is related to a functional L on the space of continuous functions on $[0, T]$ for which

$$M_t = \mathsf{E}\{L(B)|\mathcal{F}_t\},$$

with L suitably smooth and ϕ given in terms of the derivative of L. Most of the existing results in regard to stochastic control processes are presented in [101, 102, 103], and rest upon the representation of functionals of Itô processes as stochastic integrals, see [139]. In the sequel we shall outline heuristically, following [45, 4], some of the ideas on which we work out representations of the conditional optimal reward and the adjoint process.

Let us denote by $Y = (Y_t)_{\{0 \leq t \leq T\}}$ the process associated with the optimal control process β, which is driven by the SDE given by (5.8). To begin with we assume for any $\alpha \in \mathcal{A}$, the sample reward R^α is a smooth functional of the diffusion X given by (5.8) and its coefficients fulfil the following additional requirements:

A 5.7 *Further assumptions on $b^\alpha(t, x)$ and $\sigma(t, x)$*

1. For all $(t, x) \in [0, T] \times \Omega$ the inequality

$$\sigma(t, x_t) \cdot \sigma^*(t, x_t) \geq \kappa I_n \geq 0,$$

   where κ is a positive constant independent of (t, x) and I_n stands for the n-dimensional identity matrix.

2. $\mathbf{P}\left\{\int_0^T |b^\alpha(s, x_s)| ds < \infty\right\} = 1.$

3. The functions $(\sigma(t, x_t))^{(ij)}$, $i, j = 1, \ldots, n$, are Hölder continuous in x and uniformly in $(t, x) \in [0, T] \times \mathbb{R}^n$.

Under these assumptions the optimal state process driven by the SDE given by (5.8) is such that $\sup_{t \in [0,T]}\{\mathbf{E}|Y_t|^m\} < \infty$ for any integer m. Furthermore, on the conditional reward functional we need the assumptions to follow.

Let us recall first that saying the functional $L : \Omega \to \mathbb{R}$ is Fréchet differentiable at $x \in \Omega$ means, there exists a continuous linear functional $L'(x)$ such that

$$L(x + y) - L(x) = L'(x)y + o(\|y\|) \quad \text{for all } y \in \Omega.$$

Further, L is called Fréchet differentiable if it is differentiable at each $x \in \Omega$. By the Riesz representation theorem one has for each $x \in \Omega$

$$L'(x)y = \int_0^T y(s)\mu_x(ds),$$

where μ_x is a signed measure. Since $\mu_x(t) = \mu_x[0, T] = \int_0^T \mu_x(ds)$ may be taken to be right-continuous, it follows that $\|L'(x)\|_{\Omega^*} = TV(\mu_x)$, where Ω^* stands for the dual space of Ω and $\|L'(x)\|_{\Omega^*}$ for the norm of $L'(x)$ in Ω^*.

A 5.8 *Fréchet differentiability of M_t^α*

- Suppose $\beta \in \mathcal{A}$ is the optimal control process and the ($\mathcal{F}_t$-measurable random variable) functional $M_T^\beta : \Omega \to \mathbb{R}_-$, see (5.16), given by

$$M_T^\beta = \int_0^T f(s, Y_s, \beta(s, Y_s)) ds + \phi(T, Y_T) \tag{5.32}$$

  is Fréchet differentiable in x. Here Y stands for the optimal process given by (5.8), where Y is an element of the sample space Ω, i.e. the set of continuous functions from $[0, T]$ to $\mathbb{R}^n$.

- There exist positive integers m_1 and m_2 such that

$$|M_t^\beta(x)| + TV(\mu_x) \leq m_1(1 + \|x\|)^{m_2}, \tag{5.33}$$

  so that $\mathbf{E}\{M_T(y)\}^2 < \infty$ and $M_T(y)$ has certainly a representation of the type mentioned.

- The $\mathbb{R}^n$-valued Radon measure μ_x is continuous in x in the weak topology.

Lemma 5.12 *Under the assumptions A 5.7 and A 5.8, there exists a map $\Pi : \Omega \to \Omega^*$ such that for $x, y \in \Omega$*

$$M_T^\beta(x + y) = M_T^\beta(x) + \Pi(x)(y) + o(\|y\|). \tag{5.34}$$

Besides for each $x \in \Omega$, there is an $\mathbb{R}^n$-valued Radon measure $\mu_x(s)$ for $y \in \Omega$ such that

$$\Pi(x)(y) = \int_{[0,T]} y(s)\mu_x(ds), \tag{5.35}$$

where $\|\Pi(x)\|_{\Omega^*} = TV(\mu_x)$. One denotes by $\mu_x(t)$ the right-continuous bounded variation function $\mu_x([0,t])$ corresponding to the Fréchet derivative of the functional M_T^α at x.

The linearised state equations

In order to continue our search for dynamic features of the adjoint process one needs to identify the fundamental solution associated with the linearisation of the state equation. This is also a way to account for deviations from the optimal state processes resulting from various random factors.

Let $X_{t,s}(x)$ denote the position of the process X at time t, given that it started at time s from the state x and evolves according to (5.8); and further let us set $\Phi_{t,s} = \frac{\partial}{\partial x} X_{t,s}(x)$. For simplicity we write $X_{t,0} = X_t$, unless any confusion may arise. For $0 \le s \le t \le T$, here $\Phi_{t,s}$ stands thus for the matrix valued process defined by

$$d\Phi_{t,s} = A(t, Y_{t,s})\Phi_{t,s}dt + C(t, Y_{t,s})\Phi_{t,s}dB_t, \tag{5.36}$$

$$\Phi_{s,s} = I_n, \tag{5.37}$$

where $A(t, Y_{t,s})$ and $C(t, Y_{t,s})$ are matrices with entries (ij) given by the partial derivatives $\nabla_x^{(j)} b^{(i)}(t, y)$ and $\nabla_x^{(j)} \sigma^{(ij)}(t, y)$ respectively. Here $i, j = 1, 2, \ldots, n$; $k = 1, 2, \ldots, m$, I_n the identity matrix and the optimal state process Y is the random trajectory about which (5.8) is linearised. The process $\Phi_t = (\Phi_{t,s})_{s \ge 0}$ is called the *fundamental solution* and sometimes the transition matrix. It satisfies the equality $\Phi_{T,0} = \Phi_{T,t}\Phi_{t,0}$, as one easily can see from the relationship $X_{T,0}(x_0) = X_{T,t}(X_{t,0}(x_0))$ and the change rule.

The foregoing *composition rule*, as it is called in Linear Systems [32], says the solution starting at time $t_0 = 0$ from the state x_0 and moving toward the state X_T can be considered as the composition of two trajectories: first, one representing the solution starting at time $t_0 = 0$ from the state x_0 moving toward the state X_t, and second, the one starting at time t from the state x_t and moving toward the state X_T. Setting $x = X_t(x_0)$, we have $X_T(x_0) = X_{T,t}(x)$, too.

Let us stress once more that the optimal state process corresponding to the optimal control process β will be denoted by $Y = (Y_t)_{\{0 \le t \le T\}}$ and the notation $Y_{t,s}(x)$ bears the same meaning. Further, let us consider the value function (9.8) in the following form

$$v(t, x) = \mathsf{E}^\beta \left[\int_t^T f^\beta(s, Y_{s,t}(x))ds + \phi(T, Y_{T,t}(x)) \Big| \mathcal{F}_t \right],$$

where the optimal process Y restricted to the interval $[t, T]$ is such that for any $s \in [t, T]$ the state $Y_{s,t}(x)$ satisfies

$$\begin{aligned} dY_{s,t}(x) &= b(s, Y_{s,t}(x), \beta(s, Y_{s,t}(x)))ds + \sigma(s, Y_{s,t}(x))dB_s^\beta, \\ Y_{t,t}(x) &= x. \end{aligned}$$

Now we are prepared to study some path functionals of the diffusion process Y. We have the following result:

Theorem 5.13 *Assume A 5.7 and A 5.8. Then, the relationship*

$$M_T^\beta = \mathsf{E}^\beta \left[M_T^\beta \right] + \int_0^T e_t \, dB_t^\beta$$

holds.

Proof: Since β is the optimal control process, we have from (5.20) and (5.21)

$$M_t^\beta = v_0 + \int_0^t p_s \, \sigma(s, Y_s)dB_s^\beta, \tag{5.38}$$

where the value function v evaluated at (t, x) with $(t, x) \in (0, T) \times \mathbb{R}^n$ satisfies following parabolic differential equation and boundary condition

$$\begin{aligned} 0 &= \frac{\partial v}{\partial t} + \frac{1}{2} \sum_{i,j=1}^n (a(t, x))^{(ij)} \nabla_x^{(i)} v \nabla_x^{(j)} v + \\ &\quad + \sum_{i=1}^n (b^\beta(t, x))^{(i)} \nabla_x^{(i)} v + f^\beta(t, x), \tag{5.39} \\ v(T, x) &= \phi(T, x_T(x)), \quad x \in \mathbb{R}^n. \tag{5.40} \end{aligned}$$

Recall that $\sigma(t, x)\sigma^*(t, x) = a(t, x)$, where σ^* stands for the transpose of σ. Expanding $v(t, X_t(x))$ by the Itô rule and making use of the foregoing parabolic PDE yields

$$v(t, x) = \mathsf{E}^\beta \left[\int_t^T f^\beta(s, Y_{s,t}(x))ds + \phi(T, Y_{T,t}(x)) \Big| \mathcal{F}_t \right]. \tag{5.41}$$

Moreover, applying Itô's formula for change of variable to $v(s, X_{s,t}(x))$ from time t till T, see (8.17), one gets

$$\begin{aligned} v(T, Y_T(x)) &= v(t, x) + \int_t^T \left(\frac{\partial}{\partial s} + \mathsf{L}^{\sigma,b} \right) v(s, Y_{s,t}(x))ds + \\ &\quad + \int_t^T \nabla_x v(s, Y_{s,t}(x)) \sigma(s, Y_{s,t}(x))dB_s^\alpha \end{aligned}$$

which together with (5.39) and (5.40) gives

$$\phi(T, Y_T(x)) = -\int_t^T f^\beta(s, Y_{s,t}(x))ds + v(t, x) +$$

$$+ \int_t^T \nabla_x v(s, Y_{s,t}(x))\, \sigma(s, Y_{s,t}(x))dB_s^\alpha.$$

Setting $t = 0$ in the right-hand side of the foregoing equation and comparing it with (5.32) gives the following representation for M_T^β

$$M_T^\beta = v_0 + \int_0^T \nabla_x v(s, Y_{s,t}(x))\sigma(s, Y_{s,t}(x))dB_s^\beta. \tag{5.42}$$

Setting $t = T$ in (5.38), a comparison with (5.42) suggests formally a crucial relation between p and $\nabla_x v$. On the other hand, using (5.41) with $t = 0$ and (5.26) with $t = T$ in (5.38) delivers the following enlightening representation for M_T

$$M_T^\beta = \mathsf{E}^\beta \left[M_T^\beta \right] + \int_0^T \nabla_x v(s, Y_s(x))\sigma(s, Y_s(x))dB_s^\beta. \tag{5.43}$$

Furthermore, taking into account the Fréchet differentiability of M_T^β and (5.43) completes the picture of the relationship between p, $\nabla_x v$ and y, see (5.34) and (5.38). Indeed, we get comparing formally the martingale terms and using the change rule that

$$p_t = \nabla_x v(t, Y_t(x))$$

$$= \nabla_x \mathsf{E}^\beta \left\{ \int_t^T f^\beta(s, Y_{s,t}(x))ds + \phi(T, Y_{T,t}(x)) \Big| \mathcal{F}_t \right\}$$

$$= \mathsf{E}^\beta \left\{ \int_t^T \nabla_x f^\beta(s, Y_{s,t}(x))\Phi_{s,t}\, ds + \nabla\phi(T, Y_{T,t}(x))\Phi_{T,t} \Big| \mathcal{F}_t \right\}.$$

Finally, taking into account the composition rule and that for x_0 known as the σ-algebra $\sigma(Y_s : s \le t) \subset \mathcal{F}_t$, we can rewrite p_t as

$$p_t = \mathsf{E}^\beta \left\{ \int_t^T \nabla_x f^\beta(s, Y_{s,t}(x))\Phi_{s,t}ds + \nabla_x \phi(T, Y_{T,t}(x))\Phi_T \Big| \mathcal{F}_t \right\} \Phi_t^{-1}, \tag{5.44}$$

which delivers the result after setting

$$e_t = p_t\, \sigma(t, Y_t(x))$$

and writing down (5.41) for $t = 0$ and (5.38) for $t = T$.   ∎

The main line of argumentation we followed while deriving foregoing results, is deeply related to the system theoretical treatment of the control of system dynamics and focuses on the fundamental solution and associated flows as a way of accounting for perturbations around a stochastic optimal state process. A fairly readable treatment of the background of these issues is given in [14]. A clear presentation of the characterisation of the integrand in the representation of a martingale as a stochastic integral using stochastic flows and the Itô differentiation rules is given in [61].

5.3.3 Dynamics of the adjoint state process

Our next objective is to get additional insights into the structure of the integrands in (5.35), (5.42) and (5.43). To this end, one has to look at the dynamic behaviour of the process p. After having worked out the time evolution of p, we shall be able to find an expression for $d(\Phi_{s,t}\nabla_x v)$, where $\Phi_{s,t}$, with $s, t \in [0, T]$, is precisely the fundamental matrix solution of

$$
\begin{aligned}
d\eta_s \;=\;& [\nabla_x b(s, Y_s, \beta(s, Y_s)) + \nabla_\alpha b(s, Y_s, \beta(s, Y_s))\nabla_x \beta(s, Y_s)]\, \eta_s\, ds + \\
& + \nabla_x \sigma^{(k)}(s, X_s)\eta_s\, dB_s^{\beta(k)}
\end{aligned}
\tag{5.45}
$$

with $\eta_s = \Phi_{s,t}$, $0 \le t \le s \le T$. Repeated indices imply here summation from 1 to n. Let us stress, the control β considered above is state adapted or feedback. The fundamental matrix equation above reduces to the case considered before whenever $\nabla_x \beta(s, Y)$ vanishes.

Attached to the fundamental (matrix) solution $\Phi_{t,s}$ of the Itô SDE (5.45), see also (5.36), there is a fundamental adjoint solution that we denote by $\Psi_{s,t}$ which in the language of systems theory propagate the solution of the original equation backwards in time, i.e. $\Psi_{s,t} = \Phi_{t,s}^{-1}$. A simple application of the Itô rule, leads to the equation

$$
\begin{aligned}
d\Psi_{s,t} \;=\;& -\Psi_{s,t}\left[\nabla_x b^\beta(s, Y_s) - \nabla_x \sigma^{(k)}(s, Y_s)(\nabla_x \sigma^{(k)}(s, Y_s))^*\right] ds \\
& -\Psi_{s,t}\nabla_x \sigma^{(k)} dB_s^{\beta(k)},
\end{aligned}
\tag{5.46}
$$

$$
\Psi_{t,t} \;=\; I_n,
\tag{5.47}
$$

where again repeated indices indicate summation from 1 to n and $0 \le s \le t \le T$ holds. Further, we shall simply write Ψ_t for $\Psi_{t,T}$ whenever no confusion may arise. Compare with (6.74) in Chapter 6.

The derivation of the adjoint state equation to follow is an extension of the one given by Bensoussan [16].

Theorem 5.14 *The assumptions of Theorem 5.13 hold. Let us consider the adjoint state process $p = (p_t)_{t\ge0}$ introduced in Lemma 5.9. Then, there exists a $\mathcal{F}_t$-predictable process $r = (r_t)_{t\ge0}$, for which there is an increasing sequence of stopping times $\{\tau^n\}$ with $\lim \tau^n = 1$ a.s. and $\mathsf{E}\{\int_0^{\tau^n} |r_t|dt\} < 0$, so that p and r satisfy the SDE given by*

$$
- dp_s \;=\; [p_s \nabla_x b^\beta(s, Y_s) + \nabla_x f^\beta(s, Y_s) - \nabla_x \sigma^k(s, Y_s)r_s^k]\, ds + r_s dB_s^\beta,
\tag{5.48}
$$

$$
p_T \;=\; \nabla_x \phi(T, Y_T)
\tag{5.49}
$$

Proof: The objective is to prove, the process p defined by (5.44) does satisfy an SDE. To do this, rewrite (5.44) as

$$
p_t = \mathsf{E}^\beta\left[Z_t | \mathcal{F}_t\right] \Psi_t,
$$

where Z_t is simply given by the argument of the conditional expectation in (5.44), i.e. by

$$
Z_t = \int_t^T \nabla_x f^\beta(s, Y_{s,t}(x))\Phi_{s,t}\, ds + \nabla\phi(T, Y_{T,t}(x))\Phi_{T,t}.
$$

A little change in the integral expression above gives

$$
E^\beta[Z_t|\mathcal{F}_t] = -\int_0^t \nabla_x f^\beta(s, Y_s(x))\Phi_s ds +
$$

$$
+E^\beta\left[\int_0^T \nabla_x f^\beta(s, Y_s(x))\Phi_s ds + \nabla_x \phi(T, Y_T(x)\Phi_T \Big|\mathcal{F}_t\right].
$$

It is not hard to see, the expression $E^\beta[H|\mathcal{F}_t]$ is a square integrable martingale with H given by

$$
H = \int_0^T \nabla_x f^\beta(s, Y_s(x))\Phi_s ds + \nabla_x \phi(T, Y_T(x)\Phi_T.
$$

Since all square integrable $\mathcal{F}_t$-martingales are stochastic integrals of the Brownian motion $B^\beta = (B^\beta)_{t\geq 0}$, as it has been proved by Fujisaki et al. in [76], we have the stochastic integral representation

$$
E^\beta\{H|\mathcal{F}_t\} = E^\beta[H] + \int_0^t \gamma_s dB^\beta_s,
$$

where $\gamma = (\gamma_t)_{t\geq 0}$ is a uniquely defined $\mathcal{F}_t$-adapted process and for $k = 1, 2, \ldots, m$, $E\left\{\int_0^T |\gamma_t^{(k)}|^2 dt\right\} < \infty$. Setting $\zeta_t = E^\beta[Z_t|\mathcal{F}_t]$ in the representation of p above, we obtain $p_t = \zeta_t \Psi_t$ and the SDE

$$
d\zeta_t = -\nabla_x f^\beta(t, Y_t(x))\Phi_t dt + \gamma_t dB^\beta_t. \tag{5.50}
$$

Thus, a simple application of Itô's rule

$$
dp_t = \zeta_t\, d\Psi_t + d\zeta_t\, \Psi_t + d\zeta_t\, d\Psi_t,
$$

substituting (5.46) and (5.50) leads to

$$
\begin{aligned}
-dp_t = \ & p_t\left[\nabla_x b^\beta(t, Y_t) - \nabla_x \sigma^k(t, Y_t)(\nabla_x \sigma^k(t, Y_t))^*\right] dt \\
& + \left(p_t\nabla_x \sigma^k(t, Y_t) - \Psi_t\gamma_t^k\right) dB_t^{\beta(k)} + \nabla_x f^\beta(t, Y_t)dt \\
& + \nabla_x \sigma^k(t, Y_t)\Psi_t\gamma_t^k dt,
\end{aligned} \tag{5.51}
$$

after setting $r_t^k = p_t\, \nabla_x \sigma^k(t, Y_t) - \Psi_t\gamma_t^k$ and a few arrangements finish the proof. ∎

With the purpose of getting an explicit representation of the adjoint state we shall rearrange (5.51) so that we can solve it as a linear SDE. This is done by considering the form of the process r as obtained in Theorem 5.14. This is the objective of the next theorem.

Theorem 5.15 *The assumptions of Theorem 5.14 hold. Let H be a semimartingale and let Z be a continuous martingale driven by the equations*

$$
\begin{aligned}
dZ_t &= \left[\nabla_x b^\beta(t, Y_t) - \nabla_x \sigma^k(t, Y_t)(\nabla_x \sigma^k(t, Y_t))^*\right] dt + \nabla_x \sigma^k(t, Y_t)dB^{\beta(k)}, \\
dH_t &= \left[\nabla_x f^\beta(t, Y_t) + \nabla_x \sigma^k(t, Y_t)\Psi_t\gamma_t^k\right] dt - \Psi_t\gamma_t^k dB^{\beta(k)}.
\end{aligned}
$$

The semimartingales H and Z are linked by

$$p_T = H_t + \int_t^T p_{s-} dZ_s.$$

(5.52)

Then the solution $\mathcal{E}_H(Z)$ of (5.52) is given by

$$\mathcal{E}_H(Z)_t = \mathcal{E}(Z)_t \left\{ H_t + \int_{t+}^T \mathcal{E}^{-1}(Z)_s \, d(H_s - [H,Z]_s) \right\}$$

Proof: See Chapter 8.   ∎

This theorem will be very helpful in the last section. The formula is particularly simple if the diffusion term is independent of the state, since in this case the drift of the process alone gives the deterministic adjoint.

The new process $r = (r_t)_{t \geq 0}$ appearing in (5.48) gives us the opportunity of completing the definition of the functional $H(\alpha)$ introduced in Lemma 5.10. Thus, let the functional $\mathcal{H}^\alpha(t, x, p, r)$ be given by

$$\mathcal{H}^\alpha(t, X_t, p_t, r_t) = p_t b^\alpha(t, X_t) + f^\alpha(t, X_t) + \sum_{k=1}^m \sigma^k(t, X_t)^* r_t^k.$$

Since both $\sigma(t, x)$ and r_t do not depend on the control, it is easy to recognise we could have added and subtracted the expression $\sigma^k(t, x)^* r_t^k$ in our comparison formula of Lemma 5.10. Thus, the functional $H(\alpha)$ as given in (5.24) needed to be completed as above, compare with (5.24). However, we shall mention for the purpose of the present essay the functional $H(\alpha)$ will do.

Next, we shall explore the differentiability of the Hamiltonian, for which we need some additional assumptions. Let $Q = (0,T) \times G$, and $G \subseteq \mathbb{R}^n$ be an open bounded set with boundary ∂G of class $\mathbb{C}^2$, so that T is the first exit time of the process $\{(t, X_t) : t \geq 0\}$ from the open set $(0,T) \times G$, $X_0 = x_0 \in G$, and assume the following properties.

A 5.9 *Smoothness assumptions on σ, b^α, f^α and ϕ*

1. $\sigma(t, x)$ is of class $\mathbb{C}^{1,2}(\overline{Q})$. In particular, the partial derivatives $\partial_t \sigma^k(t, x)$ and $\nabla_x^{(k)} \sigma^{(ij)}(t, x)$ exist and are continuous in the arguments, $\sigma^{-1}(t, x)$ exists on $\overline{Q}$

2. $b^\alpha(t, x)$ and $f^\alpha(t, x)$ are of class $\mathbb{C}^{1,1}(\overline{Q} \times A)$. In particular, the partial derivatives $\nabla_x^{(k)} b^\beta(t, x)$ and $\nabla_x^{(k)} f^\beta(t, x)$ exist and they are continuous on $\overline{Q} \times A$.

3. $\phi(t, x)$ is of class $\mathbb{C}^2(G)$. In particular $\phi(T, x)$ is of class $\mathbb{C}^2(G)$ and $\phi(t, x)$ on $(0,T) \times \partial G$ is of class $\mathbb{C}^{1,2}(\overline{Q})$.

Under these additional assumption $v(t, x)$ becomes of class $\mathbb{C}^{1,2}(Q) \cap \mathbb{C}^{0,1}(\overline{Q})$, see [70]. Let us consider the parabolic differential equation given by (5.39) and

(5.40) which in the light of the newly introduced Hamiltonian looks like:

$$\mathcal{H}(t,x,\nabla_x v,r) \;=\; -\frac{\partial v}{\partial t} - \sum_{i,j=1}^{n} (a(t,x))^{(ij)} \nabla_x^{(i)} v\, \nabla^{(j)} v$$

$$-\sum_{k=1}^{m} \sigma^k(t,x)^* r_t^k \tag{5.53}$$

$$v(T,x) \;=\; \Phi(T,x), \tag{5.54}$$

where we have assumed, there is a uniquely defined measurable $\widehat{\beta}(t,x)$, continuous in x such that

$$\mathcal{H}(t,x,\nabla_x v,r) \;=\; \nabla_x v(t,x)^* b(t,x,\widehat{\beta}(t,x)) + f(t,x,\widehat{\beta}(t,x))$$

$$+\sum_{k=1}^{m} \sigma^k(t,x) r_t^k \qquad \mathsf{P}_{\widehat{\beta}} \;\; \text{a.s.,}$$

which is the optimal feedback control. In accordance to Hausmann we shall, in what follows, consider $\widehat{\beta}(t) = \widehat{\beta}(t,Y_t(\omega))$ for an arbitrary but fixed t. Thus, the functional $\mathcal{H}(t,x,\nabla_x v,r)$ defined above, represents the Hamiltonian.

Let $W^{1,2,p,\lambda}(Q)$ and $L_{\text{loc}}^{p,\lambda}(Q)$ be defined as follows:

$$W^{1,2,p,\lambda}(Q) \;=\; \left\{ \varphi \in L^{p,\lambda}(Q) \;\Big|\; \frac{\partial \varphi}{\partial t}, \frac{\partial \varphi}{\partial x^{(i)}}, \frac{\partial^2 \varphi}{\partial x^{(i)} \partial x^{(j)}} \in L_{\text{loc}}^{p,\lambda}(Q) \right\}$$

$$L_{\text{loc}}^{p,\lambda}(Q) \;=\; \left\{ \varphi \in L^p(Q) \;\Big|\; \mu(A) dt \le \gamma(\lambda,A) \right\},$$

$$\mu(A) \;=\; \int_0^T \int_A \left[\left|\frac{\partial \varphi}{\partial t}\right|^\lambda + \sum_{i,j=1}^{n} \left|\frac{\partial^2 \varphi}{\partial x^{(i)} \partial x^{(j)}}\right|^\lambda \right] dx$$

where $1 < \lambda < \infty$, $2 \le p < \infty$, $A \subset Q$ bounded and $\gamma(\lambda,A)$ a constant depending on λ and A.

Lemma 5.16 *The assumptions A 5.3 and A 5.5 hold. Thus, the Cauchy problem defined by (5.53) and (5.54) has one and only one solution $v \in W^{1,2,p,\mu}$ such that v and $\nabla_x v$ are continuous functions in their arguments.*

Proof: It is an adaptation of Bensoussan's proofs in [20, 17]. ∎

Lemma 5.17 *The assumptions of the foregoing lemma hold. Further, let us assume A 5.9. The Hamiltonian $\mathcal{H}$ is differentiable almost everywhere and for any $j = 1, 2, \ldots, m$, we have*

$$\nabla_x^{(j)} \mathcal{H}(t,x,\nabla_x v,r) \;=\; \sum_{i=1}^{n} \left(\nabla_x^{(i)} \nabla_x^{(j)} v \right) b^{(i)}(t,x,\widehat{\beta}(t,x))$$

$$+ \sum_{i=1}^{n} \nabla_x^{(i)} v(t,x) \left(\nabla_x^{(j)} b^{(i)}(t,x,\widehat{\beta}(t,x)) \right) +$$

$$+ \nabla_x^{(j)} f(t,x,\widehat{\beta}(t,x)) + \nabla_x^{(j)} \left(\sigma^j(t,x)^* r_t^j \right).$$

Proof: Let z denote $\nabla_x^{(k)} v(t,x)$ for $k = 1, 2, \ldots, n$. Then differentiating (5.53) and (5.54), we get

$$
\begin{aligned}
0 &= \partial_s z_s(x) + \sum_{i,j=1}^{n} a(s,x)^{(ij)} \nabla_x^{(i)} (\nabla_x^{(j)} z_s(x)) \\
&\quad + \sum_{i,j=1}^{n} \left(\nabla_x^{(k)} a(s,x)^{(ij)} \right) \left(\nabla_x^{(i)} (\nabla_x^{(j)} v) \right) + \nabla_x^{(k)} f(t,x,\widehat{\beta}(t,x)) \\
&\quad + \sum_{j=1}^{n} \left(\nabla_x^{(i)} (\nabla_x^{(j)} v) \right) b^{(j)}(t,x,\widehat{\beta}(t,x)) + \\
&\quad + \sum_{i=1}^{n} (\nabla_x^{(i)} u) \left(\nabla_x^{(k)} b^{(i)}(t,x,\widehat{\beta}(t,x)) \right) \\
z_T(x) &= \nabla_x^{(k)} \phi(T,x),
\end{aligned}
$$

from which it follows that $z \in W^{1,2,p,\lambda}(Q)$. Further, it follows, the feedback control $\widehat{\beta}$ is a continuous function of (t,x) and the differential of the Hamiltonian is continuous. ∎

To close we shall state without proofs two additional results. One concerning the functional R^β and the other concerning the Hamiltonian. We make the following assumptions on the functional R^β

A 5.10 *Assumptions on the reward functional R^β*

1. There exist positive integers n_1, n_2 such that the inequality $|R^\beta(x)| + TV(\mu_x) \le n_1(1 + ||x||)^{n_2}$ holds.

2. μ_x is continuous in x in the weak topology.

Theorem 5.18 *Suppose A 5.10 are satisfied. Then R^β has an integral representation of the form*

$$
R^\beta(x) = \mathsf{E}^\beta \{R^\beta\} + \int_0^T e_s dB_s^\beta \qquad \mathsf{P}_\beta \quad a.s., \tag{5.55}
$$

where the integrand e_s satisfies $\mathsf{E}\left\{ \int_0^T |e_s|^2 ds \right\} < \infty$ and is given by

$$
e_t = \mathsf{E}^\beta \left[\int_t^T \mu_x(ds) \Phi_{s,t} \Big| \mathcal{F}_t \right] \sigma(t, Y_t). \tag{5.56}
$$

For $0 \le t \le s \le T$ define the matrix-valued process

$$
d\Phi_{s,t} = F(s, Y_s) \Phi_{s,t} ds + E^{(k)}(s, Y_s) \Phi_{s,t} dB_s^{\beta(k)}, \tag{5.57}
$$

where

$$
\begin{aligned}
F(s, Y_s)^{(ij)} &= \nabla_x^{(j)} \widehat{b}^{(i)}(s, Y_s) \\
(E^{(k)}(s, Y_s))^{(ij)} &= \nabla_x^{(j)} \sigma^{(ik)}(s, Y_s).
\end{aligned} \tag{5.58}
$$

Proof: See [45, 103] for a rigorous proof of this theorem.   ■

Recall $R^\beta = M_T^\beta$. The Fréchet derivative of M_T^β is then

$$\mu_x(ds) = \left(\int_t^T \nabla_x \, f^\beta(s, Y_s)ds + \nabla_x \, \phi(T, Y_T) \right) \delta_1(ds), \qquad (5.59)$$

where δ_t is the Dirac measure at t. Further, one proves that $y(x)$ in (5.35) is also given by (5.56), which clearly establishes the suggested relation between the integrands in (5.35), (5.38), (5.42) and (5.43). On the basis of Malliavin Calculus one can obtain in a fairly transparent way stochastic integral representations of path functionals of the type presented above. See in this regard [61].

Properties of the Hamiltonian and final remarks

The following crucial result is due to Hausmann. The relevance of the lemma is due to the fact that no derivatives with respect to α are needed, since we are using precisely the optimal control β.

Lemma 5.19 *Assume as in A 5.9. Then the Hamiltonian H defined by (5.24) is differentiable $\mathbf{P}_\beta$-almost everywhere and for any $k = 1, 2, \ldots, n$, we have*

$$\nabla_x^{(k)} H(t, Y_t, \nabla_x v) = \sum_{i=1}^{n} \nabla_x^{(i)} \nabla_x^{(k)} v(t, Y_t) \, \widehat{b}^{(i)}(t, Y_t) + \nabla_x v \nabla_x^{(k)} \widehat{b}(t, Y_t)$$

$$+ \nabla_x^{(k)} \widehat{f}(t, Y_t) \qquad \mathbf{P}_\beta \quad a.s. \qquad (5.60)$$

Proof: See [103].   ■

Under the notation of Theorem 5.14, the expressions (5.48) and (5.49) describe the well-known equations satisfied by the adjoint variable. If σ is independent of x, then the drift term alone gives the *deterministic adjoint equation*

$$-\dot{p}_t = p_t \nabla_x \, b^\beta(t, y) + \nabla_x \, f^\beta(t, y), \qquad (5.61)$$

as one can rapidly check from the definition of the adjoint process r. Moreover, setting the diffusion coefficient equal 0, i.e. $\sigma = 0$, then (5.48) and (5.49) reduce to the deterministic adjoint equation.

The growing significance of the stochastic maximum principle is deeply connected to its contribution toward a reasonable characterisation of the adjoint processes p and r including dynamics in the form of SDE's. These processes owe their explanatory power to links with the notion of random multipliers in the framework of constrained optimisation. Stochastic techniques of optimisation are however more linked to systems and information theory, which deal with dynamic processes and problems of observation, measurement, filtering and other matters of this sort.

This enables the understanding of stochastic control issues related to the structure of reward functionals, updating of decision policies as new information becomes available and the impact of changes of control upon the probabilistic structure of controlled processes. The latter may shed additional light upon the interpretation of qualitative properties of certain processes as the result of action of uncertainty and passing of time.

Let us mention in [131], the process η is considered as a solution of the linearised form of (5.8) about x with the purpose of deriving a stochastic maximum principle relying on mathematical programming and the variational theory of Neustadt. Kushner's innovative ideas provide a fundamental framework upon which the representation of the adjoint process p rests which in turn constitutes in a certain sense the heart of the martingale approach.

Before we return to our economic problem, let us make two final comments. First of all we should point out, the semimartingale version of the stochastic maximum principle puts us in a position to deal with the optimisation problem in quite the same way as in the deterministic case, see [82, 86]. This is a great achievement of the martingale approach to optimal control which merits to be stressed because the logic of the procedure and main line of argumentation remain transparent, lucid and elegant in spite of the increasing degree of complexity of stochastic techniques involved. Second, we have preferred to argue at a basic level and proceed in an intuitive and formal manner because we are primarily interested in applications and we are confident one can give rigorous proofs of our results.

5.4   Delineating a policy for the future

Planning is together with price mechanisms and rationing a major method of coordinating resources and activities. Rarely acts of production begin without a plan, a fact, which calls for clear delineation of objectives, for working out patterns coordinating resource allocation over time. Such matters take us into questions of optimal utilisation of resources and economic growth. The growth pattern of an economy includes the movement of the whole system of quantities, flows, values and prices throughout time.

The notion of resource utilisation needs some elaboration in order to clarify various issues concealed behind orthodox qualification of scarce which, as we have already seen, is more related to the presumption of unlimited wants. In this section we shall recast ideas on social and political change in a way suitable to the aim of progress, development and evolutionary growth.

5.4.1   Stochastic accumulation and associate paths

Let us begin pointing out, the basic aim of planning is to mobilise and adapt resources toward socioeconomic objectives resulting from the attempt to meet future and present needs of society. Yet, pragmatism and reality seem to rest on the belief that policy making is a process of political bargaining directed to adjust and reconcile conflicting demands and pressures.

Therefore, independent of the ability of competing groups (classes) to articulate their interests, planning has to go beyond efficiency and priority considerations to a comprehensive assessment of social costs. With this idea in mind, we proceed to the problem of capital accumulation.

Constructing a basic accumulation path

To understand a distinguishing feature of growth, as compared with development, it is necessary to keep in mind the fact that growth processes are essentially limited by a given, more or less fixed technology, which under the sway of entropy looses structure as time passes by. Fundamentally, growth means here increases in resource utilisation and exhaustion of technological knowledge. The course of capital accumulation enters both as cause and effect, as a look at (5.1) and (5.3) shows. However, accumulation makes progress by way of quantitative changes, which in the long range may call forth changes in quality. That means, accumulation is mostly supported by replications rather than mutations.

In this sense the problem of accumulation takes income as received, i.e. output net of allowances required at any time in order to keep productive capacity intact. Further, it seeks to determine the path of capital, by selecting an investment-consumption mix, such that $Y = \dot{K} + C$ and which corresponds to an initial capital endowment K^j.

The logic of accumulation states that the higher the fraction of income invested, the higher the expansion rate of capital, output, and again future investment, consumption and so on. In a way this amounts to say, the more suitable are the investment-consumption mix $(\dot{K}, C)$ to the purpose of accumulation the higher the expansion rate of economic surplus and the wider the range open to future choice. It is in this sense that positive freedom enters the macroeconomic picture and time scale of planning begins to relate to the prosperity of a system.

At a less general level this would mean, less working-time will be necessary to provide the basic material needs of society and more working-time will remain to enhance the productive capacity of the system. This leads to rising economic surplus further and to providing potentials of improving social, economic, cultural and other standards. It is at this point, when the issue of clearly delineating objectives comes to the forefront. This is essentially a problem of decision and evaluation of manifold items on the agenda of action. Thus, before we come to deal with issues concerning priority of objectives, cultural and institutional constraints, we shall examine some technical aspects of accumulation on which the material fulfilment of key objectives depend.

For the sake of simplicity, we shall limit ourselves to consider capital and labour as a homogeneous item corresponding to the simple case where k and L are one-dimensional. However, our analysis carries over to the n-dimensional case with just a few more operations, an issue we have already explained at different points.

As a starting point, we consider a purely probabilistic process which can be visualised as deviations around a natural level of accumulation akin to the classical notion of a natural economic system. Then, we pass to consider a weak solution of the accumulation path associated with the introduction of a drift term resulting from the rise of actions aimed at generating economic surplus. Let us begin tackling the issue of the strong solution, see Lemma 5.3 and set $\sigma(t, k_t) = -\zeta_t k_t$.

Lemma 5.20 *Let us assume A 5.3 and consider the process of capital accumulation $k = (k_t)_{0 \leq t < T}$ generated by the SDE*

$$dk_t = -\zeta_t k_t dB_t, \quad k_0 > 0, \tag{5.62}$$

where k_0 is any given arbitrary positive scalar, $B = (B_t)_{0 \leq t \leq T}$ is a Brownian motion as before. Then, the process k with k_t given by

$$k_t = k_0 \exp\left\{-\frac{1}{2}\int_0^t \zeta_s^2 ds - \int_0^t \zeta_s dB_s\right\}, \quad 0 \leq t \leq T, \tag{5.63}$$

solves (5.62) for any $k_0 > 0$ given.

Proof: Since $k_0 > 0$, owing to continuity, we have that $k_t > 0$ on a certain interval. Then, setting $\eta_t = \log k_t$ and using Itô's formula, gives

$$d\eta_t = -\frac{1}{2}\frac{1}{k_t^2}\zeta_t^2 k_t^2 dt - \frac{1}{k_t}\zeta_t k_t dB_t,$$

from which we obtain

$$\eta_t = \eta_0 - \frac{1}{2}\int_0^t \zeta_s^2 ds - \int_0^t \zeta_s dB_s,$$

$$k_t = k_0 \exp\left\{-\frac{1}{2}\int_0^t \zeta_s^2 ds - \int_0^t \zeta_s dB_s\right\}.$$

∎

In the world of neoclassical economics, the process k given by (5.63) would stand for a steady-state solution associated with the time period $[0, T]$. No trend would show up due to the typically harmonious configuration assumed to prevail in the neoclassic world. The mean time-path $\mathbf{E}[k_t] = k_0 \exp\{-\frac{1}{2}\int_0^t \zeta_s^2 ds\}$ would get disturbed only by transitory oscillations around it, reflecting uncertainty captured in the Brownian motion assumption, which in a radical variant can be seen as vanishing as time passes by.

The fact that for an arbitrary but fixed $t \in [0, T]$, the state k_t is a random variable leaves little room for neoclassic thinking. The point is that the assumption of a uniform rate of profit underlying the equilibrium model under perfect competition is no longer consistent with the fact that in general k_t may take manifold values according to the prevailing probabilistic structure; see [66].

Next, let us see how a trend comes in giving rise to a weak path of accumulation.

Theorem 5.21 *Assume A 5.3 and A 5.5. Then, under the probability measure $\mathbf{P}_\alpha$, the process k is given by means of the SDE*

$$dk_t = \left[\frac{\dot{K}_t}{K_t} - (a_t - \zeta_t^2)\right] k_t dt - \zeta_t k_t dB_t^\alpha, \tag{5.64}$$

where B^α is a $\mathbf{P}_\alpha$-Brownian motion such that

$$dB_t^\alpha = dB_t - (-\zeta_t k_t)^{-1} \left(\frac{\dot{K}_t}{K_t} - (a_t - \zeta_t^2) \right) k_t dt,$$

for a given positive k_0.

Proof : From the foregoing Lemma we have a strong solution for k given by (5.62). Then, adding and substracting the term $\left[\frac{\dot{K}_t}{K_t} - (a_t - \zeta_t^2) \right] k_t dt$ to (5.62), we obtain

$$dk_t = \left[\frac{\dot{K}_t}{K_t} - (a_t - \zeta_t^2) \right] k_t dt -$$
$$-\zeta_t k_t \left\{ dB_t - (-\zeta_t k_t)^{-1} \left[\frac{\dot{K}_t}{K_t} - (a_t - \zeta_t^2) \right] k_t dt \right\}.$$

Hence, we need only to prove that B^α, as defined above, is indeed a $\mathbf{P}_\alpha$-Brownian motion.

For the sake of simplicity, let us set $\phi_t = (-\zeta_t k_t)^{-1} \left[\frac{\dot{K}_t}{K_t} - (a_t - \zeta_t^2) \right] k_t$ and define M and M_t by

$$
\begin{aligned}
M &= \exp \left\{ \int_0^T \phi_s dB_s - \int_0^T \phi_s^2 ds \right\}, \\
M_t &= \mathbf{E}[M|\mathcal{F}_t], \\
\mathbf{P}_\alpha &= M \cdot \mathbf{P}
\end{aligned}
$$

Observe that using Itô's formula, one easily gets $dM_t = M_t \phi_t dB_t$ and recalling that $dM_t \cdot dB_t = M_t \phi_t dt$, one obtains

$$
\begin{aligned}
d(M_t B_t^\alpha) &= d \left[M_t \left(B_t - \int_0^t \phi_s ds \right) \right] \\
&= d(M_t B_t) - d \left(M_t \int_0^t \phi_s ds \right), \\
&= dM_t B_t + M_t dB_t + dM_t dB_t - M_t \phi_t dt - dM_t \int_0^t \phi_s ds, \\
&= dM_t B_t^\alpha + M_t dB_t,
\end{aligned}
$$

which shows that $M_t B_t^\alpha$ is a local martingale with respect to $\mathbf{P}$.

Furthermore, let $A \in \mathcal{F}_s$. Then, from the properties of expectations, we get

$$
\begin{aligned}
\mathbf{E}^\alpha \{ B_t^\alpha \mathbb{1}_A \} &= \mathbf{E} \{ B_t^\alpha \mathbb{1}_A M \} = \mathbf{E} \{ \mathbf{E} [B_t^\alpha \mathbb{1}_A M | \mathcal{F}_t] \} = \\
&= \mathbf{E} \{ B_t^\alpha \mathbb{1}_A \mathbf{E} [M | \mathcal{F}_t] \} = \mathbf{E} \{ B_t^\alpha \mathbb{1}_A M_t \},
\end{aligned}
$$

and taking into account that $(M_t B_t^\alpha)_{t \geq 0}$ is a local martingale with respect to $\mathbf{P}$, we obtain further

$$
\begin{aligned}
&= \mathbf{E} \{ B_s^\alpha \mathbb{1}_A M_s \} = \mathbf{E} \{ B_s^\alpha \mathbb{1}_A \mathbf{E} [M | \mathcal{F}_s] \} \\
&= \mathbf{E} \{ \mathbf{E} [B_s^\alpha \mathbb{1}_A M | \mathcal{F}_s] \} = \mathbf{E} \{ B_s^\alpha \mathbb{1}_A M \} = \mathbf{E}^\alpha \{ B_s^\alpha \mathbb{1}_A \},
\end{aligned}
$$

which shows that B^α is a local martingale with respect to $\mathbf{P}_\alpha$.

A localisation argument and observing, the assumptions A 5.3 and 5.5 ensure that the stochastic integrals are well-defined, even that they imply the Novikov condition, finish the proof.   ∎

Let us point out, going from a strong to a weak solution, one switches from the (reference) probability measure $\mathbf{P}$ to $\mathbf{P}_\alpha$, which calls attention to the fact that by introducing the drift term, i.e. $b^\alpha(t, k_t) = \left[\frac{\dot{K}_t}{K_t} - (a_t - \zeta_t^2)\right] k_t$, one changes the probabilistic configuration prevailing. In other words, knowing behind the drift term hides the choice of the couple (K, C) resulting from choosing (s_k, L), it becomes apparent that any investment-consumption mix is associated with a probability measure $\mathbf{P}_\alpha$ and bears thereby its own structure of uncertainty.

It is in this sense that the drift and diffusion coefficient, i.e. $b^\alpha(t, k_t)$ and $\sigma(t, k_t)$, determine the accumulation path k^α given by (5.64) and its associated probability measure $\mathbf{P}_\alpha$. On the other hand, in order to move from the probability measure $\mathbf{P} \longrightarrow \mathbf{P}_\alpha = M\mathbf{P}$ in general and from the probability measure $\mathbf{P}_\alpha$ associated with the control α to the one associated with the control β, i.e. $\mathbf{P}_\alpha \longrightarrow \mathbf{P}_\beta = M\mathbf{P}_\alpha$, as the search of an optimal control requires, the corresponding drift and diffusion coefficient $b^\alpha(t, k_t)$ and $\sigma(t, k_t)$ need to satisfy A 5.3 and A 5.5. It is rewarding to look into the economic implications of these assumptions.

Stochastic paths of consumption and investment

Let us direct attention first to the fact that these assumptions are aimed at making sure, the stochastic integral in the exponential martingale M, i.e. $\int_0^t \phi_s dB_s$, is well-defined and bounded. For our purpose, it would be enough that ϕ be bounded; indeed $\mathbf{P}$ a.s. bounded. However, for the sake of simplicity it is convenient to require a little more. One needs $\mathbf{E}\left[\exp\left\{-\frac{1}{2}\int_0^T \phi_s^2 ds\right\}\right] < \infty$.

The foregoing formulation opens the way to a wealth of alternatives concerning the manner how $\dot{K}$, L, ζ and k relate. We like to call attention to the relationship between the drift term and ϕ defined above, which is an essential element determining the probability structure. To this end, let us write

$$-\zeta_t k_t \phi_t = \left[\frac{\dot{K}_t}{K_t} - (a_t - \zeta_t^2)\right] k_t,$$

$$s_t y_t - (a_t - \zeta^2)k_t = \theta_t(k^\circ, k_t).$$

These two expressions point out the links between the drift term and the probability structure to prevail at any moment of time. At the left-hand side of the second expression we have rewritten the right-hand of the first, in a manner putting economic surplus in a more familiar form. The right-hand side of the second expression attempts to measure the gap prevailing between the current level of k_t and a reference level of capital. Thus, requiring the drift term $b^\alpha(t, k_t)$ to close up at any time t the gap between the current and a desired level of capital k_t and k°, can be related in a transparent way to the pace of

accumulation and to the probabilistic structure to which the investment decision gives rise. Here θ_t indicates the fraction of the current gap between the
levels of current and desired capital to be closed up. It is easy to see that the
higher θ_t the sooner the system reaches the desired level. However, a high θ_t
as well as a high $b^\alpha(t, k_t)$ is associated with a high s_t, i.e. a low consumption
régime, and a particular probabilistic structure.

This sort of policy is likely to be opposed first of all due to the fact that
people usually prefer present to future consumption and second due to the
uncertainty resulting from such decisions. The latter is an issue which can be
studied on the basis of the impact of this decision upon the states of being
and doing of individuals. Examining consumption from the point of view of
characteristics and the states of being and doing to which they give rise, point
out the emptyness of any economic theorising which neglects human creativity
and freedom to function, notions on which processes of labour essentially rely.
Since the process of labour is of an undeniable evolutionary nature, shifting
focus to evolutionary features needs to become highest priority of economic
theorising.

Focusing on learning aspects of development and growth as well as on the
entropic features characterising human artefacts offers a unique opportunity to
grasp into the essence of accumulation. A serious and scientific study of an
economy from this perspective, shows how meaningless are the controversies
related to notions which can only be properly understood in a framework where
time and change are essential. This is particularly true with notions like social
value and choice, social preference and rates of discount, planning objectives
and institutional constraints, attitudes and motivation, which are beyond the
boundaries of orthodox economics.

In Part I of this monograph we have introduced and explained the most basic
aspects of notions entering our description of an evolving economy, here we shall
therefore concentrate on the elaboration of an accumulation policy which as a
byproduct leads to the elimination of labour-surplus, i.e. one which directs
labour from less productive or idleness to higher productive employment.

A 5.11 *Main requirement of the accumulation policy*

1. At the beginning of every time period one negotiates and sets accordingly
   a minimum wage rate $\overline{W}$ and an upper bound $\bar{s}_K$ which remain constant
   until the end of the period.

2. With every $\overline{W}$ one associates the amount of labour L^j available for employment. The supply of labour is infinitely elastic up to the point of full
   employment.

3. The fractions $(1-s_K)$ and $(1-s_L)$ of profit and wage income are consumed
   under the assumption that $0 < s_K \le \bar{s}_K$ holds, where s_K is chosen relying
   on optimality.

In the light of the foregoing assumptions, a mechanism of optimality comes to
supplement the instruments usually available to ends of economy policy. Hence,
to bring out a desired régime of accumulation, the government or control board

in our model may resort to a law of optimality by means of which capitalists come to save any proportion of profits as optimality desires them to save, provided it does not attempt to force capitalists to save more than $\bar{s}_K$ units of each unit of profits.

Let us stress the essential point as follows: In the face of unemployment the foregoing assumptions put certain constraints on the choice variables s_K and W, i.e. $W \geq \overline{W}$ and $0 < s_K \leq \bar{s}_K$, with the aim of preventing choices of the consumption-investment mix are made independently of the choice of the level of employment. Therefore, confronted with unemployment the only degree of freedom remaining to determine the level of social output is the choice of the level of employment. Theorem 5.21 makes clear the links between the couple (s_K, L) and the accumulation path k which emerges from that choice.

Next, we shall derive the paths of consumption and investment associated with the accumulation path k and thus with the underlying choice (s_K, L). To do that, we just need to consider $C_t = C(L_t(\omega))$ and $\dot{K}_t = \dot{K}(L_t(\omega))$ and compute the Itô differential for both with respect to L. However, the SDE (2.4) determining L is no longer driven by the Brownian motion B but rather by B^α defined in Theorem 5.21. That means, under the probabilistic structure emerging from the choice (s_K, L) the demand for labour L is governed by the SDE

$$
\begin{aligned}
dL_t &= a_t L_t \, dt + \zeta_t L_t \, dB_t^\alpha, \qquad (5.65)\\
L_0 &= L^j, \qquad t \in (0, T],
\end{aligned}
$$

where the Brownian motion B^α is the one introduced above. Let us recall, consumption and investment are defined respectively by (2.12) and (2.11). We have the following Lemma.

Lemma 5.22 *The assumptions of Theorem 5.21 hold. Let us consider the processes $C = (C(L_t, \omega))_{\{0 \leq t \leq T\}}$ and $\dot{K} = (\dot{K}(L_t(\omega))_{\{0 \leq t \leq T\}}$ with the stochastic process $L = (L_t(\omega))_{\{0 \leq t \leq T\}}$ given by (5.65). Then the stochastic functional $(C(L_t(\omega))$ and $\dot{K}(L_t(\omega))$ are governed by the SDE's*

$$
\begin{aligned}
dC_t &= \{[(s_K - s_L)W + (1 - s_K)Y_L]a_t L_t + (1 - s_K)Y_{LL}\zeta_t^2 L_t^2\} \, dt + \\
&\quad + [(s_K - s_L)W + (1 - s_K)Y_L]\zeta_t L_t \, dB_t^\alpha \\
d\dot{K}_t &= \{[(s_K - s_L)W - s_K Y_L]a_t L_t + s_K Y_{LL}\zeta_t^2 L_t^2\} \, dt + \\
&\quad + [(s_K - s_L)W - s_K Y_L]\zeta_t L_t dB_t^\alpha
\end{aligned}
$$

where Y_L and Y_{LL} stand for $(\partial Y/\partial L)$ and $(\partial^2 Y/\partial L^2)$. We set $C_t = C(L_t(\omega))$ and $\dot{K}_t = \dot{K}(L_t(\omega))$.

Proof: A simple application of Itô's rule gives

$$
\begin{aligned}
dC_t &= \frac{\partial C_t}{\partial L_t} dL_t + \frac{1}{2} \frac{\partial^2 C_t}{\partial L_t} (dL_t)^2, \\
&= \left[\frac{\partial C_t}{\partial L_t} a_t L_t + \frac{1}{2} \frac{\partial^2 C_t}{\partial L_t} \zeta_t^2 L_t^2 \right] dt + \frac{\partial C_t}{\partial L_t} \zeta_t L_t dB_t^\alpha.
\end{aligned}
$$

Computing the partial derivatives and a few arrangements proves the first result. The stochastic differential $d\dot{K}_t$ can be obtained in the same way. ∎

One can easily obtain the SDE governing the evolution of other stochastic functionals. The general philosophy of Itô's calculus should be clear by now: First, one needs a differential for the basic process and then one applies the chain rule. For the case of a product, as for instance for $c = C/L$, one needs to know dC and $d(L^{-1})$ and then apply $dc_t = dC_t \cdot L_t^{-1} + C_t \cdot d(L_t^{-1}) + dC_t \cdot d(L_t^{-1})$. In this way, we may derive SDE's for virtually all relevant stochastic functionals.

The stochastic accounting price of capital

Let us now examine the question about the amount of consumption one needs to be renounced per additional unit of investment. In order to do that, we introduce the probabilistic concept of accounting price of capital (investment), which in the deterministic case one denotes by $_*P_K$. In the probabilistic framework on which we are working, the *stochastic accounting price of capital* is a stochastic process which we denote by $\omega\text{-}_*P_K = \left(\omega\text{-}_*P_K(t)\right)_{\{0 \leq t \leq T\}}$ with $_*P_K(t,\omega)$ given by means of the following *marginal rate of transformation* (ω-MRT) defined as

$$\omega\text{-}_*P_K(t) = -\left(\frac{dC_t}{d\dot{K}_t}\right)_{T_3} = -\frac{dC(L_t(\omega))}{d\dot{K}(L_t(\omega))} \tag{5.66}$$

where $C(L_t(\omega))$ and $\dot{K}(L_t(\omega))$ are stochastic functionals with C_t and $\dot{K}_t$ as above. Setting $\zeta_t = 0$ for any $t \in [0,T]$, the stochastic accounting price of capital $\omega\text{-}_*P_K(t)$ coincides with $_*P_K$, the deterministic counterpart, as one easily checks.

For simplicity we shall refer to (5.66) as $\omega\text{-}_*P_K$, a short-hand for stochastic accounting price of capital. Unfortunately, as it stands, $\omega\text{-}_*P_K$ is a little cumbersome and does not enable us to trace out clearly the effects resulting from changes of control. This is due to the presence of the Brownian motion B^α and it is the price for the type of randomness considered. A way out of the embarrassment of having too much information is reducing it by conditional averaging. Thus, we may resort to a conditional averaging of the accounting price of capital by setting

$$
\begin{aligned}
_*P_K(t,\omega) &= \mathsf{E}^\beta\{\omega\text{-}_*P_K(t)|\mathcal{F}_t\} \\
&= \frac{[(s_K - s_L)W + (1 - s_K)Y_L]\,a_t + (1 - s_K)Y_{LL}\zeta_t^2 L_t}{[(s_K - s_L)W - s_K Y_L] + s_K Y_{LL}\zeta_t^2 L_t},
\end{aligned}
\tag{5.67}
$$

which has the advantage of keeping the random variable $_*P_K(t,\omega)$ fairly flexible and adapted to the information $\mathcal{F}_t$ available at time t; above Y_L stands for $(\partial Y/\partial L)$ and Y_{LL} for $(\partial^2 Y/\partial L^2)$. Thus, $_*P_K(t,\omega)$ compared with $\omega\text{-}_*P_K(t)$ contains in a sense only the information carried by the drift terms in the SDE's governing the evolution of C and $\dot{K}$.

Let us mention the fact that when the SDE (5.65) becomes too wild, neither $_*P_K(t,\omega)$ nor $\omega\text{-}_*P_K(t)$ exhibit unambiguously the properties of $_*P_K(t)$.

However, under the requirements introduced by the definition of stochastic integrals of the Itô type, one obtains fairly well-behaving stochastic versions of the accounting price of capital. For the sake of simplicity, we shall use in the sequel the more manageable version of the accounting price of capital given by the more simple expression

$$_*P_K^\omega = \frac{(s_K - s_L)W + (1 - s_K)Y_L}{(s_K - s_L)W - s_K Y_L} \qquad \mathbf{P}_\beta \quad \text{a.s.,} \qquad (5.68)$$

where Y_L stands for the partial derivatives of disposable income Y with respect to labour L, which depends directly on the stochastic path of accumulation k given by (5.64).

The investment régime represented by T_3 directs attention to the fact that under the assumption A (5.11) the *only choice variable remaining is L*. Here again $_*P_K^\omega$ indicates how much consumption the economy as a whole has to make available, so that $_*P_K^\omega$ defines a *social supply price of investment* in terms of consumption. Under the preceding conditions $_*P_K^\omega$ is invariably higher than one with the possible exception of null-sets with respect to $\mathbf{P}_\beta$, the probability measure associated with the prevailing control action α. This contrasts sharply with the stochastic nominal price of investment P_K^ω, $P_K^\omega(t) = 1$ $\mathbf{P}_\beta$ a.s. at any $t \in \mathbb{R}$. That means, the physical substitution of consumption for investment takes place at any time on a one-for-one basis. Again, $_*P_K^\omega$ approaches 1 only as the marginal distribution accruing to an additional unit of labour Y_L goes to zero and the nature of the constraints acting upon accumulation becomes technological rather than institutional.

The fact that $_*P_K^\omega \geq 1$ makes investment at the margin more valuable than consumption and provides the key to the proper valuation of allocations of disposable income between investment and consumption purposes which come to determine the future pace of accumulation. For investment has no value on its own other than the future consumption and employment it provides. Otherwise one cannot reasonably explain why giving up the output foregone by allowing $(L^j - \underline{L})$ workers to be idle and thus subject to entropic processes like disintegration of labour skills, know-how, organisation, state of health and nutrition, and many other features defining the potential of society to action.

Based on $_*P_K^\omega$ one calculates accounting prices, i.e. social values of profits $_*P_\pi^\omega$, wage and rental rates, $_*W^\omega$ and $_*R^\omega$ respectively, see [145]. The relation between $_*W^\omega$ and $_*R^\omega$ and the corresponding nominal wage and rental rates turns out to be

$$_*W^\omega < W, \qquad _*R^\omega > R \qquad (5.69)$$

and this may lead to greater utilisation of labour, provided the extra profit due to the premium attached to investment resulting from $_*P_K^\omega$ is directed to new investments. Needless to say, the materialisation of such a policy is linked to the availability of profitable investment opportunities and they are in turn dependent first upon the presence of an appropriate productive capacity given in terms of capital and labour characteristics which define the productive power of the system and second upon the presence of purchasing power which creates the potential for profit realisation.

These issues are crucially related with an accumulation policy aimed at creating potentials or opportunities of getting productive power and accumulating know-how on the basis of participation in labour processes and generating sources of demand by creating potentials or opportunities of getting purchasing power. It is important to get firmly in mind these observations in order to grasp the essence of our accumulation policy and understand the real message of functionals like ω-MTR and the associate accounting prices.

5.4.2   Descriptive structures at an arbitrary time

Knowing how to compute the terms of trade between consumption and investment given technology and institution, it has become natural to ask for a way of computing correspondingly the terms of trade between the benefits associated with underlying exchanges of consumption for investment. To this end we shall need also a stochastic version of the marginal rate of substitution (ω-MRS) which shall follow from prevailing social preferences and states of knowledge and information from which they emerge. As a fact of life we have assumed, institutions and the government play an active role shaping social preferences. As a representative, the government formulates a social performance index by means of $f^{\alpha}(t, k_t)$ and $\phi(t, k_t)$.

The ω-MRS shall reveal how much consumption at the macroscopic level the society and at the microscopic level individuals are willing to renounce in exchange for future consumption. At the macroscopic level we put this matter in the hand of policy-makers which act on behalf of society. At the microscopic level individuals respond in a way reflected in the subjective rate of discount introduced in Chapter 4. However, since policy making needs also to adjust democratic principles, we shall argue in terms of the preference of policy-makers on alternative investment-consumption mixes. In this sense we assume ω-MRS describes the amount of current consumption the decision-maker is willing to give up for a marginal unit of investment given its technology and institutional constraints and defines thus a *social demand price of investment* in terms of consumption. In consequence, the consumption-investment mix which equates the social supply and demand price solves partially our optimisation problem. Indeed, the ω-MRT and ω-MRS, under the assumptions in Section 2.3 and 5.1, determine unequivocally the loci of admissible consumption-investment mixes given technology, institutions and social preferences. The optimal combination of consumption and investment can to some extent be described in terms of the well-known tangency condition. However, as one may expect, the characterisation of the optimum is not simple at all.

We have purposely employed the same notation while formulating the economic problem and presenting the mathematical tools. Therefore, we shall make use of such results, functionals and variables without further economic interpretation, unless it may lead to confusion.

Linking the adjoint to accounting prices

For expository reasons, let us recapitulate a little. We have considered an arbitrary control $\alpha \in \mathcal{A}$, which for $t \in [0, T]$ is given by $\alpha_t = \alpha(s_K, L_t)$.

Further, we derived various stochastic functionals, among them the stochastic version of the accounting price of capital $_*P_K^\omega$, which hold under the probability structure determined by $\mathbf{P}_\alpha$. For the rest of this subsection we shall consider a fixed but arbitrary time t, $t \in [0,T]$. In this sense, we will not examine any explicit time dependency. Taking into account (2.11) and (2.12), which define $\dot{K}$ and C as functionals of $\{s_K(t), L_t\}$, and the SDE (5.65), which describes the labour demand L, we obtained the process of stochastic accumulation k^α given by (5.64), driven by the probability measure $\mathbf{P}_\alpha$.

At this stage it is necessary to stress the point that although the accounting prices of capital $_*P_K$ and $_*P_K^\omega$ are formally the same, there are deep differences suggested by the fact that the second holds on under the probability measure $\mathbf{P}_\alpha$ which depends on the chosen control action α.

Keeping this fact in mind we proceed to compute the optimal accumulation policy. We have the following Lemma.

Lemma 5.23 *The assumptions A 2.1, A 2.3 and A 5.1 hold. Let p be the process considered in Lemma 5.10 and let β be the optimal control characterised in Theorem 5.11. Then the process p satisfies for any $t \in [0,T]$ the relationship*

$$p_t = L_t \, f'(t, C_t^*) \,_* P_K^\omega(t) \qquad \mathbf{P}_\beta(d\omega) \otimes dt \quad a.s., \qquad (5.70)$$

where we have approximated $_*P_K^\omega(t)$ as indicated in (5.66).

Proof: We go back to Theorem 5.11 a) and assume that $\beta \in \mathcal{A}$ is the optimal control. Then, for an arbitrary but fixed $t \in [0,T]$, we know that β maximises, a.s. $\mathbf{P}_\beta(d\omega) \otimes dt$, the functional $H_t(\alpha)$ given by (5.24). Hence, we have the Hamiltonian $H_t(\beta)$ given by

$$H_t(\beta) = p_t \, b^\beta(t, k_t) + f^\beta(t, k_t) \qquad (5.71)$$

for any $t \in [0,T]$, where the drift term $b^\beta(t, k_t)$ is defined as in (5.2) which incorporate the economic features of accumulation or (5.64) with $\alpha = \beta$.

Further, $f^\beta(t, k_t)$ describes the reward functional introduced in (5.4). However, to save notation we write it also as $f^\beta(t, C_t^*)$; this is of course an abuse of notation. According to the remarks forgoing the introduction of the functional given by (5.71), we can also write the Hamiltonian (5.71) as

$$H_t(\dot{K}_t^*, C_t^*) = p_t \left[\frac{\dot{K}_t^*}{K_t^*} - (a_t - \zeta_t^2) \right] k_t^* + f^\beta(t, C_t^*), \qquad (5.72)$$

where $(\dot{K}_t^*, C_t^*)$ stands for the record at time t of the time-path of the investment-consumption mix associated with the optimal control $\beta = \beta(s_K^*, L^*)$. The total differential of the functional $H_t(\alpha)$ can be written, for $\alpha = \beta$ and any $t \in [0,T]$, fixed but arbitrarily, as

$$dH_t(\beta) = \frac{\partial H_t}{\partial \dot{K}_t^*}(\dot{K}_t^*, C_t^*)d\dot{K}_t^* + \frac{\partial H_t}{\partial C_t^*}(\dot{K}_t^*, C_t^*)dC_t^*.$$

Since the Hamiltonian $H(\beta) = (H_t(\dot{K}_t^*, C_t^*))_{\{0 \le t \le T\}}$ has a maximum at $\dot{K}^* = \dot{K}_t^*$, the control set A is all of $\mathbb{R}$, and we have $dH_t = 0$, so that

$$\frac{\partial H_t}{\partial \dot{K}_t^*}(\dot{K}_t^*, C_t^*) = \frac{\partial H_t}{\partial C_t^*}(\dot{K}_t^*, C_t^*)\left(-\frac{dC_t^*}{d\dot{K}_t^*}\right),$$

and since $f^\beta(t, C_t^*)$ is differentiable, we obtain the desired result.   ∎

We could have as well considered the Hamiltonian as a functional of (s_K, L) without altering, in principle, any result. Eq. (5.70) holds for all $s \in [0, T]$ with possible exceptions on $\mathbf{P}_\beta(d\omega) \otimes ds$-null sets. Hence it is for that reason a moment-to-moment relation, known in economic dynamics as the dynamic efficiency condition, and means that at any time s, $\mathbf{P}_\beta$-a.s. the social reward derived from the decision to invest at the rate $\dot{K}^*$ should equate the reward loss associated with the consumption the economy has to sacrifice in order to further investments as the control β requires.

The dynamic efficiency condition given by (5.70) amounts as already mentioned to the *tangency condition* between the consumption-investment transformation functional, which we denote by ω-MRT, and the consumption-investment substitution functional, denoted in turn by ω-MRS. The latter is as a rule articulated by means of the family of isoquants $\{H_s(\dot{K}, C)\}$, commonly known as Hamiltonian by analogy with a functional occurring in classical mechanics, the Hamiltonian. In this sense the functional (5.71) measures the joint reward (energy) per worker attributable to the current level of social output in terms of the characteristics attained by capital and labour: It consists of the consumption reward $f^\beta(s, C_s)$, which corresponds to a latent form of kinetic energy and $p_s\, b^\beta(s, k_s)$, the motive power inherent in net investment per worker, which corresponds to the work function usually in the form of potential energy. It is extremely important to recall that changes in the kinetic energy amounts in a sense to the work done by living force (labour) and the aforementioned changes are compensated by potential energy (negative work) provided the latter itself is recreated. This circular flow of energy is precisely the essence of the social process of reproduction, where the characteristics of capital and labour in the form of states of functioning reflect energy.

The control decision β diverts consumption from the consumption-investment mix associated with the status quo, which is then directed to investment building thereby an alternative mix. In other words the control action β diverts characteristics to function from labour to capital with the purpose of letting an additional fraction of labour – labour power or states of functioning – to come into operation; in this manner available potentials convert into kinetic energy a feature underpinning the fulfilment of states of functioning. It is noteworthy that consumption reward represents here labour characteristics to function per worker while the investment reward represents capital characteristics to contribute to the productive capacity per worker. However, from (5.70) we notice the adjoint process p measures the contribution of one additional unit of capital to the characteristics of labour as a whole.

To see how the Hamiltonian represents the joint reward per worker, it is necessary to examine the first term at the right-hand side of (5.72): there the term in brackets has no units as a relative rate of change and multiplying it by p_t keeps qualitatively the meaning given already to p_t, thus anything depends on the multiplying factor k_t^*, which represents the current level per worker of the characteristics of capital as a whole; recall $(k = (K/L)$. Thus, multiplying the first two factors by k_t^*, just brings out the whole contribution of capital to the forefront and puts the reward accruing to labour in per worker units. One can hardly overemphasise the relevance of keeping in mind the type of units involved while dealing with economic applications.

Equivalently, β releases potential energy which changes over to kinetic energy, the latter acts upon the system – actually on capital – by means of mechanisms underlying the drift $b^\beta(s, k_s)$ of the dynamics, (5.1) or more precisely (5.64). This fact deserves to be noted because it relates clearly social potentials to consumption rewards, in terms of characteristics – states of being and doing – by means of accumulation. Investments give rise to economic surplus which acting as potential energy releases manifold forces and activities. The effect of these driving forces and pervasive activities is to remove constraints releasing thereby latent potentials which in turn lead to expanding and improving the set of (potential energy) states of being and doing from which things may restart again. This self-reproducing dynamics of investments tailored to the needs of social reproduction offers a clear connection between economic development and the learning process attached to it, see [86].

Optimality conditions resulting from the Hamiltonian

At this point we have a clearly delineated picture of the Hamiltonian and the adjoint process and know how to interpret them. From this perspective it is easier to understand the role they have to play in the characterisation of optimality to follow. Let us recall that we suppress the use of the optimal control whenever it is clear that we are working with the optimal trajectories. Optimal trajectories are identified with an $*$ as superindex in the corresponding variable.

Theorem 5.24 (Optimal phases) *The assumptions of Lemma 5.23 hold. Further, we assume A 2.2 and the constraint (2.28) are fulfilled. Let us characterise the control $\beta \in \mathcal{A}$ by the formula*

$$H_s(\beta) = \underset{\alpha_s \in \mathcal{A}(s,\beta)}{\text{ess sup}} \left\{ \mathsf{E}^\beta \left[H_t(\alpha) | \mathcal{F}_s \right] \right\} \qquad \mathsf{P}_\beta(d\omega) \otimes ds \quad a.s., \quad a.e.t, \quad (5.73)$$

which is equivalent to the characterisation (5.31) given to β in Theorem 5.11. Then three phases given below characterise fully the optimal control β.
Phase I
If $L_t^* < L^j$ and $s_K^*(t) = \bar{s}_K$, then

$$\frac{p_t}{f'(t, C_t^*)} = \frac{(s_K - s_L)W + (1 - s_K)Y_L}{(s_K - s_L)W - s_K Y_L} \qquad \mathsf{P}_\beta(d\omega) \otimes dt \quad a.s. \quad a.e. \; t, \quad (5.74)$$

Phase II

If $L_t^* = L^j$ and $s_K^*(t) = \bar{s}_K$, then

$$1 \le \frac{p_t}{f'(t, C_t^*)} \le \frac{(s_K - s_L)W + (1 - s_K)Y_L}{(s_K - s_L)W - s_K Y_L} \qquad \mathbf{P}_\beta(d\omega) \otimes dt \quad a.s. \quad a.e.\ t$$

(5.75)

Phase III

If $L_t^* = L^j$ and $s_K^*(t) < \bar{s}_K$, then

$$\frac{p_t}{f'(t, C_t^*)} = 1 \qquad \mathbf{P}_\beta(d\omega) \otimes dt \quad a.s. \quad a.e.\ t \qquad (5.76)$$

Proof: Let us rewrite (5.73) in the less precise but intuitively more appealing form

$$H_s(s_K^*, L^*) = \max_{s_K, L_s} H_s(\dot{K}(s_K, L_s), C_s(s_K, L_s)) \qquad \mathbf{P}_\beta(d\omega)\, ds \quad a.s. \quad a.e.\ t,$$

(5.77)

where $\beta = (s_K^*, L^*)$ which follows the convention of using *'s to emphasise variables and parameters associated with the optimal control. Taking an arbitrary but fixed couple (ω, s) one can show, arguing in a static optimisation context on the basis of the time-decentralisation property of a functional $H(\alpha)$ of the type (5.72), the optimal control is indeed determined by the three phases above. ∎

Hence, the *static first-order conditions* which one obtains from (5.73) and the differentiability of H_s, fully describe the three phases, the economy undergoes in every time period before entering a state characterised as a relative level of full employment. The Hamiltonian together with the initial constraints determines whether the economy finds itself in Phase I, Phase II or Phase III which are fully described by (5.74), (5.75) or (5.76) respectively. A priori one may say that the economy optimally develops by moving from (5.74) toward (5.75) and then to (5.76), whenever it starts with a relatively low initial capital intensity.

The phases are to be interpreted as follows: optimality dictates either a relative level of full employment, $L_s^* = L^j$, or unemployment, $L_s^* < L^j$, and enforces either the binding savings constraints, $s_K^*(s) = \bar{s}_K$, or the nonbinding constraints $s_K^*(s) < \bar{s}_K$, accordingly. Combining these alternatives one obtains six candidates, and after sorting out those which do not fulfil various constraints there remain the three phases indicated above. Then the relative social desirability of the alternative mix of investment and consumption, which the resulting optimal policy entails, has to be measured by its effect on the Hamiltonian. That means, to evaluate the alternative $(\dot{K}^*, C^*)$ mix, one has to resort to the system of accounting values associated with $_*P_K^\omega$, since this accounting price of capital embodies crucial features on the basis of which the first-order conditions characterise the phases the system undergoes.

The control β related to $_*P_K^\omega$ stresses dependence on the $\mathbf{P}_\beta$-Brownian B^β prevailing at any time, a point Theorem 5.21 makes clear. The relations characterising the three phases hold $\mathbf{P}_\beta(d\omega) \otimes ds$ a.s. Next, let us look at the conditions of optimality imposed by the adjoint process $p = (p_t)_{\{0 \le t \le T\}}$.

A glance at the adjoint equation and transversality

Starting from the representation of the process p obtained in Theorems 9.7 and 5.18 we have proved, the adjoint process p satisfies (5.48)–(5.49) or (9.22)–(9.23) according to the properties specifying the underlying control process. Translating the qualitative features the adjoint state dynamics incorporate into the language of economic dynamics amounts to obtaining the so-called *instantaneous intertemporal consistency requirement*, an issue to which we turn.

Thus, the final part of this essay is dedicated to the interpretation and evaluation of quantitative changes described in the adjoint state equation. However, to close the first step toward visualising how optimisation works, we shall have a flashlight picture of certain sluggish features of the structure of the system at an arbitrary moment of time. The deterministic adjoint state equation given by (5.61) can be written in the following familiar form

$$-\frac{dp}{dt} = \frac{\partial H_t}{\partial k_t}. \tag{5.78}$$

This simple version of the instantaneous intertemporal consistency requirement provides a fruitful framework of analysis and means of weighing various measures of allocation efficiency and resulting rates of discount. In this sense, one may show that $-\frac{\dot{p}_t}{p_t}$, i.e. the percentage rate at which the marginal reward of investment decays over time, is the appropriate rate for discounting future investment; see [145].

The *transversality conditions* are boundary conditions linking the value function, the conditional optimal reward and the terminal reward (pay-off) and act as constraints upon various processes of change, in particular the adjoint state. At this point let us recall, the adjoint state process does not obey any initial conditions but rather terminal conditions, since it evolves backwards in time. Thus, the transversality conditions are given by

$$\phi(T, k_T) = -p_T(k_T - k^\diamond)^-, \tag{5.79}$$

where $k^\diamond$ describes the level of capital per worker which once attained can be further sustained, provided from this point on steady-state conditions prevail. We shall call an accumulation path satisfying the optimality conditions obtained above, a *stochastic optimal path* of employment and savings, to which we will return later.

Economic meaning of some static properties of control policy

Let us assume that the time horizon is given by $[0, T]$, with $T \in \mathbb{R}_+$ given. Let β denote the optimal control process given by the stochastic maximum principle and let the process $k^* = (k_t^*)_{t \geq 0}$, $C^* = (C_t^*)_{t \geq 0}$, $\dot{K}^* = (\dot{K}_t^*)_{t \geq 0}$ be the associated optimal trajectories. In Lemma 5.23 we obtained the relationship

$$p_t = L_t f'(t, C_t^*)_* P_K^\omega \quad \text{a.e. } t, \qquad \mathbf{P}_\beta \otimes dt \quad \text{a.s.} \tag{5.80}$$

which holds for all $t \in [0, T]$ with possible exceptions on $\mathbf{P} \otimes dt$−null sets. This is a moment-to-moment relationship known in dynamic economics as the *dynamic efficiency condition*, but in spite of its name it describes barely static properties of equilibrium states.

To the extent this reward functional bears the dimension of an economic value, i.e. units of reward per unit of consumption, does the adjoint variable p_t bear the dimension of a *shadow price*, i.e. reward units per unit of consumption times units of consumption per unit of investment or equivalently units of reward per units of future consumption. Furthermore, since $_* P_K^\omega(t) \geq 1$ for any $t \in [0, T]$ the relationship (5.80) tells us that to forward the economic policy represented by β, reward claims on future consumption – as described by p_t – shall be higher than the reward claims on current consumption corresponding to the fraction of current consumption that has to be postponed, i.e.

$$p_t \geq L_t f'(t, C_t^*) \quad \text{a.e. } t, \qquad \mathsf{P}_\beta \otimes dt \quad \text{a.s.} \tag{5.81}$$

where we have suppressed the use of the control process α, since the $*$ makes evident we work with the trajectories associated with the optimal control. From now on we shall adhere to this notation based on which we write $f'(t, C_t^*)$ for the derivative with respect to C of f. The dynamic efficiency condition given by (5.80) amounts to the well-known tangency condition between the investment-consumption transformation functional and the associated reward substitution functional articulated by means of the family of functionals $H_t(\dot{K}_t, C_t)$, or utility isoquants in the terminology of economic theory, which in turn define a *social demand price of investments* measured in terms of current consumption.

5.4.3   Persistent structures under the tension of change

Shifting social output from consumption to investment accentuates existing lack or deficiency in states of functioning. Associated with this operation of allocation of social output, there is another consisting in shifting today's claim on consumption characteristics into future claims on future states of functioning. Since it takes time for a policy of accumulation to materialise, a successful implementation of an accumulation is critically related to the outcomes of accumulation and the way this policy accentuates existing lack of consuming power.

A simple version of the adjoint state equation

It is at this point where the need arises for an adequate tuning of inflows and outflows of consumption characteristics. Thus besides static mechanisms the accumulation policy rests on dynamic mechanisms aimed at shaping the underlying motion according to the patterns of optimisation.

To examine some economic features of the dynamics of accumulation we shall make use of Theorem 5.14. At the risk of oversimplification, we shall recall the main results included there, see (5.48) and (5.49), as follows.

Theorem 5.25 *The assumptions of Theorem 5.24 hold. A necessary and sufficient condition for the optimality of problem (5.5) is that there exist the adjoint* *processes* $p = (p_t)_{t \geq 0}$ *and* $r = (r_t)_{t \geq 0}$ *so that they satisfy*

$$-dp = \nabla_k H_t(k_t^*, p_t, r_t)dt + r_t dB_t^\beta,$$ (5.82)
$$p_T = \nabla_k \phi(T, k_T),$$

where the Hamiltonian $H_t(k_t^*, p_t, r_t)$ *is given by the relationship*

$$H_t(k_t^*, p_t, r_t) = \max_{\alpha \in \mathcal{A}} H_t^\alpha(k_t, p_t, r_t),$$ (5.83)

which holds a.e. t, $\quad P_\beta(d\omega) \otimes dt - a.s.$ *with the functional* $H_t^\alpha(k_t, p_t, r_t)$ *defined as*

$$H_t^\alpha(k_t, p_t, r_t) = f^\alpha(t, C_t) + p_t b^\alpha(t, k_t) + r_t \sigma(t, k_t).$$ (5.84)

Proof : It is a particular case of Theorem 5.14.   ∎

Further stochastic accounting prices and shadow markets

The mechanisms governing the accumulation process take care that at a static level future and current claims achieve an appropriate balance, as we have seen in Lemma 5.23. The tangency condition, which follows easily from (5.80), can be written as

$$_*P_K^\omega(t) = \frac{p_t}{f'(t, C_t^*)} = \frac{\nabla_k H_t}{\nabla_C H_t} \quad P \otimes dt \quad \text{a.s.,} \quad \text{a.e.} t,$$ (5.85)

To any static economising problem of allocation at any time t there corresponds the dual problem of valuation over time, namely the problem of determining time paths of the adjoint variables to which we turn next.

In what follows the accounting variable related to any variable Q_t at any time t shall be denoted by $_*Q_t$, a convention that we have already used without any explicit indication. However, we shall suppress the explicit use of t, whenever no confusion may arise. In addition to the accounting price of investment $_*P_K^\omega$ defined by (5.66) a precise formula of which emerges as a byproduct of the application of the maximum principle in the specific form (5.85), there are other accounting prices. For instance, the accounting price of profit denoted by $_*P_\pi^\omega$ for which a relationship follows from the observation that a unit of profit can be allocated to investment and current consumption in the ratio $\frac{s_K}{1 - s_K}$, where s_K represents the fraction of capital income, i.e. $Y - WL$, to be reinvested.

On the other hand, due to the fact that consumption acts as a numéraire or unit of account it turns out that the accounting price of consumption $_*P_C(t) = 1$. The *accounting price of profit* is thus

$$_*P_\pi^\omega = s_K \cdot {}_*P_K^\omega + (1 - s_K) \cdot 1,$$ (5.86)

where the first term at the right-hand side represents the social value of the fraction of one unit of profit dedicated to investments and the second the social value of the corresponding fraction dedicated to consumption. Furthermore, the *accounting rental rate of capital* $_*R^\omega$ and the *accounting wage rate* $_*W^\omega$ are the social value of the marginal physical product $\partial_K Y_t$ and $\partial_L Y_t$ respectively, attributed to the allocation of one additional unit of profits to capital, i.e.

$$_*R_t^\omega = {}_*P_\pi^\omega(t)\,\partial_K Y_t \quad \text{and} \quad _*W_t^\omega = {}_*P_\pi^\omega(t)\,\partial_L Y_t, \qquad (5.87)$$

and the *accounting own rate of return to capital* is defined as the ratio ${}_*r_K^\omega = \frac{_*R^\omega}{_*P_K^\omega}$. Marglin's idea that we shall accomplish with the help of the maximum principle can be described as follows. One imagines that there exists a small project lying outside the productive sector representing an opportunity of augmenting the current social product by a certain amount using a production technique not describable by the prevailing distributional patterns that we shall characterise by means of the ratio $l = \frac{L}{K}$. To enforce the realisation of the production opportunity with a distribution scheme deviating from the one corresponding to the status-quo, one allows capitalists to get capital and labour services from a *shadow market for factors* at the shadow rates ${}_*R^\omega$ and ${}_*W^\omega$, which satisfy the inequalities (5.69) whenever they commit themselves to reinvest profits according to the institutional constraints embodied in the assumption A 5.11. This particular configuration generates different types of tensional forces given by

$$_*R_t^\omega - R_t \quad \text{and} \quad W_t - {}_*W_t^\omega, \qquad (5.88)$$

which reflect competition underlying accumulation and the struggle to close up current deficits in the states of functioning; various issues in this connection have been treated in the foregoing chapter as the aspects of learning at the bottom of development processes.

It turns out, that once labour becomes as capital scarce it is necessary to introduce a shadow price or scarcity cost of labour that we shall denote by ${}_*Z^\omega$ that has obviously to satisfy the inequality $0 \le {}_*Z^\omega \le \partial_L Y_t$. Marglin's computation of accounting prices uses the fiction that before unemployed labour gets engaged in the mentioned project it is first shifted temporarily to the productive sector and from there moves with the required capital endowment to the productive opportunity, see [145]. In this connection we assume technical progress renders unnecessarily the withdrawal of l workers from the productive sector allowing it to continue operations at the standard technique l after withdrawing one unit of capital.

A few issues concerning the dynamic properties

Let us call attention to the fact that the maximum principle enables us splitting the intertemporal economising problem of allocation (5.5) into a static and a dynamic one; the first is associated with the conservation properties of the Hamiltonian characterised by (5.83), and the second is related to the valuation, a role attached to the characterisation of the adjoint state variables as the change in the optimal value (5.5) due to changes in the initial condition of the state variables. These points have been presented in [47, 46, 89]. This duality and the resulting time decentralisation of the decision process are extremely convenient for applications.

Having mentioned the static features above we turn to some dynamic aspects of optimality associated with the adjoint state equation (5.82) and with the tranversality consistency condition. However, in order to avoid technicalities we shall obtain an approximative time path of the adjoint process; this provides us with a simple expression by means of which to calculate the adjoint variables at any time t. Thus, we shall approximate the drift coefficient in (5.82) by setting

$$- \dot{p}_t = \nabla_k H_t(k_t^*, p_t, r_t), \tag{5.89}$$

where $\dot{p}_t$ shall be understood as the mean forward derivative similarly as in (6.37) and with the notational simplification introduced there; see also in Chapter 8 the definition of the conditional forward drift.

This approximate version of the adjoint equation (5.82) allows us to concentrate in the economic issues and set aside some intrincate mathematical aspects which are treated in Theorem 5.15. Eqs. (5.89), (5.83) – (5.84) give, for any $t \in [0, T]$, the linear differential equation

$$- \dot{p}_t = \nabla_k b^\beta(t, k_t^*) p_t + \nabla_k f^\beta(t, C_t^*) - \zeta_t r_t, \tag{5.90}$$

which holds $\mathbf{P}_\beta - a.s.$ with $\mathbf{P}_\beta$ being the probability measure associated with the solution of (5.64). Recall that the derivation of (5.90) rests on the concept of mean forward derivative which is also linked to P_β, see also Subsection 4.1.4.

The relationships (5.86) and (5.87) together with a little algebra and some arrangements let us rewrite (5.90) in a way that sheds light on the dynamic interaction of the adjoint with other processes associated with some social aspects of capital efficiency. Namely, we get the following simple version of the *intertemporal consistency requirement*

$$- \frac{\dot{p}_t}{p_t} = {}_* r_K^\omega(t) - (a_t - \zeta_t^2), \tag{5.91}$$

which suggests optimality poses two additional conditions. Thus, it necessitates the accounting own rate of return to capital makes up first for the rate of decay of the premium attached to investment, see (5.81), and for the rate at which social output is allocated to the purpose of necessary productive consumption. The derivation of a more enlightening condition for intertemporal consistency demands more elaboration and is done in Chapter 6. To close this subsection let us mention that optimality involves furthermore a terminal consistency requirement known in the literature as the *transversality condition*, in this regard we refer to [47, 89, 91].

An accumulation policy that satisfies (5.74) – (5.76), (5.80), (5.91) and the transversality condition given by (5.79) as established by the stochastic maximum principle above, shall be called a *Marglin-Pontryagin path of labour allocation*, since Marglin was the first to derive a Pontryagin-path of this kind from the perspective of accounting prices.

5.4.4 Policy assessment based on the adjoint

Let us now explain shortly the logic behind our optimal accumulation policy. With this purpose in mind let us assume that at time t the economy working under the conventional framework of neoclassical distribution patterns forwards an optimal level of employment that we denote by $\underline{L}_t$ and that factors of production are rewarded at the wage and rental rates W and R respectively. Moreover, let us assume that the control board acting accordingly to an accounting criterion of the type alluded to while setting L_t^j as the level of employment to be obtained in the current planning period. That means, we shall reformulate the framework of the optimisation problem in a way that enables the economy to transfer labour from the reserve army to the advanced sector, so that the level of employment increases in the amount $L_t^j - \underline{L}_t$.

Further remarks on accounting prices

In order to assess within a framework bases on characteristics the global effect of the accumulation policy which makes affordable the aforementioned increase in the level of employment as reflected in the performance integral criterion (5.4), we have to be able to trace out the net changes in production, profits and wages attributable to the policy at stake, provided that the structural parameters – in particular those related to the patterns of consumption and investment – do not vary drastically. Furthermore, we need to evaluate the impact on the rewards for capital and labour services due to the changes set forth by the optimal policy.

According to Marglin's findings on accounting prices, we obtain the following more elaborated expressions for $_*R^\omega$, $_*W^\omega$ and $_*Z^\omega$.

$$_*R^\omega = {}_*P_\pi^\omega R + [(W - {}_*Z_t^\omega) + s_L W({}_*P_K^\omega - 1)]\, l_t, \tag{5.92}$$

$$_*W^\omega = {}_*Z^\omega - (s_L - s_K)W({}_*P_K^\omega - 1), \tag{5.93}$$

$$_*Z^\omega = {}_*P_\pi^\omega \partial_L Y + (s_L - s_K)W({}_*P_K^\omega - 1), \tag{5.94}$$

which express the social value of the corresponding variable in such a way that when the economy moves towards (a relative level of) full employment the last term in the foregoing three equations tends to zero as a consequence that the premium attached to investment falls down to unity as the marginal reward accruing to labour $\partial_L Y$ goes down to zero and unemployment becomes technological rather than institutional. The latter sort of unemployment cannot be dealt with by way of the present framework of continuous control and calls forth the introduction of impulsive techniques, see [85].

From (5.92) the accounting rental rate $_*R^\omega$ is given as the sum of the social value of (1) the nominal rental rate R and of (2) the marginal excess of wages over marginal productivity of labour times the marginal employment associated with the marginal unit of capital. The accounting wage $_*W^\omega$ as in (5.93) becomes then the difference between the scarcity cost or social value of the investment forgone in the productive sector as a result of shifting a worker together with its capital endowment to the production opportunity and the social value of the net consumption increase associated with the additional worker. Lastly, $_*Z^\omega$ given by (5.94) consists of the sum of (1) the value of the marginal output of the worker diverted to the project and of (2) the redistribution costs accompanying the transfer of income from the capitalist to the worker.

An approximate evaluation of benefits

In this section, we shall use the obtained accounting prices to get a global assessment of the social benefits at time t of one unit of social output shifted from consumption to investment at time t.

Using a little algebra and the assumption of constant returns to scale one gets easily from (5.91) and (5.87) the relationships

$$-\frac{\dot{p}_t}{p_t} = \frac{{}_*P^\omega_\pi \, \partial_K Y_t}{{}_*P^\omega_K} - (a_t - \zeta_t^2), \qquad (5.95)$$

$$= \frac{{}_*P^\omega_\pi R + {}_*P^\omega_\pi (W - \partial_L Y_t) l_t}{{}_*P^\omega_K} - \eta_t. \qquad (5.96)$$

Further, combining (5.86), (5.94) and (5.96) yields the following more explanatory version of the intertemporal consistency requirement

$$-\frac{\dot{p}_t}{p_t} = \frac{{}_*P^\omega_\pi R + [(W - {}_*Z^\omega_t) + s_L W({}_*P^\omega_K - 1)] l_t}{{}_*P^\omega_K} - \eta_t,$$

$$= \frac{{}_*P^\omega_\pi (y_t - W l_t) + [(W - {}_*Z^\omega_t) + s_L W({}_*P^\omega_K - 1)] l_t}{{}_*P^\omega_K} - \eta_t,$$

which are better understood relying on the interpretation of (5.87), (5.88) and (5.92) – (5.94); above η_t stands for $a_t - \zeta_t^2$. In Chapter 6 we present a fairly detailed computation of various structural parameters and accounting rates of the type sketched above although from a slightly different point of view. However, for the main purposes these computations can be seen as a guide; thus see Section 6.4.

A Linear Dynamic for the Adjoint

A rigorous derivation of the solution for the linear SDE (5.97) is based on Theorem 5.15; for the case of a diffusion coefficient independent of the state the solution follows quite straightforward and then taking conditional expectation delivers our approximative solution.

For that reason, we shall concentrate on the economic issues which give such solutions the interpretative power we are searching for. Thus, recalling (5.85) and a few arrangements gives the linear differential equation

$$-\dot{p}_t = f'(t, C_t^*) \{[(W - {}_*Z^\omega_t) + s_L W({}_*P^\omega_K - 1)] l_t + (1 - s_K)(y_t - W l_t)\}$$
$$+ [s_K(y_t - W l_t) - \eta_t] p_t, \qquad (5.97)$$

which can be solved by standard methods. Let us emphasise once more that the time path of p as described by (5.97) is indeed the expected time path due to the fact that the oscillations driven by the Brownian motion in (5.82) hinder the observation of the actual path. Therefore, our approximation (5.89) is not at all restrictive.

In order to appreciate better the economic relevance of (5.97) we shall recall that consumption being a flow variable should be written as a function of two

time variables, one indicating the beginning and the other the end of the period over which it is measured. On the other hand, we have based our planning
task upon the fictitious exchange of claims on present and future consumption
without introducing explicitly a market for the latter; recall however the observations concerning (5.87) and (5.88). For convenience of the argumentation
let us assume that we plan at time τ for the period $[\tau, T]$.

The maximum principle and duality enable us to work out for any time
$t \in [\tau, T]$ an expected imputed price p_t reflecting the rate of reward in terms
of increases in the flow of characteristics attributable to the unit of physical
capital installed at the time τ. However, this unit of physical capital is nothing
else but the unit of current consumption exchanged at time τ for a claim on
future consumption to be realised at time T. The adjoint p, although it is
a price of the sort just mentioned, acts as a flow and gives us in a sense the
accumulation flow of future consumption over the time period $[\tau, T]$. To see
this, let us analyse the right-hand side of (5.97) a little more.

The term in brackets represents the consumption rate per unit of additional
capital and consists of the rate of consumption out of profits, i.e. the expression
given by $(1-s_K)(y_t - Wl_t)$, and the rate of consumption net of opportunity costs
of the extra wage income, both of which are thought of as having been generated
by the additional unit of capital. The coefficient of the second term stands for
the net rate of growth of the additional unit of capital installed at time τ
attributable to the reinvestments out of extra profits. The role of $f'(t, C_t^*)$
and p_t in the first and second term is to change rates of consumption and
investment into rates of reward derived from the characteristics of consumption
and investment respectively.

Present value of future consumption streams

Now, we are better prepared to understand the solution of (5.97) which is given
by

$$p_\tau = f'(\tau, C_\tau^*) \int_\tau^T \left[(W - {}_*Z_t^\omega)l_t + s_L W({}_*P_K^\omega - 1)l_t + (1 - s_K)(y_t - Wl_t) \right]$$

$$\exp\left[-\int_\tau^t \{{}_*i_C(r) - s_K(y_r - Wl_r) + \eta_r\}dr \right] dt +$$

$$f'(\tau, C_\tau^*) \exp\left[-\int_\tau^T \{{}_*i_C(r) - s_K(y_r - Wl_r) + \eta_r\}dr \right] \frac{p_T}{f'(T, C_T^*)}, \quad (5.98)$$

where the integral of the first term at the right-hand side denotes consumption accumulated between τ and T due to the marginal unit of investment
made at time τ. The second term without the expression $f'(t, C_t^*)$ represents
the consumption value of the terminal capital accumulated from the marginal
investment. Multiplying $f'(t, C_t^*)$ just changes the units of account from consumption to reward.

Therefore, if we wish to express the benefits in terms of consumption, we can do so by rewriting (5.98) and taking (5.85) into account, i.e.

$$
{}_*P_K^\omega(\tau) = \int_\tau^T \Big[(W - {}_*Z_t^\omega)l_t + s_L W({}_*P_K^\omega - 1)l_t + (1 - s_K)(y_t - Wl_t) \Big]
$$

$$
+ \exp\Big[-\int_\tau^t \{{}_*i_C(r) - s_K(y_r - Wl_r) + \eta_r\}dr \Big] dt +
$$

$$
\exp\Big[-\int_\tau^T \{{}_*i_C(r) - s_K(y_r - Wl_r) + \eta_r\}dr \Big] \frac{p_T}{f'(T, C_T^*)}. \qquad (5.99)
$$

Finally, let us stress the fact that the reward as well as the consumption units involved in (5.98) and (5.99) have been discounted. This can be seen decomposing the exponential expression into

$$
\exp\Big[\int_\tau^T \{s_K(y_r - Wl_r) - \eta_r\}\, dr \Big],
$$

reflecting the growth of the unit of capital added at the time τ and

$$
\exp\Big[-\int_\tau^T \{{}_*i_C(r)\}\, dr \Big],
$$

acting as the discounting weights. It is easy to check that

$$
\exp\Big[-\int_\tau^T \{{}_*i_C(r)\}\, dr \Big] = \frac{f'(T, C_T^*)}{f'(\tau, C_\tau^*)},
$$

which points out that the rate at which society's weights on the increments to future consumption decay over time is given by

$$
{}_*i_C(t) = -\frac{\dot{f}'(t, C_t)}{f'(t, C_t)},
$$

so that ${}_*i_C$ is the consumption rate of discount which is known in the literature also as the *social rate of discount*.

The present study shows that stochastic analysis and control techniques may lead to realistic and operational models of capital accumulation, if appropriately supplemented with fundamental economic considerations. Further, it shows that formulating investment and employment decisions as an adequate stochastic control problem of optimisation, one gets access to a powerful version of the stochastic maximum principle. In this framework, we are able to design a control policy and derive its most basic properties.

Finally, using accounting prices à la Marglin, we obtain a formula for the evaluation of social benefits that resembles the well known formula used for asset evaluation in capital markets. However, in our approach we do not make any discounting assumption, since an equivalent factor emerges in a natural way. The discount rate and the shadow price are rather byproducts of the stochastic maximum principle. This is a crucial result.

Chapter 6

Interacting goals in a variational guise

My old concept of center and periphery was still valid, but it had to be enriched by introducing some very important consequences of the hegemony of the centers. Obviously it was not my purpose to deal theoretically with the centers. Some facts had to be clarified, however, in order to understand the other side, the periphery.

Technological progress started at the centers and its fruits remained basically there. For better or worse, they did not spread to the periphery through a general fall in prices in relation to increases in productivity. Historically, the role of the periphery had been mainly restricted to the supply of primary products. This explains why the growth of income stimulated demand and continuous technological innovations at the centers and gave impetus to industrialization. The periphery was left behind not because of malicious design but because of the dynamics of the system.

It so happened that peripheral industrialization had been greatly delayed and took place during successive crises at the centers. This accentuated the tendency of the periphery to imitate the centers—to grow in their image and likeness. We tried to adopt their technologies and life styles, to follow their ideas and ideologies, to reproduce their institutions.

All this penetrated the social structure of the periphery, which lagged considerably behind the most advanced structure of the centers, and brought significant mutations and contradictions that it is of the utmost importance to clarify. This is in fact the clue to understanding why the system tends to exclude socially those at the bottom, why it becomes more and more conflictive in the course of its evolution, and finally why it eventually tends toward a serious crisis.

Raúl Prebisch
(see [169], pp.184)

6.1 Qualitative issues of interacting ends

In this chapter we continue working as an approximation with the fiction the economic system consists, of a relatively advanced core of productive activities and a reserve army of unemployed labour. In order to examine additional issues associated with shifting a fraction of underemployed or idle labour to more highly productive activities, we shall direct attention to the interplay of the actual process of capital accumulation with a widely neglected process to which we ascribe the creation of employment.

6.1.1 On the track of potentials and goals

Looking at economic events from a system theoretical point of view and attempting to grasp their evolutionary context, impel us to focus on the informational content of the states of the economic system. Accordingly, actions and decisions in the past determine to a great extent the present state, as a configuration of actual characteristics among which income, available capital stocks, potential level of employment are irrevocably related. On the other hand, the future evolves from the present as a manifold representation of anticipations, expectations and uncertainties deeply interwoven with the ambiguously emerging knowledge of the state-dynamics of the system. In a post-Keynesian setting, this corresponds to recognising economic events come off in historical time and that at any point of time the system state is characterised by its own past and shapes its own future configuration.

The logical time-conception and hypothetical reasoning of economic orthodoxy is based on timeless constructs like Walras' auctioneer, rational economic man and the ill-conceived Say's law. Thus orthodoxy sticks by contrast to the belief that in the long run under the conditions of laissez-faire the system attains a state of equilibrium corresponding to the highest possible stage of well-being simultaneously at the macroscopic and microeconomic level. The labour market remains the most telling evidence of the coherence and efficiency of this venerable theory, as a quick examination of relationships among employment, wages, demand for social output and relative prices on the one hand and on the other savings and net investments reveals.

The fact that production of commodities generates a supply of social output and a (nonnecessarily equivalent) demand for commodities on the basis of wage income due to labour services and returns to (productive) capital, necessitates considerations aimed at elucidating the less visible requirements for finance. The relevance of the insight these issues provide is related to the need of bridging the time-lag occurring between commitment of labour and capital to production and payment of rewards, a fundamental condition linked deeply to the need of recreating system potentials at least at the same pace as they are exhausted. There is a wealth of issues on time-lags affecting production either at a distributional or financial level, which are known under labels like turnover of capital, period of turnover and relative turnover. They describe essentially dynamic features of fixed and circulating capital and will be briefly treated in [94]. Matters concerning finance capital are subject to investigation in [91].

Time, change and uncertainty interact and accentuate anomalies linked to information and complexity underpinning fulfilment and frustration of expectations. Actual substitutionality of commodities and production factors, as well as cultural and multiple individual features, decisively limit for better or for worse the system ability to adapt to the newly-emerging state of affairs. The principle of effective demand and realisation of profits represents a more realistic reformulation of Say's law. It offers therefore better insights into these questions and more adequate machinery to handle them.

Releasing potential energies and conflicts

The levels of development inherited from the past are condensed in the initial state given by $k_{\tau j}$ and the associated probability measure. The expansion of operations prescribed by long-term economic policy issues is given in our simplified version by θ_k^j which put together gives the initial condition

$$k_{\tau j} = k_{\tau j} + \theta_k^j, \tag{6.1}$$

from which our examination of the system state starts. Behind this equation there are various issues related to the choice of technique that we outlined in Subsection 2.1.4 and Section 4.1. In accordance with the principle of effective demand we let in the short-term net investments be equal to savings and postpone considerations on the demand for investment goods, marginal efficiency of capital and other to a later point; see [94].

By making net investment equal to saving, as indicated in (2.11) together with (2.13), one brings about additional changes in the capacity of production, which, when properly adjusted may help to reach the level associate with full employment relative to the current period of time. Starting at a given time from a disequilibrium state raises questions about the impact in the short-term future expectations and behavioural patterns corresponding to the initial state may have on decisions and motivation and the like. *Disequilibrium states* are primarily linked to the presence of forces inhibiting the release of development potentials of the system.

In this sense, the main objective of the present essay is to establish development paths, patterns of social valuation and an institutional framework that when acting together trigger off socioeconomic forces suitable for keeping the system moving toward a sort of golden age. A *golden age* reflects a steady-state configuration characterised by fulfilment of expectations formed by past experience and lack of immediate rise to altering it, but that together with ingredients of entropic nature render the same scenario transitory and inefficient in the long run. In an evolutionary terminology this will be a climatisation phase.

To understand this, it is necessary to keep in mind that the appearance of a certain combination of variables and building up an appropriate behavioural configuration, which as a whole exhibits some degree of coherence and self-renewal, is what we understand as *equilibrium*. Hence, the desired combination of variables and behavioural configuration, in addition to the long-term structural relations, are expected to be such as to keep the system moving toward equilibrium; this equilibrium is unlikely to be attained since entropic processes exhaust continuously available energy potentials making technology obsolete.

Various difficulties arise from our commitment to reflect facts of life. First of all, overdetermination reminds us causation bears a circular nature owing to the fact that no element in the system is completely independent of what happens to the rest. Thus, while actions may be tailored according to the spirit of a goal-seeking perspective, they are likely to include or release forces conflicting with them, so that at the outset we cannot define away the eventual presence of mechanisms counteracting goal-conducive forces. Secondly, there are also difficulties deriving from the interplay of time and change bringing about uncertainty. In this regard we refer to Chapter 3.

Hence we have to account for specific features proper of the state of affairs prevailing at any arbitrary time, the corresponding expectations concerning the future and the respectively accumulated experience. These, needless to say, affect the current decisions and behaviour of individuals since they are manifestations of their states of being and doing.

A change in the initial productive capacity of the system designed primarily to meet the goal-tailored requirements of the j-th time period, brings about changes in the state of affairs and associated expectations, which in turn have manifold consequences for the future development of the system. Since the shaping decisions and expectations to prevail in the future depend on existing pattern of individual responses in the face of uncertainty and on mechanisms available to forward information and to cope with change, it becomes of utmost importance to grasp mechanisms underpinning states of being and doing of individuals. Working, the central element of processes of labour, is the major source of human knowledge and learning; working tools are fundamentally coded information on skills and labour processes.

Processes of labour are means of acquiring, creating, storaging and updating human knowledge, while unemployment just blocks human potentials to learn. Herewith, we have to mention additional employment does not only provide the basis for a future expansion of social production and effective demand under appropriate conditions; employment does not only provide means of establishing intermediate stages on the way toward realisation of (future) profits and investment plans; employment does not only contribute to expanding the domestic market through increase in the effective demand and under the action of linkage-effects. Employment, as an intermediate stage of working, as a means of learning by doing, developing skills and know-how, is the cornerstone of states of modern existence of humankind. Thus, an economic theory that fails to grasp the essence of employment issues does not deserve any claim on scientific status.

Understanding the fiction of a golden age

Our intention to maintain the fiction of an equilibrium configuration as a frame of reference makes necessary to distinguish between human actions emerging from an *idealised state of tranquillity* – an economic equivalent of Newton's free or uniform motion associated with the concept of "nonevent" resulting from the absence of causes – and human actions typically attached to *real life*; in real life change and its perception take time, information and its processing is costly, habits and imitation – as particular elements of culture – interfere with economic rationality, anticipations and expectations of the future are based on limited knowledge of the current state of affairs, and so on.

Concerning the first we shall point out: behind many undertakings aimed at examining and assessing changes and our perceptions of them, there has emerged in the sense of a comparative dynamics [126] the idea of viewing them as events driving to an equilibrium process of the kind

$$probable\ state \xrightarrow{\ t\ } probable\ state,$$

where probable refers to a measure of the size of some as yet unspecified set or ensemble of states to which the states under consideration belong. One may think of the number of its members or some other appropriate measure. Essential is, however, the fact that the statistical features of the process remain invariant under time shifts, e.g. as time passes by and/or time reversal.

Concerning the second, i.e. human actions as happening in reality, we are led to take into account the current state of system agents and a direction of time, which distinguishes between past and future. The resulting processes are of the irreversible type and have a statistical description of the form

$$improbable\ state \xrightarrow{\ t\ } probable\ state.$$

The underlying ensembles of states are called incompletely determined states or *macroscopic states* and are usually more probable in the sense that they belong to a larger set than those of the kind *improbable state* $\xrightarrow{\ t\ }$ *improbable state*. In the language of the capability approach, the capability set of individuals as a representation of attainable states of being and doing, capability to function and positive freedom to act, is an ensemble undergoing a process of irreversible type, since expanding or contracting the capability set reflects the presence of entropic processes which unfold with time.

Thus, at any moment of time in a certain economic system, past experiences and emerging expectations of the future may generate in a natural way a new system configuration which is likely to be characterised as a relatively larger ensemble of states or potential consequences. In order to specify the *a priori* unknown ensemble of causes or as an attempt to mark off the future, one may use certain values or reference units, instances of which are natural units, normal values and production prices.

What we shall understand as normal depends on the prevailing state of affairs and is like the notion of standard of living transitory and relative. In this sense, the values certain variables assume become to be seen as *normal*, whenever the economic system enters a phase along which random events, technological changes, movements in the general price level and relative prices are such as to ensure fulfilment of expectations and smooth growth of the system. *Normal prices* and *values* pertain to an economic system evolving in historical time, in the course of which commodity production takes place, relating labour and (productive) capital, with the purpose of meeting present and future human needs both linked by agreements expressed in terms of money, see [126].

As we have already seen, a *golden age path* or *steady growth path* is a sort of fictitious dynamic equilibrium characterised by a state of affairs in which capital, labour, output, consumption and investment all grow at a constant rate, and the ratio between these variables remains constant. The equilibrium characterising a golden age is a state of tranquillity, where everybody has reached a state of acclimatisation and consequently nobody sees any advantage in changing the state of affairs prevailing, and expectations are being fulfilled. As stated before, we shall regard here golden ages just as a frame of reference imposed upon us by the omnipresence of entropy and will be primarily concerned with disequilibria, labour and paths along which an economic system can abandon such disequilibria.

Main features of the envisioned policy

To fix ideas, let us mention as main elements of a policy supporting accumulation and employment objectives: consumption characteristics aimed at maintaining productive power; the principle of effective demand as means of ensuring profit realisation and purchasing power; schemes of social reproduction as vehicle of coherence and rationality using available potentials. Within such a working context following features result.

- Ability to mobilise and discipline resources toward key objectives of society, e.g. increase economic surplus so as to ensure formation of (productive) capital as required for the elimination of certain amount of unemployment. Further it shall make sure this policy itself does not become an obstacle to further development. Recognition that at any given period of time resources available to the system may be seen as data and included in the initial conditions, but as their availability changes in the course of time qualitatively and quantitatively, data themselves have to be dealt with as variable.

- Human and physical resources may represent a barrier to the expansion of the productive capacity of society and/or become driving forces of progress, since they are essentially linked to the prevailing modus of utilisation. The modus of resource utilisation depends on the ideology behind practices of acquisition, appropriation, expropriation and annexation of them.

- Since the manner of allocating resources to production of commodities sets a boundary to productive capacity, it also limits – through learning and forgetting processes underlying their utilisation – the future productivity of active and idle recources. As a consequence the level and way of employment of available resources contribute either to unfolding or crippling productive forces of society.

These observations constitute the main message embodied in our conception of the problem of accumulation as a free boundary problem. It represents a partially known barrier to the expanding productive capacity. It moves slowly under the thrust resulting from interaction among released potential energies, e.g. idle resources, creatively blocked by lack of appropriate skills and/or means of labour, waste or wrongly conceived allocation, resignation and frustration of expectations about future states of the system; see Chapter 4.

6.1.2 Decisive ingredients of economic policy

In order to understand the essence of the ingredients building the economic policy we are looking for, let us recall processes of production rely on information and knowledge structures. They serve the aim of rearranging materials, transforming and transporting various forms of codified energy and rendering organisation available.

Technology and mechanisms of income distribution

According to our picture of the production process presented above, our analysis begins with social output $Q = (Q_t)_{t \geq 0}$ and disposable income $Y = (Y_t)_{t \geq 0}$ as given. They are seen as connected by the relationship $Q = Y + \delta K$, where the processes $K = (K_t)_{t \geq 0}$ and $\delta = (\delta_t)_{t \geq 0}$ represent the stock of social capital and its average rate of depreciation. Expansions and contractions of the system capacity to produce are essential to the mobilisation of resources both human and physical and they take place via the principle of social reproduction.

The most visible part of the process of (material) production relates to the distribution of disposable income, i.e. the social output remaining after depreciation allowances, which we denote by $Y = F(t, K, L)$, where F stands for a neoclassical distribution function as introduced in Chapter 2.

Reflections on the expansion of productive capacity and potential level of the effective demand give rise to specific expectations of profit realisation from which an appropriate pattern of labour demand, certain wage bill, purchasing power and demand for social output comes about. Since such configurations of decision and plans are associated with particular probability structures, changes expected in the relevant variables can only be probabilistically anticipated. Incorporating idle labour in the production process entails manifold costs including acquisition and updating of working skills, including training, adaptation to different organisation and working conditions. It constitutes an investment in human capital aimed at increasing the system potentials within a given period of time. Since a technology consists of a relative stable core of production techniques and a periphery, it turns out labour characteristics are likely to build a stable core of working skills and a less stable periphery. Mobility of labour between the prevailing techniques depends on characteristics like skill diversity, availability of means of financing the acquisition, updating and training of new skills as well as the pace of obsolence of prevailing techniques.

Contrasting labour and capital characteristics suggests many other sources of asymmetry among which the more striking are lack of speculative interest in the labour market, creativity and parallel tracking of several goals. However, distribution functions treat both as symmetric on the ground of property rights of capital. The distortions deriving from ill-conceived distribution mechanisms may thus jeopardise the recreation of productive power and building an appropriate demand which in turn jeopardises profit realisation.

Mechanism of income distribution as represented by $F(t, K, L)$ suggests capital and labour, called factors of production in the conventional sense of economic theory, are paid for their contribution to production. The argument t summarises changes over time in efficiency attributed to virtually any set of (productive) activities involving any aspect of knowledge, skill and means of production, since anything which contributes to efficiency of these activities roughly speaking becomes an ingredient of "technology" and as such builds decisive links to information or know-how. Distribution mechanisms associated with technology actively contribute to characterising the set of activities which determines the process of production. Hence the analysis of these mechanism gives us insight into the sort of derived motivations underlying human actions directed toward fulfilment of production aims, i.e. the inherent characteristics of human activities causally related to efficiency or effectiveness and the effect these may have upon social life. The motivation to do whatever individuals believe conducive to the achievement of ends – which they may see as valuable, satisfying, or necessary to them – is a *derived motivation*. Derived motivations are deeply connected to values of society through technical causation and its understanding leads to the study of the organisation of production, devices for owning and managing wealth, and the degree of interdependence on a large scale of the society. See Chapter 3 where we present critically various of these issues. A detailed analysis of these topics from the point of view of organisation of work, management of production and entropy is given in [94].

Human capital, learning and forgetting

Looking at the process of accumulation from the point of view of capabilities accentuates the relevance of identifying labour by a bundle of characteristics which represents *inter alia* human properties to function, to act, to create and to organise. The commodity approach avoids the limitation of the labour market of orthodoxy, which cannot distinguish between inert commodities and human beings, and enables us moreover to recover the dynamism emanating from human creativity and to formulate accumulation in an evolutionary framework; see in this regard Chapter 3, where we present various aspects concerning entropy, information, human knowledge and evolutionary processes. A detailed formulation of our evolutionary approach is given in [6].

Recognising social existence as the centre of the efforts of mankind should place the process of material production at the focus of economic research, since it provides the means of necessary consumption, i.e. the personal consumption of direct producers and productive consumption to renew and replenish used-up means of production with the aim of maintaining productive power of society at an adequate level. As a consequence, efforts undertaken and expenditures directed to improve the efficiency of productive forces of society, i.e. labour and means of production, should receive the utmost priority. In this connection it is necessary to recall, human beings have been the major suppliers of know-how and sources of information in the transformation of nature into products and/or services useful to humankind; and nature is an indispensable factor to unfolding human creativity, so that maintaining human and natural potentials of development is a necessary condition of social existence.

Therefore, no matter how complex or sophisticated a technology may appear, this should not hinder us to see it first and foremost as an instrument of labour, as the result of human labour and as the crystallisation of human knowledge. Questions related to technology adoption and innovations need to be examined with respect to eventual cultural and social implications, since these may bring about the breakdown of a whole society or large segments of it and ecological damages of irreversible character. Here it is necessary to recall the transformation of many prosperous regions and societies into conglomerates of starvation mostly on behalf of Eurocentric international corporations has been run under the name of progress, see [49].

Systems of (productive) activities are increasingly undermined by introducing a technology which openly serves Eurocentric interest, so that social structures of state nations of the periphery where these activities are embedded and their inherent system of values which provides for the necessary social coherence get threatened; natural consequences are breakdown, hunger, social conflicts and the like. Since the core of the prevailing technologies make as a rule an intensive use of energy and information, a state of affairs came into being which deprives almost two thirds of the world population of any potential access to advances supposedly attained to the benefit of mankind, a situation which places serious constraints – although some of them can be removed at cost – on human interaction and international cooperation; this is a fact of life in spite of the rhetoric with which orthodoxy still defends plundering humankind and nature.

If it is true that human beings are at the centre of interest of social research, a great deal of study on innovation and techology should focus on aspects concerning the impact on human capital, labour market, relations of production and reorganisation of production deriving from the advance of large-scale capitalist production, see [95, 229].

In the model at stake in the present essay we will deal primarily with labour as represented by a bundle of revealing states of being and doing, capability to function and act in a given social environment. In this sense human beings and in particular labour are carriers of creativity; know-how and information are considered first as a learning element acting productively in the advanced core of production and second as a (morally and physically) deteriorating factor of production, forgetting skills and dissipating human knowledge, when relegated to the reserve army. One seeks for an economic policy, which is able to support effectively the incorporation into production of a fraction of labour from the reserve army. This policy should favour an allocation of disposable income to the extent of increasing economic surplus to a level which makes affordable additional amounts of capital to the end of endowing the new incorporated workers with appropriate means of labour.

For the purpose of future reference, we shall collect here some inequalities that in one way or another make sure results remain economically meaningful. In this sense, let us recall that for any $t \in [0, T]$, in particular $t = 0$, the conditions

$$Y_t > 0, \ K_0 > 0, \ N_0 > 0, \ L_0, \ I_t > 0, \ C_t > 0 \qquad (6.2)$$

$$Y_t, \ K_0, \ N_t, \ I_t, \ C_t \in \mathbb{R}^n \tag{6.3}$$

hold, where K_0 denotes the initial capital endowment, and N_0 the initial labour available to the advanced core, including the fraction of idle labour that under the pressure of the struggle for survival moves to the secondary sector searching for employment and better life opportunities, i.e. $N_0 \geq L_0$. Moreover, let ϵ_t denote the relative rate of growth at time t of labour supply to the advanced core, with $\epsilon_t \in \mathbb{R}$. Let N_t denote labour available to productive activities at time t and assume it satisfies on average the relationship

$$N_t = N_0 \exp\left\{ \int_0^t \tilde{\epsilon}_s ds \right\}, \tag{6.4}$$

where $\tilde{\epsilon}_t$ is a kind of mean value of ϵ_t. Strictly speaking, the whole of the reserve army should constitute the labour supply to the productive sector at any time t, $t \in [0, T]$. However, due to the fact that our notion of reserve army includes also unemployed labour, uprooted handicrafts and other low-productive activities, it is natural to assume, that at any time t only a fraction of the reserve army builds the dynamics of change in the relevant labour supply, and this depends accordingly on a wage differential or the wage prevailing in the stable core of productive activities.

We are aware of the limitation of this simplifying assumption, since the reserve army as defined by us contributes greatly to the determination of wages in the stable core of productive activities. For the time being, we can only say that we are working on a more satisfactory version of this matter. [71, 141, 181, 183, 214] are detailed studies on economic growth, development, reserve armies and related issues.

6.1.3   Outline of the frame of reference

A new feature entering the present model is that accumulation will be measured with respect to available labour and not just with respect to labour actually employed. In this sense the economic policy is aimed at strengthening in the current period of time the ability of the system to accumulate capital according to the requirements posed by the current supply of labour and the goal of attaining an accumulation path along which a given relative level of employment results. Full employment in the actual sense will be a consequence in the long term of a purposeful application of an economic policy tailored by means of techniques associated with the theory of impulsive control. This is so because our analysis starts right after the corresponding impulse has been applied and ends shortly before the next impulse becomes necessary, i.e. we deal here with a special version of the control problem associated with the second phase of our global control strategy.

Variables in terms of available labour

Let us shift focus to variables in terms of available labour, since from this perspective it is easier to recognise the ability of a system to meet the requirements of social reproduction. Hence, we shall consider at the time t, with $t \in [0, T]$, consumption, capital, investment, and employment per available labour N_t and denote these variables with the corresponding small cases c_t, k_t, i_t, and λ_t which we group as follows

$$y_t = \frac{Y_t}{N_t} \qquad c_t = \frac{C_t}{N_t}, \qquad k_t = \frac{K_t}{N_t}, \qquad (6.5)$$

representing the variable on which the weight of accumulation rests, including the state variable k_t. The second group of variables is given by

$$\lambda_t = \frac{L_t}{N_t}, \qquad i_t = \frac{\dot{K}_t}{N_t}. \qquad (6.6)$$

which represent the instrument variables or control functions. The last group of variables consists of two ratios which enjoy a lot of interpretative power. Thus, let l_t and y_t represent the labour-capital and the output-capital ratio respectively:

$$l_t = \frac{L_t}{K_t} = \frac{\lambda_t}{k_t}, \qquad v_t = \frac{Y_t}{K_t}. \qquad (6.7)$$

For the reason alluded to in Chapter 2 and in this essay, we will consider mechanisms of distribution described by F which are not explicitly dependent of time, i.e. $Y = F(K, L)$. Since the function $F(K_t, L_t)$ is assumed to be linear and homogeneous, it can be written as

$$F(K_t, L_t) = K_t \, \mathfrak{f}\left(\frac{L_t}{K_t}\right), \qquad (6.8)$$

where the new map $\mathfrak{f}$, a sort of normalised function, is given by the relationship

$$\mathfrak{f}\left(\frac{L_t}{K_t}\right) = F\left(1, \frac{L_t}{K_t}\right).$$

Then from (2.10) we obtain the functional $\mathfrak{f} : t \longrightarrow y_t$, where y_t has the following simple form

$$y_t = k_t \, \mathfrak{f}(l_t), \qquad (6.9)$$

which is obviously of class $\mathbb{C}^2(\mathbb{R}_+ \longrightarrow \mathbb{R}_+)$. With the notation in (6.5) we are able to rewrite (2.12) and (2.11) in the following normalised form

$$\begin{aligned} c_t &= (1 - s_K)(y_t - W\lambda_t) + (1 - s_L)W\lambda_t, & (6.10) \\ i_t &= s_K(y_t - W\lambda_t) + s_L W\lambda_t, & (6.11) \end{aligned}$$

that added together give the corresponding version of the income identity

$$y_t = i_t + c_t.$$

Lemma 6.1 *Let us consider F and $\mathfrak{f}$ defined as in (6.8) and assume A 2.1. Then, for any $t \in [0, T]$ the marginal reward $\partial_L F(K_t, L_t)$ and $\partial_K F(K_t, L_t)$ are given by the relationships*

$$\begin{aligned} \partial_L F(K_t, L_t) &= \mathfrak{f}'(l_t), & (6.12) \\ \partial_K F(K_t, L_t) &= \mathfrak{f}(l_t) - \mathfrak{f}'(l_t)l_t. & (6.13) \end{aligned}$$

Proof: Differentiating the expression (6.8) partially with respect to L and K and applying Euler's theorem for homogeneous functions, one gets the claims of the lemma.   ∎

A distinguishing feature of our picture of labour-surplus economy is the assumption of an exogeneously determined minimum wage rate $\overline{W}$ adapted to the level of technical progress, responding to the requirement resulting from the principle of effective demand and the interplay between the upward and downward tendencies of the reserve army.

Let us introduce a pair-valued control function $\alpha : t \to \alpha_t$ with α_t given by

$$\alpha_t = (s_K(t), \lambda_t) \tag{6.14}$$

which allows us to rewrite eqs. (2.11) and (2.12) in the following more control-suggestive and simple way

$$i_t = i(k_t, s_K(t), \lambda_t) = i(k_t, \alpha_t), \tag{6.15}$$
$$c_t = c(k_t, s_K(t), \lambda_t) = c(k_t, \alpha_t). \tag{6.16}$$

Concerning the nominal wage W we shall say, it is obviously linked to a minimum wage rate $\overline{W}$ by the following inequality

$$W \geq \overline{W}. \tag{6.17}$$

Let us stress, $\overline{W}$ is not the minimum of subsistence and depends upon the state prevailing in the reserve army of unemployed labour, on the current level of technological progress and thus on the standard of living attached to productive activities associated with the stable core of techniques.

Features of the golden age and goal-conducive policies

Steady states are by definition growth paths along which a certain type of saturation results. For the purpose of the present essay we are interested in a type of saturation associated with growth paths along which capital saturation takes place and consumption per available labour attains a maximum. Thus we shall look for appropriate accumulation paths among all sustainable steady-state growth paths which satisfy our saturation criterion.

Lemma 6.2 *The assumptions of Lemma 6.1 hold. For consumption and investment paths given by (6.10) and (6.11) show the relationships*

$$c_t = k_t f(l_t) - (\epsilon_t - \pi_t^2) k_t, \tag{6.18}$$
$$\partial_\lambda c_t = f'(l_t), \tag{6.19}$$
$$\partial_K c_t = f(l_t) - f'(l_t) l_t - (\epsilon_t - \pi_t^2), \tag{6.20}$$

hold and characterise univocally an accumulation path along which capital saturate and consumption per available labour attains a maximum.

Proof: Indeed, we proceed as follows. Let us consider (2.11) and (2.12) which we rewrite according to (6.5) and (6.9) in the following normalised form

$$c_t = (1 - s_K)(k_t \mathfrak{f}(l_t) - W\lambda_t) + (1 - s_L)W\lambda_t, \qquad (6.21)$$
$$i_t = s_K(k_t \mathfrak{f}(l_t) - W\lambda_t) + s_L W\lambda_t, \qquad (6.22)$$

and combine them to obtain

$$c_t = k_t \mathfrak{f}(l_t) - i_t.$$

Foregoing equation together with (6.42) gives

$$c_t = k_t \mathfrak{f}(l_t) - D_+ k_t - (\epsilon_t - \pi_t^2) k_t.$$

Hence, setting $D_+ k_t = 0$, since capital saturation characterises our steady-state, we get (6.18).

Among the steady-state growth paths that satisfy eq. (6.18), we are interested precisely in the one that maximises consumption c_t. Differentiating eq. (6.18), we get then (6.19) and (6.20). From (6.19) results, whenever the marginal reward accruing to labour is greater than $\overline{W}$ for all techniques of production, λ_t must be equal to one. This entails the obvious result, maximum consumption requires full employment.   ∎

Let $c_t^\diamond$ and $k_t^\diamond$ stand respectively for the maximum sustainable level of consumption and capital per available labour. Let us denote by $f^\diamond$ the level of social reward corresponding to the maximum sustainable level of consumption $c_t^\diamond$, i.e. $f^\diamond = f^\alpha(t, c_t^\diamond,)$. Before continuing we want to point out once more that $D_+ k_t$, the same holds for (6.39) and (6.40), shall be interpreted as mean-conditional forward time-derivative, i.e.

$$D_+ k_t = \lim_{h \downarrow 0} \mathsf{E} \left[\frac{k_{t+h}(\omega) - k_t(\omega)}{h} \Big| \mathcal{F}_t \right],$$

where the expectation is taken with respect to the probability measure **P**. However, we stress that for the sake of simplicity and abusing the notation, we shall use the more suggestive notation k_t, λ_t and so on, unless any confusion arises.

Lemma 6.3 *Let the function* $\mathfrak{f} : \mathbb{R}_+ \longrightarrow \mathbb{R}_+$ *be defined as in (6.8) and assume that, for any* $t \in [0, T])$, *the function* $F(K_t, L_t)$ *satisfies A 2.1 and that for* c_t, *given by (6.5), the relationship (6.18) holds. Then the necessary conditions for the existence of* $c_t^\diamond$ *are given by (6.19) and (6.20). Further, the corresponding* $k_t^\diamond$ *is defined by*

$$\partial_K F(K_t, L_t) = \mathfrak{f}(l_t) - \mathfrak{f}'(l_t)l_t$$
$$= (\epsilon_t - \pi_t^2) \qquad (6.23)$$

Proof: It follows the argumentation of the preceding Lemma. Then, the level of capital saturation compatible with $c_t^\diamond$ turns out to be, according to (6.20), precisely the level of capital k_t at which the marginal reward accruing to capital $\partial_K F(K_t, L_t)$, see (6.13), renders $\partial_K c_t = 0$, which proves the Lemma.   ∎

We shall call $k_t^\diamond$, $c_t^\diamond$ and $f^\diamond$ the *golden values* of k_t, c_t and f respectively, since they describe a smooth, steady-state growth path with full employment which is known as the golden age of growth, stressing thereby its *mythical nature*, see [183, 181]. Thus defining $k^\diamond$ by means of (6.23) we obtain the so-called *golden rule of capital accumulation* which we explain shortly after the following lemma.

Lemma 6.4 *Let us assume that $k_t^\diamond$ is defined by (6.23) and the accumulation process $k = (k_t)_{t\geq 0}$ is such that $k_0 \leq k_t^\diamond$ for all $t \geq 0$. Then, it is easy to check the inequalities*

$$k_t \leq k_t^\diamond \tag{6.24}$$

$$c_t \leq \mathfrak{f}(l_t^\diamond)k_t^\diamond, \qquad with \ \ l_t^\diamond = (k_t^\diamond)^{-1} \tag{6.25}$$

$$f(t, k_t, \alpha_t) \leq f^\diamond \tag{6.26}$$

hold.

Proof: One uses the properties of $\mathfrak{f}$ and f, the income identity and the preceding lemma. ∎

Let us close this section recalling a beautiful and enlightening parable due to Marglin, see [145] concerning the question why should the maximum sustainable level of consumption per available labour $c^\diamond$ be determined according to (6.25) and hence be finite.

> Consider the head of a peasant family to whom the king makes the following offer: "You can have as much land, ready for the plough, as you like. On one condition. You must clear a sufficient amount of additional land to bequeath to your progeny an equal amount of land per capita as has been given to you. [1] If the size of this family remains constant, there need be no bound to the present generation's desire for land. But if the family is growing and it takes time and energy to clear new land, greed for the land's product will inhibit its demand for land. For the larger the initial endowment, the more land it must be cleared to maintain the per capita size of its holding. And the more time it spends in clearing the land, the less time it has to till its original holding. Intuitively, the marginal cost of satisfying the bequest constraint and the marginal benefit of a larger holding must balance at a holding of finite rather than infinite size.

In order to translate the message of Marglin's parable into our context, we first substitute the capital ratio k_t for land and then look at the golden rule as given by (6.23). This says, instead of increasing the capital level to the point at which its marginal reward $\partial_K F(K_t, L_t)$ falls to 0, the economy would optimally restrict itself to the more modest level k_t at which accumulation allows replacing capital stock used up and endowing subsequent generations at the prevailing standards, i.e. $(\epsilon_t + \delta_t)$, and provides for unpredictable fluctuations and uncertainty resulting from eventual informational confusion; see the remarks following (6.27) and (6.28) as well as Subsection 2.2.3.

---

[1] In other words, 'give to your progeny as has been given to you'. Whence the name golden rule.

Concerning social costs and terminal rewards

Let us make a few comments on the social cost f and the terminal reward ϕ. From the outset we should mention two points regarding the cost function f: (1) reflects lack of consumption characteristics associated with a commitment to the accumulation policy in implementation and (2) since through renouncing to an amount of consumption characteristics enables an expansion of the level of employment we shall often refer to f as the current rate of reward; this is a reward in the sense of bringing about a decrease in current deficit of states of functioning. Placing emphasis on social potentials like productive capacity, earning power and capabilities to function in a way that corresponds to the law of social reproduction, gives a clear suggestion how we shall evaluate the capacity of the system to function. In the context of Chapter 4 we define a performance functional as the capacity of a system to contribute effectively to the unfolding of its productive forces and in foregoing chapter we have seen how we can evaluate a performance functional corresponding to a system driven by a controlled diffusion. In this chapter we will maintain the same philosophy with minor changes. Thus the following remarks are solely made for the sake of completeness.

Thus, let us recall we assume f describes the flow rate of social cost attained by the *representative individual* and ϕ represents the terminal reward which measures the social value attached to the terminal state of capital; more explicitly, ϕ measures the social value of eventual deviations of terminal capital from the reference level k_τ^o introduced above. This term will be explained in the course of solving the control problem; as we shall see later it is connected with the transversality condition given by (6.111).

Since the problem of constructing a model of production will be treated somewhere else and we use a model of distribution akin to the orthodox one, it is extremely important to be aware the objects to be assessed are fully of different nature. First, having centred on the process of social reproduction entails that social preferences and value judgements are linked to priorities resulting from a commitment to maintain and renew society's potentials for production, creativity and innovation. Second, understanding k as determined by technical coefficients like a_k and a_l implies a commitment to maintain and renew capital and labour characteristics, in particular with respect to productive, creative and innovative potentials. But precisely this commitment links distribution mechanisms to those of effective demand. In this sense, the functionals f and ϕ measure social performance with respect to labour and capital potentials; see in this regard Sections 4.1 and 4.3 and [160] for a detailed description of technical coefficients of production and the performance functional.

6.1.4   Economic and institutional barriers

The principle of effective demand is a useful device for tuning changes in the capacity to producing and identifying the margins of action. It hints at potential bottlenecks of financial nature associated with the expansion of productive capacity and timing a gradual rise of labour income as well as wage goods corresponding to a higher level in the standard of living. Precisely at this stage, institutional interventions need to enter the picture. In this regard, we shall recall that simple economic considerations help us to focus our attention on some domains and intervals demarcating feasibilities and possibilities with respect to a given configuration of the production process in a given period of time.

Adjusting purchasing power to capacity expansion

To the aim of increasing economic surplus corresponds realisation of profits and this entails the presence of an appropriate effective demand which in turn relies on an adequate purchasing power. Since investment does not have an end in itself, purchasing power needs to rely in one way or another on the wage bill. Eventual deviations of purchasing, consuming and earning power from normal values may undermine the process of social reproduction reflected usually as bottlenecks and crises of profit realisation, unemployment and other well-known social maladies.

Knowing the expense account out of the wage bill one may anticipate the demand for social output provided the employment dynamics is adequately given. Since the basic ingredients in this framework are mostly known only probabilistically, one objective of planning and control interventions is to fix some of these elements so as to allow for a more effective anticipation of future changes. To illustrate this reflection, we shall examine briefly the information behind (6.1). The situation referred to in (6.1) describes an expansion in the productive capacity of a prescribed size; it remains to determine the size of the labour team assigned to the unit of capital endowment to prevail. That means, for the period of time under consideration the size of the capacity expansion belongs to the data of the problem, while the size of labour team which will operate with a chosen unit of capital endowment is a variable.

Let us denote the ratio of the size of the labour team to the capital unit of endowment by $l = (l_t)_{t \geq 0}$, see (6.5), which we shall understand and analyse as follows.

- The higher the size of the labour team, the higher the level of l and thus the higher the corresponding level of employment the prevailing unit of capital endowent can support. However, for a given unit of capital endowment there is a level of l beyond which marginal rewards, labour and/or the capital may get adversely affected.

- Alternatively, one may choose a unit of capital endowment which allows larger labour teams. However, concentrating capital on plant and machinery beyond certain levels may render the circular flow of production sensitive to financial bottlenecks and the like.

- A high level of l does not unambiguously mean a high level of purchasing power. It may also appear an imbalance between additions to purchasing power of workers and their contribution to social output at the prevailing capital endowment.

- For a given unit of capital endowment, operating below a certain level of l, the available capital equipment may have to operate at levels lower than the one the corresponding schedule of efficiency would indicate with respect to the expected rates of return on capital and/or various rates of turnover.

- A more efficient level of operation is commonly related to production and distributional arrangements yielding first higher output per worker with the capital endowment in question and second ensuring higher returns on capital.

In particular, let us mention the golden rule requires more capital per worker than the maximisation of surplus would indicate. Margins of choice and initial conditions referred to above, anticipate certain features of planning that once on the way entail a set of constraints and limited flexibility; they reflect, for instance, the loss of mobility in (physical) capital as soon as it gets embodied and installed in the system of material production. The alternatives remaining after the foregoing technical considerations are further limited by constraints resulting from the social relations of productions and other interacting institutions.

The relevant range of employment is then linked to the real interval $[\underline{l}, \overline{l}]$, where $\underline{l}$ is connected to the surplus-maximising level of employment and $\overline{l}$ with the level of employment at which wage income exhausts net social output. Hence, the levels $\underline{l}$ and $\overline{l}$ represent respectively an economic and a physical barrier to increases in employment and in wages. That means an increase in the level of employment or in wages remains possible, at least theoretically, as long as the associated wage bill remains less than the net social output, i.e. as long as $l \leq \overline{l}$. On the other hand, beyond $\underline{l}$ full employment is expected to entail a fall in the rate of profit and even in the rate of surplus value. The reason is that at full employment wages would increase or just tend to rise faster than productivity as a result of an eventually increased bargaining power.

Besides the customary responses of an economy based on profit, i.e. inflationary and fiscal policies, there are measures that appear as the natural vehicle to prevent the fall of the rate of profit below a certain level. We mean the so-called *rationalisation* and *automation*, i.e. replacement of human labour by machines and the like, aimed at restoring unemployment and replenishing the reserve army which in turn act as a check on any rise or even upward tendency of employment and real wages. These reflections reveal the existence of a sort of institutional barrier, i.e. the power of capital, directed toward stopping investment and increasing labour whenever the rate of profit falls too low. Under a capitalist régime this barrier can be reached well before $\underline{l}$.

Main domains of intervention

In the conventional world of neoclassical economics, we should mention, the economy would settle at $\underline{l}$, at least under perfect competition. Deviation from this optimal level may be the result of increasing labour for a given stock of capital, increasing capital for a given level of employment or both. Therefore, in order to increase employment beyond the level associated with $\underline{l}$ one needs institutional intervention, the purpose of which should be to provide accurate and fair information on the short- and long-term consequences of alternative policies available for choice.

This essay focuses in policies of capital accumulation and gives a special emphasis on processes of learning, building know-how and motivation versus forgetting, physical and moral deterioration. These processes describe natural outgrowths of the participation in and exclusion of labour from the production process. Additional features of institutional intervention are brought about indirectly through a system of social preferences and value-judgements by way of which the control board assumes a directive role in moulding social awareness and solidarity. In this fashion one supplies individuals with the fundamentals of intertemporal decisions compatible with social reproduction, in the prevailing state of affairs, individuals in turn attempt to achieve states of functioning – earning power, working skills and other characteristics – on the basis of which they may get the consumption power to maintain and improve their states of being and doing.

Issues concerning capabilities and positive freedom are expected to play thereby a major role, since they have been partially incorporated in the time factor $\exp\{-\varphi_t^s\}$ explained in the Section 4.3, a fact which gives these questions a more realistic dimension. Recall that $\exp\{-\varphi_t^s\}$ may represent an enrichment or impoverishment factor attributed to an enlargement or shortening of the life of the (representative) individual corresponding to command over a consumption bundle above/below the subsistence level attached to the current stage of development.

The standard problem of allocating disposable income for the purpose of (personal) consumption, i.e. today's investment in human capital, and for the purpose of investment, i.e. today's additions to physical capital, when viewed from the perspective of capability and positive freedom acquires its real relevance. Namely, it allows to focus more sharply on the decision of letting a fraction of available labour to deteriorate morally and physically in order to ensure the remaining fraction improves its capability and positive freedom. The presence of the factor $\exp\{-\varphi_t^s\}$ suggests these changes are to a major extent irreversible.

Consequently, policy-making cannot limit itself to look at the future and present claims associated with a feasible alternative investment-consumption mix, i.e. a path of pairs $(\dot{K}_t, C_t)$ resulting from an alternative expansion of the level of employment beyond the one associated with $\underline{l}$, but need to try to find a logic for balancing these claims. Other domains of interventions are related to capital impulses referred to in (6.1) and to the negotiation of the upper bound $\bar{s}_K(t)$ as in (2.27), in addition to issues concerning binding directives associated with the optimal policy.

6.2   Uncertainty sources in accumulation

A consequence of the fiction the economy consists of a relatively advanced core of productive activities and a reserve army of unemployed labour, is the idea unemployment comes out in the advanced core as unemployed labourers arrive searching for jobs and a better life. The purpose of this section is to show how essential ingredients of the system described above may come to interact and to explore forces at work, their dynamics and contribution to accumulation and employment.

6.2.1 Stochastic accumulation and labour

Since the attained picture of the labour-surplus economy corresponds in a way to portraying its states at any time t, $t \in [0, T]$, we shall try to bring to the forefront characteristic features of its dynamics. The sequence of the system's portraits obtained acquires more coherence and may in addition reveal various contradictory tendencies in the process of capital accumulation; further they may help us to understand the nature of permanent sources of social conflicts. Let us describe first the state of affairs in the labour market.

Random features of the labour market

Let $B^1 = (B_t^1)_{t \geq 0}$ and $B^2 = (B_t^2)_{t \geq 0}$ be two $\mathbb{R}^n$-valued Brownian motions defined on the abstract probability space $(\Omega, \mathcal{F}, \mathbf{P})$ endowed with a filtration $(\mathcal{F}_t)_{t \geq 0}$ of sub-σ-fields of $\mathcal{F}$ with the usual properties; for short the probability basis $(\Omega, \mathcal{F}, \mathbf{F}, \mathbf{P})$.

Essentially we shall assume, at any time t the average change in the supply of labour and in the demand for labour are directly proportional to the prevailing supply and demand for labour. The parameters of proportionality depend respectively on the prevailing wage rate and the anticipated level of skill chararcteristics resulting from certain types of employment, and on the prevailing demand coefficients and anticipated rate of return attached to the chosen technique; see in this regard Subsection 2.2.3. However, for the purpose of the present essay they will be considered as structural data.

Then, we shall assume changes at the time t, with $t \in [0, T]$, in the labour supply N_t and in the labour force participation L_t are governed by following SDE's of the diffusion type

$$dN_t = \epsilon_t N_t \, dt + \pi_t N_t \, dB_t^1, \qquad (6.27)$$

$$dL_t = a_t L_t \, dt + \zeta_t L_t \, dB_t^2, \qquad (6.28)$$

where $N_0, L_0 \in \mathbb{R}^n$ given and ϵ_t, π_t, a_t and ζ_t are $n \times n$ matrices specified in Subsection 2.2.3. The entries of these matrices are thus arbitrary scalars which may remain constant within the time period considered. For the sake of simplicity we shall assume that $B^1 = B^2$ unless stated otherwise. We have to point out, there is no reason other than convenience to use the same Brownian motion in (6.27) and (6.28) as way of describing the uncertainty driving the correspondence processes. A more realistic model should avoid this.

The drift term $\epsilon_t N_t$ in (6.27) will be interpreted in a sense explained in Subsection 2.2.3 as the flow of unemployed labour including underemployed workers and migrants during the time interval $[t - \Delta t, t]$ searching for a job in the core of productive activities, while the drift term $a_t L_t$ in(6.28) will be understood as the flow of job openings in the core of highly productive activities during the same interval of time. Both drifts depend upon the response of individuals, i.e. labourers and capitalists, to changes in the socioeconomic environment, attributed to the economic strategy being implemented. In Subsection 2.2.3 we have described how these coefficients are linked to the structure of the system, individual decisions and uncertainty.

The diffusion coefficients $\pi_t N_t$ and $\zeta_t L_t$ in (6.27) and (6.28) are due to unpredictable system fluctuations. They are generated by the interplay of a wealth of only partially known and/or indirectly observable decision processes and signals from interacting individuals, which often pursue conflicting goals. As a first approximation let us conceive these fluctuations as caused by a sort of confusion resulting from an inappropriate assessment of the extent and nature of these relationships as well as from the failure of anticipating properly the global/relative and the permanent/transitory character of their impact. This kind of confusion is usually attached to changes associated typically with intertemporal decisional situations under incomplete and asymmetric information and hinders individuals from ascertaining fully the socioeconomic state of affairs, its timing as well as their own position within this constellation of change.

Employment and crisis

For the sake of completeness, let us point out, the parameters of proportionality appearing in ϵ_t and a_t in (6.27) and (6.28) should be understood in a broad sense as given by relationships like $\epsilon_t = \epsilon(W_t, k_t)$ and $a_t = a(r_t, k_t)$. Here the dependence on both arguments is direct. That means, if $W^{(i)}$ and $r^{(i)}$ increase, the entries $\epsilon^{(ik)}$ and $a^{(ik)}$ increase as well, a fact which brings about accordingly changes in the supply and demand for labour. The dependence with respect to the second argument is however more subtle, since this is related to the chosen technique, the required and resulting level of skill characteristics as well as the structure of effective demand. Higher capital intensities k_t are as a rule associated with higher levels of know-how, labour skills and wages and thus jobs linked to such techniques will be attractive and give rise to increases in coefficients. At this point the probability structure of the underlying dynamics needs to be taken into account to the end of explaining fulfilling and/or frustration of expectations. Thus it appears natural to postulate increases in the second argument bring about a probability structure which amounts to lowering in the average the value of entries in ϵ_t and a_t. Furthermore, since in the reserve army, we have idle and underemployed labour, we may assume the level of skills attributed to such labour will bear probability features relatively unfavourable unless their skill characteristics are purposely improved.

In order to keep the model operational, we have expressed the corresponding drifts in such a way that they depend only on structural parameters. This amounts to suppressing any dependence on the second argument and assuming in the long-term the parameters ϵ_t and a_t depend upon the wage rate and the rate of returns on capital r_t. In the short-term $\overline{W}$ is negociated and kept fixed, so that the choice of technique and the assumption about W put a ceiling on r_t, $t \in [0, T]$. Thus, the parameters $\overline{W}$ and r_t are data for our present problem (in the language of economics they are exogenously given) and adjust to the prevailing standards of living and technical progress. See [71, 146].

Lemma 6.5 *Let us consider the process $N = (N_t)_{t \geq 0}$ with N_t driven by (6.27) and a Brownian motion $B^1 = (B_t^1)_{t \geq 0}$ defined on $(\Omega, \mathcal{F}, \mathbf{F}, \mathbf{P})$. Then for any $t > 0$ and $N_0 > 0$, N_t is given by the expression*

$$N_t = N_0 \exp \left\{ \int_0^t \left(\epsilon_s - \tfrac{1}{2} \pi_s^2 \right) ds + \int_0^t \pi_s dB_s^1 \right\}. \tag{6.29}$$

Proof: It is easy to see that the formula (6.29) is a solution of (6.27) as long as this solution does not vanish, see [82], pp. 36-38, Example 3. Indeed, setting $\eta_t = \log N_t$ and using Itô's formula, we obtain

$$d\eta_t = \frac{\partial}{\partial N_t} \{\log N_t\} dN_t + \frac{1}{2} \frac{\partial}{\partial N} \{\log N_t\} (dN_t)^2,$$

which after computing the partial derivatives, substituting (6.27) for dN_t and $\pi_t^2 N_t^2 dt$ for $(dN_t)^2$ gives

$$d\eta_t = \frac{1}{N_t} (\epsilon_t N_t dt + \pi_t N_t dB_t^1) - \frac{1}{2} \frac{1}{N_t^2} \pi_t^2 N_t^2 dt.$$

The fact that $N_0 > 0$ and that owing to continuity $N_t > 0$ on a certain interval, justifies the foregoing operations. Thus we have

$$d\eta_t = (\epsilon_t - \tfrac{1}{2}\pi_t^2) dt + \pi_t dB_t^1,$$
$$\eta_t = \eta_0 + \int_0^t (\epsilon_s - \tfrac{1}{2}\pi_s^2) ds + \int_0^t \pi_s dB_s^1,$$

from which (6.29) follows. Note that since the right-hand side of (6.29) does not vanish for $t > 0$, recall $N_0 > 0$, it represents an arbitrary solution. Therefore, (6.29) solves (6.27) for any $t > 0$. ∎

In the sequel we will need the stochastic differential of processes like N^{-1}, L^{-1}, l and l^{-1}. With the purpose of developing more feelings for that kind of operations and for future reference, we will then store some of these differentials in the following lemma.

Lemma 6.6 *Consider the processes $\lambda = (\lambda_t)_{t \geq 0}$ and $l^{-1} = (l^{-1})_{t \geq 0}$ with λ_t and l_t defined, for any $t \in [0, T]$, as in (6.5) – (6.7). Under the assumption that $B^1 = B^2$, show that λ and l^{-1} are governed by the SDE's*

$$d\lambda_t = \{(a_t - \epsilon_t) + (\pi_t - \zeta_t)\pi_t\}\lambda_t\, dt - (\pi_t - \zeta_t)\lambda_t\, dB_t, \tag{6.30}$$
$$d(l_t^{-1}) = \{i_t \lambda_t^{-1} - (a_t - \zeta_t^2)\, l_t^{-1}\}\, dt - \zeta_t l_t^{-1}\, dB_t, \tag{6.31}$$

with λ_0 and l_0 given, see (6.2) – (6.3).

Proof: Let us begin computing the stochastic differential for N^{-1} and L^{-1} under the assumption the processes N and L have dynamics given by (6.27) and (6.28). Let us set $\eta_t = N_t^{-1}$ and consider it as a function on N_t which has an Itô differential. Again, a simple application of Itô's calculus, yields

$$d\eta_t = \frac{\partial \eta_t}{\partial N_t} dN_t + \frac{1}{2} \frac{\partial^2 \eta_t}{\partial N_t^2} (dN_t)^2.$$

After substituting the partial derivatives, (6.27) and $(dN_t)^2$, one obtains

$$dN_t^{-1} = -N_t^{-1}(\epsilon_t - \pi_t^2)dt - N_t^{-1}\pi_t dB_t^1,$$

which gives the stochastic differential equation driven N^{-1}. Similarly, setting $\eta_t = L_t^{-1}$ one obtains

$$dL_t^{-1} = -L_t^{-1}(a_t - \zeta_t^2)dt - L_t^{-1}\zeta_t dB_t^2.$$

At this stage, we set $\eta_t = N_t^{-1}$ again and consider $\lambda = L\eta$ in order to compute $d\lambda$ as the stochastic differential of a product. Thus we have

$$d\lambda_t = L_t d\eta_t + dL_t \eta_t + dL_t d\eta_t,$$

where the third term of the right hand side is the distinguishing feature of Itô's calculus. Taking $B = B^1 = B^2$ and substituting the differentials $d\eta_t$, dL_t and their product $dL_t d\eta_t$, gives

$$d\lambda_t = a_t \lambda_t dt + \zeta_t \lambda_t dB_t - [\lambda_t(\epsilon_t - \pi_t^2)dt + \lambda_t \pi_t dB_t] - \lambda_t \zeta_t \pi_t dt,$$

from which (6.30) follows.

To obtain (6.31), one considers $l^{-1} = K\eta$ with $\eta_t = L_t^{-1}$. Setting $dK_t = \dot{K}_t dt$ and following closely the preceding procedure, one gets (6.31). ∎

One of the advantages of the simplifying assumption concerning the nature of the uncertainty in (6.27) and (6.28) is the transparency of (6.30). In order to provide a better feeling about the information we may lose by assuming $B^1 = B^2$, let see how (6.30) would look like without this simplifying assumption. Thus applying Itô to the process λ and using (6.27) and (6.28) as they stand, we obtain

$$d\lambda_t = [(a_t - \epsilon_t) + \pi_t^2]\lambda_t dt - [\pi_t dB_t^1 - \zeta_t dB_t^2]\lambda_t - \zeta_t \pi_t \lambda_t dB_t^1 dB_t^2,$$

which describes clearly the contribution of the uncertainty associated with N and L to one coming out in the dynamics of λ. Maintaining at least a certain degree of correlation between B^1 and B^2, as one easily recognises, would already allow a clear interaction of random features inherent in N and L. This would then play a role in the accumulation dynamics to be derived shortly.

Nevertheless, it is necessary to stress that inter-connections between the supply of labour and the accumulation of capital will remain hidden as long as the heterogeneity of capital and labour are not reckoned explicitly. However, this is beyond the scope of the present work. See [91].

A few remarks on the drift coefficient of eqs. (6.30) and (6.31) are appropriate. First, let us mention the interaction of the processes λ and l comes to bear through the term $a_t - \epsilon_t$, which we want to see as a barometer of social conflicts related to the reserve army, accumulation and employment. See [71, 146, 141].

- A large reserve army brings about a strong downward pressure on wages, due to a weak bargaining power of labour. The lower the wage rates, the stronger the tendency of the rate of profits to increase, which may lead to a greater $a_t - \epsilon_t$. However, higher rates of profit bring about by dint of tougher competition a lower uniform rate of return on capital and this puts a check on the tendency of the reserve army to shrink that accentuates deficiencies in states of functioning associated with low wage rates.

- A greater $a_t - \epsilon_t$ may increase the level of employment and reduce the reserve army that in the average may improve states of functioning. From this it may follow stronger bargaining power of labour, calling forth an upward tendency of the wage rates and downward pressure on profits, which in turn may lead to lowering $a_t - \epsilon_t$ and a replenishing the reserve army by rationalisation and automation.

- Tough competition holds sway over the advanced sector. Under the pressure of the struggle for profits and market control capitalists continuously switch to more automatised techniques associated with higher rates of accumulation and profits. The latter entails, as already suggested, lower $a_t - \epsilon_t$ replenishing again the reserve army. Intensive pace of capital formation per worker seems to require a slower pace in the creation flow of job openings.

- Finally, the uncertainty parameters may lead to an over- or underestimation of the drift of the actual dynamics with sometimes alleviating but sometimes aggravating effects on the actual state of affairs. This can be seen analysing $[\pi_t - \zeta_t]$ and $\exp\{-\varphi_t^s\}$.

It is necessary to stress that the understanding of the meaning of eqs. (6.30) and (6.31) requires in the spirit of Thesis 7.9 an adequate knowledge about what capital accumulation truly means. Here, we mention only a few features and invite the reader to take a closer look at the processes they generate. For this purpose, we write down first the solution for λ

$$\lambda_t = \lambda_0 \exp \left\{ \int_0^t [(a_s - \epsilon_s) + (\pi_s - \zeta_s)\pi_s] \, ds - \int_0^t (\pi_s - \zeta_s) \, dB_t, \right\} \quad (6.32)$$

and the two versions of the solution of (6.31). First, as the process l^{-1} with

$$l_t^{-1} = l_0^{-1} \exp \left\{ \int_0^t (s_s v_s - [a_s - \zeta_s^2]) \, ds - \int_0^t \zeta_s \, dB_s \right\}, \quad (6.33)$$

which stresses capital formation per actually employed worker and secondly as the process l with

$$l_t = l_0 \exp \left\{ - \int_0^t (s_s v_s - [a_s - \zeta_s^2]) \, ds + \int_0^t \zeta_s \, dB_s \right\}, \quad (6.34)$$

representing labour per unit of available capital, which indicates the shortcoming in the capital endowment of labour. Under the sway of competition l^{-1}

increases or l decreases. However, in order for these changes to contribute to the development goals, it is necessary to supplement them with activities able to absorb at least the uprooted labour. An elaborated analysis of these issues is given in [91].

6.2.2 Building capital and employment

An essential requirement of profit realisation is the presence of effective demand, as a dynamic mechanism providing purchasing power, renewal of productive forces and earning potentials. Thereby, the links between capital and employment acquire a vital dimension reflected partially by the processes introduced above.

The models constructed in the foregoing subsection are instances of projections of reality in the sense of Thesis 7.9. The models described by (6.27) and (6.28) represent changes in the supply and demand for labour, while (6.32) and (6.33) picture models reflecting the dynamics governing processes of employment and capital formation. Next, we will take a close look at the interaction of these processes and investigate the joint effect and compatibility of economic policies directed toward the enhancement of productive capacity and growth potentials like investment in capital stocks and those conducive to improvement and higher utilisation of available productive potentials.

The joint dynamics of interacting processes

Let us proceed to design a process of capital accumulation able to handle growth and development tasks as just mentioned. In order to see how forces of interaction behind (6.30) and (6.31) work, we shall consider k_t, i.e. capital per available labour, as given by the relationship

$$k_t = \left(\frac{K_t}{L_t}\right)\left(\frac{L_t}{N_t}\right).$$

Theorem 6.7 *Assume $l^{-1} = (l_t^{-1})_{t \geq 0}$ and $\lambda = (\lambda_t)_{t \geq 0}$ are described by the SDE's (6.31) and (6.30) in Lemma 6.6. For $B = B^1 = B^2$ and k_0 given, show the process $k = L^{-1}\lambda$ is driven by the SDE given by*

$$dk_t = \left[i_t - (\epsilon_t - \pi_t^2)k_t\right]dt - \pi_t k_t \, dB_t. \tag{6.35}$$

Proof: Once more, we shall apply a little of Itô's calculus. To be precise, we use the Itô rule for the differential of a product, which differs from the standard one only, as we already know, in the third term, see [139] pp. 122-23, Example 1, to obtain

$$
\begin{aligned}
dk_t &= d\left(l_t^{-1}\lambda_t\right) \\
&= l_t^{-1}\,d\lambda_t + d\left(l_t^{-1}\right)\lambda_t + d\left(l_t^{-1}\right)\cdot d\lambda_t,
\end{aligned}
$$

which together with (6.30) and (6.31) and some arrangements gives

$$
dk_t = \Big[\{(a_t - \epsilon_t) + (\pi_t - \zeta_t)\pi_t\} + \\
+ \Big\{s_t v_t - (a_t - \zeta_t^2)\Big\} + \zeta_t(\pi_t - \zeta_t)\Big]k_t\,dt - \pi_t k_t\,dB_t. \tag{6.36}
$$

Thus the assumption that the Brownian motions associated with (6.27) and (6.28) coincide leads to the very simple eq. (6.35).   ∎

Nevertheless, this simplifying assumption hides the interaction between the processes modelled by (6.30) and (6.31). Without this simplifying assumption (6.36) would be

$$
\begin{aligned}
k_t^{-1} dk_t = & \left\{ \left[(a_t - \epsilon_t) + \pi_t^2 \right] dt - \varsigma_t \pi_t dB_t^1 dB_t^2 - \left[\pi_t dB_t^1 - \varsigma_t dB_t^2 \right] \right\} \\
& + \left\{ \left[s_t v_t - (a_t - \varsigma_t^2) \right] dt - \varsigma_t dB_t^2 \right\} \\
& + (\pi_t dB_t^1 - \varsigma_t dB_t^2 + \varsigma_t \pi_t dB_t^1 dB_t^2)(\varsigma_t dB_t^2),
\end{aligned}
$$

which shows that alone the assumption that B^1 and B^2 are two dependent Brownian motions blur strongly the interactions.

In order to avoid losing that information, we want to stress the role of three components of the drift term in (6.36), which reflect the contribution of the processes of building capital and eliminating unemployment.

- The first one is the drift contribution due to changes in the process of eliminating unemployment represented by λ_t, the meaning of which follows from the remarks concerning (6.30).

- The second one is due to the actual economic surplus or process of capital formation per actually employed worker l_t^{-1} described by (6.31) that we have already brought forward.

- The third one accounts for the pure stochastic interaction between the processes λ_t and l_t^{-1} which may also lead to an over- or underestimation of the actual drift.

It is necessary to be aware that the mere existence of an economic surplus, as the drift of (6.35) may suggest, does not entail accumulation of capital. It is necessary that profits be realised, as we know from the principle of effective demand. Further, one needs that capitalists carry out investments, which gives evidence of the fact that the amount of employment (expected) to prevail is governed by the stock of equipment then in existence.

Since we assume in this essay that the advanced sector coexists with the reserve army without any interaction other than recruiting labour according to (6.27) and (6.28), to explore further the accumulation problem we need to understand some issues behind economic policy-making.

Economic policy and the equation of growth

The movements of k_t can be gauged by the level or relative position of the functions i_t and $\varsigma_t k_t$ at time t, i.e. the actual and required investments per available worker, where the new variable ς_t is defined as $\varsigma_t = \epsilon_t - \pi_t^2$. The difference $i_t - \varsigma_t k_t$ at time t portrays roughly the expected rate of change of the level of capital per available worker as a function of k_t, more precisely as a function of the information available at time t on the level of k_t, and

provide helpful insights into the prevailing stochastic dynamics of accumulation
of capital per available worker.

Knowing, for instance, that at time t the inequality $i_t - \varsigma_t k_t \geq 0$ holds,
means accumulation indeed takes place and that k_t increases. However, one
word of caution is in order since these statements hold only almost surely
according to the probability measure prevailing at time t. To see this, we
need to make precise how we can deal with the drift term in (6.35). In order
to keep our model operational, we have to resort to the notion of stochastic
forward derivative already introduced in Chapter 4, i.e. approximate the drift
coefficient by the mean-conditional forward derivative, see [154]. As mentioned
above, the purpose of using this notion is to eliminate too strong irregularities
of the trajectories $k_t(\omega)$ by conditioning with respect to the past $\mathcal{F}_t$ and taking
the conditional mean over all possible values $k_{t+h}(\omega)$, $h > 0$. Let us denote the
drift coefficient of (6.35) by $D_+ k_t$ and assume that the relationship

$$D_+ k_t = \lim_{h \downarrow 0} \mathbf{E}_x \left[\frac{k_{t+h}(\omega) - k_t(\omega)}{h} \bigg| \mathcal{F}_t \right], \tag{6.37}$$

holds, where $\mathbf{E}_x$ stands for the expectation operator conditioned at the level
of capital $k_t = x$. Since in this chapter we will be working only with strong
solutions of SDE's, we may as well suppress the subindex in the expectation
operator. However, it is important to keep in mind the conditional expectation
is taken always with respect to the present. To save notation, we shall use the
more suggestive expression $\dot{k}_t = D_+ k_t$ unless any confusion may arise and talk
about the forward drift.

Recalling the income identity $y_t = i_t + c_t$, and interpreting the drift as in
(6.37) one gets that accumulation obeys the relationship

$$y_t = c_t + \varsigma_t k_t + \dot{k}_t \qquad \mathbf{P} \quad \text{a.s.} \tag{6.38}$$

which states that on the average the national income per available worker y_t
will be allocated for the purpose of maintaining the level of consumption per
available worker c_t, and the level of capital per available worker, i.e. $\varsigma_t k_t$, as
well as of yielding increases in the level of capital per available worker $\dot{k}_t$.

Relying on the principle of effective demand, we may say that, at any moment of time, on the average a certain number of workers are engaged on
producing consumption goods, other fraction on producing new equipment or
replacement items of capital and the rest on producing either capital or consumption goods to meet eventualities. Hence it follows that the central role of
economic policy-making should be aimed at providing information, incentives
and binding directives that help to avoid realisation bottlenecks. Herewith, it
is necessary to observe that, for a given technique, the demand for consumption goods is deeply related to the number of workers engaged in producing
capital goods, the prevailing rate of real wage and the ratio of the output rate
of consumption goods to the wage rate.

Accumulation can go on, with the prevailing technique, as long as the output
of commodities produced by the workers employed in the consumption branch
exceeds sufficiently the level corresponding to its own wage bill, so that it can

also satisfy the demand for wage goods of workers required to keep up the productive potential of existing capital and to increase the productive power. To make sure that this is so, constitutes a further task of economic policy-making usually condensed in the constraint associated with (6.1)–(6.3) which ensure (6.35) makes sense and that simple dynamics like (6.38), known as the fundamental equation of growth, do not exhibit pathologies. Concerning various crucial issues on capital accumulation that we cannot consider here, we refer to [74, 78, 162].

Qualitative features of the accumulation dynamics

Next we shall point out a few interesting features of the qualitative behaviour of the process of accumulation of capital per available worker. With this purpose we shall look into the drift coefficient of (6.35) in the form given by (6.38) and draw a phase diagram, see Fig. 6.1. In Fig. 6.1 we depict a phase plane $(\dot{k}_t + \underline{c}_t, k_t)$, for an arbitrary but fixed time t, portraying disposable income per available worker nett of required investments per newly employed worker (vertical axis) against the level of capital per available worker (horizontal axis). It is easy to recognise that the aforementioned disposable income can be allocated to purposes of personal necessary consumption c_t and to net expansions of capital $\dot{k}_t$. Thus, the situation in Fig. 6.1 obtains after recalling $\varsigma_t = \epsilon_t - \pi_t^2$ and rewriting (6.38) as

$$\dot{k}_t + \underline{c}_t = y_t - \varsigma_t k_t,$$

where $\underline{c}_t$ stands for the minimum of necessary personal consumption per available worker, resulting from the assumption of the existence of a floor for the admissible wage rate W, i.e. from the assumption that the prevailing wage rate W has to satisfy the inequality $W \geq \overline{W}$ with $\overline{W}$ given. We shall refer to $\underline{c}_t$ as the level of *necessary personal consumption*. The latter does not have to correspond to a minimum level of subsistence and will adjust instead to the prevailing level of socioeconomic progress of the society.

The left-hand side in the foregoing equation hints to the potential allocation of economic surplus per available worker to consumption beyond the level $\underline{c}_t$ and to the expansion of the level of capital accumulation. Within certain limits, e.g. inflation barrier, bargaining power of labour, finance and the like, the decisions to carry out investment determine the rate of accumulation the system may achieve. The level of wages, in terms of consumption goods, has to be set according to the prevailing standard of living and future expectations, since changes in $\underline{c}_t$ may affect the factor $\exp\{-\varphi_t^s\}$ negatively and with it, expectation of life, motivation, know-how and skills as already mentioned in Chapter 4.

On the other hand, any increase of the level of personal consumption above $\underline{c}_t$ slows down directly the pace of capital accumulation, because it diverts economic surplus from investment. For a given technique, a set of conditions and circumstances as indicated above determines an appropriate régime of wage rates, investment decisions and ultimately a consistent rate of accumulation; such a régime may be thought to characterise a steady state toward which the system converges in the long range. Many such régimes are possible and these are distinguished according to properties exhibited by at least one of their state variables. For instance, let us mention the level of capital denoted by k_t° known as the Golden Value of capital per available worker, which represents that level of k_t capable of sustaining the maximum level of consumption denoted c_t°. This is simultaneously the saturation level of capital, i.e. the level at which $\dot{k}_t = 0$.

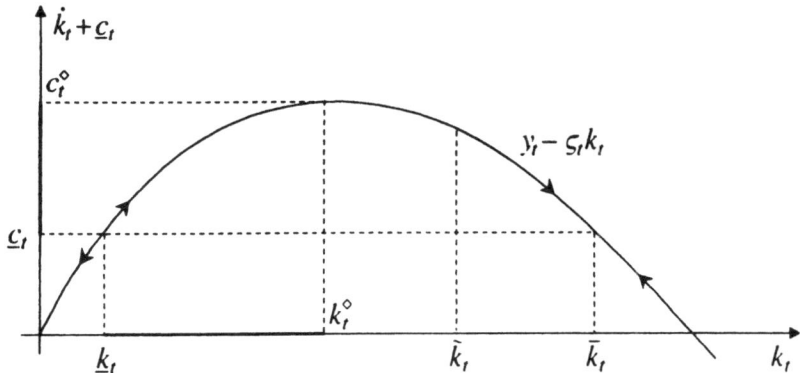

Figure 6.1: Phase diagram of accumulation of capital per available worker

Associated with $\underline{c}_t$ there are a lower and an upper level of capital denoted respectively by $\underline{k}_t$ and $\overline{k}_t$, which are also equilibrium points, the first unstable while the latter stable as the arrows in Fig. 6.1 show. The Golden Value of capital k_t° is also an unstable equilibrium state. From the point of view of economic policy the qualitative behaviour displayed in the diagram in Fig. 6.1 places some restrictions on the range of admissible initial-levels of capital per available worker.

Let us draw attention to the fact that accumulation paths starting from levels of capital at the left of $\underline{k}_t$ are driven by the dynamics of the system to the zero level of capital. On the other hand, paths starting from initial levels of capital at the right of $\underline{k}_t$ do not move necessarily towards the upper level $\overline{k}_t$, since the Brownian motion acting on the accumulation path may still drive them out of the stable region. It is easy to recognise that when the system operates at a level of capital close to and at the right of $\underline{k}_t$, i.e. when $k_t+\underline{c}_t$ barely exceeds the level of necessary personal consumption $\underline{c}_t$, the remaining product surplus is just enough to provide for slightly positive additions to the stock of capital. This situation leaves almost no room to adapt to eventual changes signalised by newly available information and to respond to contingencies coded in the coefficients ζ_t and π_t.

In passing, let us point out that the level of capital $\tilde{k}_t$ corresponds to that level associated with the average rate of return on capital beyond which further investments are no longer worthwhile. For the purpose of the present essay, for any time $t \in [0,T]$, the relevant range of capital k_t is given by $[\underline{k}_t, k_t^\circ]$.

On the dynamics steering the system

After having learned more about the ingredients of the present model and some features of their qualitative behaviour, we want to examine how to arrange for changes in the interacting processes in a manner conducive to the aims under consideration. For this purpose we want to obtain expressions of the drift terms of (6.30), (6.31) and (6.36) more amenable to our analysis and make a few comments on their economic content. Consider λ_t, l_t^{-1} and k_t as functions of time t and the canonic sample w, $w \in \Omega$. Hence, the forward drifts of λ_t, l_t^{-1} and k_t are given by

$$D_+ \lambda_t = \{(a_t - \epsilon_t) + \pi_t^2 - \zeta_t \pi_t\} \lambda_t, \tag{6.39}$$

$$D_+ l_t^{-1} = i_t \lambda_t^{-1} - (a_t - \zeta_t^2) l_t^{-1}, \tag{6.40}$$

$$D_+ k_t = [\{(a_t - \epsilon_t) + \pi_t^2 - \zeta_t \pi_t\} + \{s_t v_t - (a_t - \zeta_t^2)\}$$
$$- \zeta_t(\zeta_t - \pi_t)] k_t, \tag{6.41}$$

and according to (6.39) and (6.40), the forward drifts of (6.36) and (6.41) become equivalent and equal to

$$D_+ k_t = i_t - (\epsilon_t - \pi_t^2) k_t. \tag{6.42}$$

Let us denote the relative rate of growth of λ_t, l_t^{-1} and k_t respectively, by $\hat{\lambda}_t$, $\hat{l}_t^{-1}$ and $\hat{k}_t$, i.e.

$$\hat{\lambda}_t = \frac{D_+ \lambda_t}{\lambda_t}, \qquad \hat{l}_t^{-1} = \frac{D_+ l_t^{-1}}{l_t^{-1}}, \qquad \hat{k}_t^{-1} = \frac{D_+ k_t}{k_t}.$$

This enables us to derive from (6.41) the compact relationship

$$\hat{k}_t = \hat{\lambda}_t + \hat{l}_t^{-1} - \zeta_t(\zeta_t - \pi_t), \tag{6.43}$$

which suggests a characterisation of the continuous economic policy; i.e. the control process $\alpha = (\alpha_t, \mathcal{F}_t)_{t \geq 0}$, see (6.14). Namely, the chosen investment-consumption mix should correspond to an investment policy which brings about a process of capital accumulation proceeding at a pace that ensures in the course of time elimination of unemployment. That is, the control α should be chosen so that at any t, with $t \in [0, T]$, the inequality

$$\hat{\lambda}_t \geq 0 \tag{6.44}$$

holds. Eq. (6.44) entails that the relative rates of change of k and l be so that

$$\hat{k}_t \geq \hat{l}_t^{-1}, \tag{6.45}$$

provided that the relationship

$$\hat{\lambda}_t \geq \zeta_t(\zeta_t - \pi_t), \tag{6.46}$$

holds. That means, the relative rate of change in the process of employment formation should always be able to override the impact of uncertainty. Summing up, the choice of the control α should ensure a positive relative rate of change in the process of unemployment elimination, as indicated by (6.44), and a relative rate of change in the process of formation of capital endowment per actually employed worker so that (6.43) holds. The right-hand side of inequality (6.46) prescribes a moving lower bound of the range of values of $\hat{\lambda}_t$. Given an increasing reserve army and homogenous labour, it is not unlikely that π_t decreases toward 0 and this would entail that $\zeta_t(\zeta_t - \pi_t)$ goes up to ζ_t^2. This makes (6.46) unlikely to hold for large values of ζ_t, i.e. large uncertainty in the process of (new) job openings.

On the other hand, for a shrinking reserve army increasing π_t's are likely to materialise. As π_t moves toward ζ_t, the lower bound approaches 0 and may become even negative for π_t larger than ζ_t, which may be explained by substitution of labour for capital. Let us store for the purpose of future reference some of these reflections in the following assumption.

A 6.1 *Assumption on the control variable λ.*

- At any time t the control variable λ satisfies the inequality (6.44). Alternatively, it may satisfy (6.46) provided measures of policy-making ensure that π_t remains close to ζ_t.

The aim of the measures of policy-making alluded to is not only to ensure (6.44) holds, but also to provide unemployed labour with alternatives, to enrich information of potential migrants with information on the requirements one needs with regard to the most convenient choice of skill characteristics and diversification to diminish risk, to establish training centres and opportunities in job training.

Eq. (6.46) requires that uncertainty does not override the employment trend. This suggests also that the main source of uncertainty lies in the process of unemployment elimination and the flow describing the creation of job openings. The tails of the probability density of the process L_t and of N_t mark the level of uncertainty to prevail. Since the inequality

$$N_t \geq L_t \tag{6.47}$$

holds for any t, the inequalities

$$k_t \leq l_t^{-1} \quad \text{and} \quad 0 \leq \lambda_t \leq 1 \tag{6.48}$$

are trivial. The high degree of abstraction of our model of accumulation impedes preciser statements on the qualitative and quantitative links among λ, l^{-1} and the associated sources of uncertainty. As a matter of fact, we would need to provide information on the ratio of profit margins to wage costs, bargaining power of labour, ratio of accumulation to the stock of capital, changes in the rate of profit on capital and in the rate of real wages, and on the more likely reactions of capitalists to a fall in wages. In the remaining part of this section we will introduce some additional ingredients with the purpose of extending our analysis to some of these issues.

Since (6.36) integrates essential characteristics of the process of capital accumulation and emerges as a result of interaction of processes representing the supply of labour and the elimination of unemployment, we use it here as *system dynamics*.

A concise representation of the accumulation dynamics

For future reference and expository purposes let us denote the process of capital accumulation per available worker by $k = (k_t)_{t\geq 0}$ and assume it is governed by the stochastic differential equation

$$dk_t = b^\alpha(t, k_t)dt + \sigma^\alpha(t, k_t)dB_t, \tag{6.49}$$

where the drift $b^\alpha(t, k_t)$ measures locally the average macroeconomic tendency at time t of the rate of change of k_t, and the diffusion coefficient $\sigma^\alpha(t, k_t)$ represents the associated fluctuation about the drift.

The choice of alternative control processes α gives rise to a set of alternative drifts and diffusion coefficients $b^\alpha(t, k_t)$ and $\sigma^\alpha(t, k_t)$, and this in turn entails a set of alternative time paths of accumulation $k^\alpha = (k_t)_{t\geq 0}$. In order to understand the impact of alternative policies, one needs to introduce a criterion of social desirability that acts also as means of assessing distinguishing features of the process $k^\alpha = (k_t)_{t\geq 0}$.

The interaction of economic processes set off by the policy under consideration brings about a drift coefficient $b^\alpha(t, k_t)$, as in (6.36). A properly chosen economic policy should be capable of imposing a pace of accumulation of capital that pushes indeed the economy to higher levels of development within the degree of uncertainty attached to $\sigma^\alpha(t, k_t)$. The choice of the control α should be aimed at bringing about a drift as in (6.30) that contributes to the elimination of unemployment and a drift as in (6.31) that generates the required capital endowment per actually employed labour.

The interaction of both drift coefficients shall ultimately help to bring about a pace of accumulation of capital on the basis of which shifting idle labour into more highly productive activities can be afforded, effective demand expands and profit realisation follows. Thus we direct our attention to the evolution of the process $k = (k_t)_{t\geq 0}$ governed by the SDE (6.49) with the initial state (6.1) and on the appropriate choice of the process $\alpha = (\alpha_t)_{t\geq 0}$.

6.2.3 Further reflections on policy design

In this subsection the goal is to elaborate a little more on two notions that bear a great deal of theoretical relevance and interpretative appealing. Namely, the social supply and demand price of investment. These notions are central tools constructing an investment policy aimed at increasing productive capacity of the economy in a way that facilitates shifting effectively into production a given fraction of the reserve army. A simple way of looking at this issue leads to moral considerations underlying the distinction between capital and income. This distinction is not as obvious as it might appear, owing to the fact capital is on the one side a social relation of production, in particular it is an income-yielding property, and on the other it represents means of production. In the latter category capital yields, if it is combined with labour power, income in the form of wages. At this stage it is necessary to stress, a large part of production is therefore devoted to maintaining and increasing means of production and labour power as well with the purpose of using them in future

production. Concerning the participation in production, the underlying motive
of each individual may be understood as that of getting command over money,
purchasing power or consumption goods.

The crucial distinction between maintaining and increasing the productive
capacity as well as the distinction between capital and income can be explained
as follows:

> The morality of a peasant, who gathers his crops according to the rhythm
> of the seasons, is to put back into the soil what he takes out of it, and to
> set aside seed from each harvest, so as to preserve productive capacity
> for the future, not only for his lifetime, or his children's lifetime, but for
> the future as such. It is this morality which produces the conception of
> capital and income. Income consists in the kindly fruits of the earth, and
> capital in the fertility of the soil.

> Joan Robinson
> (see [180], pp. 33–34)

Viewing these essential issues from this side suggests first, the viability of an
economic system depends deeply upon its prevailing moral ideas and second,
the true significance of production lies in consumption, including productive
and personal necessary consumption, rather than the motive for it.

Social supply and demand price of investment

Processes underlying the generation of social output in an economic system and
the distribution of disposable income are intricate and highly interwoven with
each other. Therefore, there are great difficulties in giving precision to notions
like consumption and investment, in a way that provides an operational distinction between capital and income. At a national level (physical) capital and
labour power may be seen as consisting in natural resources, physical goods like
machines and plants, man-hours including brain and muscle, know-how, organisation, training and education. Viewing consumption goods according to the
productive and/or income-earning characteristics they support and as means
of providing the basis for the capability of individuals to function, e.g. directing attention to aspects of nutrition and/or skill acquisition and the like, offers
us however adequate instruments to work out clear-cut distinctions between
consumption and investment.

From this angle, we shall look at the problem of the social desirability of
a feasible investment-consumption mix and trace the effect on employment of
additional investments and consequently of a faster capital accumulation. In
this task the application of the peasant's morality should assist and guide our
assessment rationale. Accordingly, maintaining the productive power for the
future of natural, inherited or human resources should act as binding directive
upon decisions concerning the allocation of disposable income for consumption
and investment purposes. Further, one needs to be aware that for economic
surplus to be obtainable there must be a surplus of disposable income per
available worker over personal necessary consumption. In addition capitalists
must be impregnated with the peasant's morality so that they feel guilty of

drawing purchasing power from their stock of capital and come hence to plough profits back into production.

Taking into account the amount of employment is ultimately driven by the stock of equipment in existence and the economic surplus it generates, we can draw attention to the fact that any path $t \longrightarrow (i_t, c_t)$ gives rise to an employment path $t \longrightarrow \lambda_t$. Provided the latter represents a kind of improvement owing to actual increases of employment, learning by doing and higher capabilities of labour, then the latter path will in turn induce in the former further changes of the desired sort which we shall trace and evaluate.

Keeping technology and institutional arrangements fixed, the actual idea of our assessment methodology is to associate with the accumulation path $k = (k_t)_{t \geq 0}$ and the resulting employment process $\lambda = (\lambda_t)_{t \geq 0}$ a path $t \longrightarrow {}_*P_K(t)$. Again ${}_*P_K(t)$ represents the *accounting price of investment* and for any fixed time $t \in [0,T]$ ${}_*P_K(t)$ is given, by the relationship

$$ {}_*P_K(t) = - \left(\frac{dc}{di} \right)_{T_3} = - \left(\frac{\partial_\lambda c_t d\lambda + \partial_{s_K} c_t ds_K}{\partial_\lambda i_t d\lambda + \partial_{s_K} i_t ds_K} \right)_{T_3} . $$

Working within the framework resulting from the institutional arrangements condensed in T_3 compels capitalists to invest at the highest rate $\overline{s}_K$ whenever unemployment emerges. Therefore, the accounting price of investment ${}_*P_K(t)$ in the context of the present essay assumes the form

$$ {}_*P_K(t) = - \frac{\partial_\lambda c_t}{\partial_\lambda i_t}, \tag{6.50} $$

which indicates how much consumption $c_t = c(k_t, \alpha_t)$ the economy has to make available in order to forward an additional unit of investment $i_t = i(k_t, \alpha_t)$ along the accumulation path underlying the employment path $t \longrightarrow \lambda_t$. It is important to realise, the amount of consumption $\partial_\lambda c_t$ forgone at any time $t \in [0,T]$, with the purpose of adding one unit to investment i_t, is dictated by the prevailing technology. Parallel to these shifts of disposable income from consumption to investment one has, needless to say, shifts from current states of functioning for units of consumption forwarded to future states of functioning for units of capital added to existing plant and endowment.

Therefore, the employment process $\lambda = (\lambda_t)_{t \geq 0}$ resulting in this way requires us to consider the stream of marginal states of functioning forgone as a consequence of a marginal decrease in current level of consumption characteristics per available worker and this is represented by the path $t \longrightarrow \partial_c f(t, k_t, \alpha_t)$. This stream together with ${}_*P_K(t)$ gives the path of *functioning loss* as indicated by the path $t \longrightarrow {}_*P_K(t).\partial_c f(t, k_t, \alpha_t)$, due to a goal-adequate increment in the rate of investment or, equivalently, due to the resulting decrease in current level of consumption characteristics attributable to the higher demand for disposable income accompanying a higher level of employment, e.g. resources to cover required additions to capital endowment and plant. This path represents a *social supply price of investment*, because it indicates at any time $t \in [0,T]$ the amount of consumption to be given up, in order to afford a desired path accumulation and the resulting path of employment.

For that reason we have to face the problem of finding an appropriate *social demand price of investment*. The role of the social demand price of investment is to reflect adequately the (social) value in terms of consumption characteristics one may be willing to pay for a goal-conducive investment increase. To better appreciate the fairness of a (social) price it is necessary to assess the states of functionings (social reward) one may expect to derive from future states of consumption that the investment increase at isssue will make possible. It is to this question that we turn next.

On the role of the performance index

At this point we shall recall some central features of systems evolving to major extent on the basis of mechanisms of autoreplication, in changing environment; see Subsection 2.1.4. These central features are adjusted to the goal of self-reproducing, self-replacing and self-excelling, which in the case of the human society acquires a particular character owing to the social and biological nature of humankind. These features make available elaborate social and individual devices with the purpose of coping with time and change.

- First, to the end of reproducing social structures and patterns of organisation on which human interactions may take place and to maintain the basis on which the interplay between society and nature rests.

- Second, to the end of creating anew and replenishing social and individual potentials to function, to adapt to new emerging situation, to anticipate and prepare to irreversible exhaustion of moving forces and the like.

- Third, to the end of releasing blocked potentials by removing material, cultural and social constraints which stimulate creativity, learning and progress. The latter features are better understood in the face of evolutionary processes of growth, development and structural changes.

Recasting acts of consumption in terms of operations aimed at renewing, replenishing and creating potentials to function in and interact with the process of social reproduction, gives an accumulation policy its true relevance, a feature which contrasts heavily with the one resulting from the cake-eating metaphors of orthodoxy.

From this perspective investment decisions acquire a fundamental character. Namely, allocating available resources to removing constraints on the potentials to function of certain fraction of workers currently in the reserve army or to the improvement of the productive power of capital and labour currently in production, with the perspective of being later in the future in a better shape to cope with the problem of removing constraints. The design of the performance index has to take into account the need of providing objective mechanisms to weigh future rewards against current costs.

Let us recall the control process is given by $\alpha = (\alpha_t)_{t \geq 0}$ with $\alpha_t = (s_K(t), \lambda_t)$. Since any path of pairs $(s_K(t), \lambda_t)$ uniquely determines a path of the investment-consumption mix (i_t, c_t), we introduce a social welfare criterion J^* by means of a map $\alpha \longrightarrow J^*(\alpha.)$ with J^* given by the function

$$J^*(\alpha) = \mathsf{E}\left\{ \int_0^T f(t, k_t, \alpha_t)\, dt + \phi(T, k_T) \right\}, \qquad (6.51)$$

where $f(t, \cdot, \alpha_t)$ and $\phi(T, \cdot)$ are strict concave functionals of class $\mathbb{C}^1(\mathbb{R}^n \to \mathbb{R}_-)$. In (6.51) above, E denotes the mathematical expectation with respect to the probability measure P; the time horizon $[0, T]$, is considered as given. The functional $J^*(\alpha)$ conveys a measure of social desirability of the control decision α. The terminal reward ϕ reflects the increases in the range of the capability set of individuals which a successfully accomplished accumulation policy makes affordable in the future.

Thinking in terms of individual capability to function under a given accumulation régime, frees the performance criterion (6.51) from the irrealistic and scientifically ill-founded utility theory advanced by economic orthodoxy. In this sense the cost reflected by $f(t, k_t, \alpha_t)$ has an objective nature. It represents the social burden attached to a deficit in the "states of being and doing" characterising an individual. It conveys information on the dynamics of change reflecting learning and skill acquisition only in connection with employment opportunities; provided such opportunities materialise a process of catching up initiates at a pace determined by the earning power and/or the consuming power underlying sustainable income at the state k_t. Social cost is thus attributable to the presence of individuals deprived of consumption characteristics along the course of the process k up to the time t. Since in our setting the existence of unemployment is characterised by deficits in the level of individuals' capability in the sense described above, it becomes apparent that the rationale behind the performance criterion (6.51) is of the sort of "the less deficit the better".

Although any path of the *continuous control* α uniquely determines an accumulation process k and in turn a path c, we shall write for the sake of clarity the continuous control α explicitly as an argument of the functional f. However, we shall omit it sometimes in order to save notation, if no confusion may arise. On the other hand, the integrand in (6.51) may be denoted by $f(t, k_t, \dot{k}_t)$ in order to stress either the relation between the (state-position in the sense of) capital endowment and (momentum in the form of) changes in the means of labour or to recall connections with Hamiltonian mechanics on which most of our analysis rests. However, a look at (6.5), (2.12), (2.11), (6.14), (6.15) and (6.16) convinces us of the notational convenience and equivalence of the alternative notations of the integrand in (6.51).

From economic policy-making to optimisation

In order to guarantee boundedness of the integral in (6.51) we follow the more natural alternative of considering capital saturation. In a loose manner, we define a level of *capital saturation* as the level beyond which further accumulation is no longer worthwhile according to the rationale underlying the golden-age dynamics in their various guises. From this point on, we expect the capital-labour ratio k_t, in (6.5), to remain constant at a level depending of the golden-age variety prevailing which we seek to identify with a particular steady-state growth path. As in the conventional neoclassical growth theory, we shall associate with particular paths $t \longrightarrow (s_K(t), \lambda_t)$ specific steady-state growth paths and translate preference for particular growth paths into preference for the corresponding specific paths $t \longrightarrow (s_K(t), \lambda_t)$.

Eqs. (6.19)–(6.20) and (6.23)–(6.26) characterise the rationale chosen in the present essay in order to sort out the golden path $k^\diamond$. The economic policy-making philosophy is then to design and negotiate a framework following which an accumulation régime results capable of putting the economic system on the way towards $k^\diamond$. Lemmas 6.4 and 6.48 give already conditions that the sought régime of capital accumulation has to fulfil. Our reflections concerning (6.43) through (6.48) are directed to single out further conditions. Additional requirements have been pointed out in connection with qualitative behaviour of the process of accumulation of capital per available labour depicted in Fig. 6.1. The following assumption is intended to summarise some of these requirements.

A 6.2 *Assumptions of admissibility of the control variables.*

- For any time t, the control variables s_K and λ satisfy the assumptions involved in A 2.2, (2.26)–(2.28) and (6.48), and they are compatible with the associate restriction on the state variables.

- The initial conditions (2.17)–(2.19) hold and k_0 belongs to the range of relevance $[k_t, k_t^\diamond]$. Furthermore, the initial state k_0 is such that k never leaves the aforementioned interval.

As we have already explained in Chapter 1 one can relax some constraints on the state variables. For instance, in order to avoid that the process k leaves the range of relevance, one can introduce a reflection or a jump mechanism close to $\underline{k}_t$ which hinders an outcrossing of this boundary state. This will be treated in [94].

The family of control processes $\alpha = (\alpha_t)_{t \geq 0}$ that satisfy, for any time t, the assumption in A 6.1 and A 6.2 and do not lead to values of the state that violate the constraints imposed on it, is called the family of *admissible control processes* and it is denoted by $\mathcal{A}$.

Lemma 6.4 gives the conditions characterising the paths of capital accumulation k dominated by the path $k^\diamond$ underlying the maximum sustainable level of consumption per available labour $c_t^\diamond$, where $c_t^\diamond$ is taken as a function of $k_t^\diamond$ and λ_t, i.e. $c_t^\diamond = c(k_t^\diamond, \lambda_t)$. It is natural to try to characterise the path $k = (k_t)_{t \geq 0}$ along which one can come as close as possible to $k^\diamond$ using control processes $\alpha \in \mathcal{A}$. Thus, we can, following [124], take the golden rule path as a suitable reference path from which to assess the system performance by means of the following meaningful criterion on the basis of the functional $\alpha \longrightarrow J(\alpha.)$, where J is given by

$$J(\alpha) = \mathsf{E} \left\{ \int_0^T [f(t, k_t, \alpha_t) - f^\diamond] \, dt + \phi(T, k_T) \right\}, \qquad (6.52)$$

where the expectation is taken with respect to the probability measure P. The integrand at the right-hand side measures the rate at which the path k fills the gap in the level of the "states of being and doing" that it can support. This notion is relative to the corresponding k°. The second term is to be understood as before. Now we are prepared to formulate the question of an economic policy presented above as the following problem of optimisation.

Optimising problem 6.1 (Control with complete information) *Let the assumption A 6.1, A 6.2 and those of Lemma 6.4 hold. Find among the $\alpha \in \mathcal{A}$, an optimal control β such that*

$$J(\beta) = \sup_{\alpha \in \mathcal{A}} J(\alpha), \tag{6.53}$$

subject to (6.1) and (6.35).

A precise mathematical treatment of this problem will be given in the following section, where we deal in some detail with various aspects of the solution of (6.53), prove existence and uniqueness and obtain the adjoint process using a stochastic maximum principle due to Bensoussan [19]. In the final section of this chapter we will come back to essential issues of economic policy-making and work out the economic content of the solution.

6.3* Variational control of diffusions

In this section we will deal with the problem of controlling diffusions by variational methods and consider the case of control entering into the diffusion term. We will give a rigourous derivation of the stochastic maximum principle drawing heavily on results due to [16, 19]. Bensoussan's powerful and transparent approach relies mainly on variational methods which are very similar to those used in deterministic theory and recovers most of the results which exist in the literature by more elementary techniques. The fact the control variable enters explicitly into the diffusion term represents an improvement against the approach followed in the preceding chapter; we should mention however that the underlying information structure would be one generated by the Brownian motion considered. In the next subsection we give a fairly general statement of the problem to be treated and introduce various notions, specially that of the adjoint state, and discuss some basic properties of optimally controlled diffusions.

6.3.1 Control setting with complete information

Although stochastic control with partial information on observations of the state would be better suited to economic reality and econometric practice, we restrict ourselves to the case of controlling a diffusion process with complete information. Since the optimal control of partially observed diffusions is mathematically more involved and needs an infinite dimensional setting, we shall treat this case elsewhere; see [18, 68, 69].

Notation and setting of the problem

Let $(\Omega, \mathcal{F}, \mathbf{P})$ be a probability space, endowed with a filtration $(\mathcal{F}_t)_{t \geq 0}$, with the usual properties on which we shall consider a standard $\mathcal{F}_t$-Brownian motion $B = (B_t, \mathcal{F}_t)_{t \geq 0}$ with values in $\mathbb{R}^n$, where $t \in [0, T]$ with $T \in \mathbb{R}_+$. In particular this means the process B is a $\mathcal{F}_t$-martingale.

A 6.3 *Assumption on the case with complete information*

- Let us assume the σ-algebras $\mathcal{F}_t$ are given by the relationship $\mathcal{F}_t = \sigma(B_s, s \leq t)$, which entails they content the information available at the instant of time t.

In the case of partial information, one would work with a filtration $(\mathcal{G}_t)_{t \geq 0}$ such that $\mathcal{G}_t \subseteq \mathcal{F}_t$ for any t. In order to define the state equation and to assure that it possesses a unique solution for a given initial condition, we define the drift vector b and the diffusion matrix σ, and introduce certain assumptions.

Let $b(t, x, \alpha) : \mathbb{R}^n \times \mathbb{R}^m \longrightarrow \mathbb{R}^n$ and $\sigma(t, x, \alpha) : \mathbb{R}^n \times \mathbb{R}^m \longrightarrow \mathcal{L}(\mathbb{R}^n; \mathbb{R}^n)$ be, for any t, two Borel random functions defined on $(\Omega, \mathcal{F}, \mathbf{P})$ be such that the assumptions to follow hold.

A 6.4 *Assumptions on the drift vector b and diffusion matrix σ.*

a) $b(t, \cdot, \cdot)$ is continuous and continuously differentiable with respect to x, and α_t.

b) $\sigma(t, \cdot, \cdot)$ is continuous and continuously differentiable with respect to x and α_t, and

c) the partial derivatives $\partial_x b$, $\partial_\alpha b$, $\partial_x \sigma^j$ and $\partial_\alpha \sigma^j$ are bounded functions, where $j = 1, 2, \ldots, n$ and σ^j denotes column vectors of σ.

Let us point out that from A 6.4 a) – c), it follows that:

$$\begin{aligned}
|b(t, x, \alpha)| &\leq c(1 + |x| + |\alpha|), \\
|\sigma(t, x, \alpha)| &\leq c(1 + |x| + |\alpha|),
\end{aligned} \tag{6.54}$$

where c is a deterministic constant. Next we introduce the control function α and impose conditions under which the system under consideration shall be optimally controlled. Let us assume that the control process $\alpha = (\alpha_t)_{t \geq 0}$ takes values on a set A_{ad} such that the following assumptions are fulfilled.

A 6.5 *Assumptions on the control process α.*

a) A_{ad} is a non-empty closed convex subset of $\mathbb{R}^m$ representing a prescribed constraint set which ensure the states and control variables fulfil desired properties.

b) For $\alpha = (\alpha_t(\omega), \mathcal{F}_t)_{t \geq 0}$ the space of control random functions $L^2_{\mathcal{F}}(0,T)$ defined as

$$L_{\mathcal{F}}(0,T) = \Big\{ \alpha \in L^2(\Omega \times (0,T), \mathbf{P}(d\omega) \otimes dt; \mathbb{R}^m) :$$

$$\text{a.e. } t, \ \alpha_t \in L^2(\Omega_t, \mathcal{F}_t, \mathbf{P}; \mathbb{R}^m) \Big\},$$

is a sub-Hilbert space of L^2.

c) The set $\mathcal{A} = \big\{ \alpha \in L^2_{\mathcal{F}}(0,T) : \alpha_t \in A_{ad}, \text{ a.e., a.s.} \big\}$ is a closed convex sub-set of $L^2_{\mathcal{F}}(0,T)$.

For any $t \in [0,T]$ with finite $T \in \mathbb{R}_+$ any process $\alpha = (\alpha_t)_{t \geq 0}$, viewed as an element $\alpha : t \longrightarrow \alpha_t$ of $\mathcal{A}$ will be called an *admissible control* and $\mathcal{A}$ stands for the *space of admissible controls*. The use of admissible controls assures now that given an initial state $x_0 \in \mathbb{R}^n$, one can define a state equation which possesses a unique solution.

State equation and performance functional

Let $L^2_{\mathcal{F}}(0,T;\mathbb{R}^n)$ represents the sub-space of $L^2(\Omega \times (0,T), \mathbf{P}(d\omega) \otimes dt; \mathbb{R}^n)$ consisting of processes $z = (z_t)_{t \geq 0}$ such that for a.e. t, $z_t \in L^2(\Omega, \mathcal{F}_t, \mathbf{P}; \mathbb{R}^n)$. With any admissible control $\alpha \in \mathcal{A}$, characterised by A 6.5 a) – c), we associate a state process $X = (X_t, \mathcal{F}_t)_{t \geq 0}$ according to the theorem to follow.

Theorem 6.8 *Under the assumptions A 6.3, A 6.4 and A 6.5, there exists, for any admissible control $\alpha \in \mathcal{A}$ one and only one process X satisfying, for any $t \in [0,T]$, the equation*

$$dX_t = b(t, X_t, \alpha_t)\, dt + \sigma(t, X_t, \alpha_t)\, dB_t, \tag{6.55}$$

where the process $X \in L^2_{\mathcal{F}}(0,T;\mathbb{R}^n) \cap L^2(\Omega, \mathcal{F}, \mathbf{P}; \mathbb{C}(0,T;\mathbb{R}^n))$ with $X_0 = x_0 \in \mathbb{R}^n$.

Proof: One easily checks that the requirements for the application of Itô's theorem are satisfied. See Chapter 8; alternatively see [17]. ∎

Let us introduce a cost function $f(t, X_t, \alpha_t) : [0,T] \times \mathbb{R}^n \times \mathbb{R}^m \longrightarrow \mathbb{R}_-$ and a terminal reward function $\phi(t, X_t) : [0,T] \times \mathbb{R}^n \longrightarrow \mathbb{R}_-$ be two Borel random functionals such that for any $t \in [0,T]$ the assumptions to follow hold.

A 6.6 *Assumptions on the cost and terminal reward functionals*

a) f and ϕ are Borel, strictly concave, continuously differentiable with respect to (X_t, α_t) and X_t respectively, and

b) the partial derivatives $\partial_x f$, $\partial_\alpha f$ and $\partial_x \phi$ are bounded functions.

For the sake of simplicity we shall indicate partial differentiation also as sub-index, whenever no confusion arises. Note that from the assumption above follows that

$$|f_x(t, X_t, \alpha_t)| \leq c(|x| + |\alpha| + 1), \qquad (6.56)$$
$$|f_\alpha(t, X_t, \alpha_t)| \leq c(|x| + |\alpha| + 1), \qquad (6.57)$$
$$|\phi_x(t, x)| \leq c(|x| + 1), \qquad (6.58)$$
$$f(t, 0, 0) \in L^\infty(0, T), \qquad (6.59)$$

are fulfilled, where c is again a deterministic constant. Let us define a goal functional or performance index $J : \alpha \longrightarrow J(\alpha)$ as

$$J(\alpha) = \mathsf{E}\left\{ \int_0^T f(t, X_t, \alpha_t)\, dt + \phi(T, X_T) \right\}. \qquad (6.60)$$

Our task is to characterise an admissible control function by means of which $J(\cdot)$ attains a supremum. We are now ready to state the control problem as follows.

Optimising problem 6.2 *The assumptions in A 6.4 a) – c), A 6.5 a) – c), and A 6.6 a) – b) hold. Find, among the $\alpha \in \mathcal{A}$, an optimal control β such that*

$$J(\beta) = \sup_{\alpha \in \mathcal{A}} J(\alpha), \qquad (6.61)$$

given the state equation as in Theorem 6.8.

We shall denote by β an optimal control satisfying A 6.5 a) – c) for which the underlying state and performance structure satisfy A 6.4 a) – c) and A 6.6. Further, the associated optimal state will be denoted by Y, i.e. the process Y satisfies the corresponding state equation

$$dY_t = f(t, Y_t, \beta_t)\, dt + \sigma(t, Y_t, \beta_t)\, dB_t , \qquad (6.62)$$

where $Y_0 = x_0 \in \mathbb{R}^n$, the initial state x_0 given and the state process $Y = (Y_t)_{t \geq 0}$ is such that $Y \in L^2_{\mathcal{F}}(0, T; \mathbb{R}^n) \cap L^2(\Omega, \mathcal{F}, \mathsf{P}; \mathbb{C}(0, T; \mathbb{R}^n))$.

At this stage let us mention, the fact that the filtration $(\mathcal{F}_t)_{t \geq 0}$ is a priori given and it does not depend on the state is not as restrictive as it may appear. This is particular true when the diffusion term does not depend on the control.

6.3.2 Gâteaux differentiability and perturbations

Next we shall use perturbation methods with the purpose of obtaining some preliminary results on conditions of extremality. Perturbation theory provides powerful techniques for the study of equations close to equations of a specific form. These are extremely useful, since most differential equations admit neither an exact analytical solution nor a complete qualitative description. In Chapter 7 we present this perturbation idea in a deterministic setting.

Perturbed controls and associated trajectories

Let t_0 be fixed in $[0, T)$ and ϵ a parameter which will tend to 0. Further, let β be an optimal control and α an arbitrary admissible control. Let us denote by $\beta^\epsilon = (\beta_t^\epsilon)_{t \geq 0}$ the control process with β_t^ϵ given, for any $t \in (0, T)$ by

$$\beta_t^\epsilon = \begin{cases} \beta_t, & t \in (0, t_0] \\ \alpha_t, & t \in (t_0, t_0 + \epsilon] \\ \beta_t, & t \in (t_0 + \epsilon, T], \end{cases}$$

provided that $t_0 + \epsilon < T$. Clearly, the control β^ϵ is admissible. Let Z denote the solution of the linearised state equation 6.64. We will denote the state process associated with β^ϵ by $Y^\epsilon = (Y_t^\epsilon)_{t \geq 0}$ and write for the sake of simplicity $\beta^\epsilon = \beta + \epsilon \alpha$. Further, let us set

$$\tilde{Y}^\epsilon = \frac{Y^\epsilon - Y}{\epsilon} - Z \quad \text{or} \quad Y^\epsilon - Y = \epsilon(Z + \tilde{Y}^\epsilon),$$

from which it follows that

$$\begin{aligned} d\tilde{Y}_t^\epsilon &= \left[\frac{1}{\epsilon}(b(t, Y_t + \epsilon(Z_t + \tilde{Y}_t^\epsilon), \beta_t + \epsilon\alpha_t) - b(t, Y_t, \beta_t)) \right. \\ &\quad \left. - (b_x'(t, Y_t, \beta_t)Z_t + b_\alpha'(t, Y_t, \beta_t)\alpha_t) \right] dt \\ &\quad + \sum_{j=1}^n \left\{ \frac{1}{\epsilon}\left(\sigma^{j'}(t, Y_t^\epsilon + \epsilon(Z_t + \tilde{Y}_t^\epsilon), \beta_t + \epsilon\alpha_t) - \sigma^{j'}(t, Y_t, \beta_t) \right) \right. \\ &\quad \left. - (\sigma_x^{j'}(t, Y_t, \beta_t)Z_t + \sigma_\alpha^{j'}(t, Y_t, \beta_t)\alpha_t) \right\} dB_t^{(j)}, \\ \tilde{Y}_0^\epsilon &= 0, \end{aligned}$$

hold. Setting $Y_t^\epsilon = Y_t + \lambda(Y_t^\epsilon - Y_t)$ and $\beta_t^\epsilon = \beta_t + \lambda(\beta_t^\epsilon - \beta_t)$ with $\lambda \in [0, 1]$, the foregoing equation can be written as

$$\begin{aligned} d\tilde{Y}_t^\epsilon &= \int_0^1 b_x'(t, Y_t + \lambda(Y_t^\epsilon - Y_t), \beta_t + \lambda\epsilon\alpha_t)\tilde{Y}_t^\epsilon \, d\lambda \, dt \\ &\quad + \sum_{j=1}^n \int_0^1 \sigma_x^{j'}(t, Y_t + \lambda(Y_t^\epsilon - Y_t), \beta_t + \lambda\epsilon\alpha_t)\tilde{Y}_t^\epsilon \, d\lambda \, dB_t^{(j)} \\ &\quad \int_0^1 (b_x'(t, Y_t + \lambda(Y_t^\epsilon - Y_t), \beta_t + \lambda\epsilon\alpha_t) - b_x'(t, Y_t, \beta_t))Z_t \, d\lambda \, dt \\ &\quad + \int_0^1 (\beta_\alpha'(t, Y_t + \lambda(Y_t^\epsilon - Y_t), \beta_t + \lambda\epsilon\alpha_t) - b_\alpha'(t, Y_t, \beta_t))\alpha_t \, d\lambda \, dt \\ &\quad + \sum_{j=1}^n \int_0^1 (\sigma_x^{j'}(t, Y_t + \lambda(Y_t^\epsilon - Y_t), \beta_t + \lambda\epsilon\alpha_t) - \sigma_x^{j'}(t, Y_t, \beta_t))Z_t \, d\lambda \, dt \, dB_t^{(j)} \\ &\quad \sum_{j=1}^n \int_0^1 (\sigma_\alpha^{j'}(t, Y_t + \lambda(Y_t^\epsilon - Y_t), \beta_t + \lambda\epsilon\alpha_t) - \sigma_\alpha^{j'}(t, Y_t, \beta_t))\alpha_t \, d\lambda \, dB_t^{(j)}. \end{aligned}$$

Let us point out, computing $E|\tilde{Y}_t^\epsilon|^2$ after integrating the foregoing equation on the interval $[0,t]$ one easily recognises that except for the first two all other terms tend to vanish as $\epsilon \to 0$. This is due to the continuity properties of the derivatives involved. Using this observation, we get the inequality

$$E|\tilde{Y}_t^\epsilon|^2 \leq c\, E \int_0^t |\tilde{Y}_s^\epsilon|^2 ds + \rho^\epsilon,$$

where $\rho^\epsilon \to 0$, as just mentioned, as $\epsilon \to 0$, and is given by

$$\rho^\epsilon = E\left(\int_0^t |Z_s|\left|\int_0^1 [b_x'(s, Y_s + \lambda(Y_s^\epsilon - Y_s)\beta_s + \lambda\epsilon\,\alpha_s) - b_x'(s, Y_s, \beta_s)]\right| ds\right)^2 + \cdots$$

Further, one can prove that

$$E\left\{\sup_{0\leq t\leq T} |Y_t^\epsilon - Y_t|^2\right\} \leq c\epsilon^2,$$

so that at least for a subsequence one has

$$\sup_{0\leq t\leq T} |Y_t^\epsilon - Y_t|^2 \longrightarrow 0 \quad \text{a.s.}$$

Knowing these features of a perturbation of the optimal state process we can begin to explore the extremal properties of the control $\beta^\epsilon = \beta + \epsilon\alpha$.

Gâteaux derivative of the performance index

Let us apply the preceding result to the problem of computing the derivative of the performance functional for a perturbed control.

Lemma 6.9 *The assumptions of Theorem 6.8 and A 6.6 hold. Then, the functional $J(\alpha)$ is Gâteaux differentiable and the following formula holds*

$$\frac{d}{d\epsilon} J(\beta + \epsilon\alpha)|_{\epsilon=0} =$$

$$E\left\{\int_0^T [f_x'(t, Y_t, \beta_t)\, Z_t + f_\alpha'(t, Y_t, \beta_t)\alpha_t]\, dt + \phi_x'(T, x_T)Z_T\right\}, \quad (6.63)$$

where Z is the solution of the linearised version of the state eq. (6.62), i.e.

$$dZ_t = [b_x'(t, Y_t, \beta_t)Z_t + b_\alpha'(t, Y_t, \beta_t)\alpha_t]\, dt$$

$$+ \sum_{j=1}^n (\sigma_x^{j\,'}(Y_t, \beta_t)Z_t + \sigma_\alpha^{j\,'}(Y_t, \beta_t)\alpha_t)\, dB_t^{(j)}), \quad (6.64)$$

with the initial state holding the condition $Z_0 = z_0 = 0$ and the linearised process $Z = (Z_t)_{t\geq 0}$ such that $Z \in L_\mathcal{F}^2(0, T; \mathbb{R}^n) \cap L^2(\Omega, \mathcal{F}, P; \mathbb{C}(0, T; \mathbb{R}^n))$.

Proof: It is an adaptation of [19]. One considers

$$\frac{1}{\epsilon}[J(\beta^{\epsilon}) - J(\beta)] =$$

$$\frac{1}{\epsilon}\mathbf{E}\left\{\int_0^T [f(t, Y_t^{\epsilon}, \beta_t^{\epsilon}) - f(t, Y_t, \beta_t)]\, dt + [\phi(T, Y_T^{\epsilon}) - \phi(T, Y_T)]\right\},$$

from which one obtains, following the perturbation procedure described above, the expression

$$J(\beta^{\epsilon}) - J(\beta) - \epsilon\,\mathbf{E}\left\{\int_0^T [f_x'(t, Y_t, \beta_t)Z_t + f_\alpha'(t, Y_t, \beta_t)\alpha_t]\, dt + \phi_x'(T, Y_T)Z_T\right\}$$

$$= \epsilon\,\mathbf{E}\left\{\int_0^T \left[\int_0^1 \left[f_x'(t, Y_t + \lambda(Y_t^{\epsilon} - Y_t), \beta_t + \lambda\epsilon\,\alpha_t) - f_x'(t, Y_t, \beta_t)\right](\tilde{Y}_t^{\epsilon} + Z_t)d\lambda\right.\right.$$

$$+ \int_0^1 \left[f_\alpha'(t, Y_t + \lambda(Y_t^{\epsilon} - Y_t), \beta_t + \lambda\epsilon\,\alpha_t) - f_\alpha'(t, Y_t, \beta_t)\alpha_t\right]d\lambda$$

$$+ \int_0^1 \left[\phi_x'(T, Y_T + \lambda(Y_T^{\epsilon} - Y_T)) - \phi_x'(T, Y_T)\right](\tilde{Y}_T^{\epsilon} + Z_T)d\lambda$$

$$\left.\left. + f_x'(t, Y_t, \beta_t)\tilde{Y}_t^{\epsilon}\right]dt + \phi_x'(T, Y_T)\tilde{Y}_T^{\epsilon}\right\}.$$

Here one easily sees the integrand in the first three terms tends to vanish as $\epsilon \to 0$, owing to the assumption on the partial derivatives. Therefore the right-hand side of this equality is of the type $o(\epsilon)$, where $o(\epsilon) \to 0$, as $\epsilon \to 0$. This entails the desired result.   ∎

6.3.3   Adjoint state and optimality conditions

In order to understand the extremality properties of the control β, it is necessary to be able to write (6.63) in a form where the control α appears explicitly. However, to do this one needs to introduce the adjoint variables to which we now turn.

The adjoint variables and the Hamiltonian

With the purpose of introducing adjoint variables we examine the effect of some changes in the equation of first variation about the (optimal) solution Y on the behaviour of (6.63). Hence, let us consider the functions φ and $\psi^1, \psi^2, \psi^1, \dots, \psi^n$ in $L_{\mathcal{F}}^2(0, T; \mathbb{R}^n)$ and $\zeta = (\zeta_t)_{t \geq 0}$, where ζ_t is given by

$$d\zeta_t = (b_x'(t, Y_t, \beta_t)\zeta_t + \varphi_t)\, dt + \sum_{j=1}^n \left(\sigma_x^{j\,'}(t, Y_t, \beta_t)\zeta_t^j\right) + \psi_t^j)\, dB_t^{(j)}, \quad (6.65)$$

where $\zeta_0 = 0$ and the process $\zeta = (\zeta_t)_{t \geq 0}$ is defined so that it is characterised by $\zeta \in L_{\mathcal{F}}^2(0, T; \mathbb{R}^n) \cap L^2(\Omega, \mathcal{F}, \mathbf{P}; \mathbb{C}(0, T; \mathbb{R}^n))$. It is easy to check that the map

$\varphi, \psi^1, \psi^2, \ldots, \psi^n \longrightarrow \zeta_t$ defined from the space of functions $L^2_{\mathcal{F}}(0, T; \mathbb{R})^{n+1}$ in the set of functions $L^2_{\mathcal{F}}(0, T; \mathbb{R}^n) \cap L^2(\Omega, \mathcal{F}, \mathbf{P}; \mathbb{C}(0, T; \mathbb{R}^n))$ is linear and continuous. We consider then the functional

$$\varphi, \psi^1, \psi^2, \ldots, \psi^m \longrightarrow \mathbf{E}\left\{\int_0^T f'_x(t, Y_t, \beta_t)\zeta_t \, dt + \phi'_x(T, Y_T)\zeta_T\right\}$$

is in turn linear and continuous from $(L^2_{\mathcal{F}}(0, T; \mathbb{R}^n))^{m+1}$ into $\mathbb{R}$. Therefore, we can define in a unique way stochastic processes $p(\cdot)$ and $r^1(\cdot), r^2(\cdot), \ldots, r^n(\cdot)$ in $L^2_{\mathcal{F}}(0, T; \mathbb{R}^n)$ such that the following relation holds

$$\mathbf{E}\left\{\int_0^T p_t \varphi_t \, dt + \sum_{j=1}^n \int_0^T r^j_t \psi^j_t \, dt\right\} = \mathbf{E}\left\{\int_0^T f'_x(t, Y_t, \beta_t)\zeta_t \, dt + \phi'_x(T, Y_T)\zeta_T\right\}.$$

$$(6.66)$$

We like to point out that the adjoint variables emerge here in a natural way, build a unique adjoint state and are independent of the control.

On the conditions of optimality

Now, we are better prepared to derive the conditions of optimality. To this end we present first a lemma, which enables us to get a stochastic version of Pontryagin's maximum principle. Eq. (6.69), together with Lemma 6.9, gives the following lemma.

Lemma 6.10 *The assumptions of Lemma 6.9 hold. We have the formula*

$$\frac{d}{d\epsilon}J(\beta + \epsilon\alpha)|_{\epsilon=0} =$$

$$\mathbf{E}\left\{\int_0^T \left[f'_\alpha(t, Y_t, \beta_t) + p_t b'_\alpha(t, Y_t, \beta_t) + \sum_{j=1}^n r^j \sigma^{j\,'}_\alpha(t, Y_t, \beta_t)\right] \alpha_t \, dt\right\},$$

for short,

$$\frac{d}{d\epsilon}J(\beta + \epsilon\alpha)|_{\epsilon=0} = \mathbf{E}\left\{\int_0^T H'_\alpha(t, Y_t, \beta_t, p_t, r_t)\alpha_t \, dt\right\}, \qquad (6.67)$$

where

$$H(t, X_t, \alpha_t, p_t, r_t) = f(t, X_t, \alpha_t) + p_t b(X_t, \alpha_t) + \sum_{j=1}^n r^j_t \sigma^{j\,'}(t, X_t, \alpha_t). \quad (6.68)$$

Proof: Applying the result condensed in (6.66), but considering (6.64) instead of (6.65) and comparing coefficients we obtain

$$\mathbf{E}\left\{\int_0^T f'_x(t, Y_t, \beta_t)Z_t \, dt + \phi'_x(T, Y_T)Z_T\right\} =$$

$$\mathbf{E}\left\{\int_0^T \left[p_t b'_\alpha(t, Y_t, \beta_t)\alpha_t + \sum_{j=1}^m r^j \sigma^{j\,'}(t, Y_t, \beta_t)\alpha_t\right] dt\right\}. \quad (6.69)$$

Then substituting this in (6.63) proves the lemma.   ■

Lemma 6.10 provides a derivative of the performance index in which the control
α appears explicitly in each of its terms. This leads to the introduction of the
Hamiltonian functional, where the process r accounts for (uncertainty) changes
in the system performance owing to changes in the control and the constraints.

Pontryagin's conditions for optimality

Before we go over to formulate the stochastic maximum principle let us point
out that due to the fact we have assumed strict concavity, see Assumption A
6.6, the optimality conditions stated below are both necessary and sufficient.
Further, let us mention that at the present stage we get only one of the equations building the Hamiltonian system known from classical mechanics.
    One immediately verifies (6.62) may be written with the aid of the functional
H, introduced in Lemma 6.10, in the form of the canonical equation of the
Hamiltonian corresponding to the state Y. The canonical equation associated
with the adjoint processes p and r will be obtained in Theorem 6.13.

Theorem 6.11 (The stochastic maximum principle) *Let assumptions A
6.3, A 6.4 a) – c), A 6.5 a) – c), A 6.6 a) – b) hold. Then for the control
process β to be a solution of the optimal control problem defined by (6.61) and
(6.62), and denoting by Y the corresponding trajectory, and by $p = (p_t)_{t \geq 0}$
and $r^j = (r^j_t)_{t \geq 0}$ the processes uniquely defined in (6.66), it is necessary and
sufficient that the condition*

$$\frac{d}{d\epsilon} J(\beta + \epsilon\alpha)\big|_{\epsilon=0} \leq 0, \qquad (6.70)$$

*holds. Taking into account Lemma 6.10, the expression (6.67) and condition
(6.70) entail, for any admissible control process $\alpha = (\alpha_t)_{t \geq 0}$ as above, the
following condition*

$$\mathbf{E}\left\{ \int_0^T H_\alpha(t, Y_t, \beta_t, p_t, r_t) \cdot (\alpha - \beta_t) dt \right\} \leq 0, \qquad (6.71)$$

*or equivalently for any admissible control process α such that 6.5 a) – c) holds,
the inequality*

$$H_\alpha(t, Y_t, \beta_t, p_t, r_t) \cdot (\alpha_t - \beta_t) \leq 0, \ a.e. \ t, \qquad \mathbf{P} \otimes dt \quad a.s. \qquad (6.72)$$

follows.

Proof: It follows from (6.70) and a classical localisation argument. See [15], for
instance.   ■

To make sure that the obtained maximum principle becomes a powerful working
tool, one still needs to obtain the dynamics governing the processes p and r.
This is the subject of the next subsection.

6.3.4   The equation of the adjoint state

Let us proceed to the formulation of the theorem which should yield the missing canonical equation of the Hamiltonian. With this aim we shall recall some additional properties of the linearised version of the state equation.

Some properties of the fundamental solution

The basic idea in the control theory of systems governed by differential equations is that of translating the space of trajectories of the system, say $\mathfrak{X}_0$, associated with the initial condition x_0, to the space $\mathfrak{X}_1$ corresponding to the trajectories emanating from x_1. The state x_1 corresponds in turn to the transfer of x_0 to state $X_{t_1} = x_1$, so that we may think of the transformation under consideration as that which transfers the initial state of $\mathfrak{X}_0$ to that of $\mathfrak{X}_1$.

Assuming for a moment that given a certain parameter β, the equation (6.62) has a solution passing through x_0 at time t_0, let us denote its value at later time t as $\phi(t, x_0, t_0)$ or for the sake of simplicity $\phi(t, t_0)$. If the initial data and/or the parameter β are changed slightly then one expects that the solution of (6.62) will also change slightly. The fundamental solution of (6.64), represents a linear mapping with the property of generating a family of spaces $\mathfrak{X}$ which are obtained from z_0 by a transfer along $X = (X_t)_{t \geq 0}$. The fundamental solution is also called the fundamental matrix of (6.64) and it is associated in a natural way with the fundamental solution of the adjoint equation. Thinking in terms of matrices, one can write $\Phi(t, t_0) = [\phi_1(t, t_0), \phi_2(t, t_0), \ldots, \phi_n(t, t_0)]$, where $\phi_i(t, t_0)$ with $i = 1, \ldots, n$, stands for the value at time t of the i-th solution emanating from the initial state x_0.

The fundamental solution of the adjoint equation will be denoted by Ψ and runs from t towards t_0. We write it $\Psi(t_0, t)$ with the purpose of stressing the fact, it propagates the solution of the original equation backwards in time. However, we will suppress the use of t_0 whenever it does not lead to confusion. Further, thinking of Φ and Ψ as processes, we will write $\Phi = (\Phi_t)_{t \geq 0}$ and $\Psi = (\Psi_t)_{t \geq 0}$ with $\Phi_t = \Phi(t, t_0)$ and $\Psi_t = \Psi(t, T)$. The following property of Φ and Ψ is of great importance.

Lemma 6.12 *Let Φ and Ψ be the fundamental solution of (6.64) and its adjoint equation. Then for any $t \in [0, T]$ one has*

$$\Phi_t \Psi_t = \Psi_t \Phi_t = I.$$

Proof: It is sufficient to prove that $d\Phi_t \Psi_t = d\Phi_t \Psi_t = 0$. Thus, let us denote by Φ, with $\Phi \in \mathcal{L}(\mathbb{R}^n, \mathbb{R}^n)$ for any $t \in [0, T]$, the fundamental solution of eq. (6.64). Then, Φ_t satisfies the stochastic matrix equation

$$d\Phi_t = b_x(t, Y_t, \beta_t)\Phi_t \, dt + \sum_{j=1}^{n} \sigma_x^{j'}(t, Y_t, \beta_t)\Phi_t \, dB_t^j, \qquad (6.73)$$

$$\Phi_0 = I,$$

where I denotes the identity matrix. Since Ψ_t represents the adjoint of Φ_t, it is uniquely defined by

$$
d\Psi_t = [-\Psi_t b'_x(t, Y_t, \beta_t) + \Psi_t \sum_{j=1}^{n} \sigma_x^{j'}(t, Y_t, \beta_t) \sigma_x^{j'}(t, Y_t, \beta_t)] \, dt -
$$

$$
\sum_{j=1}^{n} \Psi_t \, \sigma_x^{j'}(t, Y_t, \beta_t) \, dB_t^{(j)}, \tag{6.74}
$$

$$
\Psi_0 = I. \tag{6.75}
$$

The lemma easily follows using the Itô rule to compute the differential $d\,\Psi_t\Phi_t$ and $d\,\Phi_t\Psi_t$. Let us note that Φ and Ψ can only be considered once the optimal control β and the corresponding state process Y has been chosen, since β and Y enter into the right-hand side of the associated differential system, i.e. the eqs. (6.64) and (6.74).

The canonical equation associated with the adjoint state

To complete the formulation of the stochastic maximum principle, we need to derive the canonical equation associated with the adjoint state. The two canonical equations are independent and meaningful. They are called in classical mechanics the *Hamilton equations of motion*.

Theorem 6.13 (The adjoint state equation) *Let the assumptions of Theorem (6.11) hold. Then the processes p_t and $r_t^{(j)}$ satisfy*

$$
- dp_t = H'_x(t, Y_t, \beta_t, p_t, r_t) dt - \sum_{j=1}^{n} r_t^j dB_t^{(j)}, \tag{6.76}
$$

$$
p_T = \phi'_x(T, Y_T), \tag{6.77}
$$

where the process $p = (p_t)_{t \geq 0}$ so that $p \in L^2_{\mathcal{F}}(0, T; \mathbb{R}^n) \cap L^2(\Omega, \mathcal{F}, \mathbf{P}; \mathbb{C}(0, T; \mathbb{R}^n))$, and the process $r^j = (r_t^j)_{t \geq 0}$, $r^j \in L^2_{\mathcal{F}}(0, T; \mathbb{R}^n)$, such that

$$
\sum_{j=1}^{m} \lambda^j \mathbf{E} \int_0^T |r_t^j| dt < \infty.
$$

Proof: Let us consider the following particular case of (6.65)

$$
d\zeta_t = (b'_x(t, Y_t, \beta_t)\zeta_t + \varphi_t) \, dt + \sum_{j=1}^{n} \sigma_x^{j'}(t, Y_t, \beta_t)\zeta_t \, dB_t^{(j)}, \tag{6.78}
$$

$$
\zeta_0 = 0.
$$

Then, expression (6.66) becomes

$$
\mathbf{E}\left\{ \int_0^T p_t \varphi_t \, dt \right\} = \mathbf{E}\left\{ \int_0^T f'_x(t, Y_t, \beta_t)\zeta_t \, dt + \phi'_x(T, Y_t)\zeta_T \right\}. \tag{6.79}
$$

Let us compute the Itô differential

$$d\,\Psi_t\zeta_t = (d\Psi_t)\zeta_t + \Psi_t d\zeta_t + d\Psi_t \cdot d\zeta_t,$$

which together with (6.74) and (6.78) gives

$$d\,\Psi_t\zeta_t = \Psi_t\zeta_t\,dt$$

and further

$$\Psi_t\zeta_t = \int_0^t \Psi_s\varphi_s\,ds.$$

Hence, we have the following representation for ζ_t

$$\zeta_t = \Phi_t \int_0^t \Psi_s\varphi_s\,ds. \tag{6.80}$$

Now, substituting (6.80) in (6.79), we obtain

$$\mathsf{E}\left\{\int_0^T p_t\varphi_t\,dt\right\} = \mathsf{E}\left\{\int_0^T dt\,f_x'(t,Y_t,\beta_t)\Phi_t\int_0^t \Psi_s\varphi_s\,ds + \right.$$
$$\left. +\phi_x'(T,Y_T)\Phi_T\int_0^T \Psi_s\varphi_s\,ds\right\}. \tag{6.81}$$

But (6.81) can be rewritten as

$$\mathsf{E}\left\{\int_0^T p_t\varphi_t\,dt\right\} = \mathsf{E}\left\{\int_0^T ds\int_s^t dt\,f_x'(t,Y_t,\beta_t)\Phi_t\Psi_s\,\varphi_s + \right.$$
$$\left. \int_0^T \phi_x'(T,Y_T)\Phi_T\Psi_s\,\varphi_s\,ds\right\}. \tag{6.82}$$

The right-hand side of (6.82) can be rewritten further using the properties of expectation as

$$\mathsf{E}\left\{\int_0^T p_t\varphi_t\,dt\right\} =$$
$$= \mathsf{E}\left\{\int_0^T ds\left(\mathsf{E}\left[\phi_x'(T,Y_T)\Phi_T + \int_s^T f_x'(t,Y_t,\beta_t)\Phi_t\,dt\Big|\mathcal{F}_t\right]\Psi_s\right)\varphi_s\right\}. \tag{6.83}$$

Further, (6.83) suggests the following representation for p_s

$$p_s = \left(\mathsf{E}\left[\phi_x'(T,Y_T)\Phi_T + \int_s^T f_x'(t,Y_t,\beta_t)\Phi_t\,dt\Big|\mathcal{F}_s\right]\right)\Psi_s \qquad \mathsf{P}(d\omega)\otimes dt \quad \text{a.s..} \tag{6.84}$$

One can easily check that p_s is a local $\mathcal{F}_s$-semimartingale, square integrable and right-continuous. Now, we set

$$X = \phi_x'(T,Y_T)\Phi_T + \int_0^T f_x'(t,Y_t,\beta_t)\Phi_t\,dt.$$

Therefore, using the Kunita-Watanabe integral representation for local semi-martingales, we obtain for $i = 1, 2, \ldots, n$

$$\mathsf{E}\{X^{(i)}|\mathcal{F}_s\} = \mathsf{E}\{X^{(i)}\} + \int_0^s \sum_{j=1}^n \eta_t^{(ij)} dB_t^{(j)}, \tag{6.85}$$

where the $\eta^{(ij)}$ are $\mathcal{F}_s$-processes such that

$$\int_0^s |\eta_t^{(ij)}|^2 dt < \infty \quad \text{a.s.} \tag{6.86}$$

Thus, setting

$$\Lambda_t = \mathsf{E}\{X\} + \int_0^t \eta_s \, dB_s - \int_0^t f_x'(s, Y_s, \beta_s)\Phi_s \, ds, \tag{6.87}$$

we get from (6.84) for the process p the expression

$$p_s = \Lambda_s \Psi_s. \tag{6.88}$$

Since (6.87) satisfies the stochastic differential

$$d\Lambda_t = -f_x'(t, Y_t, \beta_t)\Phi_t \, dt + \sum_{j=1}^n \eta_t^j \, dB_t^{(j)}. \tag{6.89}$$

We apply Itô's calculus to (6.88) in order to obtain a stochastic differential for p. Thus, it follows

$$dp_t = (d\Lambda_t)\Psi_t + \Lambda_t \, d\Psi_t + d\Lambda_t \cdot d\Psi_t, \tag{6.90}$$

and substituting (6.87), (6.89) and (6.74) in (6.90), and recalling $\Phi_t \Psi_t = I$, we get

$$-dp_t = \left(f_x'(t, Y_t, \beta_t) + p_t b_x'(t, Y_t, \beta_t) + \sum_{j=1}^n r^j \sigma_x^{j\,'}(Y_t, \beta_t) \right) dt - r_t \, dB_t, \tag{6.91}$$

where $r_t^j = \eta_t^j \Psi_t - p_t \sigma_x^{j\,'}(Y_t, \beta_t)$. Under the notation introduced in Lemma 6.10 one obtains

$$-dp_t = H_x'(t, Y_t, \beta_t, p_t, r_t) \, dt - \sum_{j=1}^n r_t^j \, dB_t^{(j)},$$

$$p_T = \phi_x'(T, Y_T).$$

Hence, it remains to verify

$$\mathsf{E}\left\{ \int_0^T \left(p_t \varphi_t + \sum_{j=1}^n r_t^j \psi_t^j \right) dt \right\} = \mathsf{E}\left\{ \int_0^T f_x'(t, Y_t, \beta_t)\zeta_t \, dt + \phi_x'(T, Y_T)\zeta_T \right\}.$$

To this end, we compute the Itô differential

$$dp_t \zeta_t = (dp)\zeta_t + p_t \, d\zeta_t + dp_t \cdot d\zeta_t. \tag{6.92}$$

Then, after substituting (6.65) and (6.91) in (6.76) and some arrangements, we obtain

$$\mathsf{E}\{\phi_x'(T, Y_T)\zeta_T\} = \mathsf{E}\left\{\int_0^T \left(p_t\varphi_t - f_x'(t, Y_t, \beta_t)\zeta_t + \sum_{j=1}^n r_t^j \psi_t^j\right) dt\right\}$$

and this concludes the proof.   ■

6.4 Qualitative trends and interpretation

In this section we return to the study of the interaction of the processes l^{-1} and λ, which reflect capital and employment formation, with the aim of applying obtained results to the stochastic control problem OP 6.1, see (6.53). A sound interpretation of the role of several variables and equations with the purpose of visualising meaning and power of the solution shall prepare the ground for a clear treatment of issues concerning attainability and reversibility of essential features of underlying policy.

6.4.1 The background of control actions

In order to understand the role and impact of control actions we are compelled to identify the correspondence between ingredients, variable links and philosophy of the control and economic problem. Thus, we begin making precise some issues on the respective background intended to identify the explanatory power of the various notions building the core of the maximum principle presented above. Only in this sense the mathematical results established in the foregoing section come to characterise an optimal control process and the appertaining paths of accumulation, employment and consumption among others corresponding to the stochastic control problem formulated in (6.53).

To grasp the economic relevance and power of the stochastic maximum principle condensed in Theorems 6.11 and 6.13, let us proceed to identify the links between the economic problem condensed in (6.53) and the control problem given by (6.61) and make sure the drift term, diffusion coefficient and other functions involved fulfilled the assumptions. To facilitate this, we have used almost the same notation throughout this chapter. In this section we return to the one-dimensional framework of the first two sections. However, we shall emphasise that due to the structural form of the state and associate equations the same analysis can be extended component by component to the n-dimensional case; see in this regard Subsection 2.2.3. For the drift and diffusion coefficients we obtain the following correspondence

$$b(t, X_t, \alpha_t) = i_t - (\epsilon_t - \pi_t^2)k_t \tag{6.93}$$

$$\sigma(t, X_t, \alpha_t) = -\pi_t k_t. \tag{6.94}$$

As we have already suggested, our analysis will depend heavily on the drift term, while the diffusion coefficient is primarily used to explain sources of uncertainty. Thus, to stress the dynamic thrust resulting from the drift we shall call attention to the following points which can be read from (6.93) and (6.94) taking into account the discussion on these topics presented in Sections 6.1 and 6.2.

- The expansion of productive capacity increases with the rate of investment and the income/capital ratio, and decreases with the level of available opportunities to develop working capacities. As a first approximation, this can be seen setting the drift equal to $\dot{k}_t$, recalling $i_t = s_t v_t$ and (6.5) – (6.6).

- At a given rate of capacity utilisation, the greater the rate of investment the lower the relative share of income going to workers as a consequence of diverting resources into capital formation. Further, the rate of labour supply increases as the prospect of employment opportunities becomes higher and with it the underlying uncertainty tension increases.

Beside these two points there are others which reveal the impact upon accumulation per available worker resulting from the interaction of capital building and employment creation suggested by the manifold relationships we worked out in Sections 6.1 and 6.2.

Furthermore, the functionals $f(t, x, \alpha)$ and $\phi(t, x)$ characterised by Assumption A 6.6 in Section 6.3 correspond to those appearing in (6.52) which represent the current rate of cost and the terminal reward. The reason for resorting to the performance index (6.52) rather than to the one defined by (6.51), which would appear as the most natural, goes back to our intention to measure deviations from the steady-state chosen as reference state, see Lemma 6.4.

Hence, the deviation of the instantaneous rate of reward from its golden value described by $f(t, k_t, \alpha) - f^\circ$ cannot simply be understood on the grounds of deviations from a bliss value f°. Focusing on the principle of social reproduction and characteristics of consumption goods as nourishment of states of being and doing of social agents compel us to assess the performance of an economic system on the basis of the capability set its state of evolution can offer individuals. In this sense from the interaction of capital building and employment creation results the need to weigh the system ability to expand its productive capacity, to deliver material goods, to forward employment opportunities and to afford an adequate effective demand against the system ability to increase appropriately the options open to individuals to function and to act in accordance to the state of progress prevailing. Following Eichner's lines of argumentation in [58], we advance three points relating critically the system and individuals' ability to function.

- As long as employment remains the primary source of purchasing and consuming power in an economic system, the appropriate functioning of the system and individuals will remain critically related.

- Since renewing earning and productive power of individuals and the system are essential to the continuity of social existence, the capacity to

provide opportunities to acquire and develop know-how as well as to renew and expand productive power needs to be critically reflected in the
performance index.

- Processes of time and change remind us with appalling evidence of the omnipresence of the law of entropy and stress human creativity as quintessen-
tial to the advancement of social existence. This is a feature relating
critically the performance of a system to the degree up to which human
dynamism, ingenuity, intrinsic willingness and ability to set and fulfil
goals are free of material, cultural, economic, social, political or whatsoever constraints.

In the light of the paradigm of social reproduction these three points contribute
to elucidate the meaning the aim of developing the productive forces of society
actually carries and the sharp contrast between freedom of action and the
vacuous notion of freedom of will.

It remains to identify qualitative and quantitative changes in the individual
characteristics underlying capabilities to function and the range of options expected to materialise in the future attributable to changes in the allocation of
currently available resources. In particular those changes serving the purpose
of strengthening the productive capacity and/or the pace of accumulation, see
for instance (1.10). However, since individual entitlements and capability sets
are far from following an egalitarian distribution, there is a sense of arbitrariness focusing merely on an average increase in options, as for instance (1.10)
suggests, and assigning certain worth to unreleased potentials associated with
the presence of unemployment; see in this regard [5].

The social value of investment is described by the process p which evolves
according to (6.76). To understand the rôle played by this equation, we would
need further elaboration. For the time being, we shall understand the social
value of investment in the sense explained above and represented quantitatively by expressions like (1.10), (9.19) or (5.98). The process r appearing in
(6.76) and more explicitly in (6.91) describes the uncertainty underlying the
materialisation of the social value of investment.

Now we are prepared to complete the search for the correspondence of the
main ingredients envolved in the economic and control problem, a necessary
step toward a proper understanding of the design of an optimal policy. The
functional H was introduced in Lemma 6.10. Thus in the light of the given
correspondence the functional H defined by (6.69) assumes for any $t \in [0, T]$
the form

$$H(t, k_t, \alpha_t, p_t, r_t) = [f(t, k_t, \alpha_t) - f^\circ] + p_t \, b(t, k_t, \alpha_t) + r_t \, \sigma(t, k_t, \alpha_t) \quad (6.95)$$

which taking into account (6.93) and (6.94), and the comment following these
equations can be rewritten as

$$H_t(k_t, i_t, c_t, p_t, r_t) = [f(t, c_t) - f^\circ] + p_t \, b(t, k_t, i_t) - r_t \pi_t k_t. \quad (6.96)$$

This simplified notation, which is particular convenient for our purpose, stresses the fact that the functional H depends on the control and thus on the pair

(i_t, c_t) and that the processes p and r are control independent. As we know already, the functional H is called the Hamiltonian, whenever the control process α is the optimal one. At any time t the function H_t represents a potential describing in the light of the capability approach first, the capability of individuals to function, given that at time t a consumption bundle becomes available embodying the characteristics c_t and second, the productive power available to the end of accumulation as condensed in k_t and i_t. The term p_t translates units of productive power at time t into future consumption characteristics measured in terms of units prevailing at time t, while r_t accounts for eventual deviations resulting from changes in resource allocation effectuated at time t.

At this point let us recall, the construction of the function H_t obeys a rationale operating at the vertical level, which accounts for its eminently static character. The term in brackets indicates the capability of available labour to function within the context determined by the technology prevailing at time t and in the event of optimality this term is linked to the bundle of consumption characteristics needed to maintain it. The second term plays the same rôle in regard to capital per available labour, while the third term accounts for provisions associated with uncertainty. In a loosely way, we may write (6.96) simply as $H(i_t, c_t)$, which has been assumed to articulate in the form of isoquants the existing system of social preferences and value-judgements and orders completely alternative combinations of investment i and consumption c. [199, 200, 201] deal admirably with questions related to the determination of such social orderings.

The Hamiltonian analogy revisited

To understand the meaning of economic dynamics, Leibniz's dual distinction between dead force (*vis mortua*) and living force (*vis viva*) is extremely important. The first points to a phenomenal motion which has only a potential existence and hence is an unobserved motive power. The second is on the other hand a moving power susceptible to quantitative assessment only in conjunction with the first and further, it is not necessarily conserved in the process of phenomenal motion due to the presence of irreversible processes. A striking feature worth noting is, motive power requires for its realisation the presence of a tension underpinned by potential reflected latent, profit, opportunities, which in turn may give rise to the actualisation of moving power, see [5].

From this perspective, we shall let the functional H represent the action power of the system, which can be defined in per head, per worker terms or the like. Moreover, its first term shall represent the moving power and its second the motive power. This interpretation is akin to the one resulting from the capability approach and gives accumulation a rationale naturally tied to the process of social reproduction. Thus, under the guide of the optimal control the action power H given by (6.95) measures the total social energy associated with the system at any time $t, t \in [0, T]$. We shall call the first and second term in (6.96), i.e. $f(t, k_t, c_t) - f^\circ$ and $p_t\, b(t, k_t, i_t)$ the kinetic and potential social energy respectively, since they resemble the concepts of kinetic and potential energy in classical physics. The last term $-r_t \pi_t k_t$ accounts for the social energy deviations due to the uncertainty of the system.

The logic underlying any accumulation policy cannot fully be explained on the basis of the foregoing analogy, since various basic elements of the learning process building the heart of development and the growth process have not been explicitly integrated. However, since we have considered these issues in Chapter 4 we refer to it, in particular see Section 4.3.

To conclude let us just mention, the drift term of the system $b(t, k_t, \alpha_t)$ indicates the ability of the system to generate economic surplus, which is then allocated to expanding the moving and motive power of the system. The crystallisation of the allocation objective generates among other things, opportunities for profit realisation in the form of capabilities, know-how, means of production, effective demand, expectations of higher standard of living and actualisation of moving power in terms of employment opportunities. The last term is unpredictable, it may favour or inhibit the contribution of the first two terms depending on π which accounts for unpredictable fluctuations and confusions mostly related to the information flow on the configuration of changes resulting from the accumulation policy.

6.4.2   Static analysis of the optimal policy

Let $\beta = (\beta_t)_{t \geq 0}$ denote the optimal control and $k^* = (k_t^*)_{t \geq 0}$, $c^* = (c_t^*)_{t \geq 0}$, $i^* = (i_t^*)_{t \geq 0}$ the optimal trajectories corresponding to k, c and i respectively. Thereby the function (6.96) gives at time t the value of the Hamiltonian $H_t(k_t^*, i_t^*, c_t^*, p_t, r_t)$ which we will write for simplicity as

$$H_t(i_t^*, c_t^*) = [f(t, c_t^*) - f^\circ] + p_t\, b(t, i_t^*) - r_t \pi_t k_t^*, \tag{6.97}$$

and sometimes we shall simply use $H_t(\beta)$. We have the following lemma.

Lemma 6.14 *The assumption of the optimising problem 6.1 hold. Let $p = (p_t)_{0 \leq t \leq T}$ denote the adjoint state process characterised in Theorem 6.13. Then for any $t \in [0, T]$ the adjoint state satisfies the relationship*

$$p_t = f_c'(t, c_t^*) * P_K(t) \qquad P(d\omega) \otimes dt \quad a.s. \quad a.e.\ t, \tag{6.98}$$

where the accounting price of capital $*P_K(t)$ is given by (6.50).

Proof: Here we use the fact that c_t and i_t are functions of k_t, see (2.12) and (2.11). Since the Hamiltonian has a maximum at $i^* = i_t^*$, the control set $R^1 A$ is all of $\mathbb{R}_+$ and since H_t is differentiable in i^*, we must have

$$0 = \frac{dH_t}{di_t^*} = -\frac{\partial f}{\partial c} \left(-\frac{dc_t^*}{di_t^*} \right) + p_t,$$

where we use the property that the Hamiltonian associated with an optimal policy is constant. Now, taking into account (6.50), (6.67) and (6.72) we get the result.   ∎

The relationship (6.98) holds for all $t \in [0, T]$ with possible exceptions on $P(d\omega) \otimes dt$-null sets. For that reason it is a moment-to-moment relation known in dynamic economics as the *dynamic efficiency conditions*.

The dynamic conditions make sure the characteristics to function of living force given up with the purpose of furthering capital accumulation is recovered at any time with probabilistically negligible exceptions. The characteristics of individuals to function is measured here in terms of characteristics to function per available worker, which are associated with a certain consumption bundle.

Taking into account that $_*P_K(t) \geq 1$ for any $t \in [0, T]$ as a simple computation shows, (6.98) tells us that to forward the accumulation policy represented by β amounts of future characteristics to function per unit of capital turning back to living labour as described by p_t have to be higher than the amount of current characteristics to function given up at time t per unit of consumption, i.e.

$$p_t \geq f'(t, c_t) \qquad P(d\omega) \otimes dt \quad a.s. \quad a.e.\ t, \qquad (6.99)$$

The dynamic efficiency condition given by eq. (6.98) amounts to the well-known tangency condition between the transformation functional of investment and consumption characteristics represented by $\mathcal{I}_3$ and the substitution functional of the associated characteristics to function given by the Hamiltonians $H(i_t^*, c_t^*)$, which in turn defines a *social demand price of investment in terms of current consumption*. The tangency condition, which follows easily from eq. (6.98), can be written as

$$_*P_K(t) = \frac{p_t}{f'(t, c_t^*)} = \frac{\partial_i H_t}{\partial_c H_t}. \qquad (6.100)$$

Eq. (6.100) becomes evident recalling, $_*P_K(t)$ is given by (6.50) and taking into account (6.95) and (6.93) as well as (6.96).

Theorem 6.15 *Let the assumption of Lemma 6.14 hold. On the basis of the stochastic maximum principle formulated in Theorem 6.11 let us characterise the optimal control β by*

$$H_t(k_t^*, \beta_t, p_t, r_t) = \max_{\alpha \in A} H(t, k_t, \alpha_t, p_t, r_t) \qquad P(d\omega) \otimes dt \quad a.s. \qquad (6.101)$$

Taking into account the differentiability of $H(t, k_t, \alpha_t, p_t, r_t)$ in Lemma (6.9), one obtains static first-order conditions of optimality which fully describe the following three phases the system undergoes in every time period before entering the golden age era:

Phase I
If $\lambda_t^* < 1$, and $s_K^*(t) = \overline{s}_K$, then

$$\frac{p_t}{f'(t, c_t^*)} = \frac{(s_K - s_L)W + (1 - s_K)f'(l_t)}{(s_K - s_L)W - s_K f'(l_t)} \qquad P(d\omega) \otimes dt \quad a.s. \quad a.e.\ t, \quad (6.102)$$

Phase II
If $\lambda_t^* = 1$, and $s_K^*(t) = \overline{s}_K$, then

$$1 \leq \frac{p_t}{f'(t, c_t^*)} \leq \frac{(s_K - s_L)W + (1 - s_K)f'(l_t)}{(s_K - s_L)W - s_K f'(l_t)} \qquad P(d\omega) \otimes dt \quad a.s. \quad a.e.\ t,$$

$$(6.103)$$

Phase III
If $\lambda_t^* = 1$, *and* $s_K^*(t) < \bar{s}_K$, *then*

$$\frac{p_t}{f'(t, c_t^*)} = 1 \qquad \mathbf{P}(d\omega) \otimes dt \quad a.s. \quad a.e. \ t. \qquad (6.104)$$

Proof: The proof follows for the main purposes the steps of the one given in [94]; see also Marglin [145]. ∎

The Hamiltonian, together with the initial conditions and constraints, determines whether the economy finds itself in Phase I, II or III. As we shall see below, the system optimally develops by moving from Phase I to Phase II and from Phase II to Phase III, when it starts from a capital intensity k_0 which is low enough. However, one has not always to begin with Phase I; a sufficiently large initial endowment of capital k_0 may put the economy also in Phase II or even in Phase III.

The phases obtained as the necessary conditions of (6.101) shall be interpreted as follows. First of all, the combination of events like full employment, i.e. $\lambda_t^* = 1$, or unemployment, i.e. $\lambda_t^* < 1$, and a binding investment policy requiring to direct profits back to production at the highest rate $\bar{s}_K$, i.e. $s_K^*(t) = \bar{s}_K$, or a nonbinding investment policy which allows for reinvestment at a lower rate, i.e. $s_K^*(t) < \bar{s}_K$, characterise the case to hold and this is indicated at the left-hand side under the corresponding phase. Then, according to the phase, if optimality dictates full employment or unemployment and a binding or a nonbinding investment policy, the relative social desirability of the couple $(s_K(t), \lambda_t)$ or alternatively of the resulting investment-consumption mix (i_t, c_t) has to be measured by means of the corresponding weight $_*P_K(t)$ associated with (6.101) and (6.100), resulting from the first-order conditions; see (6.102) to (6.104).

6.4.3   Dynamic study of the adjoint state

Let us point out at this stage, the maximum principle condensed in Theorem 6.11 and Theorem 6.13 enables us to split the intertemporal optimisation problem (6.51) into a static, i.e. eq. (6.101), and a dynamic one, i.e. eqs. (6.110) and (6.111). In other words, the maximum principle allows a time decentralisation of the decision process and this is extremely convenient for applications.

Since we have considered the static features in the foregoing subsection, let us look at the dynamic aspects of optimality. From eq. (6.76) in Theorem 6.13, we obtain under the notation of the present section

$$D_+ p_t = \partial_k H_t(i_t^*, c_t^*), \qquad (6.105)$$

where the left-hand side in eq. (6.105) has to be interpreted as the mean forward conditional derivative with respect to time. That means $D_+ p_t$ is given by

$$\lim_{h \downarrow 0} \mathbf{E} \left[\frac{p(t+h, \omega) - p(t, \omega)}{h} \Big| \mathcal{F}_t \right].$$

Taking into account eqs. (2.12) through (2.11), (6.93) through (6.94) , and (6.95) through (6.71), and after a few arrangements, eq. (6.105) becomes

$$\hat{p}_t + (\epsilon_t - \pi_t^2) + r_t\pi_t = \frac{(1 - s_K) + s_K(\frac{p_t}{f'(t,c_t^*)})}{\frac{p_t}{f'(t,c_t^*)}}(f(l_t) - f'(l_t)l_t)$$

$$= \frac{{}_*P_\pi(f(l_t) - f'_t(l_t)l_t)}{{}_*P_K} \tag{6.106}$$

where ${}_*P_\pi$ is the accounting price of a unit of profits, i.e.

$$_*P_\pi = (1 - s_K) + s_K \cdot {}_*P_K.$$

The accounting price of profit ${}_*P_\pi$ measures the social value of a unit of profit which is allocated between consumption and investments according to the fractions $1 - s_K$ and s_K respectively. Since the social value of a unit consumption is the unity and that of investment is ${}_*P_K$, the meaning of ${}_*P_\pi$ becomes apparent. Recall that ${}_*P_K$ is given by (6.50) and see (6.102) to convince yourself that it is greater than 1. The reason why investment is valued higher than consumption is because investment generates future consumption which entails increases in the future states of functioning. This higher social valuation is a consequence of the institutional constraints introduced in Section 6.1 and Section 6.2.

The expression ${}_*P_\pi(f(l_t) - f'(l_t)l_t)$ in (6.106) measures the social value of increasing marginal rewards accruing to capital F_K, see (6.13). It is central to the materialisation of this social value that the increases in profit income are indeed directed back to production in the fraction indicated by the optimal policy. The technical conditions of this materialisation of value are condensed in $P_K(t)$. Then, since ${}_*P_K$ is the social value of investment, the ratio in the right-hand side of (6.106) stands for the investment rate of return which we denote as customary by ${}_*r_K$. Then, (6.106) acquires the more simple form

$$- \hat{p}_t + (\epsilon_t - \pi_t^2) + r_t\pi_t = {}_*r_K. \tag{6.107}$$

On the other hand, the relative rate of decay of the premium attached to investment is given by the logarithmic derivative of ${}_*P_K = (p_t/f'(t,c_t^*))$

$$\frac{{}_*\dot{P}_K}{{}_*P_K} = \hat{p}_t - \frac{\partial_c f''(t,c_t^*)}{\partial_c f'(t,c_t^*)} - \frac{s_K(1 - s_K)(f'(l_t))^2 \zeta_t l_t^2 \pi_t \eta_t}{{}_*P_K^2}, \tag{6.108}$$

where η stands for the elasticity of marginal reward, i.e. $\eta_t = -\frac{d(\log f'(t,C_t^*))}{d(\log C)}$.

Let $\frac{\partial f(t,C_t^*)}{\partial C}$ denote the marginal cost attributed to one unit less of social consumption C and let ${}_*i_C$ denote its relative rate of decay which we call the *consumption rate of discount*. Similarly, the quantity ${}_*i_C$ is given by the relationship

$$_*i_C - \frac{s_K(1 - s_K)(f'(l_t))^2 \zeta_t l_t^2 \pi_t \eta_t}{{}_*P_K^2} = -\frac{\frac{d}{dt}(\partial_c f(t,c_t^*))}{\partial_c f(t,c_t^*)}.$$

Due to the fact that the argument of the cost functional f is $c = \frac{C}{N}$, it results

$$_*i_C = -\frac{\frac{d}{dt}(\partial_c f(t,c_t^*))}{\partial_c f(t,c_t^*)} + \epsilon_t. \tag{6.109}$$

Therefore, combining (6.107), (6.108) and (6.109) it follows

$$_*i_C + (r_t - \pi_t)\pi_t - \frac{s_K(1-s_K)(f'(l_t))^2 \zeta_t l_t^2 \pi_t \eta_t}{_*P_K^2} = {_*r_C}, \qquad (6.110)$$

where $_*r_C$ is called the *consumption rate of return* is given by

$$_*r_C = {_*r_K} + \frac{_*\dot{P}_K}{_*P_K}.$$

The compact relation (6.110) reveals the economic content of Theorem 6.9, in particular of (6.76). It says, in order to motivate postponing claims on current consumption as a policy directed to strengthening capital accumulation entails, the consumption rate of return $_*r_C$ has to make up for the following three components:

1) The relative rate of decay of consumption characteristics per available labour. This entails a relative decay of individual states of functioning the consumption rate of return has to compensate to the end of getting an appropriate level of characteristics to function;

2) The relative rate of decay of consumption attributable to consumption from the increased margins of income accruing to capitalists within the scope of optimal economic policy. This term tends to vanish as s_K approaches unity and increases as s_K becomes small. This term undergoes changes due to the passing of time, uncertainty, accumulation and the accounting price of capital;

3) The decay due to uncertainty attributable to accumulation itself as well as to constraints and control interventions acting upon accumulation, as the second term in the left-hand side of (6.110) indicates.

The optimality requirement stated by (6.110) shall be called the *instantaneous intertemporal consistency conditions*. One can hardly overemphasise the need to account for the tendency of the characteristics to function to deteriorate, a fact which has to be counteracted by improving adequately the characteristics of the current consumption bundle. These features are the essence of our modelling endeavour in Section 4.3 and reveal the omnipresence of entropy see in particular (4.24). The instantaneous intertemporal consistency conditions remind us of pervasive effects of the law of entropy, acting upon the earning power of living labour, the productive power of capital and technology.

Since from the perspective of the capability approach deficiency in the characteristics to function jeopardise earning power and thus the action power of the system, we let the terminal conditions force the system to avoid capital levels which deviate from the less possible deficiency. This reflects the idea, the accumulation path along which the system optimally evolves is the one which allows individuals to traverse on consumption paths exhibiting the less possible deficiency in their characteristics to function. The *transversality condition* then is given by

$$\phi(T, k_T^*) = -p_T(k_T - k^\circ)^- \qquad (6.111)$$

where the notation indicates that for any variable z, the expression z^- stands for the negative part of z, i.e. $z^- = -\min(0, z)$.

Summing up, we shall call a control path $t \longrightarrow \alpha_t$ or equivalently a policy path of investment and employment $t \longrightarrow (s_K(t), \lambda_t)$ optimal, if and only if it satisfies: the static first-order conditions of optimality as given by (6.98) and (6.102) through (6.104), the instantaneous intertemporal consistency condition represented by (6.110), the transversality condition (6.111) and the dynamic efficiency condition (6.98). In the next section we deal with questions of attainability and reversibility of optimal paths.

6.4.4   A brief qualitative analysis

Our next step is to look for conditions which guarantee attainability of the three phases obtained above and prevent that once higher-numbered phases have been attained, the system moves back to lower-numbered ones.

The rationale behind this issue shall be intuitively visualised as follows. First, we choose an appropriate level of capital per actually employed worker l_t^{-1} from which we shall work out certain capital endowment. This endowment of capital is thought of as given to any worker in the production process as well as to those who are expected to enter into production during the time period under consideration. Choosing an appropriate l^{-1} is tantamount to finding out, at any time t, an amount of labour per unit of capital endowment l_t so as to allow capital accumulation. Second, since consumption provides for means of maintaining, renewing and improving the characteristics of individuals to function, one identifies on the way to determine λ, the associated paths of earning and consuming power which provide in turn the means of fulfilling the aims of accumulation. Let us recall

$$k_t = l_t^{-1} \lambda_t, \tag{6.112}$$

from which we may infer the main alternative of interplay of the basic processes of accumulation which we briefly sketch. Throughout Phase I, which lasts as long as (6.102) holds, l_t shall remain constant while λ_t and k_t grow up. That means that in Phase I, we are mainly eliminating unemployment and widening capital.

In Phase II, we shall hold λ_t constant, since according to (6.103) we have attained the level $\lambda = 1$, and let l_t^{-1} grow with the aim of bringing economic surplus close to its optimal level. That means, Phase II is one of capital deepening under saving binding constraints. The fact that $_*P_K$ falls down to unity makes sure that capital deepening comes to an end. In Phase III, which is characterised by (6.104), the savings constraint is no longer binding, what enables consumption to increase and the states of functioning improve so that the economy moves towards the golden age. Increasing consumption serves the objective of enlarging the capability set as a way of affording a higher degree of freedom to act. After this quick sketch of the meaning of the optimal policy, we shall start examining its qualitative properties.

Qualitative analysis of attainability

Let $\bar{s}$ denote the ratio s given by (2.19) corresponding to the case when $s_K(t) = \bar{s}_K$ and let v_t° stand for the process $\mathfrak{f}(l_t)$ when $l_t = l_t^\circ$, see (6.7) and (6.9). Further, let $\underline{l}_t$ denote the level of l_t at which the economic surplus given by (6.42), or alternatively by (6.116), attains a maximum. A simple computation shows that at $\underline{l}$ the marginal income $\mathfrak{f}'(l_t)$ is given by

$$\mathfrak{f}'(l_t) = \left(\frac{s_K(t) - s_L(t)}{s_K(t)} \right) W \tag{6.113}$$

which we denote $\underline{\mathfrak{f}}'(l_t)$.

Lemma 6.16 *Let the assumptions of Theorem 6.15 hold. A necessary and sufficient condition for the attainability of k_t° is that the inequality*

$$\bar{s}_t v_t^\circ > \epsilon_t - \pi_t^2 \tag{6.114}$$

holds.

Proof: Let us recall that the mean forward conditional-derivative $D_+ k_t$ is given by

$$D_+ k_t = i_t - (\epsilon_t - \pi_t^2) \tag{6.115}$$

Then, substituting (2.11) in (6.115) we get

$$\hat{k}_t = s_K(t)(\mathfrak{f}(l_t) - Wl_t) + s_L(t)Wl_t - (\epsilon_t - \pi_t^2) \tag{6.116}$$

or, equivalently, see (2.19),

$$\hat{k}_t = s_t \mathfrak{f}(l_t) - (\epsilon_t - \pi_t^2) . \tag{6.117}$$

Hence, positive capital accumulation amounts to positivity of the right-hand side of (6.117). Thus, setting $s_t^* = \bar{s}$ in (6.117) gives the accumulation feasibility condition (6.114) at the golden-rule values $l_t = l_t^\circ$ and $\mathfrak{f}(l_t) = v_t^\circ$. Then, applying a continuity argument completes the proof; see [145]. Indeed, the continuity of $\mathfrak{f}(l_t)$ and (6.114) entails positive accumulation of capital, regardless of the level of k_t, in a neighbourhood of l_t° as well as at l_t° itself. Therefore, l_t can be brought sufficiently close to l_t° within this neighbourhood by sacrificing employment.

Furthermore, let us observe that attaining l_t° and correspondingly c° is tantamount to attaining k_t°. Finally, knowing the feasibility of attaining and sustaining k_t° this must optimally happen in finite time. Otherwise, from (6.52) it follows easily that such a path would be dominated by any path attaining c_t° in finite time, and therefore cannot be optimal. ∎

Lemma 6.17 *Let the assumptions of Theorem 6.15 hold. Let $\underline{l}_t$ be the labour-capital ratio at which the economic surplus, the right-hand side of (6.116) or*

(6.117), attains a maximum. The golden-rule level of the capital per available labour ratio $k_t^\diamond$ associated with $l_t^\diamond$ is necessarily higher than the level $\underline{k}_t$ associated with the level $\underline{l}_t$, i.e. the inequalities

$$l_t^\diamond < \underline{l}_t \quad and \quad \underline{k}_t < k_t^\diamond \tag{6.118}$$

hold.

Proof: Let us recall that $k_t^\diamond$ by definition, i.e. (6.23), satisfies

$$v_t^\diamond - f'^\diamond(l_t)l_t = \epsilon_t - \pi_t^2 \tag{6.119}$$

where $f'^\diamond(l_t)$ stands for $f'(l_t)$ at

$$l_t = l_t^\diamond.$$

Eq. (2.23) together with the attainability condition (6.114) gives

$$\overline{s}_t v_t^\diamond > v_t^\diamond - f'^\diamond(l_t)l_t^\diamond. \tag{6.120}$$

Rewriting the left-hand side of (6.120), see (6.38), (2.19) and (6.116), we obtain

$$\overline{s}_K \left[v_t^\diamond - \left(\frac{s_K(t) - s_L(t)}{s_K(t)} \right) W l_t^\diamond \right] > v_t^\diamond - f'^\diamond(l_t)l_t^\diamond.$$

Since $\overline{s}_K \leq 1$, we obtain further

$$v_t^\diamond - \left(\frac{s_K(t) - s_L(t)}{s_K(t)} \right) W l_t^\diamond > v_t^\diamond - f'^\diamond(l_t)l_t^\diamond. \tag{6.121}$$

From (6.121) one obtains easily

$$\underline{f}'(l_t) = \left(\frac{s_K(t) - s_L(t)}{s_K(t)} \right) W < f'^\diamond(l_t)$$

and this inequality delivers the first inequality in (6.118). Furthermore, since at the golden-rule level $\lambda = 1$, it provides the second inequality in (6.118) also. ∎

Lemma 6.18 *The relative rate of growth of the aggregate consumption is given by the following relation*

$$\frac{\partial_t C_t}{C_t} = s_t f(l_t) + \frac{(1 - s_K)f'(l_t) + (s_K - s_L)W}{(1 - s_K)f(l_t) + (s_K - s_L)W l_t} \left(\frac{\partial l_t}{\partial t} \right) \tag{6.122}$$

$$- \frac{(f(l_t) - W l_t)}{(1 - s_K)f(l_t) + (s_K - s_L)W l_t} \dot{s}_K + \frac{\frac{1}{2}(1 - s_K)f''(l_t)\zeta_t^2 l_t^2}{(1 - s_K)(f(l_t) - W l_t) + (1 - s_L)W l_t}.$$

Proof: It follows easily from (2.12), some arrangements and Itô's calculus. ∎

Lemma 6.19 *The relative rate of growth of consumption per available labour* c_t *is given by*

$$\frac{\frac{\partial c_t}{\partial t}}{c_t} = \frac{\frac{\partial C_t}{\partial t}}{C_t} - \epsilon_t + \pi_t^2. \tag{6.123}$$

Proof: The computations leading to (6.123) go via Itô's calculus and are similar to these preceding (6.36) and (6.39) through (6.41).   ∎

Similarly one can compute the conditional mean forward derivative for the accounting price of capital from which we get the following lemma.

Lemma 6.20 *The following relation for* $*P_K$ *holds*

$$\frac{\frac{\partial *P_K}{\partial t}}{*P_K} = \frac{(1 - s_K)f''(l_t)\frac{\partial l_t}{\partial t} + [W - f'(l_t)]\dot{s}_K - \frac{1}{2}(1 - s_K)f'''(l_t)\zeta_t^2 l_t^2}{(s_K - s_L)W + (1 - s_K)f(l_t)'} \tag{6.124}$$

$$+ \frac{s_K f''(l_t)\frac{\partial l_t}{\partial t} - [W - f'(l_t)]\dot{s}_K - s_K f'''(l_t)\zeta_t^2 l_t^2 [(s_K - s_L)W - s_K f'(l_t)]^{-1}}{(s_K - s_L)W - s_K f'(l_t)}.$$

Proof :   Starting from (6.100), the procedure is as before. A little Itô calculus, one approximates the conditional mean forward derivative and a few arrangements gives the lemma.   ∎

It is necessary to recall the time derivative of random processes are to be interpreted as conditional mean forward derivative as in Sections 6.1 and 6.2. One computes on the basis of an already known SDE using stochastic calculus. However, approximations based on the mean drift are used only to keep the number of operations fairly limited and are in no way essential.

Let us now go over to analyse the question related to the once-and-for-all transition from Phase I to Phase II, and from Phase II to Phase III, we would like to make a few remarks intended to facilitate the argumentation.

First of all, we shall call attention to the fact that (6.109) can be written as

$$*i_c(t) = \eta_t \frac{\dot{c}_t}{c_t} + \epsilon_t \tag{6.125}$$

where η is the elasticity of the marginal reward $(\partial f / \partial c)$ and therefore measures the relative rate of change of marginal reward with an increase in consumption c_t of one percent. The strict concavity of f entails that $\eta_t > 1$. Now, combining (6.122), (6.123) and (6.126), one obtains (6.126) that we state in the following Lemma.

Lemma 6.21 *We have the formula*

$$*i_c(t) = \eta_t \left\{ [s_t f(l_t) - \epsilon_t] + \frac{[1 - s_K(t)]f'(l_t) + [s_K(t) - s_L(t)]W}{[1 - s_K(t)]f(l_t) + [s_K(t) - s_L(t)]Wl_t} \left(\frac{\partial l_t}{\partial t} \right) \right.$$

$$- \frac{(f(l_t) - Wl_t)}{[1 - s_K(t)]f(l_t) + [s_K(t) - s_L(t)]Wl_t} \dot{s}_K(t) \tag{6.126}$$

$$\left. + \frac{\frac{1}{2}[1 - s_K(t)]f''(l_t)\zeta_t^2 l_t^2}{[1 - s_K(t)](f(l_t) - Wl_t) + [1 - s_K(t)]Wl_t} + \pi_t^2 \right\} + \epsilon_t.$$

Lemma 6.22 *If the path variable l is such that the relationships*

$$\frac{{}_*P_\pi(t)[v_t - \mathfrak{f}'(l_t)]}{{}_*P_K(t)} = \eta_t\{s_t v_t - (\epsilon_t - \pi_t^2)\} + \epsilon_t \tag{6.127}$$

and

$$\left(\frac{s_K(t) - s_L(t)}{s_K(t)}\right) W > \mathfrak{f}'(l_t) \tag{6.128}$$

are simultaneously satisfied, then it must follow that $l_t \leq l_\epsilon(t)$, where l_ϵ is defined by

$$l_\epsilon(t) = \inf\{l_t : l_t > \underline{l}, \quad s_t \mathfrak{f}(l_t) = \epsilon_t - \pi_t^2\}. \tag{6.129}$$

Proof: Our proof goes indirectly. Therefore, let us assume l satisfies (6.127) and (6.128), while $l_t > l_\epsilon$ holds. From (6.129) follows that

$$s_t v_t \leq \epsilon_t - \pi_t^2$$

and since $\eta_t > 1$, we obtain from (6.127)

$$\frac{{}_*P_\pi(v_t - \mathfrak{f}'(l_t)l_t)}{{}_*P_K} \leq s_t v_t. \tag{6.130}$$

On the other hand,

$$\frac{{}_*P_\pi}{{}_*P_K} = \frac{(s_K - s_L)W}{(s_K - s_L)W + (1 - s_K)\mathfrak{f}'(l_t)}$$

which together with (6.128) gives

$$\frac{{}_*P_\pi}{{}_*P_K} > s_K. \tag{6.131}$$

Now, using (6.131) in (6.130) provides the inequality

$$s_K[v_t - \mathfrak{f}'(l_t)l_t] < s_t v_t \tag{6.132}$$

or equivalently,

$$s_K[v_t - \mathfrak{f}'(l_t)l_t] < s_K\left[v_t - \left(\frac{s_K - s_L}{s_K}\right)W l_t\right]. \tag{6.133}$$

But this inequality contradicts (6.128).   ∎

Next, let us list properties and requirements which characterise the three phases.

P 6.1 *Characterising properties of the optimal phases*

State values	Time		Behaviour	
	invariance	varying	equilibrating	Hamiltonian

a) Phase I

State values	invariance	varying	equilibrating	Hamiltonian
$\lambda_t^* < 1$ $k_t = \lambda_t l_t^{-1}$	$\dot{l}_t = 0$ $\dot{s}_K(t) = 0$	$\dot{\lambda}_t > 0$ $\dot{k}_t > 0$	$s_K^*(t) = \bar{s}_K$ $_* P_K(t) \geq 1$ $\left(\frac{s_K(t) - s_L(t)}{s_K}\right) W \geq f'(l_t)$	$\frac{\partial H}{\partial \lambda} = 0$

b) Phase II

State values	invariance	varying	equilibrating	Hamiltonian
$\lambda_t^* = 1$ $k_t = l_t^{-1}$	$\dot{\lambda}_t = 0$ $\dot{s}_K(t) = 0$	$\dot{k}_t > 0$ $\ddot{k}_t > 0$	$s_K^*(t) = \bar{s}_K$ $_* P_K(t) \geq 1$ $_* \dot{P}_K < 0$ $\left(\frac{s_K(t) - s_L(t)}{s_K}\right) W \leq f'(l_t)$	$\frac{\partial H}{\partial \lambda} \geq 0$ $\frac{d}{dt}\left(\frac{\partial H}{\partial \lambda}\right) > 0$

c) Phase III

State values	invariance	varying	equilibrating	Hamiltonian
$\lambda_t^* = 1$ $k_t = l_t^{-1}$	$\dot{\lambda}_t = 0$ $_* \dot{P}_K(t) = 0$	$\dot{k}_t > 0$ $\ddot{k}_t \gtreqless 0$ $\dot{s}_K(t) < 0$	$s_K^*(t) < \bar{s}_K$ $_* P_K(t) = 1$ $\left(\frac{s_K(t) - s_L(t)}{s_K}\right) W < f'(l_t)$	$\frac{\partial H}{\partial \lambda} \geq 0$ $\frac{d}{dt}\left(\frac{\partial H}{\partial \lambda}\right) > 0$

For the sake of simplicity we have relaxed the notation of the time derivative of stochastic processes in the foregoing list. The following lemma will be needed.

Theorem 6.23 *Let us denote the optimal value of l_t by l_t^*, i.e. the value of l_t associated with the optimal control β. Then, it follows that in Phase I, l_t^* satisfies the inequality*

$$l_t^* < l_\epsilon(t). \tag{6.134}$$

Proof: We proceed here again in an indirect way. Hence, we assume that $l_t^* > l_\epsilon(t)$ and show that this leads to a contradiction. We consider the following three cases:

- $\dot{l}_t^* = 0$, which turns out to be inconsistent with the first-order optimality conditions. In particular, it precludes the possibility of obtaining full employment.

- $\dot{l}_t^* < 0$, this inequality results to be inconsistent with the instantaneous intertemporal consistent condition, otherwise the foregoing case applies directly.

- $\dot{l}_t^* > 0$, in this last case it turns out the transversality conditions is violated.

Then, taking into account the intertemporal consistency condition (6.110) as well as Lemmas 6.18 through 6.21, and the characterisation of Phase I, one proves that for the first two cases (6.128) holds. Thus, since in Phase I the relationship

$$\left(\frac{s_K - s_L}{s_K}\right) W \geq f'(l_t) \, ,$$

then (6.128) holds also and application of Lemma 6.22 delivers the contradiction.

The last case, i.e. $l_t^* > l_\epsilon$ and $\dot{l}_t^* > 0$, can be reduced to one of the first two cases or it leads to a contradiction of the attainability of the golden-rule values. Therefore, we can safely conclude that in Phase I the relationship $l_t^* < l_\epsilon$ must hold, which entails that Phase I has to come to an end in finite time. ∎

Let us point out, optimality dictates unemployment whenever the social value of the net contribution of the marginal worker falls to zero before full employment can be reached.

Qualitative analysis of reversibility

Further, we like to mention that $\frac{\partial H}{\partial \lambda} = 0$ is a necessary condition characteristic of Phase I. We like to state the following.

Lemma 6.24 *Once the economy reaches full employment with $\dot{k}_t > 0$, then $\dot{k}_t$ has to remain positive as long as the economy remains in Phase I and maintains full employment.*

Proof: The fact that in Phase II the relationship $l_t = k_t^{-1}$ holds and a small calculation deliver the lemma. ∎

Lemma 6.25 *We have the formula*

$$\frac{\partial H}{\partial \lambda} = p_t s_K \left[f'(l_t) - \left(\frac{s_K - s_L}{s_K}\right) W \right]$$
$$+ f'(t, c_t) \left[f'(l_t) - s_K \left(f'(l_t) - \left(\frac{s_K - s_L}{s_K}\right) W \right) \right]. \quad (6.135)$$

Proof: The lemma follows from (6.93), (6.95), (2.11), and (2.12) after a small calculation and few arrangements.

Lemma 6.26 *We also have the formula*

$$\frac{d}{dt}\left(\frac{\partial H}{\partial \lambda}\right) = f'_c(c)\left\{ (1 - s_K)[v_t - f'(l_t)l_t]\dot{k}_t \right.$$

$$\left. + \left[f'(l_t) - s_K \left(f'(l_t) - \left(\frac{s_K - s_L}{s_K}\right) W + \right) \right] \dot{\lambda}_t \right\}$$

$$\cdot \left[f'(l_t) - s_K \left(f'(l_t) - \left(\frac{s_K - s_L}{s_K}\right) W \right) \right]$$

$$+ f'(t, c_t) \left[(1 - s_K) f''(l_t)l_t \right] + p_t s_K f''(l_t)\dot{l}_t$$

$$-\left\{ f'(t,c_t)(1-s_K)(v_t-f'(l_t)l_t)+p_t[s_K(v_t-f'(l_t)l_t)-(\epsilon_t-\pi_t^2]-r_t\pi_t\right\}$$

$$\cdot\left[f'(l_t)-\left(\frac{s_K-s_L}{s_K}\right)W\right]$$

$$+\frac{1}{2}(1-s_K)f''(t,c_t)\left\{ \left[f'(l_t)-s_K\left(f'(l_t)-\left(\frac{s_K-s_L}{s_K}\right)\right)W\right]\right.$$

$$\left.\cdot\left[(f'(l_t)-f'(l_t))\lambda_t\pi_t^2+f'(l_t)l_t\lambda_t(\zeta_t-\pi_t)^2+f''(l_t)\zeta_t^2 l_t^2\right]\right\}. (6.136)$$

Proof: One applies Itô's formula to $\frac{\partial H_t}{\partial\lambda}$. Recall that $H_t = H_t(k_t,\alpha_t,p_t,r_t)$ is given by (6.95) and that $\alpha_t = \alpha(\lambda_t,s_K(t))$. Furthermore, the processes k_t, λ_t, p_t are random and their differentials are given respectively by (6.36), (6.30) and (6.76). See also (6.105).

It remains to show that when the economy enters Phase II and Phase III, that Phase I and Phase II respectively, can never recur. In this regard, the following theorems hold unambiguously only for the fairly restrictive case $\bar{s}_K = 1$. However, it is likely that the theorems also hold without this assumption.

Theorem 6.27 *Whenever the economy has entered Phase II, then Phase II can never recur.*

Sketch of the proof: One has to prove that once Phase II has lasted for a positive interval of time, $\frac{\partial H}{\partial\lambda}$ given by Theorem 6.25, can never again become zero. Hence, one has to prove that for small values of $\frac{\partial H}{\partial\lambda}$, its time derivative given by Lemma 6.26

$$\frac{d}{dt}\left(\frac{\partial H}{\partial\lambda}\right)$$

must be positive. Recall that in Phase II conditions like $\lambda_t = 1$, $s_K(t) = \bar{s}_K$, $\dot{\lambda}_t = 0$ and so on hold. See the list of properties characterising Phase II. Furthermore, we know that Phase II comes to an end when $_*P_K$ falls down to unity. ∎

In Phase III the following Lemma is valid.

Lemma 6.28 *In Phase III the intertemporal consistency condition (6.110) becomes simply*

$$_*i_C(t) + (r_t - \pi_t)\pi_t - \frac{s_K(1-s_K)(f'(l_t))^2\zeta_t l_t^2\pi_t\eta_t}{_*P_K^2} = v_t - f'(l_t)l_t. \qquad (6.137)$$

Proof: It follows from (6.106), (6.107) and (6.110) and the fact that Phase III the relationship $_*P_K(t) =_* P_\pi(t) = 1$ holds. ∎

Theorem 6.29 *Show that the transition from Phase II to Phase III is irreversible.*

Sketch of the proof: Combining (6.137) and (6.126) one gets a formula for $\dot{s}_K$ and proves that for $_*P_K = 1$ and s_K close to $\bar{s}_K$ its time derivative $\dot{s}_K$ has to be negative. See the characteristic features of Phase III in the corresponding table. Hence, once the economy has lived in Phase III for a while, then s_K can never become equal to $\bar{s}_K$. Therefore, the economy would remain by itself in Phase III for ever.   ∎

We have modelled the process of capital accumulation as the interplay of the processes of unemployment elimination and formation of economic surplus and we have thereby gained some insights into qualitative and quantitative aspects of their interaction as the economy moves towards full employment. Furthermore, we have obtained conditions for the attainability and reversibility of the three phases the system undergoes before entering its optimum state.

However, we still have to learn more about the time the economy spends in the different phases in order to be able to use the model in economic policy issues.

Part III

Background and Tools

1492: The Spirit of Eurocentrism

There now begins a reign of terror on Hispaniola for which I can find no proper historical parallel. Our world has a long and cruel history, and the word "unprecedented" should perhaps not ever be used. But the unique horror of Columbus' new state was that even the blindest obedience could not save the people. What was demanded of them was the impossible. Here was indeed created a new and unheard of society. ... To fill the empty ships going back to Castile, to stop his detractors from taking, to prove his success, Columbus needed gold. And the following system was adopted for the end.

Every man and woman, every boy or girl of fourteen or older, in the province of Cibao (of the imaginary gold fields) had to collect gold for the Spaniards. As their measure, the Spaniards used those same miserable hawks' bells, the little trinkets they had given away so freely when they first came "as if from Heaven." Every three months, every Indian had to bring to one of the forts a hawks' bell filled with gold dust. The chiefs had to bring in about ten times that amount. In the other provinces of Hispaniola, twenty-five pounds of spun cotton took the place of gold.

Copper tokens were manufactured, and when an Indian had brought his or her tribute to an armed post, he or she received such a token, stamped with the month, to be hung around the neck. With that they were safe for another three months while collecting more gold.

Whoever was caught without a token was killed by having his or her hands cut off. ... There were no gold fields, and thus, once the Indians had handed in whatever they still had in gold ornaments, their only hope was to work all day in the streams, washing out gold dust from the pebbles. It was an impossible task, but those Indians who tried to flee into the mountains were systematically hunted down with dogs and killed, to set an example for the others to keep trying. ... Thus it was at this time that the mass suicides began: the Arawaks killed themselves with cassava poison.

During those two years of the administration of the brothers of Columbus, an estimated one half of the entire population of Hispaniola was killed or killed themselves. The estimates run from 125,000 to one-half million.

Then, in 1496, when there was obviously not one grain of gold left, the gold tribute system was changed to that of the *repartimientos*, later known as *encomiendas*. The Spaniards cut out estates for themselves; the Indians still living on this land became their property. They could be used to work the land for the owner or could be hired out indefinitely as labor gangs anywhere else.

Hans Koning
(see [123], pp. 83–84)

Chapter 7

Essentials on systems and control

... there is a tendency to forget that all science is bound up with human culture in general, and that scientific findings, even those which at the moment appear the most advanced and esoteric and difficult to grasp, are meaningless outside their cultural context. A theoretical science unaware that those of its constructs considered relevant and momentous are destined eventually to be framed in concepts and words that have a grip on the educated community and become part and parcel of the general world picture – a theoretical science, I say, where this is forgotten, and where the initiated continue musing to each other in terms that are, at best, understood by a small group of close fellow travellers, will necessarily be cut off from the rest of cultural mankind; in the long run it is bound to atrophy and ossify however virulently esoteric chat may continue within its joyfully isolated groups of experts.

Erwin Schrödinger
(in [171], pp. 18–19)

7.1   The concepts of system and model

The aim of systems theory is to investigate a variety of phenomena regardless
of their specific nature but rather focusing on a set of essential relationships
linking the objects involved and on questions concerning changes emanating
from interactions among aforementioned objects and from their interplay with
the environment. Systems theory places emphasis on certain features of the
above set of essential relationships which come to characterise structure, organisation and patterns of order interrelating the objects under investigation;
in particular in the way these objects function.

   Therefore, the subject of study in systems theory is a rather formal entity
reflecting distinguishing relationships and features attributable to the set of
objects involved in observed phenomena; this entity is called a *system*. It is
crucial to note systems theory does not attempt to scrutinise explicitly the nature of phenomena under study – e.g. physical, chemical or social phenomena
– but rather, it aims at conceptualising such phenomena in a manner appropriate to interdisciplinary research and to facilitate communication among the
various fields. A feature of central importance in systems theory is its concern
with goal-conducive behaviour and decision making.

   By now, a large number of scientific fields use the concept of system to
describe its objects of study. In general, one uses the concept of system to
represent a great variety of living objects in nature, to describe various types of
relations emerging from innumerable economic, productive and organisational
activities of society, to refer to certain parts of our universe, to characterise
artificial mechanisms, and so on. Since it is not easy to give a precise and
general definition of the concept of system, we shall just mention a few notions
based on which one usually characterises a system.

   First, let us say that the main characteristic of a system is its *structure*. Any
system is made of components, elements and members. Individual elements
interact and interrelate in a specific way. However, the determination of its
structure is for any system relative, since according to the purpose at hand
one may need or be willing to observe the system at different structural levels.
Thus, any system is susceptible to a variety of arrangements of its elements;
so the same system may be seen as consisting of a large or small number
of elements, which may build themselves systems called then *subsystems* and
comes to exhibit accordingly a more or less complex structure.

   Second, associated with any system there corresponds its *environment* with
which it interacts in various ways. The separation of a system from its environment as well as the nature of the interaction ascribed to them are rather
arbitrary. Nevertheless, this arbitrariness should serve mainly the purpose of
ensuing functionality and efficiency in systems theoretical studies.

   Third, any system can be characterised *quantitatively* and *qualitatively* by
dint of *variables* of various sorts, i.e. dependent and independent variables, state
variables and system parameters, essential and unessential variables, and so on.
The number of such quantitative variables itself is a quantitative characteristic
of the system.

   Fourth, the *time evolution* of the system including change of either all or some of its charateristics is of crucial importance; we shall focus on changes in available potentials to function. Thereby, the characteristics of the system vary as a result of interactions among its elements and between the system with its environment.

Finally, the complexity of a system is a more basic feature rather than its size. Complexity is essentially related to information without which systems are unable to keep functioning. In the face of limited perception and/or deficiency in information flows, the operation of actual systems undergoes a variety of *random actions*. As a result of this, the state variables of the system develop in the course of time into random processes as a consequence of various kinds of related processes of change.

Thus, it is safe to say that virtually all existing actual systems are stochastic in nature, in particular those associated with the use of observations resulting from processes undermined by limited perception.

7.1.1   Systems characterisation

At this point, it may be clear systems theory breaks with the classical concept of the world consisting of matter and energy. The system-theoretical picture of the world rests instead on matter, energy and information, three components capable of explaining the existence of mechanisms underpinning structure, organisation and order which in turn are essential for understanding systems.

Summing up foregoing considerations, we shall say that facing the task of studying a phenomenon from the point of view of systems theory, we shall look at the following main particular features:

- Description of the structure of the system and in a certain sense the *choice of the structural level* at which the study shall take place;

- Influences from the system environment and actions of the system upon its environment, for short *system-environment interactions* ;

- Availability of quantitative characteristics capable of determining at any time the state of the system;

- Time evolution of the system state and its links with *entropic processes*;

- *Information content* and influence of certain random factors upon the functioning of the system.

We are mainly interested in *socioeconomic systems*; that means, in those systems describing activities – which are directly related with production, distribution and consumption of objects – aimed at fulfilling social needs in a way the process of social reproduction becomes basic to any activity of society. At the bottom of our enquiry is the visualisation of an appropriate network through which social systems draw energy and information from their environment (outside world) with the purpose of maintaining their structures, organisation and patterns of order. A key role in this undertaking is ascribed to the principle of maximum entropy.

In general, the investigation of socioeconomic systems pursues particular objectives. In many cases, an objective consists in ascertaining a *system structure* and a *system dynamics* (evolution) following which the system attains a state at which an a priori given criterion satisfies certain properties. The most common criteria refer to efficiency, stability, optimality, viability or a mixture of them. However, since arrangements of system elements and determination of its environment are arbitrary, it is of paramount importance to keep in mind the range of applicability of additional knowledge is limited to a chosen point of perspective. Further, the recognition of the approximative nature of insights and transitory character of system descriptions is an essential virtue of systems theory, no matter how detailed and accurate the study of the behaviour of an arbitrary system may be.

At this stage let us mention *systems theory* deals with the relationships between the elements of a system, with the relations between system structure and functioning, with the interaction of its various components, subsystems and environment. For this purpose systems theory applies mathematical techniques from a variety of fields like differential equations, topology, graphs, matrix analysis, Boolean and Lie algebras, etc. *Control theory* is concerned furthermore with modifying the behaviour of the system so as to achieve desired goals. Diverse arrays of problems of this sort are encountered in many fields of human endeavour. However, control theory focuses on those susceptible to precise mathematical description and seeks to understand fundamental principles for the guidance of processes taking place in the system at stake.

7.1.2   Methodological basis of model building

A natural outgrowth of recognising the relative nature of knowledge is the attempt to approach reality on the basis of model building. This leads to a storehouse of consciously codified representations of certain salient characteristics of systems under consideration. The human brain is a remarkably active model builder, it continuously captures, stores, creates, updates, rejects, destroys and recreates pictures of reality. The fact that model building bases on cognitive processes, learning, operations of a semantic and/or syntactic type, information processing and the like makes it very difficult to give a short and useful presentation of the methodology underlying model building. Thus, we limit ourselves to mention a few points.

Let us point out, a common feature of most interdisciplinary research undertakings in applied sciences is the quest for the fundamental understanding of their systems. In this sense, it has become customary to construct simple systems that replicate conceptually the actual ones in order to facilitate their understanding and analysis. Under certain conditions regarding the correspondence of system characteristics like structure, environment, state variables, functioning, information content, dynamics, random factors and so on, we may say two systems S_1 and S_2 are *similar*. Then, it is usual to call the more simple system a *model* of the other one.

Models are usually classified first according to the *analogy* in its structure, functioning and behaviour. Then, we talk about structural, functional and be-

haviorual models. Secondly, according to the *intended purpose* which motivates
labels like demonstration and experimental models. Thirdly, according to the
style of representation which brings about labels like verbal, graphical, mathematical and physical models. In this regard, we shall say we are interested in
mathematical modelling of socioeconomic systems. Modelling builds a bridge
between the actual world and mathematical theories and enables thereby techniques of simulation, optimisation, control, and so on, to come into practice.

7.2 Information, knowledge and modelling

In the foregoing section we have called attention upon some characteristics
based on which we may start thinking of a *system* as a set of interacting subjects
of any nature. The whole external world, any of its parts, any sector of a
national economy, a plant, an animal, an organism of a man, an airplane, a
car may serve as examples of a system. Since the application of mathematical
methods for the investigation of systems assumes a precise account of it, for
instance in the form of a mathematical model, we are led to the conclusion that
systems description, model building, and the design of control mechanisms are
just the system-theoretical version of human observation of reality, perception
of the world, becoming aware of interrelations of various kinds that precede
most actions in society and nature. Therefore, it shall not be surprising that
the cornerstones of model building are very much alike to those of knowledge,
cognition and learning, see [163].

7.2.1 Models as building elements of knowledge

Any model serves for the purpose of reflecting, in a simplifying manner, certain
aspects of a system and in a sense of storing the knowledge of the model designer
on this regard. Thus, models act as communication means and fulfil a sort of
linguistic function the semantic part of which is represented by the knowledge
content they convey while the syntax is reflected by the way of coding or storing
this information.

At a fairly general level, *learning* may be viewed as consisting of acquisition of knowledge on the basis of processes of communication, transmission,
recognition, association, rejection, mutation, imitation, comparison and doing,
where transmission and doing are basic. *Processes of learning* are particular
manifestations of cognitive operations underpinning human behaviour which
from the perspective of orthodoxy are dichotomised into mental and physical,
a consequence of the ill-conceived separation of mind and matter (brain); see
[190, 177]. A formalisation of model building parallels human learning as the
theses to follow forwarded by Peschel in [163] suggest.

Thesis 7.1 *Any model represents stages of a learning process the knowledge
content of which is coded in the human brain or stored in an external medium.*

It is necessary to stress the temporary and relative character of a model as well
as the requirement of being susceptible of extension and improvement. As an
intermediate result of a learning process a model shall be seen as a provisional

picture of some aspects or manifestations of reality and as a simplified substitute of it, as a more adequate and an efficient object of research conducive to
a better knowledge of objective reality.

Thesis 7.2 *A model shall be characterised by following qualitative features:*

- *The basic information, a sort of frame of key-concepts supposed to avoid
  redundancy and to enable the reconstruction of the information contained
  in the original object or system.*

- *Fundamental relationships representing a semantic network of concepts
  that reflects a whole of dependencies and connections in objective reality.*

- *A hierarchical structure drawing an analogy with an associative memory according to which information is stored in various space- and time-
  dependent interconnected levels.*

- *Capability of extension by means of widening and deepening of information which give rise to model operations like aggregation and structural
  enlargement.*

The information content of a model can be used for syntactic purpose. This
holds in particular for those models whose information content has been stored
in an external medium, e.g. think of robotics and artificial intelligence. The
use of the information content for semantic purpose is associated with the in-
tercalation of eventually formalised intermediate stages as in the human brain.
In other words, the latter requires a highly efficient process of communication,
e.g. think of paintings, a piece of music and literature writings.

7.2.2   Goal-oriented modelling

System theory builds on a variety of dialectic relationships. Let us mention as
a way of example dialectics like analysis-synthesis, perception-action, essential-
unessential, dynamic-static, deterministic-stochastic about which we shall have
later something to say. It turns out, a great majority of models have not been
designed just for the sake of improving and increasing human knowledge but
obey other intentions mostly specified in advance. The latter models cannot
be sharply distinguished from the former, however.

Thesis 7.3 *Models are often tailored to support clearly stated goals. In such
cases the setting of goals determines in a sense the goal-adequate cut of intervening dialectic relationships so that the meaning within the model at issue of
concepts like essential, static and so on are to some extent settled in advance.
Therefore, a goal-tailored model shall be understood mainly as a projection of
objective reality from a point of view chosen by the model designer.*

Let us stress the point that the construction of a goal-tailored model is not itself
an objective but only an intermediate step towards it; it reflects the right to
choose that framework appearing to the model maker as the most appropriate
vehicle for approaching reality. Therefore, in order to achieve the envisaged

goal the model maker shall commit himself to undertake the research work
necessary so that the knowledge stored in the model actually serves the goal
chosen at the outset.

It is natural to see complex systems as consisting of a variety of more simple
systems. As a consequence models of complex systems shall be designed as
composed of a set of goal oriented models everyone reflecting a specific aspect
of the complex model. This leads to view goal-tailored models as coarse models
of the original system, since any of them describes particular aspects and only
as a whole they provide a more accurate picture of it.

Thesis 7.4 *According to the purpose at hand and the tools used a model maker
may obtain a variety of diverse projections corresponding roughly to a manifold
of essential aspects characterising as a whole a complex system. However, it
may be that putting together some of these projections gives rise to contradictory
and discording features which do not seem to appear in the original.*

It may be annoying and sometimes even embarrassing having to account for
contradictions and dissonances of the type mentioned in Thesis 7.4. However,
from a dialectic point of view this cannot be considered as totally unexpected. It is known quantitative changes give rise to qualitative ones and vice
versa. Many instances of this type are well-known in quantum physics and
chemistry. In particular, in connection with socioeconomic systems changes in
the behaviour of individuals or social groups of the sort at stake have been
recognised and investigated extensively.

7.2.3   A few remarks on model representation

According to the role of being communication means and to the linguistic
function attributed as a consequence to models, it becomes necessary to make
a precise distinction of its semantic and syntactic features.

Concerning the *semantic* features of a model, we should say they comprise
all these elements conveying its knowledge content and meaning, which when
properly activated makes the model appear similar to the original. However,
in order to establish an effective communication between model designer and
its user, it is necessary the latter is able to access the associated semantics.
The *syntactic* features of the model refer to the medium on which one codes as
well as the formal tools used to represent basic information, fundamental relationships and hierarchical structures characterising the model. In an intuitive
sense, the syntax of a model may be understood as the vehicle of its semantics.

Since any model has in a sense the role of being a substitute for the original
system, it becomes natural to require the process of modelling to satisfy certain
criteria of adequacy embodied in Thesis 7.5. The process by means of which
a user tries to understand a model on the basis of her/his own inner models
generates a virtual picture of it. This virtual picture is in a way the user's own
replica or (model) interpretation of the model.

Thesis 7.5 *A model is considered to fulfil the adequacy requirement, whenever
its interpretation corresponds to a virtual picture that to great extent duplicates
precisely the original system. In particular, it should mimic those traits of*

the system that have been stressed during the process of model building. In general, this quality of being similar with the original or replicating essential charateristics needs quantitative appraisal.

This adequacy requirement is essential to our conception of model building as a learning process and lies at the heart of the dialectic foundations of the theory of knowledge. However, we have to recall models bear the character of projections of reality and as such can never reproduce the original in full.

7.3 Basic features of model design

One can never overemphasise the role models played supporting the interaction of humankind with nature aimed at transforming natural objects, at humanising natural environments and so on with the purpose of satisfying social needs. A purposeful employment of models entails a well-balanced setting of goals and a consequently goal-oriented model design. Nevertheless, this is a difficult matter.

One trouble and cause of bias we like to mention in this regard, is the subordination of goals that occurs quite often when using mathematics for modelling purposes. It seems no few researchers tempted by the advantages of dealing with tools easy to handle, seduced by the beauty of mathematical theories and methods, unwilling to give up attained familiarity with techniques already available in the own toolkit sacrifice essential system components and decisive features of modelling endeavour to the prospect of getting models amenable to favoured techniques and views.

The abuse and misuse of mathematics may be avoided by means of a system theoretical application of model building, a deep understanding of the semantic and syntactic features of a model and the acquisition of knowledge to such an extent as to ensure access to the semantics of the model at hand.

7.3.1 The qualitative features

In the application of modelling methods to technical, economic, social, biological investigations behavioural models play an outstanding role. *Behavioural models* are, roughly speaking, those models focusing essentially on behavioural aspects of the system they attempt to replicate.

Thesis 7.6 *Under given conditions behavioural models should reflect the behaviour of actual systems in a way such as to establish adequately a sort of correspondence between cause and effect.*

It is usual to formalise this relationship between *cause* and *effect* thinking of *external actions* on the system and *responses* of the system to its environment. The variables describing the external actions are called *inputs* which we denote by Y_i and those describing the responses or actions of the system on its environment are called *outputs* denoted in turn by X_i. Besides this set of variables one may include others characterising actions of various components of the system on each other.

The relevant information on the current correspondence between the causes Y_i and the effects X_i is coded in variables characterising the state of the system at any time and these are called *state variables* of the system. Evolutionary and adaptive elements of a system in the face of change in its environment bring about responses aimed at attaining characteristics – like states of functioning – which ensure system reproduction. These essential features of system behaviour rest on mechanisms of various types – structure preserving, self-organising, homeostatic, cognitive, entropic, anti-entropic among others – from which result in general an average trend within acceptable limits of viability and an enormous number of random deviations.

In social and living systems energy and information flows are of paramount importance determining patterns of behaviour, functionings, organisations and states of order. Since the capacity of self-reproduction of a system depends critically on its ability to absorb energy and order from its environment, it is at this level where the introduction of the principle of maximum entropy in social discourse acquires crucial significance. In particular, because the entropy of the environment will decrease by sucking energy and order from it, so that its potentials for changes decrease which may lead to irreparable damages.

Thesis 7.7 *The design of behavioural models is shaped to a great extent by the need to take into account basic tasks of systems theory like system estimation, prediction and realisation. One expects thereby the information gathered on the system behaviour enables the prediction of its behaviour so that a purposeful guidance and control of the future performance of system operations become effective even in the face of inputs not previously considered.*

An essential feature of any learning process is a continuously alternating change of focus from the detail to the totality. This corresponds to switching from a microscopic to a macroscopic view which relates deeply to the dialectic relationship between analysis and synthesis. This observation is associated with following thesis.

Thesis 7.8 *The representation of a system rests upon a deep understanding and detailed description of its elements or items. However, the system itself turns out to be more than the mere sum of its items due to mutual interdependence and interactions among its elements which are to a great extent responsible for change of qualitative and quantitative nature that one perceives alternating focus from the micro to macro level.*

On the description of the items of the system relies its basic information and its conceptualisation so that it is safe to assert that system details are coded in the form of concepts. Yet, concepts represent a wealth of details describing elements and have thus the character of class.

Next, we shall turn to quantitative features associated with the notion of concept. However, we shall stress there cannot be a sharp distinction between qualitative and quantitative properties underpinning a concept due to the dialectic principle according to which there is no quality without quantity and vice versa.

7.3.2   Quantitative features

At this stage, we would like to come back to the requirement on the model user of being able to access the semantics attached to the model. This would mean the model user is expected to decode the information behind any concept and to reconstruct the wealth of details condensed in it with the purpose of recognising the totality. On the other hand, in the process of building a model one attempts to keep redundancy at a minimum and to bring about concepts carrying only class-specific information, and one appeals to the user's ability to interpret the model properly. It goes without saying that this modelling rationale bears some fragility.

Thesis 7.9 *The logical basis of modelling rests upon the following features which clearly point to potential difficulties and sources of misinterpretations.*

- *The whole of a system is to be disentangled in basic information and fundamental relationships to such a degree that its knowledge content can be suitably coded in an appropriate medium.*

- *The model representation of the knowledge content of a system has to be so that when required it enables a reconstruction of the system that reflects virtually the totality.*

- *The role of a model as a projection of the objective reality and of concepts as suitably coded system information together with the notion of virtual reconstructionability of the totality of a system give rise to the operation of model interpolation.*

Dealing with systems in general it is useful to build up on analogies with classical physics. However, a word of caution is appropriate in particular regarding socioeconomic systems. Because in such systems there are quite many properties that cannot be satisfactorily grasped, specially in quantitative terms.

Any theory is a process in society and as such the set of concepts building it undergoes continuous changes. On the other hand, how the theory operates, how and in which direction it develops depends to great extent on what its practitioners understand the theory to be. This leads to a problem of epistemological nature known as *overdetermination*, see in this connection [175].

Thesis 7.10 *A system variable is the model description of a particular phenomenon which emerges as an attribute of an itemised property representing the class attached to the conceptualisation of a more complex phenomenon. As a consequence, it is necessary to stress that a system variable describes a phenomenon that may occur with varying intensities and it should be amenable to observation, directly or indirectly, and to quantitative appraisal.*

The fact that the conceptualisation of a system undergoes continuous changes becomes crucial when one has to face the problem of evaluating quantitatively a system variable. This issue relates on the one hand to the dialectic relation between rest and motion, and on the other to the eventually qualitative changes induced in the behaviour of a system element when put in isolation with the purpose of quantitative assessment.

A problem of this sort is known in quantum mechanics as *Heisenberg's Principle of Uncertainty*. At a fairly general level, this principle says that when energy and information are part of a system one cannot receive energy/information from its environment without changing the environment and the system as well. It seems that the mere act of inquiring (measuring/observing) changes the potential answer, giving rise to interesting questions concerning interactions (information/knowledge).

Thesis 7.11 *The quantitative assessment of the variables of a system relies on the fulfilment of the conditions to follow.*

- *In order to evaluate quantitatively a system variable it is necessary that there is a value it can assume and that this remains constant as long as the assessment process lasts.*

- *There exists a system potential and an associated general principle based on which one can split the study of the behavioural aspects of the system into static and dynamic analysis that together build a complete picture of the system including quantitative and qualitative features as well.*

A well-known principle of the kind just mentioned is *Hamilton's Principle of Least Action*; see [5, 91] as a way of example. The dialectic unity of motion and rest as well as necessity and randomness plays here a decisive role.

7.4 State representation and systems control

It is usual to represent the state of a system on the basis of n state variables $X^{(1)}, X^{(2)}, \ldots, X^{(n)}$ condensing in the sense of Thesis 7.6 relevant information on the underlying type of causation. The state of the system is then given by the vector $X = (X^{(1)}, X^{(2)}, \ldots, X^{(n)})'$ by means of which one expresses the system dynamics when acting as a process $X : [0, T] \longrightarrow \mathbb{R}^n$. This system dynamics certainly includes also its statics when it assigns to any moment of time $t \in [0, T] \subseteq \mathbb{R}$ a state value $X_t \in \mathbb{R}^n$ summarising in this way crucial properties of the dynamic behaviour of the system. In this sense the state of a system describes phenomena undergoing change in the course of time and a system is viewed as an object in which processes take place.

The state representation needs to reflect also specific patterns of structure and order as exhibited by the system. As a part of the system description we shall consider a set of parameters $\alpha^{(1)}, \ldots, \alpha^{(k)}$ and they build the vector $\alpha = (\alpha^{(1)}, \ldots, \alpha^{(k)})$. The vector α acts as the process $\alpha : [0, T] \longrightarrow A$, where A is termed the control set and reflects the constraints to which the control process is subject. In general $A \subseteq \mathbb{R}^k$. The control process α is taken from a prescribed class of functions denoted by $\mathcal{A}$ which is commonly assumed to contain the piecewise continuous functions.

7.4.1 State equation and the problem of control

It is convenient to think of state variables $X^{(i)}$ as described by differential equations. To see this let us introduce a drift function b, a rate of reward f

and terminal payoff g characterised by the assumptions to follow, where $\bar{b}$, $\bar{f}$ and $\bar{g}$ are constants.

A 7.1 *Assumption on the drift, reward and payoff functions*

- Let $b(x,\alpha) : \mathbb{R}^n \times \mathbb{R}^k \longrightarrow \mathbb{R}^n$ be a continuously differentiable function such that $|b(x,\alpha)| \leq \bar{b}(1 + |x| + |\alpha|)$ holds and b_x is bounded.

- Let $f(x,\alpha) : \mathbb{R}^n \times \mathbb{R}^k \longrightarrow \mathbb{R}$ be a continuously differentiable function such that $|f(x,\alpha)| \leq \bar{f}(1 + |x|^2 + |\alpha|^2)$, $|f_x| \leq \bar{f}(1 + |x| + |\alpha|)$ and $|f_v| \leq (1 + |x| + |\alpha|)$ hold.

- Let $g(x) : \mathbb{R}^n \longrightarrow \mathbb{R}$ be a continuously differentiable function such that $|g(x)| \leq \bar{g}(1 + |x|^2)$ and $|g_x| \leq \bar{g}(1 + |x|)$ hold.

The expressions f^α and b^α are shorthand for $f(x,\alpha)$ and $b(x,\alpha)$. An *admissible control* shall be a function $\alpha \in L^2(0,T;\mathbb{R}^k)$ such that $\alpha_t \in A_{\text{ad}}$ a.e., where A_{ad} is a non-empty closed convex subset of $\mathbb{R}^k$. Let $\mathcal{A}_{\text{ad}}$ denote the family of admissible controls.

Lemma 7.1 *The assumption A 7.1 holds.*

(1) *Let $X = (X_t)_{t \geq 0}$ be the (deterministic) process associated with an arbitrary control $\alpha \in \mathcal{A}$ and assume that for any $t \in (0,T]$ the corresponding state $X_t = X_t(\alpha)$ is given by*

$$dX_t = b(X_t, \alpha_t), \tag{7.1}$$

and $X_0 = x_0$ with $x_0 \in \mathbb{R}^n$ given. Then the process X is well-defined and belongs to $H^1(0,T;\mathbb{R}^n)$.

(2) *Let $J = (J_t)_{t \geq 0}$ be the process associated with X, where for any $t \in [0,T)$ the state $J_t = J_{T_t}(\alpha)$ is defined by*

$$dJ_t = -f(X_t, \alpha_t)dt$$

and $J_T = g(X_T)$. Let us set $J(\alpha) = J_0 (= J_T(\alpha))$. Then the process J is well-defined and for any admissible control $\alpha \in \mathcal{A}$ the functional $J(\alpha)$ given by

$$J(\alpha) = \int_0^T f(X_t, \alpha_t)dt + g(X_T)$$

is well defined.

The mapping $J(\alpha)$ introduced above is known as the *performance functional* of the system driven by (7.1) which is in turn known as the *state equation*. A natural question is that of finding an admissible control for which the performance functional may attain properties to be viewed as optimal in a certain sense. Provided one optimal control exists, we shall denote it by β and by Y the corresponding optimal state process. The state process $X = (X_t)_{0 \leq t \leq T}$ associated with an admissible control α is an absolutely continuous function defined on $[0, T]$ with range in $\mathbb{R}^n$ which we shall call sometimes the trajectory corresponding to $\alpha \in \mathcal{A}_{\mathrm{ad}}$.

A central concern of the theory of optimal control is the choice of an appropriate functional framework and a family of admissible controls. The purpose of this choice is a qualitative and quantitative characterisation of the optimal control β and the performance functional $J(\beta)$, according to various issues resulting from the type of questions under considerations.

In general one may require that at the initial time t_0 the initial state x_0 belongs to a pre-assigned set such that $(t_0, x_0) \in \mathcal{T}_0$, where $\mathcal{T}_0$ is in the (t, x)-space. In addition one may require that at the terminal time t_T the terminal state X_T is such that (t_T, X_T) is the first entrance point of a pre-assigned set $\mathcal{T}_T$ in the (t, x)-space. However, for the sake of simplicity we set $t_0 = 0$ and $t_T = T$ with $T > 0$. The sets $\mathcal{T}_0$ and $\mathcal{T}_T$ are called the *initial* and *terminal sets* for the problem considered. In this case one says that a control α transfers a system driven by (7.1) from $x_0 \in \mathcal{T}_0$ to $x_T \in \mathcal{T}_T$ abusing the notation a little.

Let us denote the corresponding trajectory by $\phi(t_T, t_0)$ or $\phi(t_T, t_0, x_0)$ whenever it is necessary to emphasise the initial state. Let us assume further that going from x_0 to x_T the system goes through the state x^* at time t^*. Then the trajectory satisfies the *composition rule*

$$\phi(t_T, t_0, x_0) = \phi(t, t^*, \phi(t^*, t_0, x_0)),$$

which means that the trajectory going from (t_0, x_0) to (t_T, x_T) is the composition of the trajectory from (t_0, x_0) to (t^*, x^*) and the trajectory from (t^*, x^*) to (t_T, x_T).

The set of states the trajectory may attain is called the *state space* for the system driven by a given state equation. The notation $\mathcal{A}_{\mathrm{ad}}$ is short-hand for the requirement that going from the initial to the terminal set the control α and the corresponding trajectory ϕ satisfy a certain set of constraints. For the purpose of emphasising eventual links between the initial and terminal set one introduces the set $\mathcal{T}_{0T} = \mathcal{T}_0 \times \mathcal{T}_T$ which focuses on trajectories ϕ starting from $(t_0, x_0) \in \mathcal{T}_0$ and terminating at $(t_T, x_T) \in \mathcal{T}_T$. Thus the set $\mathcal{T}_{0T}$ defines the end data of a control problem.

Note that to a given admissible control $\alpha \in \mathcal{A}_{\mathrm{ad}}$ there may correspond more than one admissible trajectory ϕ as a result of different choices of admissible end data. Now we are prepaired to state the following control problem.

Optimising problem 7.1 *Let $\mathcal{A}_{\mathrm{ad}}$ denote the family of admissible controls and let $\mathcal{A}_{\mathrm{ad}}$ be non-empty. Further let us assume that the assumptions in A 7.1 hold and that $J(\alpha)$ is defined as in Lemma 7.1 with $\alpha \in \mathcal{A}_{\mathrm{ad}}$ and end data in $\mathcal{T}_{0T}$. Let $\mathcal{A}^\epsilon$ be a non-empty subset of $\mathcal{A}_{\mathrm{ad}}$. Find a control $\beta \in \mathcal{A}^\epsilon$ and a state process Y satisfying (7.1) and with end data in $\mathcal{T}_{0T}$ such that*

$$J(\alpha) \leq J(\beta) \quad \text{for all } \alpha \in \mathcal{A}^\epsilon$$

holds.

The precise formulation of the above control problem is rather lengthy and beyond the scope of our present undertaking. A pair (β, Y) that solves the above control problem is often called an *optimal control pair*.

7.4.2 Necessary conditions for optimality

In control theory a great deal of effort is directed to finding necessary conditions for optimality. To this end let us begin with an extremum condition. Let $\alpha \in L^2(0,T,\mathbb{R}^k)$ be any arbitrary admissible control, β be the optimal control as before and set $\beta^\epsilon = \beta + \epsilon\alpha$ with β_t^ϵ given by

$$\beta_t^\epsilon = \beta_t + \epsilon\alpha_t,$$

for any $t \in [0,T]$. Further, we consider the *linearised equation* about the optimal trajectory Y or equation of first variation, with $\delta x = z$, defined by

$$\frac{dz_t}{dt} = b_x(Y,\beta)z_t + b_\alpha(Y,\beta)\alpha_t, \quad z_0 = 0$$

which is well defined in $H^1(0,T;\mathbb{R}^n)$. The coefficients above are short-hand for

$$\frac{\partial b^{(i)}(Y_t,\beta_t)}{\partial x_t^{(j)}}, \quad \text{and} \quad \frac{\partial b^{(i)}(y_t,\beta_t)}{\partial \alpha_t^{(l)}},$$

where $i,j = 1,2,\ldots,n$, and $l = 1,2,\ldots,k$. Then the aim is to calculate the Gâteaux derivative of $J(\beta^\epsilon)$ with respect to ϵ for $\epsilon = 0$. The state process corresponding to β^ϵ shall be denoted by Y^ϵ which gives rise to the approximating processes

$$\tilde{Y}^\epsilon = \frac{Y^\epsilon - Y}{\epsilon} - Z \quad \text{or} \quad Y^\epsilon - Y = \epsilon(Z + \tilde{Y}^\epsilon).$$

Lemma 7.2 *Let (β,Y) be the optimal control pair. Under the assumption A 7.1 the following formula holds.*

$$\frac{d}{d\epsilon}J(\beta^\epsilon)|_{\epsilon=0} = \int_0^T (f_x(y_t,\beta_t) \cdot z_t + f_\alpha(y_t,\beta_t)\alpha_t)\,dt + g_x(y_T)z_T. \quad (7.2)$$

Proof: The state process $\tilde{Y}^\epsilon$ satisfies the condition $\tilde{Y}_0^\epsilon = 0$ and the equation

$$\begin{aligned}
\frac{dY_t^\epsilon}{dt} &= \frac{1}{\epsilon}\left(b(Y_t^\epsilon,\beta_t^\epsilon + \epsilon\alpha_t) - b(y_t,\beta_t)\right) - (b_x(y_t,\beta_t)z_t + b_\alpha(y_t,\beta_t)\alpha_t) \\
&= \int_0^T b_x(y_t + \lambda(y_t^\epsilon - y_t),\beta_t + \lambda\epsilon\alpha_t)\tilde{y}_t^\epsilon\,d\lambda + \\
&\quad + \int_0^T [b_x(y_t + \lambda(y_t^\epsilon - y_t),\beta_t + \lambda\epsilon\alpha_t) - b_x(y_t,\beta_t)]\,z_t\,d\lambda + \\
&\quad + \int_0^T [b_x(y_t + \lambda(y_t^\epsilon - y_t),\beta_t + \lambda\epsilon\alpha_t) - b_\alpha(y_t,\beta_t)]\,\alpha_t\,d\lambda.
\end{aligned}$$

Taking into account that $y_t^\epsilon \to y_t$ for all t and that b_x and b_α are continuous it turns out that as $\epsilon \to 0$ the sum of the last two integrals converges in $L^2(0,T,\mathbb{R}^n)$ towards 0, which entails that $\tilde{y}^\epsilon \to 0$ in $H^1(0,T,\mathbb{R}^n)$.

Applying the same arguments to the expression

$$J(\beta^\epsilon) - J(\beta) = \int_0^T [f(y_t^\epsilon,\beta_t^\epsilon + \epsilon\alpha_t) - f(y_t,\beta_t)\alpha_t]\,dt + g(y_T^\epsilon) - g(y_T),$$

we obtain

$$J(\beta^\epsilon) - J(\beta) - \epsilon\left\{\int_0^T [f_x(y_t,\beta_t)z_t + f_\alpha(y_t,\beta_t)\alpha_t]\,dt + g_x(y_T)z_T\right\} = o(\epsilon),$$

from which the lemma follows. ∎

The fact that in formula (7.2) the control α does not appear explicitly in each of its terms poses certain disadvantages which one tries to remedy by introducing adjoint variables. These are characterised by the so-called adjoint state equation the key property of which is propagating the solution of the state equation (7.1) backwards in time. To this end let us consider $\psi \in L^2(0,T;\mathbb{R}^n)$ and introduce the process ζ defined by

$$\frac{d\zeta_t}{dt} = b_x(y_t,\beta_t)\zeta_t + \psi_t,$$

where $\zeta_0 = 0$. Then the map

$$\psi \longrightarrow \int_0^T f_x(y_t,\beta_t)\cdot\zeta_t\,dt + g_x(y_T)\cdot\zeta_T$$

is linear and continuous $L^2(0,T;\mathbb{R}^n) \longrightarrow \mathbb{R}$ so that there is a unique process p in $L^2(0,T;\mathbb{R}^n)$ such that

$$\int_0^T f_x(y_t,\beta_t)\cdot\zeta_t\,dt + g_x(y_T)\cdot\zeta_T = \int_0^T p_t\cdot\psi_t\,dt,$$

for all $\psi \in L^2(0,T;\mathbb{R}^n)$. The process p is called the *adjoint state process* and enables us to rewrite (7.2) in the desired form

$$\frac{d}{d\epsilon}J(\beta^\epsilon)|_{\epsilon=0} = \int_0^T [f_\alpha(y_t,\beta_t)\cdot\alpha_t + p_t\cdot b_x(y_t,\beta_t)\alpha_t]\,dt$$
$$= \int_0^T H_\alpha(y_t,\beta_t,p_t)\cdot\alpha_t\,dt,$$

where we have set
$$H(x,\alpha,p) = f(x,\alpha) + p\cdot b(x,\alpha).$$

The function H is often called the *Hamiltonian* in analogy with a quantity appearing in classical mechanics. However, this should be reserved for the case where the process α becomes the optimal control.

Chapter 8

Semimartingale calculus

The preceding selection of arguments from Kant and nineteenth-century sources reveals at least three distinct sorts of intellectual reasons for the slow rise of probabilism: (1) the fusion of an ancient epistemological tradition concerning the nature of knowledge with more recent successes of mechanical science; (2) the inscrutable and profuse depths of causal thinking; and (3) the difficulty of understanding the nature of statistical laws. Reason (1) helped to entrench deterministic beliefs. ... Reason (2) is a notorious and persistent source of philosophical confusion. ...

Reason (3) for the slow rise of probabilism can obviously only be overcome by transforming deeply rooted views about the nature of cause and law, indeed of scientific knowledge in general. Provided that a theoretical explanation of statistical statements and methods is sought at all, the required transformation of scientific thinking appears to involve a double ontological price; two pieces of traditional science will have to be abandoned: (a) reductionism and (b) determinism. The rejection of (a) is suggested by the fact that the properties that can explain the regular behaviour of a statistical aggregate cannot meaningfully be ascribed to the single parts—they characterize the aggregate as a whole, or at least sizable portions of it. This feature is at variance with the analytical spirit of previous science according to which the behavior of any system ought to be explained in terms of its composition and the behavior of its elementary parts. Theoretical pluralism had not thus far been an official part of modern science. The rejection of (b) is suggested as a consequence. If the apparent contingency of a single instance is explained by placing it within a whole whose relevant properties *cannot* be reduced to properties of the parts, even the hypothetical assumption of a complete determination of the instance in question will appear doubtful, because incompatible with a novel explanatory scheme, indeed the only scheme that eventually does the work of explaining the phenomena.

Lorenz Krüger
(see [127], p. 83)

8.1 Elements of semimartingale theory

The notion of stochastic integral has entered a wide variety of fields and has become a versatile working tool in many applied areas of investigation. The Brownian motion and its mathematical description, the Wiener process, are indeed still widely used as a starting-point for the conceptualisation of stochastic models, however the increasing interest to account for discontinuity, dynamic features of systems and more general evolutions direct attention to the powerful theory of semimartingales.

We shall present a short introduction to semimartingales and stochastic calculus with two purposes in mind: First to make easily available notions we have used in the present essays and second to improve the ability of non-probabilistically trained readers to interpret and understand real-world phenomena dominated by the presence of uncertainty.

8.1.1 The probabilistic setting and basic terminology

The usual setup for the general theory of processes is a complete probability space $(\Omega, \mathcal{F}, \mathbf{P})$ endowed with a filtration $(\mathcal{F}_t)_{t \geq 0}$, i.e. an increasing family of sub-σ-fields of $\mathcal{F}$, which is assumed to satisfy the usual conditions of right-continuity and inclusion of all $\mathbf{P}$-null sets. The notion of right-continuity, written as $\mathcal{F}_t = \mathcal{F}_{t+}$, where $\mathcal{F}_{t+} = \bigcap_{\epsilon > 0} \mathcal{F}_{t+\epsilon}$, says that $t \longrightarrow \mathcal{F}_t$ is increasing and right-continuous. The arrangement $(\Omega, \mathcal{F}, (\mathcal{F}_t)_{t \in \mathbb{T}}, \mathbf{P})$ is called a *filtered probability space*, where $\mathbb{T}$ is an interval of $\mathbb{R}_+$. For simplicity of notation we shall often write $\mathbf{F}$ for the filtration $(\mathcal{F}_t)_{t \in \mathbb{T}}$ and refer to the filtered probability space $(\Omega, \mathcal{F}, \mathbf{F}, \mathbf{P})$ as to the *probability basis*.

A *stochastic process* is a function $X : \mathbb{R}_+ \times \Omega \longrightarrow \mathbb{R}$ such that for each $t \geq 0$, $X_t = X(t, \cdot)$ is a measurable function from $(\Omega, \mathcal{F})$ into $(\mathbb{R}, \mathcal{B})$, where $\mathcal{B}$ is the Borel σ-field. The functions $t \longrightarrow X_t(\omega)$ mapping $\mathbb{R}_+$ into $\mathbb{R}$ are called the *sample paths* of the stochastic process X. The process X is said to be *adapted* to $(\mathcal{F}_t)_{t \geq 0}$ if for each t, X_t is a measurable function from $(\Omega, \mathcal{F}_t)$ into $(\mathbb{R}, \mathcal{B})$. Further, X is said to be *càdlàg* (after the French: continu à droite avec des limites à gauche) if its sample paths $\{t \longrightarrow X_t(\omega), \omega \in \Omega\}$ are right-continuous with finite left limits. The *predictible σ-field* $\mathcal{P}$ on $\mathbb{R}_+ \times \Omega$ is the σ-field generated by the left continuous, adapted stochastic processes. Alternatively, we may assume the predictable σ-field is generated by the sets of the form $(s, t] \times F_s$, for $F_s \in \mathcal{F}_s$, $0 \leq s \leq t < \infty$, together with the sets of the form $\{0\} \times F_0$, for $F_0 \in \mathcal{F}_0$. A process is said to be predictible if it is measurable with respect to this σ-field.

For notational simplicity most of the notions involved are given for real-valued processes; thus, whenever a process takes values in $\mathbb{R}^n$ it should be understood the definitions apply component by component unless stated otherwise.

A stopping time relative to the filtration $(\mathcal{F}_t)_{t \geq 0}$ is a map τ on Ω with values in $[0, \infty]$, such that for every t the event $\{\tau \leq t\} \in \mathcal{F}_t$. This is equivalent to saying the random variable $\tau : \Omega \longrightarrow [0, \infty]$ is called a *stopping time* if the process $X_t = \mathbb{1}_{(0,\tau]}(t)$ is adapted. If the process just defined is predictible, the

τ is said to be a *predictible time*. With a stopping time τ and a process X, we can associate the *stopped process* $X_|$ defined for any t by $X_{|t}(\omega) = X_{t \wedge \tau}(\omega)$; with an arbitrary sequence of stopping times (τ_n) and a process X, we can associate the sequence of stopped processes $(X_|^n)$ where for any fixed n and any $t \in \mathbb{T}$ we have $X_{|t}^n = X_{t \wedge \tau_n}$. A *change of time* or time-change is a positive, continuous, increasing process A, not necessarily adapted, but such that each random variable A_t is a stopping time. Under the action of a time-change a new filtration $(\widetilde{\mathcal{F}}_t)_{t \geq 0}$ is defined by setting $\mathcal{F}_{A_t} = \widetilde{\mathcal{F}}_t$. Two important examples of time changes are $A_t = \tau \wedge t$ and $A_t = \tau + t$, where τ is a stopping time. The first gives rise to the stopped process $X_|$ which is adapted to the new and initial filtration. The second transforms X into $\widetilde{X}$ with $\widetilde{X}_t = X_{\tau + t}$ which means the process $\widetilde{X}$ starts at τ anew and thus forgets everything lagging behind τ. If A is an open set in $\mathcal{B}$ and X is a stochastic process in Ω, viewed as the space of continuous paths from $\mathbb{R}_+$ to A, then the time τ_A defined by $\tau_A = \inf\{t > 0 : X_t \in A\}$ is a stopping time known as the *hitting time of A for X*.

A real-valued, adapted process $X = (X_t)_{t \in \mathbb{T}}$ is called a *supermartingale* with respect to $(\mathcal{F}_t)_{t \in \mathbb{T}}$ if

(1) For each $t \in \mathbb{T}$ one has $X_t \in L^1(\Omega, \mathcal{F}, \mathbf{P})$, i.e. $\mathbf{E}\{|X_t|\} < \infty$;

(2) For $s, t \in \mathbb{T}$ such that $s \leq t$, it follows $\mathbf{E}\{X_t | \mathcal{F}_s\} \leq X_s$.

A process X for which $-X$ is a *supermartingale* is called a submartingale, and a process which is both a super- and submartingale is called a *martingale*. More precisely, a martingale is an adapted family of integrable random variables such that

$$\int_A X_s \, d\mathbf{P} = \int_A X_t \, d\mathbf{P},$$

for every pair s, t with $s < t$ and $A \in \mathcal{F}_s$. A continuous *local martingale* is a continuous adapted process X such that for any fixed n, the process $X_|^n$ with $X_|^n \mathbf{1}_{\{\tau_n > 0\}}$ and $t \in \mathbb{T}$ is a martingale, where τ_n is the stopping time given by $\inf\{t : |X_t| \geq n\}$. A martingale X is said to be *closed* by a random variable Y if $\mathbf{E}\{|Y|\} < \infty$ and $X_t = \mathbf{E}\{Y | \mathcal{F}_t\}$, where $t \in \mathbb{T}$.

Two stochastic processes X and Y are *modifications* if $X_t = Y_t$ a.s., for each t. If X and Y are modifications there exists a null set N_t such that $X_t(\omega) = Y_t(\omega)$ whenever $\omega \notin N_t$; the null set N_t depends on t. Two processes X and Y are *indistinguishable* if a.s., for all t, $X_t = Y_t$, i.e. there exists a null set N independent of t such that for $\omega \notin N$ and for all t one has $X_t(\omega) = Y_t(\omega)$. An additional remark concerns exceptional sets with regard to certain properties of sample paths of processes, which are defined on $\mathbb{T} \times \Omega$ while the probability measure $\mathbf{P}$ is defined only on Ω. To this end one calls a random set $A \subset \mathbb{T} \times \Omega$ such that its projection on Ω has a probability measure 0, an *evanescent* set. In this sense for two indistinguishable random processes X and Y, the random set $\{(\omega, t) : X_t(\omega) \neq X_t(\omega)\}$ is an evanescent set.

Theorem 8.1 *Let X be a supermartingale. The function $t \longrightarrow \mathbf{E}\{X_t\}$ is right-continuous if and only if there exists a unique modification Y of X and Y is càdlàg.*

Proof: See [176], pp. 61–62.   ∎

All martingales have right-continuous modifications. The set of martingales
with respect to a given filtration is a vector space; if the stochastic process X
is a martingale such that for every t and for $p \geq 1$ the condition $\mathsf{E}\{|X_t|^p\} < \infty$
holds, then owing to Jensen's inequality the process $|X|^p$ turns out to be a
submartingale. An important remark is that if Y is an integrable random
variable and $(\mathcal{F}_t)_{t \in \mathbb{T}}$ a given filtration, by choosing for each t one random
variable from the equivalence class $\mathsf{E}\{Y|\mathcal{F}_t\}$, one defines a martingale $(Y_t)_{t \in \mathbb{T}}$
with $Y_t \in \mathsf{E}\{Y|\mathcal{F}_t\}$.

Any stochastic process X is adapted to its *natural* filtration consisting of
σ-fields $\mathcal{F}_t^0 = \sigma(X_s, s \leq t)$ and $(\mathcal{F}_t^0)_{t \in \mathbb{T}}$ is the minimal filtration with respect
to which X is adapted. Thus to say that X is adapted to $(\mathcal{F}_t)_{t \in \mathbb{T}}$ amounts
to say that $\mathcal{F}_t^0 \subset \mathcal{F}_t$ for each t, note natural filtrations are not assumed to
contain all the P-null sets of $\mathcal{F}$. It is the presence of a filtration which gives the
parameter t the nature of time and in this sense saying X is adapted to $(\mathcal{F}_t)_{t \in \mathbb{T}}$
means that it is a functional on its history, i.e. the set of possible "pasts", up
to t; see [176], pp. 40. A stochastic process X is *progressively measurable* or
simply progressive with respect to the filtration $(\mathcal{F}_t)_{t \in \mathbb{T}}$, if for every t the map
$(s, \omega) \longrightarrow X_s(\omega)$ from $[0, t] \times \Omega$ into $(\mathbb{R}, \mathcal{B})$ is $\mathcal{B}([0, t]) \otimes \mathcal{F}_t$-measurable. A
subset A of $\mathbb{R}_+ \times \Omega$ is *progressive* if the process $X = \mathbf{1}_A$ is progressive.

A process X is increasing (resp. of finite variation) if it is adapted and
the paths $t \longrightarrow X_t(\omega)$ are finite, right-continuous and nondecreasing (resp. of
finite variation on each compact interval of $\mathbb{R}_+$) for almost every ω. A $\mathcal{B} \otimes \mathcal{F}$-
measurable process A with values in $\mathbb{R}_+$ is said to be an *increasing process* if
almost every sample path $t \longrightarrow A_t(\omega)$ is right continuous and nondecreasing.
Since increasing functions have left limits, an increasing process is thus càdlàg.
Let $\mathbb{W}^+$ denote the class of increasing processes. Let us consider an increasing
process $A \in \mathbb{W}^+$ and fix an ω such that $t \longrightarrow A_t(\omega)$ is right-continuous and
nondecreasing. This path induces a measure on $\mathbb{R}_+$ denoted for simplicity by
$dA_s(\omega)$ for $s \leq t$ and each $t > 0$. Then the process I, with $I_t(\omega) = \int_0^t H_s dA_s(\omega)$,
is well defined for a bounded Borel function H on $\mathbb{R}_+$. For $H_s = H(s, \omega)$
bounded and jointly measurable, the process I – defined ω-by-ω – is then
right-continuous in t and jointly measurable; see [115] pp. 27–28.

Further let $\mathbb{W} = \mathbb{W}^+ \ominus \mathbb{W}^+$ denote the class of càdlàg processes for which
almost every sample path is of finite variation on each compact subset of $[0, \infty)$.
Indeed, a process is in $\mathbb{W}$ if and only if it is the difference of two members of
$\mathbb{W}^+$. The members of $\mathbb{W}$ are called *processes of bounded variation*. Note that
if $A \in \mathbb{W}^+$ it admits a *terminal variable* A_∞ which takes its values in $\overline{\mathbb{R}}_+$, i.e.
it converges a.s. to a limit A_∞ as $t \uparrow \infty$. If $A \in \mathbb{W}$ then $D \in \mathbb{W}^+$, where
$D_t(\omega) = \int_0^t |dA_s(\omega)|$ is the *total variation* of A. The process $D \in \mathbb{W}^+$ is unique
and so that for a.e. ω, the measure $dD_t(\omega)$ on $\mathbb{R}_+$ is the absolute value of the
signed measure $dA_t(\omega)$.

Let $\mathbb{V}^+$ be the class of real-valued processes A which are càdlàg and adapted,
so that each path $t \longrightarrow A_t(\omega)$ is nondecreasing. Further, let $\mathbb{V} = \mathbb{V}^+ \ominus \mathbb{V}^+$
denote the class of all real-valued processes which are càdlàg and adapted, so
that almost every sample path is of *finite variation* over each finite interval

$[0, t]$. The set of integrable increasing processes shall be denoted by $\mathbb{A}^+$, i.e. $A \in \mathbb{A}^+$ if and only if $A \in \mathbb{V}^+$ and $\mathsf{E}\{A_\infty\} < \infty$. The set $\mathbb{A} = \mathbb{A}^+ \ominus \mathbb{A}^+$ denotes all the processes of *integrable variation*, i.e. processes $A \in \mathbb{V}$ such that $\mathsf{E}\{D_\infty\} = \mathsf{E}\{\int_0^\infty |dA_s|\} < \infty$.

Finally, let us recall that if $\mathbb{C}$ denotes a class of processes, then $\mathbb{C}_{loc}$ stands for the class of processes obtained by *localisation*, i.e. $X \in \mathbb{C}_{loc}$ if there exists an increasing sequence of stopping times τ_n depending on X which generates for any n a stopped process $X^n_\tau \in \mathbb{C}$. The sequence (τ_n) is called the *localising sequence* of X relative to $\mathbb{C}$. By the way, a class $\mathbb{C}$ of processes is said to be *stable under stopping* if for any stopping time τ the stopped process $X_\tau \in \mathbb{C}$.

8.1.2   Basics on stochastic integration

To the end of getting a fairly general notion of differential applicable to a wide class of stochastic processes, one resorts to integrals which use semimartingales as integrators, predictible processes as integrands and which, as result of the integration, give a random variable defined a.e. A process H is said to be *simple predictible* if H has for $(t, \omega) \in \mathbb{R}_+ \times \Omega$ a representation of the form

$$H_t(\omega) = H_0(\omega)\mathbb{1}_{\{0\}}(t) + \sum_{i=1}^n H_i(\omega)\mathbb{1}_{(\tau_i, \tau_{i+1}]}(t),$$

where $0 = \tau_0 \leq \tau_1 \leq \cdots \leq \tau_{n+1} < \infty$ is a finite sequence of stopping times, $H_i \in \mathcal{F}_{\tau_i}$ with $|H_i| < \infty$ a.s. and $0 \leq i \leq n$. The collection of simple predictible processes is denoted by $\mathbb{S}$.

Let $\mathbb{S}_u$ denote $\mathbb{S}$ endowed with the topology of uniform convergence in (t, ω) and $\mathbb{L}^0$ denote the space of finite-valued random variables topologised by convergence in probability. Let X be a stochastic process, then we introduce the operator I_X on the basis of a linear mapping from $\mathbb{S}$ into $\mathbb{L}^0$ given by

$$I_X(H) = H_0 X_0 + \sum_{i=1}^n H_i(X_{t \wedge \tau_{i+1}} - X_{t \wedge \tau_i}),$$

where $H \in \mathbb{S}$. The process X above is called an *integrator* so that the operator I_X can be written $I_X(H) = \int H_s dX_s$. The operator I_X is linear and satisfies a weak form of the bounded convergence theorem, according to which uniform convergence of processes H^n to H implies only the convergence in probability of $I_X(H^n)$ to $I_X(H)$.

The operator I_X maps processes into random variables. Further, since this operator is based on a path by path definition for step functions $H.(\omega)$, it does not depend on the choice of the representation of H in $\mathbb{S}$. The convergence in probability has the advantage of remaining unaltered under an equivalent change of measure.

A process X is said to be a *total semimartingale* if X is càdlàg, adapted, and $I_X : \mathbb{S}_u \longrightarrow \mathbb{L}^0$ is continuous. A process X is called a *semimartingale* if, for each $\tau \in \mathbb{T}$, the stopped process X_τ is a total semimartingale. Following stability properties of semimartingales are relevant; see [172].

Theorem 8.2 *The set of (total) semimartingales is a vector space.*

Theorem 8.3 *If Q is a probability measure which is absolutely continuous with respect to P, then every (total) P-semimartingale X is a (total) Q-semimartingale.*

Theorem 8.4 (Stricker) *Let X be a semimartingale for the filtration $(\mathcal{F}_t)_{t \in \mathbb{T}}$. Let $(\mathcal{G}_t)_{t \in \mathbb{T}}$ be a subfiltration of $(\mathcal{F}_t)_{t \in \mathbb{T}}$, such that X is adapted to the $\mathcal{G}$-filtration. Then X is a $\mathcal{G}$-semimartingale.*

Let $\mathbb{D}$ denote the space of adapted processes with càdlàg paths, $\mathbb{L}$ denote the space of adapted processes with càglàd paths (left continuous with right limits) and $b\mathbb{L}$ denote processes in $\mathbb{L}$ with bounded paths. Let us in addition introduce a third type of convergence. A sequence of processes $(H^n)_{n \geq 1}$ converges to a process H *uniformly on compacts in probability* (abbreviated ucp) if, for each $t > 0$, $\sup_{0 \leq s \leq t} |H_s^n - H_s|$ converges to 0 in probability.

Theorem 8.5 *The space $\mathbb{S}$ is dense in $\mathbb{L}$ under the ucp topology.*

Let us introduce an operator that maps processes into processes. This is the *stochastic integral operator* J_X induced by the process X. For $H \in \mathbb{S}$ and X a càdlàg process in $\mathbb{D}$, we define the operator J_X on the basis of the linear mapping from $\mathbb{S}$ into $\mathbb{D}$ given by

$$J_X(H) = H_0 X_0 + \sum_{i=1}^{n} H_i(X_{|t}^{i+1} - X_{|t}^i),$$

where $H \in \mathbb{S}$, $H_i \in \mathcal{F}_{\tau_i}$ and for the stopping time τ_i we have $X_{|t}^i = X_{t \wedge \tau_i}$ as above. For $H \in \mathbb{S}$ and $X \in \mathbb{D}$, we call $J_X(H)$ the *stochastic integral of H with respect to X*. Following three notations are used interchangeably for the stochastic integral:

$$J_X(H) = \int H_s \, dX_s = H \cdot X.$$

Theorem 8.6 *Let X be a semimartingale. The mapping $J_X : \mathbb{S}_{ucp} \longrightarrow \mathbb{D}_{ucp}$ is continuous.*

Proof: See [172], pp. 50.   ∎

Since $\mathbb{S}_{ucp}$ is dense in $\mathbb{L}_{ucp}$ and $\mathbb{D}_{ucp}$ is a complete metric space, we are able to extend the stochastic integral operator J_X from $\mathbb{S}$ to $\mathbb{L}$ by continuity. For X a semimartingale, the continuous linear mapping $J_X : \mathbb{L}_{ucp} \longrightarrow \mathbb{D}_{ucp}$ obtained as an extension of $J_X : \mathbb{S} \longrightarrow \mathbb{D}$ is called the *stochastic integral*.

8.1.3   Semimartingale characteristics and Lévy processes

The end of this subsection is to present three well-known processes on the basis of which one can illustrate various aspects of the theory of semimartingales. The class of Lévy processes makes up the major part of them and includes the other two as particular cases. Let us begin introducing two fundamental processes.

A *process with independent increments* on $(\Omega, \mathcal{F}, (\mathcal{F}_t)_{t \in \mathbb{T}}, \mathbf{P})$ is a $\mathbb{R}^n$-valued process X adapted to the filtration $(\mathcal{F}_t)_{t \in \mathbb{T}}$ and with càdlàg sample paths such that (1) $X_0 = 0$, (2) for all $0 \leq s \leq t$, the random variable $X_t - X_s$ is independent of the σ-field $\mathcal{F}_s$. A *process with stationary independent increments* on $(\Omega, \mathcal{F}, (\mathcal{F}_t)_{t \in \mathbb{T}}, \mathbf{P})$ is a process with independent increments for which the distribution of the variables $X_t - X_s$ depends solely on the difference $t - s$. For X a càdlàg process, the processes $X_- = (X_{t-})_{t \in \mathbb{T}}$, the *process of left limits*, and $\Delta X = (\Delta X_t)_{t \in \mathbb{T}}$, the *process of jumps*, are defined for any $t \in \mathbb{T}$ by

$$X_{t-} = \lim_{s \uparrow t} X_s, \text{ for } s < t$$

$$\Delta X_{t-} = X_t - X_{t-} \text{ and } X_{0-} = X_0.$$

The process of left limits X_- and the process of jumps ΔX are predictible, see [115] pp.7 and 17. If $\sup\{|\Delta X_t| < c; t \in \mathbb{T}\} < \infty$, where c is a nonrandom constant, then X is said to be a process with bounded jumps. A time $t \in \mathbb{T}$ is called a *fixed time of discontinuity* if $\mathbf{P}(\Delta X_t \neq 0) > 0$.

Let X be an adapted process on $(\Omega, \mathcal{F}, (\mathcal{F}_t)_{t \in \mathbb{T}}, \mathbf{P})$ with $X_0 = 0$ a.s., independent and stationary increments. Then X is said to be a *Lévy process*, if X_t is continuous in probability, i.e. it holds $\lim_{t \to s} X_t = X_s$, where the limit is taken in probability. Let us recall at this point, a process X with independent increments, $X_0 = 0$ and with fixed times of discontinuity has an infinitely divisible distribution. A probability measure μ on $\mathbb{R}$, or a real-valued random variable Y with law μ, is said to be *infinitely divisible* if, for any $n \geq 1$, there is a probability measure μ_n such that $\mu = \mu_n^{*n}$ or equivalently if Y has the law of the sum of n independent identically distributed random variables.

For simplicity, we consider X as real-valued. Then, its characteristic function is of the form $\mathbf{E}\{\exp(iuX_t)\} = \exp[\psi_t(u)]$, with the function $\psi_t(u)$ given by the Lévy-Khintchine formula

$$\psi_t(u) = iub_t - \frac{u^2}{2}c_t + \int (e^{iux} - 1 - iuh(x))F_t(dx), \tag{8.1}$$

where $b_t \in \mathbb{R}$, $c_t \in \mathbb{R}_+$, F_t is a positive measure which integrates $1 \wedge x^2$, and h is any bounded Borel function with compact support which behaves like x near the origin. From the property of independent increments follows that the process $M^u = (M_t^u)_{t \in \mathbb{T}}$ with M_t^u given by

$$M_t^u = \exp(iuX_t)/exp[\psi_t(u)]$$

is a martingale for each $u \in \mathbb{R}$. The distribution of the process X is then said to be characterised by the three terms: (1) the drift given by b_t, (2) the variance of the Gaussian part represented by c_t and (3) the Lévy measure $F_t(dx)$.

When X is a semimartingale the idea is to replace this triple by another consisting of two processes $\beta = (\beta_t)_{t \in \mathbb{T}}$ and $\sigma = (\sigma_t)_{t \in \mathbb{T}}$, and a random measure $\nu([0,t] \times dx)$, so that the process M defined above is a martingale. The triple (β, σ, ν) is then called the *characteristics* of X. This brief observation makes clear how in order to examine certain notions concerning semimartingales, it is helpful to understand the structure of Lévy processes.

Theorem 8.7 *Let X be a Lévy process. There exists a unique modification Y of X which is càdlàg and which is also a Lévy process.*

Proof: See [172], pp. 21–22.   ∎

On the basis of this result we shall always assume we use the càdlàg version of any given Lévy process. The following theorem shows that Lévy processes provide filtrations which satisfy the usual conditions.

Theorem 8.8 *Let X be a Lévy process and $\mathcal{G}_t = \mathcal{F}_t^0 \vee \mathcal{N}$, where $(\mathcal{F}_t^0)_{t \in \mathbb{T}}$ is the natural filtration of X, and $\mathcal{N}$ are the P-null sets of $\mathcal{F}$. Then $(\mathcal{G}_t)_{t \in \mathbb{T}}$ is right-continuous.*

Proof: See [172], pp. 22–23.   ∎

Theorem 8.9 *Let X be a Lévy process and τ a stopping time. On the set $\{\tau < \infty\}$ the process $Y = (Y_t)_{0 \leq t < \infty}$ defined by $Y_t = X_{\tau+t} - X_\tau$ is a Lévy process adapted to $\widetilde{\mathcal{F}}_t = \mathcal{F}_{\tau+t}$, Y is independent of $\mathcal{F}_\tau$ and Y has the same distribution as X.*

Proof: See [172], pp. 23–24   ∎

This theorem makes evident, Lévy processes start anew at stopping times. Before we come to study briefly the behaviour of the jumps of Lévy processes, let us introduce two particular and well-known cases of Lévy processes.

Let $B = (B_t)_{t \in \mathbb{T}}$ be a process with independent increments defined as before on $(\Omega, \mathcal{F}, (\mathcal{F}_t)_{t \in \mathbb{T}}, \mathsf{P})$ with values in $\mathbb{R}$ and $B_0 = 0$, then B is said to be a *Brownian motion*, if $\mathsf{E}\{B_t\} = 0$ and $\mathsf{E}\{B_t^2\} < \infty$ for each $t \in \mathbb{T}$. Further, the function $\sigma^2(t) = \mathsf{E}\{B_t^2\} < \infty$ is called the *variance function* of B and for $\sigma^2(t) = t$, the process B is called a *standard* Brownian motion. Since a standard Brownian motion is a Lévy process, the foregoing theorem describes the strong Markov property for the Brownian motion. It is easy to check that a Brownian motion is a martingale as long as $\mathsf{E}\{|B_0|\} < \infty$, and thus B has a version with right-continuous paths a.s.; see Theorem 8.1.

Theorem 8.10 *Let B be a Brownian motion. Then there exists a modification of B which has continuous paths a.s.*

Theorem 8.11 *Let B be a real-valued Brownian motion with $B_0 = 0$. Then the process M defined for any $t \in \mathbb{T}$ by $M_t = B_t^2 - t$ is a martingale.*

The second example is introduced as follows. Let $N = (N_t)_{t \in \mathbb{T}}$ be a process with independent increments defined as above on $(\Omega, \mathcal{F}, (\mathcal{F}_t)_{t \in \mathbb{T}}, \mathsf{P})$ with values in $\mathbb{N}$ and $N_0 = 0$, then N is said to be an *extended Poisson process* if $\mathsf{E}\{N_t\} < \infty$ for each $t \in \mathbb{T}$ and $N_t - N_s$ is Poisson distributed with mean $\lambda(t) - \lambda(s)$ for all $0 \leq s < t$, $s, t \in \mathbb{T}$. The function $\lambda(t) = \mathsf{E}\{N_t\}$ is called the *intensity* of N.

For $\lambda(t)$ continuous we say that N is a Poisson process, and for $\lambda(t) = t$, the process N is a *standard* Poisson process. With the process N one associates a family of stopping times τ_n given by

$$\tau_n = \inf\{t : N_t = n\},$$

so that $\tau_0 = 0$ and $\tau_n < \tau_{n+1}$ on the set $\{\tau_n < \infty\}$, and $\lim_{(n)} \tau_n = \infty$. On the basis of this sequence of stopping times, one characterises N by

$$N_t = \sum_{n=1}^{\infty} \mathbf{1}_{\{\tau_n \leq t\}},$$

which has relevance on its own. Since the process N defined above gives at any time t the number of jumps occurring in the interval $(0, t]$, it is called the *counting process* associated with the sequence $(\tau_n)_{n \geq 1}$. At any time t the jump process ΔN takes by definition only the values 0 and 1. If we set $\tau = \sup_{n \geq 1} \tau_n$, we have $[\tau_n, \infty) = \{N_t \geq n\} = \{(t, \omega) : N_t(\omega) \geq n\}$ as well as $[\tau_n, \tau_{n+1}) = \{N_t = n\}$ and $[\tau, \infty) = \{N_t = \infty\}$. Thus, the random variable τ is called the *explosion time* of the process N. The compensator of N is $N_t^p = \lambda(t)$, see [115], p. 34.

Theorem 8.12 *Let X be a Lévy process with bounded jumps. Then $\mathsf{E}\{|X_t|^n\} < \infty$ for all $n = 1, 2, 3, \ldots$.*

Proof: From the definition of bounded jumps and the right-continuity of sample paths it follows that for $n = 1, 2, \ldots$ the family of random times defined by $\tau_n = \inf\{t > \tau_{n-1} : |X_t - X_{\tau_{n-1}}| \geq c\}$ with $X_0 = 0$ and $\tau_0 = 0$, forms a strictly increasing sequence of stopping times. Further, Theorem 8.8 entails the random variable $\tau_n - \tau_{n-1}$ is independent of $\mathcal{F}_{\tau_{n-1}}$ and has the same distribution as τ_1. Thus, for some α, $0 \leq \alpha < 1$, we have

$$\mathsf{E}\{e^{-\tau_n}\} = (\mathsf{E}\{e^{-\tau_1}\})^n = \alpha^n,$$

and further

$$\mathsf{P}\{|X_t| \geq 2nc\} \leq \mathsf{P}\{\tau_n < t\} < e^t \alpha^n,$$

which entails, X_t has an exponential moment and hence moments of all orders.   ∎

This theorem provides a quick insight into the role of Poisson measures and the Brownian motion in the construction of Lévy processes. Considering for instance an arbitrary homogeneous stochastically continuous process X with independent increments and finite moments of second order such that $\mathsf{E}\{X_t\} = 0$, we have the representation

$$
\begin{aligned}
X_t &= M_t + A_t, \\
M_t &= \sigma_t B_t, \\
A_t &= \int_{\mathbb{R}} x \tilde{N}_t(\cdot, dx), \\
\tilde{N}_t(\cdot, \Lambda) &= N_t(\cdot, \Lambda) - t\nu(\Lambda),
\end{aligned}
$$

where $t\nu(\Lambda) = \mathsf{E}\{N_t(\omega,\Lambda)\}$ and $\Lambda \in \mathcal{B}_0$. The role of the Poisson measure $N_t(\cdot,\Lambda)$ is that of counting the number of jumps of the process A with values in the set Λ. If the closure $\bar{\Lambda}$ of the set Λ does not contain the point 0, then $N_t(\cdot,\Lambda)$ takes on a finite value with probability 1. Since sample functions of the process B possess with probability 1 an unbounded variation on any interval $[0,t]$, the sample paths of X will exhibit in general the same property. The σ-field $\mathcal{B}_0$ is a σ-subfield of $\mathcal{B}$ consisting of Borel sets in $\mathbb{R}$ such that their closure does not contain the point 0.

8.1.4   The bracket process and finite variation

The next objective is to introduce the notion of bracket or quadratic covariation of two processes. This process is very useful and owns its name to a certain analogy with the covariance function of a Gaussian process.

Let X and Y be two semimartingales. The *bracket process* (or quadratic covariation) of X and Y, denoted by $[X,Y] = ([X,Y]_t)_{t\geq 0}$, is defined uniquely up to an evanescent set by

$$[X,Y] = XY - X_0Y_0 - \int X_- \, dY - \int Y_- \, dX,$$

which is an integration by the parts formula. In differential form this is written as

$$d(XY) = X_- \, dY + Y_- \, dX + d[X,Y].$$

If either X or Y is continuous the bracket process $[X,Y]$ is itself continuous, a result which is a natural consequence of a formula linking the bracket of X and Y and their jumps, i.e.

$$\Delta[X,Y] = \Delta X \, \Delta Y.$$

The operation $(X,Y) \longrightarrow [X,Y]$ is bilinear and symmetric and is characterised by the following properties:

$$[X,Y]_0 = 0, \ [X,Y] = [X - X_0, Y - Y_0],$$
$$[X,Y] = \frac{1}{4}([X+Y, X+Y] - [X-Y, X-Y]).$$

The bracket process (or quadratic variation) of X is defined by

$$[X,X] = X^2 - X_0^2 - 2\int X_- \, dX.$$

On the basis of the properties of the bracket process, one easily recognises that $[X,Y]$, as the difference of two increasing processes, has paths with finite variation on compacts which are càdlàg. Since the process $[X,Y]$ itself is adapted, then it is a semimartingale, too. Further, since the stochastic integrals are semimartingales, one concludes from the integration by the part formula that XY is also a semimartingale. Since the process $[X,X]$ is nondecreasing with continuous paths, and since $\Delta[X,X] = (\Delta X)^2$, one can decompose $[X,X]$ path by path into its continuous and its pure jump part. For a semimartingale X, the process $[X, X]^c$ denotes the path by path continuous part of $[X, X]$.

At this stage we shall mention the existence of the *angle bracket* (or conditional quadratic covariation) of the semimartingales X and Y. It is denoted by $\langle X, Y \rangle = (\langle X, Y \rangle_t)_{t \geq 0}$ with $\langle X, Y \rangle_t$ defined formally by

$$\langle X, Y \rangle_t = \langle X, Y \rangle_0 + \int_0^t \mathsf{E}\{dX_s dY_s | \mathcal{F}_s\}.$$

Let us mention also that for any semimartingale X there is a (unique) continuous local martingale part denoted by X^c and that $[X, X]^c = [X^c, X^c] = \langle X^c, X^c \rangle$.

Theorem 8.13 *If X and Y are semimartingales with continuous martingale parts X^c and Y^c, then*

$$[X, Y]_t = \langle X^c, Y^c \rangle_t + \sum_{s \leq t} \Delta X_s \, \Delta Y_s.$$

Proof: See [115], p. 55. ∎

A semimartingale is called a *quadratic pure jump semimartingale* if $[X, X]^c = 0$. Let us close this subsection presenting a class of processes with bounded stochastic variation known as quasimartingales. Let X be an adapted process with respect to $\mathbf{F}$ such that for any $t \in \overline{\mathbb{R}}_+$ the random variables X_t are integrable. Then X is said to be of *bounded stochastic variation* if the mean conditional variation $\mathsf{E}\{D_t(\omega)\}$ remains bounded over all finite partitions $\{t_i\}$ of $[0, t]$. Here $D_t(\omega) = D_t^+(\omega) + D_t^-(\omega)$, where $D_t^+(\omega)$ and $D_t^-(\omega)$ are the positive and negative parts of the mean conditional variations, i.e.

$$D_t^+(\omega) = \sum_{i=0}^{n} \left(\mathsf{E}\{X_{t_{i+1}} - X_{t_i} | \mathcal{F}_{t_i}\} \right)^+,$$

$$D_t^-(\omega) = \sum_{i=0}^{n} \left(\mathsf{E}\{X_{t_{i+1}} - X_{t_i} | \mathcal{F}_{t_i}\} \right)^+,$$

An adapted process X with càdlàg paths is said to be a *quasimartingale* on $[0, \infty]$ if $\mathsf{E}\{|X_t|\} < \infty$ for each t and it has bounded stochastic variation.

8.2   Operating with semimartingales

In this section we shall collect some formulas for operation with stochastic integrals and discuss briefly properties of a few semimartingales which appear in various ways in the present essays.

8.2.1   A remarkable formula for change of variables

A fundamental formula which enables us to operate on the class of semimartingales is the well-known Itô formula, which can be easily derived from the integration by the parts formula; see for instance [50], pp. 350–351 and [115], pp. 57–58. For the purpose of this chapter we shall limit ourselves to state a change of variables formula valid for semimartingales and make some explanatory remarks.

Theorem 8.14 (Itô's formula) *Let X be a semimartingale and f be a function of class $\mathbb{C}^2(\mathbb{R}^n \longrightarrow \mathbb{R})$. Then $f(X)$ is a semimartingale and we have*

$$f(X_t) = f(X_0) + \sum_{i=1}^{n} D_i f(X_-) \cdot X^{(i)} + \frac{1}{2} \sum_{i,j=1}^{n} D_{ij} f(X_-) \cdot [X^{(i)}, X^{(j)}]^c$$

$$+ \sum_{0 < s \leq t} \left[f(X_s) - f(X_{s-}) - \sum_{i=1}^{n} D_i f(X_{s-}) \Delta X_s^{(i)} \right],$$

where $D_i f$ and $D_{ij} f$ stand for the first and second partial derivatives of f and $X^{(i)}$ represents the i-th component of X.

The fact that X is a semimartingale with values in $\mathbb{R}^n$ means, each component of X is a real-valued semimartingale, so that for any $i = 1, 2, \ldots, n$ the real-valued process $f \circ X^{(i)}$ is again a semimartingale. The last two terms at the right-hand side of the formula above are processes of finite variation associated with the jumps of the semimartingale X. Since $\sum \|\Delta X_s^2\|$ is convergent over any finite interval of time, the third term represents the continuous while the fourth term reflects the purely discontinuous nature of the semimartingale. In this sense, it becomes essential to render predictible the integrands appearing in the formula, a point which is visualised by writing X_-, an exception to this is however the third term where we work with the continuous part and are thus allowed to dispense with it.

At the heart of the semimartingale theory is the observation that as long as integrators generate predictible and optional σ-fields which conceal the historical nature of the filtrations underlying the integration, the need of predictible integrands cannot be essential. This contrasts greatly with stochastic calculus based on the Brownian motion for which one has $\mathcal{P} = \mathcal{O}$, a deep consequence of the strong Markov property according to which at any time future and past are independent. Putting future and past on the same footing allows the underlying process to start always anew, however it renders information irrelevant and deprives time of its arrows.

8.2.2 The Doléans-Dade exponential

A highly interesting application of the change of variables formula refers to the study of the stochastic exponential. This is a peculiar instance of a process, driving a class of stochastic differential equations invariant by change of law, which is a discontinuous function of bounded variation. To stress the power of the semimartingale theory, let us mention that it is possible to solve, for a prescribed Z_0, SDE's of the type

$$dZ_t = f(t, Z_{t-}) dX_t,$$

where the unknown process Z and the driving process X are semimartingales, and f is an appropriate matrix which satisfies the same Lipschitz conditions as in the theory of ODE's. Let us turn to the Doléans-Dade exponential formula.

Theorem 8.15 *Let X be a semimartingale with $X_0 = 0$. Then there exists for $t \geq 0$ a unique semimartingale Z solving the stochastic integral equation*

$$Z_t = 1 + \int_0^t Z_{s-} dX_s,$$

where Z is given by

$$Z_t = \exp(X_t - \frac{1}{2}[X,X]_t^c) \prod_{0 < s \leq t} (1 + \Delta X_s) \exp(-\Delta X_s + \frac{1}{2}(\Delta X_s)^2),$$

and the infinite product converges.

Proof: See [172], pp. 77-78.   ∎

The stochastic exponential of X given in the preceding theorem is denoted by $\mathcal{E}(X)$ and $Z = \mathcal{E}(X)$ is known as the *Doléans-Dade exponential*. For a continuous semimartingale X with $X_0 = 0$, the stochastic exponential is given by the simple formula

$$\mathcal{E}(X) = \exp(X - \frac{1}{2}[X,X]),$$

which for a multiple λ of the standard Brownian motion $B = (B_t)_{t \geq 0}$ gives for any $t \geq 0$

$$\mathcal{E}(\lambda B)_t = \exp(\lambda B_t - \frac{\lambda^2}{2}[B,B]_t) = \exp(\lambda B_t - \frac{\lambda^2}{2}t).$$

Since for the standard Brownian motion we know that $[B,B]_t = t$, the process $\mathcal{E}(\lambda B)$ is known as the *geometric Brownian motion*.

8.2.3   Compensation of supermartingales

A property which is quite helpful working with processes is that of stability with respect to the operation of stopping. In particular the class of all uniformly integrable martingales, i.e. all martingales X such that the family of random variables $(X_t)_{t \in \mathbb{T}}$ is uniformly integrable which we denote by $\mathfrak{M}$, and the class of all square integrable martingales, i.e. all martingales X such that $\sup_{t \in \mathbb{T}} \mathsf{E}\{X_t^2\} < \infty$ which we denote by $\mathfrak{H}^2$, are stable under stopping. Let us recall, a family of random variables $(X_t)_{t \in \mathbb{T}}$ is uniformly integrable means that $\sup_{t \in \mathbb{T}} \mathsf{E}\{X_t^2\} < \infty$ and for any $A \in \mathcal{F}_t$, $\sup_{t \in \mathbb{T}} \mathsf{E}\{X_t \mathbb{1}_A(\omega)\} \longrightarrow 0$ as $\mathsf{P}(A) \longrightarrow 0$. Further, for $X \in \mathfrak{M}$ (resp. $X \in \mathfrak{H}^2$) X_t converges a.s. and in $L^1(\Omega, \mathcal{F}, \mathbf{F}, \mathsf{P})$ (resp. in $L^2(\Omega, \mathcal{F}, \mathbf{F}, \mathsf{P})$) to the terminal variable X_∞. It is evident that $\mathfrak{H}^2 \subset \mathfrak{M}$, and $\mathfrak{H}^2$ as well as $\mathfrak{M}$ are stable under the operation of stopping.

A real valued adapted process X is said to be of class (D) (respectively (DL)) if the family of random variable $X_\tau \mathbf{1}_{\{\tau < \infty\}}$ where τ ranges through all stopping times (respectively bounded stopping times) is uniformly integrable. On the other hand, a uniformly integrable martingale is of class (D). If Y is integrable on $(\Omega, \mathcal{F}, \mathbf{F}, \mathbf{P})$ and $\mathcal{G}$ ranges through the σ-fields of $\mathcal{F}$, the family of conditional expectations $\mathsf{E}\{Y|\mathcal{G}\}$ is uniformly integrable. The following result is at the heart of the decomposition of supermartingales.

Theorem 8.16 *Let Z be a positive supermartingale of class (D). Then there exists a unique (up to an evanescent set) predictible process $A \in \mathbb{A}^+$ with $A_0 = 0$ such that for any $t \in \mathbb{T}$ the relationship*

$$Z_t = \mathsf{E}\{A_\infty - A_t | \mathcal{F}_t\} \qquad a.s.$$

holds. Inversely, if Z admits the foregoing representation, it is of class (D).

Proof: See [50], pp. 211–212. ∎

The process A is said to be the *increasing process associated* with Z and when Z is a potential, then it is said to be the *potential generated* by A. A process $X \in \mathbb{D}$ is a *potential* if it is a positive supermartingale such that $\lim_{t \to \infty} \mathsf{E}\{X_t\} = 0$.

Conditional expectations are the more natural objects on the basis of which updating takes place as new information arrives. This observation directs attention to a slightly more general class of processes, namely the processes with conditionally independent increments. As a natural object in this vein, we shall consider the notion of (instantaneous) conditional forward (or backward) drift of a process X given by

$$D^+ X_t = \lim_{h \downarrow 0} \mathsf{E}\left\{ \frac{X_{t+h} - X_t}{h} \bigg| \mathcal{F}_t \right\},$$

assuming the random variables X_t at any time t are integrable, so that the conditional expectation has a meaning.

An intuitive manner of viewing the operation of compensating, may be that of adding to a given process (say a supermartingale X) a smoother one (say a differentiable, here $D^+ X$) so that one gets a martingale, here $X_t + \int_0^t D^+ X_s \, ds$. Compare with Meyer in [62], pp. 138–139 from which we take the following result, where the differentiability requirement is replaced by that of bounded variation. However, let us mention that under suitable integrability conditions one should understand $\mathsf{E}\{X^*\} > -\infty$, where X^* stands for $\sup_t |X_t|$.

Theorem 8.17 *Every supermartingale X can be compensated, i.e. there exists a unique increasing process which is locally natural with $A_0 = 0$, known as the compensator, such that $X + A$ is a local martingale. If the decreasing process $\mathsf{E}\{X\}$ is bounded, the random variable A_∞ is integrable. Under suitable integrability conditions, A_t can be expressed as*

$$A_t = \lim_{h \downarrow 0} \int_0^t \mathsf{E}\left\{ \frac{X_{s+h} - X_s}{h} \bigg| \mathcal{F}_s \right\} ds,$$

and the process $X + A$ is a true martingale. For a positive supermartingale the result is true without any additional integrability condition.

For the bracket process of a square integrable martingale X, the compensator is the angle bracket process $\langle X, X \rangle$, defined for any $t > 0$ by

$$\langle X, X \rangle_t = \lim_{h \downarrow 0} \int_0^t \mathsf{E} \left\{ \frac{(X_{s+h} - X_s)^2}{h} \Big| \mathcal{F}_s \right\} ds.$$

In particular, if $A \in \mathbb{A}_{\mathrm{loc}}^+$ there is up to an evanescent set a unique predictible process $A^p \in \mathbb{A}_{\mathrm{loc}}^+$, which is said to be the compensator of A, such that $A - A^p$ is a local martingale.

Let us draw attention to a special feature of martingales obtained by compensation. Namely, a purely discontinuous local martingale X is generally not the sum of its jumps. First, the series $\sum_{s \le t} \Delta X_s$ may diverge and even when it does converge, it may still differ from X_t. As a way of example, recall the extended Poisson process with an intensity function $\lambda(t)$, which is continuous as we have previously seen. Then the process $M = N - \lambda$ is a purely discontinuous local martingale and $N_t = \sum_{s \le t} \Delta M_t$ which obviously differs from M_t, see [115], pp. 32, 34 and 40.

8.2.4   On the notion of semimartingale

The space of all semimartingales exhibits very pleasant properties and remains stable under a wide range of operations. We have already mentioned stability under stopping, absolutely continuous change of measure, and various types of stochastic integration. Semimartingales stay stable also under change of filtration and constitute the largest possible class of processes with respect to which one may reasonably integrate all bounded predictible processes. According to the type of integrands and integrators considered, one obtains various sorts of semimartingales. To record the relationship between them we need additional definitions.

An adapted, càdlàg process X is *decomposable* if there exist processes M and A with $M_0 = A_0 = 0$, such that for any $t \in \mathbb{T}$ the relationship

$$X_t = X_0 + M_t + A_t,$$

holds, where $M \in \mathfrak{H}_{\mathrm{loc}}^2$ and $A \in \mathbb{V}$. An adapted càdlàg process Y is called a *classical semimartingale* if there exist processes N and B with $N_0 = B_0 = 0$ such that, for any $t \in \mathbb{T}$, the relationship

$$Y_t = Y_0 + N_t + B_t$$

holds, where $N \in \mathfrak{M}_{\mathrm{loc}}$ and $B \in \mathbb{V}$. Let $A \in \mathbb{A}$ with $A_0 = 0$, then A is said to be a *natural process* if

$$\mathsf{E}\{[M, A]_\infty\} = 0$$

for all bounded martingales M. Further, let $A \in \mathbb{A}_{\mathrm{loc}}$ with $A_0 = 0$, then A is a *locally natural process* if

$$\mathsf{E}\{[M, A_|]_\infty\} = 0,$$

for all bounded martingales M, where $A_|$ stands for the stopped process with respect to any stopping time τ such that $\mathsf{E}\{\int_0^\tau |dA_s|\} < \infty$. By definition there

exists a sequence of stopping times (τ_n) increasing to ∞ and with it one associates a sequence (A_1^n), where for any n the process A_1^n is in $\mathbb{A}$. The following
theorems summarise the relation between the above notions of semimartingales.

Theorem 8.18 *Let X be an adapted, càdlàg process. The following statements
are equivalent:*

  (1) X is a semimartingale;

  (2) X is decomposable;

  *(3) given $\epsilon > 0$, there exist processes M and A with $M_0 = A_0 = 0$ such that
  for any $t \in \mathbb{T}$*

$$X_t = X_0 + M_t + A_t,$$

  *with $|\Delta M_t| \leq 2\epsilon$ and $A_t = \sum_{0 < s \leq t} \Delta X_s \, 1_{\{|\Delta X_s| > \epsilon\}}$, where $M \in \mathfrak{M}_{loc}$
  and $A \in \mathbb{V}$;*

  (4) X is a classical semimartingale.

Proof: See [172], pp. 88 and 99–102.   ∎

Theorem 8.19 *Let $A \in \mathbb{A}$ with $A_0 = 0$. Then A is natural if and only if*

$$\mathsf{E}\left\{\int_0^\infty M_{s-} dA_s\right\} = \mathsf{E}\{M_\infty A_\infty\}$$

for any bounded martingale M.

Proof: See [172], p. 89.   ∎

Theorem 8.20 *Let X be a semimartingale. If X has a decomposition $X_t =
X_0 + M_t + A_t$, where $M \in \mathfrak{M}_{loc}$ and $A \in \mathbb{V}$ locally natural with $M_0 = A_0 = 0$,
then such a decomposition is unique.*

Proof: See [172], p. 107.   ∎

Let us close with two more definitions. Let X be a semimartingale. If X has a
decomposition $X_t = X_0 + M_t + A_t$, where $M \in \mathfrak{M}_{loc}$ and $A \in \mathbb{V}$ locally natural
with $M_0 = A_0 = 0$, then X is said to be a *special semimartingale*. If X is a
special semimartingale, then the unique decomposition $X_t = X_0 + M_t + A_t$,
with A locally natural is called the *canonical decomposition*. Let us mention at
this point that if A is bounded and natural, then it is predictible.

8.3   Predictible integrands and representation

In this section we shall mainly deal with two additional operations under which the class of all semimartingales remains stable. Regarding the first, one examines an operation which generates a semimartingale whenever one replaces an original probability measure, viewed as part of the data, by another measure which is absolutely continuous with respect to the first. Regarding the second, one studies the map $X \longrightarrow H \cdot X$ where X is a semimartingale and H is a predictible process and looks for conditions under which the stochastic integral $H \cdot X$ represents a semimartingale.

Furthermore, we are interested in rewriting semimartingales as stochastic integrals, a question known as representation theorems which constitute the entry point toward the identification of the underlying density process. We shall conclude solving a general linear equation on the basis of a stochastic exponential. Let us begin with the Girsanov transformation.

8.3.1   The Girsanov transformation

Let us consider two probability measures $\mathbf{P}$ and $\mathbf{Q}$ defined on a filtered space $(\Omega, \mathcal{F}, \mathbf{F})$. Further, let X be a semimartingale on the filtered probability space $(\Omega, \mathcal{F}, \mathbf{F}, \mathbf{P})$ defined as before. The basic assumption refers either to absolute continuity or a localised version of it. We shall assume $\mathbf{Q}$ is absolutely continuous with respect to $\mathbf{P}$, for short $\mathbf{Q} \ll \mathbf{P}$.

Let Z denote a version of the density process of $\mathbf{Q}$ with respect to $\mathbf{P}$, which is a martingale on $(\Omega, \mathcal{F}, \mathbf{F}, \mathbf{P})$. Indeed, for each $t \in \mathbb{T}$, one chooses a random variable $Z_t \in \mathbf{E}\{Z_\infty | \mathcal{F}_t\}$ where Z_∞ stands for the density of $\mathbf{Q}$ with respect to $\mathbf{P}$ restricted to $\mathcal{F}_\infty$ and for any random time τ, Z_τ is similarly the density of $\mathbf{Q}$ with respect to $\mathbf{P}$ restricted to $\mathcal{F}_\tau$. The following result holds:

Theorem 8.21 *Any semimartingale with respect to* $\mathbf{P}$ *is a semimartingale with respect to* $\mathbf{Q}$.

Proof: It is a consequence of Theorem 8.18.   ∎

To avoid confusion, we shall write $\widehat{\mathbf{E}}$, $\widehat{\mathcal{F}_t}$ and so on, in order to distinguish those objects associated with $\mathbf{Q}$, whenever it becomes necessary. An issue of great relevance is related to the question of the impact upon the characteristics and the decomposition of a semimartingale when the measure $\mathbf{Q}$ substitutes $\mathbf{P}$. In order to avoid technicalities, we shall illustrate this issue under the assumption that $\mathbf{P}$ and $\mathbf{Q}$ are equivalent, i.e. when both $\mathbf{Q} \ll \mathbf{P}$ and $\mathbf{P} \ll \mathbf{Q}$ hold. Then, we have:

Lemma 8.22 *For the adapted process* X *with càdlàg paths to be a local martingale with respect to* $\mathbf{Q}$, *it is necessary and sufficient that* XZ *be a local martingale with respect to* $\mathbf{P}$.

Proof: It suffices to prove the lemma for martingales, since the localisation procedure uses for both measures the same sequences of stopping times going to infinity. Let t be an arbitrary time with $t \in \mathbb{T}$. Then if XZ is a martingale with respect to $\mathbf{P}$, we have

$$\int |X_t| \mathbf{Q}(d\omega) = \int |X_t| Z_t \mathbf{P}(d\omega) < \infty.$$

Further, let $s \in \mathbb{T}$ be so that $s < t$ and assume $A \in \mathcal{F}_s$. Then we have

$$\int_A X_s \mathbf{Q}(d\omega) = \int_A X_s Z_s \mathbf{P}(d\omega) = \int_A X_t Z_t \mathbf{P}(d\omega) = \int_A X_t \mathbf{Q}(d\omega),$$

since Z_s (resp. Z_t) is the density of $\mathbf{Q}$ with respect to $\mathbf{P}$ restricted to $\mathcal{F}_s$ (resp. $\mathcal{F}_t$). This shows X is a martingale with respect to $\mathbf{Q}$.

To prove the converse, it is enough to observe that from $\mathbf{P} \ll \mathbf{Q}$, it follows the existence of a random variable $U_\infty \in L^1(\Omega, \widehat{\mathcal{F}}, \widehat{\mathbf{F}}, \mathbf{Q})$ such that $\mathbf{P} = U_\infty \mathbf{Q}$ and $\widehat{\mathbf{E}}\{U_\infty\} = 1$. Then, for any $t \in \mathbb{T}$, one may select a $U_t \in \widehat{\mathbf{E}}\{U_\infty | \widehat{\mathcal{F}}_t\}$ from which one gets a right continuous version of the process $U = (U_t)_{t \in \mathbb{T}}$ and thus U is a local martingale with respect to $\mathbf{Q}$. Further, since $\mathbf{P}_t$ and $\mathbf{Q}_t$ are the restrictions to $\mathcal{F}_t$, it turns out $Z_t \mathbf{P}_t = \mathbf{Q}_t$ and $\mathbf{P}_t = U_t \mathbf{Q}_t$ so that $Z_t = U_t^{-1}$. To conclude, assume X is a martingale with respect to $\mathbf{Q}$. Then we have

$$\int_A X_t Z_t \mathbf{P}(d\omega) = \int_A \frac{X_t}{U_t} \mathbf{P}(d\omega) = \int_A X_t \mathbf{Q}(d\omega) = \int_A X_s \mathbf{Q}(d\omega)$$
$$= \int_A \frac{X_s}{U_s} \mathbf{P}(d\omega) = \int_A X_s Z_s \mathbf{P}(d\omega).$$

Thus XZ is a martingale with respect to $\mathbf{P}$.   ∎

Before we state the main result on change of law, we have some remarks. The bracket process $[X, Y]$ of two semimartingales X and Y remains the same under the probability measures $\mathbf{P}$ and $\mathbf{Q}$. However, the angle bracket $\langle X, Y \rangle$, which represents the predictible compensator of $[X, Y]$ whenever it exists, does not in general coincide under the laws $\mathbf{P}$ and $\mathbf{Q}$. Let us also recall that the process Z_- never vanishes; see for instance [50], pp. 256.

Theorem 8.23 (Girsanov-Meyer) *Let X be a local martingale with respect to $\mathbf{P}$ with $X_0 = 0$.*

a) *The process X is a special semimartingale with respect to $\mathbf{Q}$ if and only if the angle bracket process $\langle X, Z \rangle$ exists so that the canonical decomposition of X becomes $X = M + A$, where M is a local martingale with respect to $\mathbf{Q}$ and A a predictible process in $\mathbb{V}$ given respectively by*

$$M_t = X_t - \int_0^t \frac{1}{Z_{s-}} d\langle X, Z \rangle_s,$$

$$A_t = \int_0^t \frac{1}{Z_{s-}} d\langle X, Z \rangle_s.$$

b) *In general, the process A given by*

$$A_t = \int_0^t \frac{1}{Z_s} d[X, Z]_s$$

is a local martingale with respect to $\mathbf{Q}$.

Proof: See [50], pp. 255–261 or [172], pp. 109–110.   ∎

Lenglart [135] presents a similar version of the result above directed to the case where the probability measure $\mathbf{Q}$ is only absolutely continuous with respect to $\mathbf{P}$ to which we refer. Let us close this subsection stating a theorem according to which semimartingales in the sense defined above are equivalent to classical semimartingales or semimartingales usually encountered in the literature.

Theorem 8.24 (Bichteler-Dellacherie) *An adapted process X with càdlàg path is a semimartingale if and only if it is a classical semimartingale. That is, the process X is a semimartingale if and only if it has a unique canonical decomposition $X = X_0 + M + A$, where $M \in \mathfrak{M}_{loc}$ and $A \in \mathbb{V}$ with $M_0 = A_0 = 0$.*

Proof: See Protter [172], pp. 114–115.   ∎

Let us mention at this stage restriction to processes X with $X_0 = 0$ does not involve loss of generality, since taking $Y_t = X_t - X_0$ gives the opportunity to recover the original process.

8.3.2   Predictible integrands

The predictible σ-field $\mathcal{P}$ on $\mathbb{T} \times \Omega$ is the smallest σ-field making all processes in $\mathbb{L}$ measurable. Recall that $\mathbb{L}$ is the space of adapted processes having left continuous paths with right limits. Before we pass to outline the stochastic integral with predictible integrands, let us mention that for bounded processes of finite variation the notions of a natural process and a predictible process are equivalent. We shall record this in the following theorem.

Theorem 8.25 *Let the process A be in $\mathbb{A}_{loc}$ and $A_0 = 0$. Then A is predictible if and only if A is locally natural.*

In this regard we refer to [172] pp. 117–121 and [115] pp. 27–30. Here a central tool is the integration by parts formula

$$\int_0^\infty M_{s-}\, dA_s = M_\infty A_\infty - M_0 A_0 - \int_0^\infty A_{s-}\, dM_s - [M, A]_\infty,$$

where M is an arbitrary bounded martingale. Observing that $N_t = \int_0^t A_{s-}\, dM_s$ builds a local martingale and

$$[A, M]_\infty = \sum_{0 < s \le \infty} \Delta A_s \Delta M_s = \int_0^\infty \Delta A_s\, dM_s = \int_0^\infty (A_s - A_{s-})\, dM_s$$

$$= \int_0^\infty A_s\, dM_s - \int_0^\infty A_{s-}\, dM_s$$

directs attention to the main line of argumentation. We close our remarks on natural processes with a lemma.

Lemma 8.26 *Let A be a natural process in $\mathbb{V}$ and let H be in $\mathbb{L}$ such that $\mathbf{E}\left\{\int_0^\infty |H_s|\,|dA_s|\right\} < \infty$. Then the process $H \cdot A$ in $\mathbb{V}$ is natural.*

Proof: For any bounded martingale M, the martingale $H \cdot M$ is locally bounded. Thus one easily checks that $\mathsf{E}\{[H \cdot M, A]_\infty\} = 0$ since A is natural. Observing further that

$$[M, H \cdot A]_\infty = \int_0^\infty H_s \, d[M, A]_s = [H \cdot M, A]_\infty$$

gives the lemma.   ∎

Next, we shall extend the definition of stochastic integral to the most general class of predictible integrands, so that integration delivers a semimartingale, indeed a class of semimartingales.

The next step towards a more general stochastic integral centres on the map $H \longrightarrow H \cdot X$ where $H \in \mathfrak{bL}$ and X is a special semimartingale living in the space $\mathcal{H}^2$ to be defined in short. This map has an extension denoted for simplicity also by $H \longrightarrow H \cdot X$ which is then defined on $\mathfrak{bP}$, i.e. the set of bounded processes that are $\mathcal{P}$-measurable, and with X belonging to the particular class of semimartingales aforementioned. The extension of the stochastic integral process visualised by the map $H \longrightarrow H \cdot X$ is essentially related to smoothness and measurability requirements on the integrands H, which are then chosen from the family of predictible processes, and to the need for the integrators X to be semimartingales, a feature which relies heavily on stability properties revealed by Bichteler-Delacherie's theorem. One can further enlarge the class of integrands and integrators on the basis of approximations and uniform convergence operations under which predictibility and semimartingale features are conserved.

Let $\mathcal{H}^2$ be the space of semimartingales consisting of all special semimartingales X distinguished first, by the canonical decomposition $X = \bar{N} + \bar{A}$ where $\bar{N}$ is a local martingale and $\bar{A}$ a locally predictible process in $\mathbb{V}$ and second, by the finite norm $\| \cdot \|_{\mathcal{H}^2}$ defined for $X \in \mathcal{H}^2$ as

$$\|X\|_{\mathcal{H}^2} = \|[\bar{N}, \bar{N}]_\infty^{1/2}\|_{L^2} + \left\| \int_0^\infty |d\bar{A}_s| \right\|_{L^2},$$

under which the *space $\mathcal{H}^2$ of semimartingales* is a Banach space. The task of extending the stochastic integral so as to allow for predictible integrands is based on two essential points. First, for $X \in \mathcal{H}^2$ the space $\mathfrak{bL}$ is dense in $\mathfrak{bP}$ under an appropriate $d_X(\cdot, \cdot)$ to be introduced later. Second for $X \in \mathcal{H}^2$ and $H^n \in \mathfrak{bL}$, such that H^n is Cauchy under d_X, it follows $H^n \cdot X$ is Cauchy in $\mathcal{H}^2$. Then for $H \in \mathfrak{bP}$ such that $\lim_{n \to \infty} d_X(H^n, H) = 0$, the *stochastic integral* $H \cdot X \in \mathcal{H}^2$ is then defined by the unique semimartingale in $\mathcal{H}^2$ given by $\lim_{n \to \infty} H^n \cdot X$, which we shall write as

$$H \cdot X = \left(\int_0^t H_s dX_s \right)_{t \geq 0}.$$

For $H, J \in \mathfrak{bP}$ and $X \in \mathcal{H}^2$ with a canonical decomposition as above, we define $d_X(H, J)$ by

$$d_X(H, J) = \left\| \left(\int_0^\infty (H_s - J_s)^2 \, d[\bar{N}, \bar{N}]_s \right)^{1/2} \right\|_{L^2} + \left\| \int_0^\infty |H_s - J_s|^2 \, |d\bar{A}_s| \right\|_{L^2}.$$

Let us call attention to the following points: (a) if $H \in b\mathbb{L}$ and $X \in \mathcal{H}^2$ then the stochastic integral $H \cdot X \in \mathcal{H}^2$; (b) if $X = \bar{N} + \bar{A}$ is the canonical decomposition of X, then $H \cdot \bar{N} + H \cdot \bar{A}$ is in turn the canonical decomposition of $H \cdot X$; (c) the approximations of the stochastic integral above converge uniformly.

Let us collect some properties of the stochastic integral.

Theorem 8.27   *(1) Let $X, Y \in \mathcal{H}^2$ and $H, K \in b\mathbb{P}$. Then*

$$
\begin{aligned}
(H + K) \cdot X &= H \cdot X + K \cdot X, \\
H \cdot (X + Y) &= H \cdot X + H \cdot Y.
\end{aligned}
$$

(2) Let τ be an arbitrary stopping time. Then $(H \cdot X)_| = H \mathbf{1}_{[0,\tau]} \cdot X = H \cdot X_|$.

(3) Let $X \in \mathcal{H}^2$ and $H, K \in b\mathbb{P}$. Then $K \cdot X \in \mathcal{H}^2$ and $H \cdot (K \cdot X) = (HK) \cdot X$.

(4) Let $X \in \mathcal{H}^2$ be a (square integrable) martingale, and $H \in b\mathbb{P}$. Then $H \cdot X$ is a square integrable martingale.

(5) Let $X, Y \in \mathcal{H}^2$ and $H, K \in b\mathbb{P}$. Then, for $t \geq 0$ we have

$$
\begin{aligned}
[H \cdot X, K \cdot Y]_t &= \int_0^t H_s K_s \, d[X,Y]_s, \\
[H \cdot X, H \cdot X]_t &= \int_0^t H_s^2 \, d[X,X]_s.
\end{aligned}
$$

Proof: See [172], pp. 128–130.   ∎

To conclude we shall say that restricting integrands to $b\mathbb{P}$ and integrators to $\mathcal{H}^2$ is not essential, but a matter of convenience. A more general stochastic integral can be defined for arbitrary semimartingales as integrators and predictible processes as integrands which need not be bounded. Moreover, many of the properties are simple extensions of those already mentioned.

8.3.3   Martingale representation

In this subsection we shall briefly present an issue, which has great relevance for a wide variety of applications in engineering, meteorology, medicine, physics and many other fields, concerning the representation of semimartingales as a stochastic integral. Here we shall focus on the representation of martingales or local martingales defined on a probability basis $(\Omega, \mathcal{F}, \mathbf{F}, \mathbf{P})$. The problem consists in characterising a class of (local) martingales in a manner that it attains the property according to which any member of this class admits a unique stochastic integral representation with predictible integrands and an integrator which belongs itself to this class. Whenever a class of (local) martingales exhibits the representation property one can explicitly compute the density process of any other measure $\mathbf{Q}$ such that $\mathbf{Q} \ll \mathbf{P}$ locally with respect to $\mathbf{P}$.

Since there is a bijective correspondence between the elements X of $\mathfrak{H}^2$ and their terminal variables X_∞, then it becomes natural to endow $\mathfrak{H}^2$ with a

Hilbertian structure on the basis of the *scalar product* $(X,Y)_{H^2}$ and the *norm* $\|X\|_{L^2}$ defined by

$$(X,Y)_{H^2} = \mathsf{E}\{X_\infty Y_\infty\} \quad \text{and} \quad \|X\|_{H^2} = \|X_\infty\|_{L^2}$$

for any X and Y in $\mathfrak{H}^2$. It is easy to check, $\mathfrak{H}^2$ is a *Hilbert space* and its *dual space* is also $\mathfrak{H}^2$. For $X,Y \in \mathfrak{H}^2$ we have $\langle X,Y\rangle \in \mathbb{A}$, which is known as the predictible quadratic covariance and $XY - \langle X,Y\rangle \in \mathfrak{M}$. Furthermore,

$$(X,Y)_{H^2} = \mathsf{E}\{\langle X,Y\rangle_\infty\} + \mathsf{E}\{X_0 Y_0\}.$$

Two local martingales X and Y are said to be *strongly orthogonal* if their product XY is a local martingale and X is called a *purely discontinuous local martingale* if $X_0 = 0$ and it is strongly orthogonal to all continuous local martingales.

For $X,Y \in \mathfrak{H}^2$ we say that X and Y are *orthogonal* (in the Hilbertian sense) if $\mathsf{E}\{X_\infty Y_\infty\} = \mathsf{E}\{X_0 Y_0\} = 0$. A closed subspace F of $\mathfrak{H}^2$ is said to be a *stable subspace* if it is stable under the operations of stopping and multiplication with $\mathbb{1}_A$ for $A \in \mathcal{F}_{0-} (= \mathcal{F}_0)$. Further, the intersection of all closed, stable subspaces containing F is called the *stable subspace generated* by F and is denoted by $\sigma(F)$.

Theorem 8.28 *Let $F = \{X^i : X^i \in \mathfrak{H}^2 \text{ with } i = 1,2,\ldots,n\}$ so that for any $i \neq j$, X^i and X^j are strongly orthogonal. Then $\sigma(F)$ is given by the set $\mathfrak{I}(F) = \{Y = \sum_{i=1}^n H^i \cdot X^i : H^i \in \mathcal{L}_p(X^i) \text{ and } X^i \in F\}$, where $\mathcal{L}_p(X^i)$ is the space of predictible integrands H^i such that*

$$\|H^i\|_{X^i} = \left(\mathsf{E}\left\{\int_0^\infty (H^i_s)^2 d[X^i, X^i]_s\right\}\right)^{\frac{1}{2}} < \infty, \quad 1 \le i \le n.$$

Proof: See [172], pp. 149–150. ∎

A finite set F in $\mathfrak{H}^2$ such that $\mathfrak{I}(F) = \mathfrak{H}^2$ is said to have the (*predictible*) *representation property*. Let us mention that stable subspaces and predictible representation are linked in a sense by the probability measure P. Indeed, this link may be attributed to a property of the probability measure P when this is viewed as one element among the collection of probability measures rendering all elements of F square-integrable martingales with respect to the underlying probability basis.

8.3.4 Linear stochastic equations

The end of this subsection is to present a closed form solution of a SDE driven by a semimartingale with an exogenous semimartingale term which prevents it to be strictly linear.

Theorem 8.29 *Let H be a semimartingale and let X be a continuous semimartingale and consider, for $t \in \mathbb{T}$ the process Z given for any $t \ge 0$ by*

$$Z_t = H_t + \int_0^t Z_{s-} dX_s. \tag{8.2}$$

Then the solution $Z = \mathcal{E}_H(X)$ of (8.2) is given by

$$\mathcal{E}_H(X)_t = \mathcal{E}(X)_t \left\{ H_0 + \int_{0+}^{t} \mathcal{E}(X)_s^{-1} d(H_s - [H, X]_s) \right\},$$

where $U = \mathcal{E}(X)$ is the solution of the homogeneous equation considered in Theorem 8.15.

Proof: By analogy with the method of variation of constants, let us assume the solution is of the form $Z = CU$, where C is càdlàg and U continuous. From the integration by parts formula we get

$$
\begin{aligned}
dZ_t &= C_{t-} \, dU_t + U_t \, dC_t + d[C, U]_t \\
&= C_{t-} U_t \, dX_t + U_t dC_t + U_t \, d[C, X]_t \\
&= Z_{t-} \, dX_t + U_t \, d\{C_t + [C, X]_t\}.
\end{aligned}
$$

Since Z solves (8.2), the left-hand side of the foregoing equation can be arranged so as to get

$$
\begin{aligned}
dH_t + Z_{t-} \, dX_t &= Z_{t-} \, dX_t + U_t \, d\{C_t + [C, X]_t\} \\
\frac{1}{U_t} dH_t &= d\{C_t + [C_t, X]_t\}.
\end{aligned}
$$

Calculating at both sides of the last equation the quadratic covariation with X and observing that $[[C, X], X] = 0$ yields

$$\frac{1}{U_t} dH_t = dC_t + \frac{1}{U_t} d[H, X]_t,$$

since

$$\left[\frac{1}{U} \cdot H, X \right] = [C, X].$$

Thus, we get

$$C_t = \int_0^t \frac{1}{U_s} d(H_s - [H, X]_s)$$

which completes the theorem. Recall $Z = CU$. ∎

For a continuous semimartingale X, we have the following worthwhile relationships

$$d\left(\frac{1}{\mathcal{E}(X)} \right) = \frac{dX - d[X, X]}{\mathcal{E}(X)} \quad \text{and} \quad \frac{1}{\mathcal{E}(X)} = \mathcal{E}(-X + [X, X]).$$

8.4   Stochastic calculus of the Itô type

The aim of this section is to survey a few issues of the theory of SDE's appropriate to construct and study a diffusion in $\mathbb{R}^n$, i.e. a mathematical description of the phenomenon concerning propagation of energy, as time passes by, e.g. in the form of capability to function, from one part of a system to another as a result of interactions of a random nature.

A classical example describes a system consisting of a tall cylindrical vessel with its lower part filled with a iodine solution and a column of clear water poured carefully and slowly on its top so as to avoid building of convention currents. Since at first the coloured part appears to be separated from the clear by a sharp, well-defined boundary and after a certain time the upper part begins to exhibit colour – the colour getting fainter towards the top while becoming less intense at the bottom –, it is evident that a transfer of iodine molecules from the lower part of the vessel to the upper part takes place. The iodine is said to diffuse into the water and a diffusion process is a mathematical description of the transfer of individual molecules of iodine from the lower to the upper part of the vessel viewed from an average point of view.

8.4.1 Path functionals and diffusions

The classical picture of diffusions has been dominated by a mathematical model representing a diffusion as a strong Markov process with continuous paths. Indeed, the classical picture was inspired by the study of PDE's underlying Fick's description of a diffusion, giving a solid quantitative basis to the phenomenon of heat transfer by conduction due to random molecular motion. Fick adopted the mathematical equation of heat conduction derived some years earlier by Fourier and provided the framework for a mathematical theory of diffusion in isotropic substances. This is based on the hypothesis that the rate of transfer of diffusing substances through a unit area of a section is proportional to the concentration gradient measured normal to the section, i.e.

$$F = -D\frac{\partial C}{\partial x}, \tag{8.3}$$

where F stands for the rate of transfer per unit area of section, C for the concentration of the diffusion substance, x for the space coordinate measured normal to the section, and D for the so-called diffusion coefficient. The negative sign suggests that diffusion occurs in the direction opposite to that of increasing concentration. If F, the amount of material diffusing, and C, the concentration, are both expressed in terms of the same unit of quantity, e.g. gram or gram molecule, then it is clear from (8.3) that D is independent of this unit and has dimension $(\text{length})^2(\text{time})^{-1}$, e.g. cm^2s^{-1}.

The basic hypothesis underlying the mathematical model expressed by (8.3) is in general consistent only for an isotropic medium, whose structure and diffusion properties in the neighbourhood of one point are the same relative to all directions. As a consequence of this symmetry, the direction of the flow of the diffusion substance at any point is normal to the surface of constant concentration through the point. However, anisotropic media have different diffusion properties in different directions so that (8.3) needs to be modified with the purpose of reflecting a preferential direction of orientation. For the time being, let us mention that the fundamental differential equation of diffusion in an isotropic medium resulting from (8.3) is given by

$$\frac{\partial C}{\partial t} = D \frac{\partial^2 C}{\partial x^2} \tag{8.4}$$

if the diffusion is one-dimensional. The expressions (8.3) and (8.4) are known as Fick's *first* and *second laws of diffusion*, and represent a mathematical model of the phenomenon of heat conduction; for more details see [43].

As a first approximation, a mathematical description of diffusion which places more emphasis on sample paths and average behaviour can be given as follows. Let a diffusion be a continuous process X in $\mathbb{R}^n$ such that

$$\mathsf{E}\left\{X_{t+\Delta t}^{(i)} - X_t^{(i)} | \mathcal{F}_t\right\} = b^{(i)}(X_t)\Delta t + o(\Delta t),$$

$$\mathsf{E}\left\{(X_{t+\Delta t}^{(i)} - X_t^{(i)})(X_{t+\Delta t}^{(j)} - X_t^{(j)}) | \mathcal{F}_t\right\} = a^{(ij)}(X_t)\Delta t + o(\Delta t),$$

for $i, j = 1, 2, \ldots, n$, t and $t + \Delta t$ to points of time in $\mathbb{T}$, and $(\mathcal{F}_t)_{t \in \mathbb{T}}$ the underlying filtration. Here $b^{(i)}$ and $a^{(ij)}$ are measurable functions with $a(x) = (a^{(ij)}(x))$ representing a matrix for each x which is an element of S_n^+, the set of all real $n \times n$ nonnegative-definite symmetric matrices. The measurable functions $a : \mathbb{R}^n \longrightarrow S_n^+$ and $b : \mathbb{R}^n \longrightarrow \mathbb{R}^n$ are called the *covariance* and *drift* of the process X. Then a *diffusion* in $\mathbb{R}^n$ with covariance a and drift b, for short an (a, b)-diffusion, is a continuous $\mathbb{R}^n$-valued semimartingale $X = (X^{(1)}, \ldots, X^{(n)})$ defined on a probability basis such that for each $i = 1, 2, \ldots, n$, one has

$$M_t^{(i)} = X_t^{(i)} - X_0^{(i)} - \int_0^t b^{(i)}(X_s)ds$$

is a continuous local martingale, and for $i, j = 1, 2, \ldots, n$ one has

$$[M^{(i)}, M^{(j)}]_t = \int_0^t a^{(ij)}(X_s)ds.$$

Essential in the preceding definition is the absence of any direct link between an (a, b)-diffusion X and any Markov property. However, if a and b satisfy certain conditions of great practical relevance any (a, b)-diffusion becomes a process satisfying a strong Markov property.

Itô's calculus is based on the idea of constructing sample paths of a diffusion X directly from those of a Brownian motion B. To see how we can describe X in terms of B via SDE's, let us assume that for any positive integer m we can find a measurable map $\sigma : \mathbb{R}^n \rightarrow \mathcal{L}(\mathbb{R}^m, \mathbb{R}^n)$ for which we have $\sigma(x)\sigma(x)^* = a(x)$ for each $x \in \mathbb{R}^n$; above $\mathcal{L}(\mathbb{R}^m, \mathbb{R}^n)$ stands for all $n \times m$ real matrices. Further, it is necessary to assume that we can find a probability basis $(\Omega, \mathcal{F}, \mathbf{F}, \mathbf{P})$ carrying both the diffusion X and the Brownian motion B. Under these circumstances the diffusion X can be described by

$$dX_t = \sigma(X_t)dB_t + b(X_t)dt,$$

where Itô's differential dB_t can be interpreted as a white noise by writing $b_t = \int_0^t \xi_s \, ds$. Rewriting the foregoing equation as

$$\sigma(X_t)dB_t = dX_t - b(X_t)dt,$$

setting $dM_t = \sigma(X_t)dB_t$, and observing $[dM, dM]_t = [\sigma(X)dB, \sigma(X)dB]_t = a(x)dt$, one easily recognises the link between the foregoing SDE and the diffusion process introduced above; see [187] pp. 110–114 and [172] pp. 187–188. Since dB and dt can be seen as semimartingale differentials, studying SDE's one may resort to some extent to analogies with the Brownian motion and Lebesgue measure. However, SDE's with coefficients that depend not only on the current value of the state but on the past history are more natural and interesting.

Path uniqueness and strong solutions

To the aim of stressing basic path features, let us write SDE's or more properly stochastic integral equations in the form

$$X_t = x_0 + \int_0^t \sigma(s, X.)dB_s + \int_0^t b(s, X.)ds, \qquad (8.5)$$

where B is again a Brownian motion on the probability basis $(\Omega, \mathcal{F}, \mathbf{F}, \mathbf{P})$, the initial condition x_0 is $\mathcal{F}_0$-measurable, and σ and b are suitably defined coefficients so as to ensure the stochastic integral and the (continuous) process X enjoy certain properties. More precisely, the coefficients $\sigma(s, X.)$ and $b(s, X.)$ need to be predictible and adapted respectively. Both requirements are fulfilled whenever they are predictible path functionals, i.e. if the coefficients σ and b act on the path space consisting on the set $\mathbf{W}^n = \mathbb{C}(\mathbb{T} \longrightarrow \mathbb{R}^n)$ of continuous paths $w : t \longrightarrow w(t)$ and the filtration $\mathcal{P}_W$ which renders such paths predictible.

Let Y be an arbitrary continuous process defined on the probability basis $(\Omega, \mathcal{F}, \mathbf{F}, \mathbf{P})$ with the initial condition x_0 which satisfies (8.5) as above, then we say the SDE (8.5) has a solution which is *pathwise uniquely defined* if we have $\mathbf{P}(\{\omega : X_t(\omega) = Y_t(\omega), \forall t\}) = 1$. If pathwise uniqueness holds in any set-up as described above, then it holds on the particular set-up generated by the initial condition x_0 and the driving Brownian motion B.

Moreover, it turns out that there is a deterministic function $F(x_0, B)$ of x_0 and B which determines a pathwise unique solution $X.$ of (8.5) on the canonical set-up (or in a sense the minimal set-up) generated by x_0 and B, i.e. there is a function $F : \mathbb{R}^n \times \mathbf{W}^m \longrightarrow \mathbf{W}^n$ such that $X. = F(x_0, B)$ solves (8.5) on the canonical set-up, $\mathbf{P}(\{\omega : X. = F(x_0, B.)\}) = 1$ and the finiteness condition

$$\int_0^t \{|\sigma(s, X.)|^2 + |b(s, X.)|\}ds < \infty \qquad \text{a.s.} \qquad (8.6)$$

holds for each t, where $|\sigma|^2 = \text{trace}(\sigma\sigma^*)$. Let us mention that even when the above condition fails to hold, it is possible to work on the basis of localisation with the solution of (8.5) up to the explosion time, i.e. the time ζ defined as $\zeta = \inf\{t : \text{the integral in (8.6) becomes infinity}\}$.

For any fixed but arbitrary probability measure μ characterising probabilistically the initial condition $x_0 \in \mathbb{R}^n$, we understand the *canonical set-up with initial law μ* as the arrangement consisting of the probability basis $(\Omega, \mathcal{F}, \mathbf{F}, \mathbf{P})$ with $\mathbf{P} = \mu \times \mathbf{P}_W$, the initial $\mathcal{F}_0$-measurable random variable x_0 and the $\mathbb{R}^m$-valued Brownian motion defined as above such that for any

$(x_0, w) \in \mathbb{R}^n \times \mathbf{W}^m$ and $t \in \mathbb{T}$ we have $B_t(x_0, w) = w(t)$ and the uncompleted
σ-field $\mathcal{F}_t^0 = \sigma(\{B_s : s \leq t\})$ generate in the usual sense the filtration $\mathbf{F}$. Above
$\mathbf{P}_W$ stands for the Wiener measure on $\mathbf{W}^m$ and (B_t) for the canonical process.
Further, the deterministic function F is said to be a *strong solution* of (8.5)
for every initial law μ if the process $X_\cdot = F(x_0, B_\cdot)$ solves (8.5) with any x_0 as
above on any set-up $(\Omega, \mathcal{F}, \mathbf{F}, \mathbf{P})$; for a rigorous and complete presentation of
these notions see [187] pp. 124–127 and [112] pp. 145–153.

Weak solutions and uniqueness in law

On the basis of SDE's one constructs a wide variety of (a, b)-diffusions which
cover most of the standard types of applications arising in practice, whenever
the coefficients σ and b are Lipschitz (recall $\sigma\sigma^* = a$). For various reasons it is
often appropriate to get solutions of SDE's which not necessarily hold on any
but only on a particular set-up; in such cases we shall say the SDE has a *weak
solution*. The notion of uniqueness corresponding to weak solutions is that of
uniqueness in law. We shall say that the SDE (8.5) has a *weak solution* with
initial (probability) law μ if there exists a probability basis $(\Omega, \mathcal{F}, \mathbf{F}, \mathbf{P})$, a $\mathbb{R}^n$-
valued continuous semimartingale X and a $\mathbb{R}^m$-valued $\mathcal{F}_t$-measurable Brownian
motion B such that x_0 has the law μ, the finiteness condition (8.6) holds and
X solves (8.5). Let X and Y be two arbitrary weak solutions of (8.5) (perhaps
on different set-ups) such that the laws of X_0 and Y_0 coincide, then we say the
SDE (8.5) has the property of *uniqueness in probability law* if the laws of X
and Y are the same.

The heart of the methodology of weak solution is shifting of probability
structures. This serves the purpose of associating two different weak solutions
with Brownian motions which are not only the same in law but even pairwise
identical. Hereby, one may transfer the structure of a weak solution X to a
canonical set-up in which $\mathbf{W}^n$ becomes the path space for X and $\mathbf{W}^m$ that for
B. The weak-solution methods rely on weak convergence techniques, properties
of regular conditional probabilities and localisation procedures.

Martingale problems and the strong Markov property

Next we shall outline the method known as the martingale problem which
constitutes a powerful alternative to the study of existence of weak solutions
and uniqueness in law. At a fairly intuitive level, the martingale problem is the
method by means of which one establishes links between an (a, b)-diffusion X
starting at time s from x and the measure $\mathbf{P}_{s,x}$ associated with the underlying
Brownian motion.

Indeed, a simple application of Itô's formula for all functions of class $\mathbb{C}^\infty$
let us identify a process X_φ given for any t by

$$X_\varphi(t) = \varphi(X_t) - \int_0^t \mathbf{L}_s \varphi(X_s) ds, \qquad (8.7)$$

where $\mathbf{L}$ stands for the differential operator appearing in Itô's formula, i.e.
$\mathbf{L}_t = \frac{1}{2} a^{(ij)}(t, \cdot) D_i D_j + b^{(i)}(t, \cdot) D_i$ where D_i represents the partial derivative

with respect to the i-th component of x and any index which appears twice in any product entails a summation over that index. One easily recognises, the process X_φ is a local martingale, and if in addition the coefficients a and b are bounded, we obtain the process X_φ is even a martingale for each $\varphi \in \mathbb{C}_0^\infty$.

The *martingale problem* is a methodology resulting from the interest to know (1) if there is for each $x \in \mathbb{R}^n$ a probability measure $\mathbf{P}_{s,x}$ on $\mathbf{W}^n$ such that $\mathbf{P}_{s,x}\{X_s = x\} = 1$ for $s \in \mathbb{T}$ and X_φ is a local martingale (resp. a martingale) for each $\varphi \in \mathbb{C}^\infty(\mathbb{R}^n)$ (resp. $\varphi \in \mathbb{C}_0^\infty(\mathbb{R}^n)$), (2) if the probability measure starting from (s, x) is unique, and (3) the implications deriving from the unique existence of the measure $\mathbf{P}_{s,x}$. The probability measure $\mathbf{P}_{s,x}$ on $\mathbf{W}^n$ is said to be a *solution to the martingale problem* for the (a, b)-diffusion starting from x at the time s. It is said that the martingale problem for a and b is *well-posed* whenever there exists exactly one solution to the martingale problem for the (a, b)-diffusion starting from x at time s.

To conclude, let us recall the issue of existence of a weak solution of (8.5) starting from $(0, x_0)$ is equivalent to that of existence of a solution to the martingale problem for the (a, b)-diffusion starting from $(0, x_0)$. Similarly, the question of uniqueness in law for (8.5) is equivalent to the uniqueness of the corresponding martingale problem. Finally, in the diffusion context $a(t, X_.) = a(X_t)$ and $b(t, X_.) = b(X_t)$, it turns out that an (a, b)-diffusion is a strong Markov process if the corresponding martingale problem is well-posed. For details see [187] pp. 158–163 and [213], Ch. 6.

8.4.2   Change of measure and representation

In addition to the various methods of constructing diffusions we have already outlined, there are other alternatives of great relevance for applications and of highly interpretative value. Three general methods which rely on the idea of transforming a SDE into another more amenable to available techniques, are known under the labels: change of time scale, change of state space and change of measure. In this connection we shall briefly describe how one can change the drift of the SDE by changing the measure of the underlying Brownian motion.

Absolutely continuous change of measure

Let $(\Omega, \mathcal{F}, \mathbf{F}, \mathbf{P})$ be a complete probability basis and assume the filtration $\mathbf{F}$ possesses the following consistency property:

P 8.1 *Extension property of absolutely continuous measures*

- If μ_t is an absolutely continuous probability measure on $(\Omega, \mathcal{F}_t)$ with respect to $\mathbf{P}$ such that μ_t restricted on $\mathcal{F}_s$ coincides with μ_s for any $t > s \geq 0$, then there exists a probability measure μ on $(\Omega, \mathcal{F})$ such that μ restricted on $\mathcal{F}_t$ coincides with μ_t for every $t \geq 0$.

Let $\mathfrak{c}\mathfrak{H}_{\mathrm{loc}}^2$ denote the set of all locally square integrable $\mathcal{F}_t$-martingales on $(\Omega, \mathcal{F}, \mathbf{F}, \mathbf{P})$ with a.s. continuous paths. Let $X = (X_t)_{t \geq 0}$ be an element of

$c\mathfrak{H}_{loc}^2$ and $X_0 = 0$ a.s. for $X \in c\mathfrak{H}_{loc}^2$, we define a process M by setting for any $t \in \mathbb{T}$

$$M_t = \exp(X_t - \frac{1}{2}\langle X \rangle t) \tag{8.8}$$

where $\langle X \rangle$ denotes the *quadratic variational process* corresponding to X. For simplicity we assume that M is a martingale; however let us mention that in the present context we do not need this result since the properties of the coefficients σ and b ensure Novikov's condition, i.e. $\mathsf{E}\{\exp(\frac{1}{2}\langle X \rangle_t\} < \infty$ for every t. Now, we define a measure $\widehat{\mathsf{P}}_t$ on $(\Omega, \mathcal{F}_t)$, for each $t \geq 0$ and $A \in \mathcal{F}_t$, by the formula

$$\widehat{\mathsf{P}}_t(A) = \mathsf{E}[M_t \, \mathbb{1}_A]. \tag{8.9}$$

Then it can easily be proved that for any $t > s \geq 0$

$$\widehat{\mathsf{P}}_t|_{\mathcal{F}_s} = \widehat{\mathsf{P}}_s,$$

since from the properties of conditional expectations we have for $A \in \mathcal{F}_s$

$$\mathsf{E}\{M_t \, \mathbb{1}_A\} = \mathsf{E}\{\mathsf{E}[M_t | \mathcal{F}_t] \, \mathbb{1}_A\} = \mathsf{E}\{M_s \, \mathbb{1}_A\}.$$

Further, under the assumption given by P 8.1, there exists a probability measure $\widehat{\mathsf{P}}$ on $(\Omega, \mathcal{F})$ such that

$$\widehat{\mathsf{P}}|_{\mathcal{F}_t} = \widehat{\mathsf{P}}_t.$$

$\widehat{\mathsf{P}}$ is called the probability measure *which has density M with respect to* P; for more details see for instance [112] pp. 176–177 or [17], pp. 42–43. We denote $\widehat{\mathsf{P}}$ as

$$\widehat{\mathsf{P}} = M \cdot \mathsf{P}$$

Let us recall the following theorem obtained by Girsanov in the case when $X = (X_t)_{t \geq 0}$ is a Brownian motion.

Theorem 8.30 (Girsanov)   *1. Let $Y \in c\mathfrak{H}_{loc}^2$, if we define $\tilde{Y}$ by*

$$\tilde{Y}_t = Y_t - \langle Y, X \rangle_t \tag{8.10}$$

then $\tilde{Y} \in c\mathfrak{H}_{loc}^2$. $\langle Y, X \rangle$ denotes the quadratic variational process corresponding to Y and X or, loosely speaking, the cross-variation process.

2. Let $Y^1, Y^2 \in c\mathfrak{H}_{loc}^2$ and define $\tilde{Y}^1$ and $\tilde{Y}^2$ by eq. (8.10), then

$$\langle Y^1, Y^2 \rangle = \langle \tilde{Y}^1, \tilde{Y}^2 \rangle. \tag{8.11}$$

The transformation of measures $\mathsf{P} \longrightarrow \widehat{\mathsf{P}}$ is called the Girsanov transformation of drift, since it induces a drift $\langle Y, X \rangle$ for every local martingale Y. Precisely, the Girsanov Theorem implies that every continuous local martingale Y with respect to P is transformed under the probability $\widehat{\mathsf{P}}$ into a process $Y = a$ consisting of a continuous local martingale $+ \langle Y, X \rangle$ as a result of the transformation of probability measures $\mathsf{P} \longrightarrow \widehat{\mathsf{P}} = M \cdot \mathsf{P}$.

A final point we want to mention refers to a martingale representation deriving from the martingale characterisation of the Brownian motion. To this end we consider the Brownian motion B defined on the probability basis $(\mathbf{W}^n, \mathcal{B}(\mathbf{W}^n), \mathbf{P}_W)$ and consider the filtration $(\mathcal{F}_t^B)$ generated by B and completed in the usual way. Then for $T > 0$ and $Y \in L^2(\mathbf{W}^n, \mathcal{F}_T^B, \mathbf{P}_W)$ there exists a $\mathcal{F}^B$-predictible process H with values in $\mathbb{R}^n$ and $\mathbf{E}\left\{\int_0^T H_s^2 ds\right\} < \infty$ such that

$$Y_t = \mathbf{E}\{Y | \mathcal{F}_0^B\} + \int_0^t H_s dB_s,$$

where $Y_t \in \mathbf{E}\{Y | \mathcal{F}_0^B\}$ and H is uniquely determined up to $\mathbf{P}_W \otimes t$ null-sets. In this connection we shall say that the integrand H can be described explicitly by setting $H_t = (d/dt)[Y, B]_t$, see [187] pp. 73–74.

8.4.3   Miscellany on Itô calculus

In the rest of this chapter we outline various features of stochastic calculus which is based on in a certain sense less probabilistic interpretation of basic ingredients like martingales, filtrations and related notions.

We shall consider a probability basis $(\Omega, \mathcal{F}, \mathbf{F}, \mathbf{P})$ with the usual properties and a standard Brownian motion defined on it such that $\mathcal{F}_t^0 = \sigma(B_s : s \leq t)$. The σ-algebras $(\mathcal{F}_t)$ are augmented in the usual sense. Let $L^p(\Omega, \mathcal{F}, \mathbf{P}; \mathbb{R}^n)$ denote for $1 \leq p < \infty$ the space of (equivalence classes of) random variables X for which X^p are integrable. For $n = 1$, one writes simply $L(\Omega, \mathcal{F}, \mathbf{P})$.

The $\mathbb{R}^n$-valued Brownian motion is said to be standard if for any $t_1, \ldots, t_n$ the vector $(B_{t_1}^1, B_{t_2}^2, \ldots, B_{t_n}^n)$ is a Gaussian vector with mean 0 and $\mathbf{E}\{B_t B_s^*\} = I\,t \wedge s$, where I stands for the identity matrix.

Stochastic integral in the sense of Ito

Let us consider a $\mathbb{R}^n$-valued standard Brownian motion B and let H be a function in $L_{\mathcal{F}}^2(0, T; \mathbb{R}^n)$, the space of processes H with $H_t = H(t, \omega)$ such that $\mathbf{E}\left\{\int_0^T |H_t|^2 dt\right\} < \infty$ and $H_t \in L^2(\Omega, \mathcal{F}_t, \mathbf{P}; \mathbb{R}^n)$ for almost all t. The stochastic integral denoted by

$$I_B(H) = \int_0^T H_t \cdot dB_t$$

is a continuous linear operator from $L_{\mathcal{F}}^2(0, T; \mathbb{R}^n)$ into $L^2(\Omega, \mathcal{F}_t, \mathbf{P})$. The operator I_B is an isometry, $\mathbf{E}\{I(H)\} = 0$ and

$$\mathbf{E}\{I_B(H) I_B(H')\} = \mathbf{E}\left\{\int_0^T H_t \cdot H_t' dt\right\},$$

where H and H' are arbitrary processes in $L_{\mathcal{F}}^2(0, T; \mathbb{R}^n)$. Let us consider the partition $\{t_0 = 0 < t_1 < \cdots < t_n = T\}$ and a process $H \in L_{\mathcal{F}}^2(0, T; \mathbb{R}^n)$, then we associate with H a step process such that $H(t, \omega) = H^k$ in $[t_k, t_{k+1})$ for $k = 0, 1, \ldots, n - 1$, where H^k is $\mathcal{F}_{t_k}$-measurable and $\mathbf{E}\{|H^k|^2\} < \infty$. Let us mention that the set of step processes is dense in $L_{\mathcal{F}}^2(0, T : \mathbb{R}^n)$. Next, the integral $I(H)$ is given by the Riemann sum

$$I_B(H) = \sum_{i=1}^{n-1} H^i \cdot (B_{t_{i+1}} - B_{t_i}). \tag{8.12}$$

The sum converges on $L^2(\Omega, \mathcal{F}_t, \mathbf{P}; \mathbb{R}^n)$ and the limit is called a stochastic integral in the sense of Itô. The sequence $H^n \longrightarrow H$ in $L^2(0, T; L^2(\Omega, \mathcal{F}, \mathbf{P}, \mathbb{R}^n))$ so that the integral is defined on the basis of Riemann sums converging on $L^2(\Omega, \mathcal{F}, \mathbf{P}; \mathbb{R}^n)$, which have then for any H and H' in $L^2_{\mathcal{F}}(0, T, \mathbb{R}^n)$ the following properties:

$$\mathbf{E}\left\{ \int_0^T H_t \, dB_t \right\} = 0$$

$$\mathbf{E}\left(\int_0^T H_t \, dB_t \int_0^T H'_t \, dB_t \right) = \mathbf{E}\left\{ \int_0^T H_t \cdot H'_t dt \right\}.$$

The norm of $L^2_{\mathcal{F}}(0, T; \mathbb{R}^n)$ is then given by $\|H\|^2_{L^2_{\mathcal{F}}} = \mathbf{E}\{(I_B(H)^2\}$; complete details are given in [20] pp. 29–42. For $[\tau, \sigma) \subset [0, T]$ let us set

$$\int_\tau^\sigma H_t \cdot dB_t = \int_0^T \mathbf{1}_{[\tau,\sigma)}(t) H_t \cdot dB_t$$

and we have the properties

$$\mathbf{E}\left\{ \int_\tau^\sigma H_t \cdot dB_t | \mathcal{F}_t \right\} = 0$$

$$\mathbf{E}\left\{ \left(\int_\tau^\sigma H_t \cdot dB_t \right)^2 \Big| \mathcal{F}_t \right\} = \mathbf{E}\left\{ \int_\tau^\sigma |H_t|^2 dt \, \Big| \, \mathcal{F}_t \right\}.$$

From the foregoing considerations it follows the process $I_B(H)$ can be shown to be a continuous martingale with respect to the filtration $\mathbf{F}$.

Stochastic differential equations

Let us consider the $\mathcal{F}_t$-adapted functions b and σ

$$b \; : \; [0, T] \times \mathbb{R}^n \longrightarrow \mathbb{R}^n$$
$$\sigma \; : \; [0, T] \times \mathbb{R}^n \longrightarrow \mathcal{L}(\mathbb{R}^m, \mathbb{R}^n)$$

which satisfy the conditions

$$|b(t, x) - b(t, x')| \leq K|x - x'|$$
$$|\sigma(t, x) - \sigma(t, x')| \leq K|x - x'|$$
$$|b(t, x)|^2 + |\sigma(t, x)|^2 \leq K\left[1 + |x|^2\right]$$

The problem is to find a process $X = (X_t)_{t \geq 0}$ satisfying for any $t \in \mathbb{T}$

$$dX_t = b(t, X_t)dt + \sigma(t, X_t)dB_t, \tag{8.13}$$

where the process B is the standard Brownian motion with values in $\mathbb{R}^m$ and $\mathcal{F}_t$-adapted, and $X_0 = x_0 \in \mathbb{R}^n$ and $\mathcal{F}_0$-adapted. That means, the process X solves the SDE above, which is indeed a shorthand for

$$X_t = x_0 + \int_0^t b(s, X_s)\, ds + \int_0^t \sigma(s, X_s)\, dB_s \qquad (8.14)$$

One can show that if $\mathsf{E}\{|X_0|^2\} < \infty$ and the conditions

$$|b(t, x) - b(t, x')| + |\sigma(t, x) - \sigma(t, x')| \leq k|x - x'|$$
$$|b(t, x)|^2 + |\sigma(t, x)|^2 \leq k_0^2(1 + |x|^2)$$

are satisfied, where $|\sigma|^2 = \mathrm{tr}(\sigma\sigma^*)$, then the SDE above has one and only one solution such that

$$\mathsf{E}\left\{ \sup_{0 \leq t \leq T} |X_t|^2 \right\} \leq k(1 + \mathsf{E}\{|X_0|^2\})$$

holds. Setting $\sigma = 0$, one easily recognises the analogy with ODE's. A case extremely useful in applications is the one where the initial conditions are random. One considers a filtration generated by the Brownian motion, a stopping time τ and a random variable ξ which are both $\mathcal{F}_t$-measurable such that $X_\tau = \xi$. Then, multiplying the coefficients of the SDE above component by component with $1 - \mathbf{1}_{\{t < \tau\}}$, one obtains a SDE starting at the random time τ from ξ and behaving for $t > \tau$ as before.

Let b and σ be $\mathcal{F}$-adapted processes with values in $\mathbb{R}^n$ and $\mathcal{L}(\mathbb{R}^m, \mathbb{R}^n)$ respectively. Let us assume $b \in L_{\mathcal{F}}^1(0, T; \mathbb{R}^n)$, i.e almost surely integrable, $\int_0^T |b_t|\, ds < \infty$ a.s., and for every $j = 1, 2, \ldots, m$, $\sigma^j \in L_{\mathcal{F}}^2(0, T; \mathbb{R}^n)$, where σ^j is the j-th column vector of the matrix σ. The cornerstone of stochastic calculus is the Itô formula, which we present in the following theorem

Theorem 8.31 *Let $\psi : [0, T] \times \mathbb{R}^n \longrightarrow \mathbb{R}$ be a functional of class $\mathbb{C}^{1,2}(\mathbb{T} \times \mathbb{R}^n)$. Let the process $X = (X_t)_{t \in \mathbb{T}}$ be the solution of the SDE*

$$dX_t = b_t\, dt + \sigma_t\, dB_t,$$

where $X_0 = x_0 \in \mathbb{R}^n$ is $\mathcal{F}_0$-measurable, and the coefficients b and σ are defined as above. Then the following formula holds

$$
\begin{aligned}
d\psi(t, X_t) =\ & \frac{\partial \psi}{\partial x}(t, X_t)\sigma_t\, dB_t \\
& + \left\{ \frac{\partial \psi}{\partial x}(t, X_t)b_t + \frac{1}{2}\mathrm{tr}\frac{\partial^2 \psi}{\partial x^2}(t, X_t)\sigma_t\sigma_t^* + \frac{\partial \psi}{\partial t}(t, X_t) \right\} dt
\end{aligned}
$$

where $\mathrm{tr}A$ stands for the trace of the matrix A.

An easy way of understanding the appearance of the correction term with second order derivatives is as follows. We shall consider for simplicity only the scalar case. To this end, let us integrate from t to $t + \Delta t$ and take expectation at both sides to obtain

$$
\begin{aligned}
& \mathsf{E}\left\{\psi\left(t + \Delta t, X_{t+\Delta t}\right)\right\} \\
& = \quad \mathsf{E}\left[\psi(t, X_t)\right] + \Delta t\, \mathsf{E}\left[\frac{\partial \psi}{\partial t}(t, X_t) + \frac{\partial \psi}{\partial x}(t, X_t)b_t + \frac{1}{2}\frac{\partial^2 \psi}{\partial x^2}(t, X_t)\sigma_t^2\right],
\end{aligned}
$$

where the additional term appears. Then, let us make the following approximate computation of the term $\psi(t + \Delta t, X_{t+\Delta t})$,

$$
\begin{aligned}
\psi\left(t + \Delta t, X_{t+\Delta t}\right) & \simeq \quad \psi\left(t + \Delta t, X_t + b_t\Delta t + \sigma_t \Delta B_t,\right) \\
& = \quad \psi(t, X_t) + \Delta t\left[\frac{\partial \psi}{\partial t}(t, X_t) + \frac{\partial \psi}{\partial x}(t, X_t)b_t\right] \\
& \quad + \frac{\partial \psi}{\partial x}(t, X_t)\sigma_t \Delta B_t + \frac{1}{2}\frac{\partial^2 \psi}{\partial x^2}(t, X_t)\sigma_t^2 \Delta^2 B_t + \dots
\end{aligned}
$$

Taking expectation again and after a comparison, we obtain $\mathsf{E}\{\Delta^2 B_t\} = \Delta t$ which makes clear the properties of the Brownian motion give rise to the correction term appearing in the Itô formula.

We shall recall a special version of the Itô rule applicable to stochastic processes with continuous paths. Let τ be a Markov time, σ and b two processes such that $\sigma \in L^2_{\mathcal{F}}(0, T; \mathcal{L}(\mathbb{R}^m; \mathbb{R}^n))$ and $b \in L^1_{\mathcal{F}}(0, T; \mathbb{R}^n)$, where $\mathcal{L}(\mathbb{R}^m, \mathbb{R}^n)$ denotes the set of $n \times m$ real matrices. Let us assume further that $\mathbb{1}_{\{t \leq \tau\}}\sigma_t$ and $\mathbb{1}_{\{t \leq \tau\}}b_t$ are progressively measurable, where $\mathbb{1}_{\{t \leq \tau\}}$ is an indicator function, i.e. $\mathbb{1}_{\{t \leq \tau\}} = 1$ if $\omega \in \{t \leq \tau\}$, otherwise it is 0. Then, let $X = (X_t, \mathcal{F}_t)_{\{t \geq 0\}}$ be a process with continuous paths such that the relationship

$$
\mathsf{P}\left\{\sup_{\tau \geq t}\left|X_t - x - \int_0^t \mathbb{1}_{\{s \leq \tau\}}\sigma_s\,dB_s - \int_0^t \mathbb{1}_{\{s \leq \tau\}}b_s\,ds\right| = 0\right\} = 1, \qquad (8.15)
$$

and the initial condition $X_0 = x$ hold. The process X shall also be represented for $i = 1, 2, \dots, n$ by the stochastic differential equation

$$
\begin{aligned}
dX_t^{(i)} & = \quad \sum_{k=1}^m \sigma_t^{(ik)}dB_t^{(k)} + b_t^{(i)}dt, \quad \text{for } t > 0, \quad \text{and} \qquad (8.16) \\
X_0^{(i)} & = \quad x^{(i)}.
\end{aligned}
$$

Eqs. (8.16) are only a formal representation of eq. (8.15). Here, we shall deal indeed with strong Markov processes with continuous paths which are known as *diffusion processes* under certain additional smoothness requirements. The following lemma shall be used quite often in the sequel. Stochastic differential equation shall be denoted by SDE for short.

Lemma 8.32 *Assume the process X satisfies the stochastic differential equation (8.16) and let $h(t, X_t)$ be a functional of class $\mathbb{C}^{1,2}(\mathbb{T} \times \mathbb{R}^n \longrightarrow \mathbb{R})$. Then the process $h(t, X)$ satisfies, for any $0 \leq t \leq \tau$, the following SDE*

$$
dh(t, X_t) = \quad \frac{\partial h(t, X_t)}{\partial t}dt + \sum_{i=1}^n D^i h(t, X_t)dX_t^i
$$

$$+\frac{1}{2}\sum_{i,j=1}^{n} D^i D^j h(t, X_t) d\langle X^{(i)}, X^{(j)}\rangle_t, \qquad (8.17)$$

$$h(0, X_0) = h(0, x_0),$$

where D^i stands for the partial derivative with respect to the i-th component of X.

Eq. (8.17) is an analogue of Taylor's formula with the first two terms and it is known as the *chain rule for stochastic differentials*. Applying the usual rules for removing parentheses and for the product of stochastic differentials, i.e. $(dB_t^{(i)})^2 = dt$, $dB_t^{(i)} dB_t^{(j)} = 0$ for $i \neq j$, $dB_t^{(i)} dt = 0$ and $(dt)^2 = 0$, one obtains for the third term in (8.17), the expression

$$d\langle X^{(i)}, X^{(j)}\rangle_t = \sum_{k=1}^{m} \sigma_t^{(ik)}(\omega)\sigma_t^{(jk)}(\omega)dt,$$

which we shall write for short as

$$d\langle X^{(i)}, X^{(j)}\rangle_t = S_t^{(ij)} dt,$$

with $S^{(ij)}$ given by

$$S_t^{(ij)} = \sum_{k=1}^{m} \sigma_t^{(ik)}\sigma_t^{(jk)}.$$

Let us recall that for a diffusion process the characteristic operator or more precisely the differential operator coincides with the second order differential operator $\mathsf{L}_t^{\sigma,b}$ given by

$$\mathsf{L}_t^{\sigma,b}(\omega) = \frac{1}{2}\sum_{i,j=1}^{n} S_t^{(ij)} D^i D^j + \sum_{i=1}^{n} b_t^{(i)} D^i. \qquad (8.18)$$

Then, (8.17) can be written for any $t \in \mathbb{T}$ in its integral version as

$$h(t, X_t) = h(0, x) + \int_0^t \mathbf{1}_{\{s \leq \tau\}} \nabla_x h(s, X_s) \cdot \sigma_s dB_s \qquad (8.19)$$

$$+ \int_0^t \mathbf{1}_{\{s \leq \tau\}} \left(\frac{\partial}{\partial t} + \mathsf{L}_t^{\sigma,a}\right) h(s, X_s) ds, \quad t \geq 0,$$

which is known as the *change of variable formula*. For more details see [82, 139, 20] which contain generalisations of Itô's formula in various forms quite useful for control theoretical applications.

Chapter 9

Stochastic control theory

The uncertainty principle had profound implications for the way in which we view the world. Even after more than fifty years they have not been fully appreciated by many philosophers, and are still the subject of much controversy. The uncertainty principle signaled an end to Laplace's dream of a theory of science, a model of the universe that would be completely deterministic: one certainly cannot predict future events exactly if one cannot even measure the present state of the universe precisely! We could still imagine that there is a set of laws that determines events completely for some supernatural being, who could observe the present state of the universe without disturbing it. However, such models of the universe are not of much interest to us ordinary mortals. It seems better to employ the principle of economy known as Occam's razor and cut out all the features of the theory that cannot be observed. This approach led Heisenberg, Erwin Schrödinger, and Paul Dirac in the 1920s to reformulate mechanics into a new theory called quantum mechanics, based on the uncertainty principle. ...

In general, quantum mechanics does not predict a single definite result for an observation. Instead, it predicts a number of different possible outcomes and tells us how likely each of these is.

Stephen W. Hawking
(see [104] p. 59)

9.1   Semimartingale features of control

In this chapter we shall review various concepts of the theory of stochastic control, which support the application of control techniques to various problems related to capital accumulation. We shall restrict ourselves to martingale techniques with the purpose of illustrating the clarity of probabilistic interpretations, the powerful insight and flexibility the martingale approach provides. Preparing this quick survey we have heavily relied on [59] to which we refer for a careful and detailed study of stochastic control with emphasis on probabilistic analysis.

Let us consider a stochastic process X defined on the measurable space $(\Omega, \mathcal{F}_\infty)$ which evolves in the course of time. Let us assume we are given a current of σ-algebras $\mathcal{F}_t$ of $\mathcal{F}_\infty$ which describes the history of the process X up to the time ζ. The time ζ is known as the *life* or *terminating time* of the process X and it may be finite or infinite. The history of the evolution of the process X is said to be given by the filtration $(\mathcal{F}_t)_{0 \le t \le \zeta}$ with $t \in \mathbb{R}_+$. We are interested in steering the evolution of the process $X = (X_t, \mathcal{F}_t)_{0 \le t \le \zeta}$ in a manner to become apparent later.

9.1.1   A brief outline of the question of control

From the outset we shall assume the process X can be observed at certain random times τ which are called *times of observation*. It is natural to assume moreover that at any time τ an observer gets only partially access to the information associated with the history of the evolution of the process X up to time τ.

Let $\mathcal{O}$ denote the σ-algebra generated by the information accessed up to time ζ, which is obviously a sub-algebra of $\mathcal{F}_\infty \times \mathcal{B}(\mathbb{R}_+)$. A process X measurable with respect to $\mathcal{O}$ will be called an *observable process*; as short-hand we will then write $X \in \mathcal{O}$. We shall limit ourselves to the class of observable processes $\mathcal{O}$. The class of observable processes $\mathcal{O}$ is a sub-σ-algebra of $\mathcal{F}_\infty \times \mathcal{B}(\mathbb{R}_+)$ characterised by a family of processes defined on $\Omega \times \mathbb{R}_+$, measurable with respect to $\mathcal{F} \times \mathcal{B}(\mathbb{R}_+)$ so that

- it is generated by the right-continuous processes with left-limits, $\mathcal{F}_t$-adapted,

- it contains the deterministic algebra $\{\emptyset, \Omega\} \times \mathcal{B}(\mathbb{R}_+)$, and

- it remains stable under the operation of stopping at fixed times, i.e. for any $X \in \mathcal{O}$ and $s \in \mathbb{R}_+$, then the process $\tau \longrightarrow X_{s \wedge \tau}$ is also in $\mathcal{O}$.

Let us denote by $\mathcal{T}$ the class of observation times, i.e. the class of $\mathcal{O}$-stopping times or random variables τ such that for any $t < \zeta$ the family of random variables $\mathbf{1}_{\{\tau \le t\}}$ builds a $\mathcal{O}$-measurable process. The class $\mathcal{T}$ contains 0, remains stable under sup and inf operations, and contains a sequence $\{\tau_n\}$ for which $\sup \tau_n = \infty$. Given an observation time $\tau \in \mathcal{T}$, we shall say that the random variable Z is observable at time τ, whenever there is a $\mathcal{O}$-measurable process X such that $X_\tau = Z$. The σ-algebra generated by the events observable at τ

shall be denoted by $\mathcal{G}_\tau$. Let O_G be a variable signalling the observation time of
an event $G \in \mathcal{G}_\tau$ in the sense that on G it holds $O_G = \tau$, while on $G^\complement$ one has
$O_G = \infty$. Then the σ-algebra $\mathcal{G}_\tau$ can be characterised as the σ-algebra with
respect to which all the variables O_G are $\mathcal{O}$-stopping times. By convention one
sets $\mathcal{G}_\infty = \sigma(\bigcup_\tau \mathcal{G}_\tau)$. It is convenient to assume first, the life time $\zeta \in T$ and
second, if $S \in T$ then $\{\zeta \leq S\} \subseteq \{S = \infty\}$, i.e. observation times after ζ are
set equal to infinity.

The idea of intervening in or *steering the evolution* of the process X refers to
eventual attempts to influencing the probability of occurrence of given events
by purposely choosing certain actions or interventions. That means, to describe
this type of operations we need to introduce a family $\mathfrak{P}$ of probability measures
indexed by a set $\mathcal{A}$ representing the available actions or interventions. Thus,
let $\mathfrak{P}$ be given by

$$\mathfrak{P} = \{\mathsf{P}_\alpha \mid \alpha \in \mathcal{A}\},$$

where P_α is a probability measure defined on $(\Omega, \mathcal{F}_\infty)$. Since the process X
may be expected to satisfy certain conditions, it turns out the set $\mathcal{A}$ needs in
turn to fulfil requirements aimed at sorting out any action which may result
in violations of the conditions imposed on X. A set $\mathcal{A}$ fulfilling requirements
of that sort qualifies for building the family of admissible controls, actions or
interventions. At a fairly general level, the set $\mathcal{A}$ is said to be the *family
of admissible controls* if $\alpha \in \mathcal{A}$ entails $\alpha \in \mathcal{O}$, i.e. α is an observable or $\mathcal{O}$-
measurable process, and α takes values on a Lusin space $(C, \mathcal{B}(C))$ where $\mathcal{B}(C)$
stands for the Borel σ-algebra of C. Let us recall, a Lusin space is a Hausdorff
space that is the image of a Polish space under continuous bijection. It is
customary to add to C a cementery point $\natural$ such that for any $t > \zeta$ one gets
$\alpha_t = \natural$. For the sake of simplicity, one assumes all controls take the same
value for $t = 0$. For any $\alpha \in \mathcal{A}$, let us denote by X^α a meaningful process
resulting from the application of the control action α and this will be termed
the *controlled process* while P_α shall be called the *controlled object*. We shall
make these notions precise in what follows.

For the time being let us just mention that in the face of several admissible
controls it is necessary to introduce a performance index to be used as a criterion on the basis of which to discriminate among the available controls. Summing
up, let us say steering the evolution of the process X means, determining a prob-
ability measure P_α by choosing an admissible control action $\alpha \in \mathcal{A}$ from which
a meaningful process X^α emerges fulfilling certain desired properties not avail-
able in the original process X. The arrangement $(\Omega, \mathcal{F}, (\mathcal{F}_t)_{t \geq 0}, \zeta, \mathcal{O}, \mathsf{P}_\alpha, \mathcal{A})$
is often referred to as the stochastic base of the process X. This preliminary
outline of the question of control is intended to support the understanding of
heuristic argumentation.

The set of all admissible controls which coincide with α at time t shall be
denoted by $\mathcal{A}(\alpha, t)$. With a time super- or subindex, i.e. $\mathcal{A}^t$, $\mathcal{A}_t$, we shall
distinguish admissible control processes the observations of which before, re-
spectively after t, are neglected. For $s \leq t$, one defines similarly $\mathcal{A}_t^s$, i.e. a
set of admissible control chosen on the basis of observations made between the
times s and t. The set $\mathcal{A}$ of admissible controls is said to be *compatible* with
$(\Omega, \mathcal{F}_t, \mathcal{O}, \zeta)$, if for any admissible control $\alpha \in \mathcal{A}$, any observation time $s \in T$

and any $C \in \mathcal{F}_s$, we have $\mathbf{P}_\alpha(C) = \mathbf{P}_\beta(C)$ for any $\beta \in \mathcal{A}(\alpha, s)$. The probability measure $\mathbf{P}_\alpha$ is said to be $\mathcal{G}_t$-*stable*, whenever $\beta \in \mathcal{A}(\alpha, s)$ and $\alpha = \beta$ on the event $A \in \mathcal{G}_t$ entail $\mathbf{P}_\alpha(A \cap C) = \mathbf{P}_\beta(A \cap C)$ for any $C \in \mathcal{F}_\infty$. The following Lemma holds.

Lemma 9.1 *Let* $\beta \in \mathcal{A}(\alpha, t)$ *be an arbitrary control action, then the set of probability measures* $\mathbf{P}_\beta$ *is* $\mathcal{G}_t$-*stable.*

Proof. See El Karoui [59], p. 87.   ∎

To the end of ensuring a certain degree of operationality, one needs to endow the set $\mathcal{A}$ of admissible controls with an appropriate structure, so as to hinder changes in control interventions taking place at observation times in a given event drive states of the system out of its admissible range. To this purpose we introduce three basic control operations. At the heart is the idea that splitting a control action at an observation time one should maintain the admissibility of the action.

Let us consider an observation time $\tau \in T$ and an event $G \in \mathcal{G}_\tau$. Further, we consider an arbitrary control $\beta \in \mathcal{A}$ and its associate family of controls $\mathcal{A}(\beta, \tau)$. Then for a control $\alpha \in \mathcal{A}(\beta, \tau)$, a control γ which acts on G as the control α and on $G^\complement$ as the control β is also an admissible control. Indeed, the control γ changes at the observation time O_G from the action α to β for $G \in \mathcal{G}_\tau$ given. For $F \in \mathcal{F}_\infty$ we have $\mathbf{P}_\gamma(F) = \mathbf{P}_\alpha(F \cap G) + \mathbf{P}_\beta(F^\complement \cap G)$. Following operations make sure a control consists of suitably chosen control actions.

P 9.1 *Operations among control processes* $\alpha \in \mathcal{A}$

- *Restriction:* For any $s \le \tau \le T$ and $\alpha \in \mathcal{A}_T^s$, the restriction of the control process α to the interval $[s, \tau]$ defined by the map $\alpha \longrightarrow \alpha|_{[s,\tau]}$ is again a control process in $\mathcal{A}_\tau^s$.

- *Concatenation:* For $\alpha^1 \in \mathcal{A}_\tau^s$ and $\alpha^2 \in \mathcal{A}_T^\tau$, the control process $\beta = (\beta_t)_{t \ge 0}$ defined by

$$\beta_t(\omega) = \begin{cases} \alpha_t^1(\omega), & \text{if } t \in [s, \tau] \\ \alpha_t^2(\omega), & \text{if } t \in (\tau, T] \end{cases}$$

  is a control process in $\mathcal{A}_T^s$.

- *Finite mixing:* For $\alpha^1, \alpha^2 \in \mathcal{A}_T^s$ and $G \in \mathcal{G}_s$, the control process β given by

$$\beta_\tau(\omega) = \begin{cases} \alpha_\tau^1(\omega), & \text{if } \omega \in G \\ \alpha_\tau^2(\omega), & \text{if } \omega \in G^\complement \end{cases}$$

  is again in $\mathcal{A}_T^s$.

Since the application of control actions may be aimed at steering during the whole time period $[0, T] \subset \mathbb{R}_+$, the operations among control actions introduced above allow a flexible tuning of interventions according to the response of the system and the information becoming available at any moment of time. From this perspective, we may speak about adapting the evolution of the process X

by means of arbitrarily chosen control $\alpha \in \mathcal{A}$ so that some functionals of X attain desired properties. For instance, one may think of the reward derived from (respectively cost or time required to) running a system represented by the process X and we may wish these features satisfy certain conditions of optimality.

At this stage the basic steps of controlling a dynamic system are more explicit:

(1) capturing the system essentials in a model that accurately reflects the impact of time, change and basic elements of interaction;

(2) identifying means of intervening or interacting with the process describing the system evolution;

(3) defining the admissible operations on the basis of which one attempts to steer the model dynamics; and

(4) introducing an adequate criterion capable of selecting a desired succession of interventions and outcomes.

To close this brief outline of the framework of stochastic control, let us stress the fact that by endowing the family $\mathcal{A}$ of admissible controls with a rich and flexible structure one identifies in a certain sense the structure of the class $\mathfrak{P}$ of associate probability measures or controlled objects. Further characterisations of controls are problem specific, so we may resort to nonanticipative, stationary and Markovian features of controls and these lead in turn to a distinguishing class of controlled objects and solution techniques. Concerning the controlled process we just mentioned, it may be characterised on the basis of a strong or weak formulation, where the control process acts merely as a parameter; see Chapter 8. Next, we pass to study the structure underlying reward and/or functionals of the control process.

9.1.2   Performance criteria and optimality

Let us proceed to make more precise the criterion guiding the choice of "the best control policy". It is intuitive and appealing to resort to a performance index measuring *total reward* or *utility* derived from the process X under the control α acting along the time interval $[0, T]$. This is reflected by introducing the functional $R(\omega) : \alpha \longrightarrow R^{\alpha}(\omega)$ with $R^{\alpha}(\omega)$ defined as

$$R^{\alpha}(\omega) = \int_{0}^{T} f(t, X_t, \alpha_t)dt + \phi(T, X_T), \tag{9.1}$$

where for any time $t \in [0, T]$ the functions $f(t, x, \alpha)$ and $\phi(T, x)$ shall be assumed to be concave, negative bounded measurable with respect to the underlying filtration. These functions may be thought of as describing the instantaneous rate of reward and terminal pay-off. The functional $R^{\alpha}(\omega)$ shall be $\mathcal{F}_T$-measurable, negative, $\mathbf{P}_{\alpha}$-integrable and control-consistent in the sense that on observable sets on which two controls α and α° coincide we should have $R^{\alpha}(\omega) = R^{\alpha^{\circ}}(\omega)$ $\mathbf{P}_{\alpha} - a.s.$ and $\mathbf{P}_{\alpha^{\circ}} - a.s.$ At this point we may say, the question of control can be seen as the problem of maximising the expectation of the functional $R^\alpha(\omega)$ over a prescribed set of admissible controls.

The functional $R^\alpha(\omega)$ acts indeed on $\mathbb{D} \times \mathbb{P}$, where $\mathbb{D}([0,T] \longrightarrow \mathbb{R}^n)$ is the space of right-continuous functions on $[0,T]$ with left-limits and $\mathbb{P}([0,T] \longrightarrow A)$ the space of predictible functions with values in C. Further, we shall require f and ϕ are of class $\mathbb{C}_b([0,T] \times \mathbb{R}^n \times C \longrightarrow \mathbb{R}_-)$ and $\mathbb{C}_b([0,T] \times \mathbb{R}^n \longrightarrow \mathbb{R}_-)$ respectively. However, the characterisation of f and ϕ is in a sense determined by the control problem under consideration. Since the time horizon needs to suit the type of control questions risen, it is natural to choose smoothness of reward and pay-off functions f and ϕ so as to render the total reward $R^\alpha(\omega)$ meaningful. Corresponding to the type of questions at issue, one may hence formulate optimising problems focusing (1) in the terminal (total) reward without specifying its structure, (2) in observations of the controlled process up to the time, it first exits a region $G \subseteq \mathbb{R}^n$ where $T = \inf\{t \geq 0 \mid X_t \notin \overline{G}\}$, (3) in time-averaged asymptotic behaviour or ergodic control, and the like, all requiring an appropriate control-theoretical framework; see for instance [131, 132, 133].

A central objective of the theory of stochastic control is to devise means of choosing an admissible process β for which the mean value or mathematical expectation of the functional $R^\alpha(\omega)$ attains a supremum and forwarding techniques helpful in finding the attached *expected total reward* $\mathbf{E}^\beta\{R^\beta(\omega)\}$. It is important to note the mean reward depends on the chosen control and the associate trajectory of the controlled process which in turn is determined by the controlled object. The optimising process β will be called the *optimal control*. We have then the following control problem.

Optimising problem 9.1 (Global optimality) *Let $\mathcal{A}$ be the set of admissible controls associated with the system dynamics condensed in the process X, which in particular can be thought of being governed by the SDE (5.8). Further let R^α, given by (9.1), be the chosen performance index. Find an admissible control $\beta \in \mathcal{A}$ and the attached function R^β such that*

$$R^\beta = \mathbf{E}^\beta\{R^\beta(\omega)\} = \sup_{\alpha \in \mathcal{A}} \mathbf{E}^\alpha\{R^\alpha(\omega)\}. \tag{9.2}$$

The basic questions of the theory of control emerge already in the just formulated simple problem. They consist in:

- existence of an optimal control β,

- characterisation of β,

- designing of methods of determining β,

- in the case of failure of the above steps, one attempts to go through all these steps again, resorting to ϵ-optimality, and

- determining the optimal reward functional R^β.

The foregoing notion of optimality, though very general, is not quite appropriate, since it does not allow the control process to update flexibly as new information on the evolution of X arrives. The mean reward is based on the whole information on the controlled process X which is not always available. From a practical point of view it seems more reasonable to let the control process $\alpha = (\alpha_t)_{t \geq 0}$ take values on the basis of observations of the controlled process X until time t. In other words, the control α_t may be a functional of the trajectory $X_{[0,t]} = \{(s, X_s) : 0 \leq s \leq t\}$ which means that $\alpha_t = \alpha(t, X_{[0,t]})$ or $\alpha_t = \alpha(t, X.)$. This form of the control process is called *feedback control*. In general an admissible control α is said to be a feedback control if it is progressively measurable with respect to the natural filtration generated by the controlled process X.

A natural way of extending the notion of (global) optimality introduced above is obtained by defining for any $s \in T$, the functional $R_s : \alpha \longrightarrow R_s^\alpha$, known as the *expected conditional reward* of the process X^α corresponding to the control policy α, updated with the information $\mathcal{F}_s$ available at the time $s \in [0,T]$. The just introduced functional R_s^α can also be called the total expected conditional reward. The functional R_s^α is given by means of the formula

$$R_s^\alpha = \mathsf{E}^\alpha \left\{ \int_0^T f(t, X_t, \alpha_t)dt + \phi(T, X_T) \middle| \mathcal{F}_s \right\} \qquad \mathsf{P}_\alpha \quad \text{a.s.,} \qquad (9.3)$$

builds up the notion of conditional optimality and relies on the information generated by $X_{[0,s]}$. We shall call *conditional reward at the time s* the set of random variables R_s^α with $s \in T$ and $\alpha \in \mathcal{A}(\tilde{\alpha}, s)$ with $\tilde{\alpha} \in \mathcal{A}_s$, i.e. conditional reward at time s is based on control policies with values restricted to the time interval $[0, s]$ and s an observation time. Equivalently, we can say the choice of the control $\alpha \in \mathcal{A}^s$ cannot be based on the anticipation of the values it may take during the time interval $[s, T]$. This set fulfils various interesting properties among which the one of being a lattice shall in the sequel play a crucial role. Thus, we record it in the following Lemma.

Lemma 9.2 *The set of random variables $\{R_s^\alpha, \ s \in T, \ \alpha \in \mathcal{A}\}$ is stable with respect to* sup *and* inf *operations.*

Proof: See El Karoui [59], p. 93.   ∎

Let us briefly recall some basic properties characterising the conditional reward $\{R_s^\alpha, \ s \in T, \ \alpha \in \mathcal{A}\}$ at the observation time s. Here it is essential to note the notion of conditional reward opens the way for adaptive control actions.

P 9.2 *Properties of the conditional reward R_s^α*

1. The relationship $R_s^\alpha = R_t^\alpha$ P_α a.s. holds on the set $\{\omega : t = s\}$.

2. For any $\alpha \in \mathcal{A}$ and $s \in T$, the random variable R_s^α is $\mathcal{G}_s$-measurable.

3. Whenever $\alpha^\circ \in \mathcal{A}(\alpha, t)$ and $\alpha = \alpha^\circ$ on $A \in \mathcal{G}_t$, it follows that the relationship $R_t^\alpha = R_t^{\alpha^\circ}$ holds $\mathsf{P}_\alpha - a.\,s.$ on the set A.

A set of random variables $\{R_s^\alpha : \ s \in T, \ \alpha \in \mathcal{A}\}$ that satisfy the properties 1.–3. above is called a $(T, \mathcal{A})$-*system*. This gives rise to the question of finding for fixed but arbitrary initial data (s, x), a function $v(s, x) : [0, T] \times \mathbb{R}^n \longrightarrow \mathbb{R}_+$ and an optimal control process $\beta = \beta(s, X_{[0,s]})$ so that it solves again an optimality problem with the property that when $s = 0$, then the relationship

$$v(x) = R^\beta, \qquad (9.4)$$

holds with $v(x) = v(0, x)$ and $X_s = x$. However, the flexibility gained by turning to the notion of conditional reward calls for a little more sophisticated optimising procedure. This is because conditioning amounts to a switching of informational structure akin to changes of probability measure. To this end we introduce the concept of essential supremum.

Let $(\Omega, \mathcal{F}, \mathbf{F}, \mathbf{P})$ be a probability basis as before. Let $\{f_\gamma(\omega) : \gamma \in C\}$ be a family of measurable, a.s. non-positive (possibly $-\infty$) functions defined on $(\Omega, \mathcal{F}, \mathbf{F}, \mathbf{P})$. Then the function $f(\omega)$ given by

$$f(\omega) = \text{ess} \sup_{\gamma \in C} f_\gamma(\omega) \qquad \mathbf{P} \quad \text{a.s.}$$

is said to be the *essential supremum* of the family defined above provided,

(1) $f(\omega)$ is $\mathcal{F}$-measurable

(2) for all $\gamma \in C$
$$f(\omega) \geq f_\gamma(\omega) \qquad \mathbf{P} \quad \text{a.s.}$$
   if $g(\omega)$ satisfies (i) and (ii), then
$$g(\omega) \geq f(\omega) \qquad \mathbf{P} \quad \text{a.s.}$$

   hold.

A property of the essential supremum which plays a decisive role in the sequel, is that of interchangeability of the operations of essential supremum and conditional expectations. To this point we shall shortly come back and turn next to the concept of conditional optimality.

Optimising problem 9.2 (Conditional optimality) *For any time $s \in T$ and any control $\alpha \in \mathcal{A}$, find a functional $J : (s, \alpha) \longrightarrow J(s, \alpha)$ and an admissible control process $\beta \in \mathcal{A}^s$ such that the $(T, \mathcal{A})$-system defining the conditional reward holds the relationships*

$$J(s, \beta) = \text{ess} \sup_{\alpha \in \mathcal{A}(\beta,s)} \mathbf{E}^\beta \left[\int_0^T f(t, X_t, \alpha_t) dt + \phi(T, X_T) \Big| \mathcal{F}_s \right] \qquad \mathbf{P}_\beta \quad \text{a.s.}$$

$$J(s, \beta) = \text{ess} \sup_{\alpha \in \mathcal{A}(\beta,s)} R_s^\alpha \qquad \mathbf{P}_\beta \quad \text{a.s.} \qquad (9.5)$$

where the supremum is taken over all controls restricted to the class $\mathcal{A}(\beta, s)$.

A control process $\tilde{\beta}$ shall be called (s, β)-*conditionally optimal*, if $\tilde{\beta} \in \mathcal{A}(\beta, s)$ and the relationship

$$J(s, \beta) = R_s^{\tilde{\beta}} \qquad \mathbf{P}_\beta \quad \text{a.s.} \qquad (9.6)$$

holds. The set $\{J(s, \alpha) | \ s \in T, \alpha \in \mathcal{A}(\beta, s)\}$ builds a $(T, \mathcal{A})$-system and is called the *conditional maximal reward*. Let us add the two following properties which make a $(T, \mathcal{A})$-system a supermartingale.

P 9.3 *Properties making a $(\mathcal{T}, \mathcal{A})$-system a supermartingale*

4. For any $s \in \mathcal{T}$ and $\alpha \in \mathcal{A}$, the random variable R_s^α is integrable with respect to $\mathbf{P}_\alpha$.

5. For $s, t \in \mathcal{T}$ such that $s \le t$, it holds $\mathbf{E}^\alpha\{R_t^\alpha \mid \mathcal{G}_s\} \le R_s^\alpha$ $\mathbf{P}_\alpha$-a.s..

6. For $\beta \in \mathcal{A}(\alpha, s)$, it follows $R_a^\alpha = R_s^\beta$ $\mathbf{P}_\alpha$-a.s..

A set of random variables $\{R_s^\alpha : s \in \mathcal{T}, \alpha \in \mathcal{A}\}$ which is a $(\mathcal{T}, \mathcal{A})$-system and satisfies in addition the properties in P 9.3 is said to be a $(\mathcal{T}, \mathcal{A})$-*supermartingale*. We have the following lemma.

Lemma 9.3 *The set $(\mathcal{T}, \mathcal{A})$-system characterising the conditional maximal reward $J(s, \beta)$ is a $(\mathcal{T}, \mathcal{A})$-supermartingale.*

Proof: It is an adaptation of El Karoui [59], p. 94.   ■

For any $s \in [0, T]$ and $\alpha \in \mathcal{A}^s$, let us define the *conditional remaining reward* at time s by means of the functional $v^\alpha : (s, x) \longrightarrow v^\alpha(s, x)$ with $v^\alpha(s, x)$ given by

$$v^\alpha(s, x) = \mathbf{E}^\alpha \left[\int_s^T f(t, X_t, \alpha_t) dt + \phi(T, X_T) | \mathcal{F}_s \right] \qquad \mathbf{P}_\alpha \quad \text{a.s.} \qquad (9.7)$$

where we have assumed that at time s the condition $X_s = x$ holds. Further, taking the supremum over all control processes $\alpha \in \mathcal{A}_T^s$, one obtains the function $v(s, x)$ by means of the formula

$$v(s, x) = \operatorname*{ess\,sup}_{\alpha \in \mathcal{A}_T^s} v^\alpha(s, x) \qquad \mathbf{P}_\alpha \quad \text{a.s.} \qquad (9.8)$$

The process $v : (s, x) \longrightarrow v(s, x)$, where $X_s = x$, is then called the *value function* or *value process* and determines which strategies shall hold our attention. At this point let us recall, the value function is usually brought into play on the basis of a PDE known as the Hamilton-Jacobi-Bellman equation, for short HJB equation. Resorting to the HJB equation to obtain the value function is the central idea of dynamic programming, a control method developed by Bellman rooted in analogies with the Hamilton-Jacobi approach to classical mechanics. For the time being we shall say, one of the merits of the martingale approach is the introduction of the value function without loosing sight of the underlying controlled process.

Since we are not saying anything concerning the dynamics governing the controlled process, we shall limit ourselves to stress that the definition of the value function rests on the filtration generated by the controlled process, the controlled objects determined by the control actions and the reward structure we have already presented. That $v^\alpha(s, x)$ and $v(s, x)$ exist indeed and are well-defined will be seen in Section 5.2.1. Note that $v(s, x) = v^\beta(s, x)$ for the optimal control β defined in (9.5). Thus, for the sake of notational simplicity, we shall suppress the superindex β in $v(\cdot, \cdot)$, whenever it is clear that we are using the optimal control β. The controlled system $(\Omega, \mathcal{F}, \mathbf{F}, T, \mathcal{O}, \mathcal{T}, \mathbf{P}_\alpha; \alpha \in \mathcal{A})$ has the ϵ-*lattice property* for the the family $\{v_t^\alpha : \alpha \in \mathcal{A}\} \subset L(\Omega, \mathcal{F}, \mathbf{P})$ provided for $\epsilon > 0$, an arbitrary but fixed $t \in \mathcal{T}$ and given $\alpha^1, \alpha^2 \in \mathcal{A}$, there exists $\alpha^3 \in \mathcal{A}$ that satisfies

$$ v_t^{\alpha^3} \geq v_t^{\alpha^1} \bigvee v_t^{\alpha^2} - \epsilon \qquad \mathbf{P}_\alpha \quad \text{a.s.} $$

where the symbol $\vee$ stands for the supremum lattice operation in (5.12) and (9.8). Drawing on results obtained by Striebel in [211, 212] it turns out, a sufficient condition for a legitimate interchange of the operations of lattice supremum and conditional expectation is that the controlled system satisfies the ϵ-lattice property for the associated family of value functions.

9.1.3 Bellman's principle of optimality

In order to see intuitively how works the philosophy behind the idea of stochastic optimisation, we need to specialise a little the notation. Hence, let α° denote the control process resulting from the application of an arbitrary control process α along the time interval $[0, s]$ and then switching to the optimal control β to which we stick until the terminal time T. That means, the control process α° arises from the concatenation of the controls α and β, an operation we have introduced above. Then, by construction $\alpha^\circ \in \mathcal{A}(\beta, s)$ and according to (9.3) the expected conditional reward $R_s^{\alpha^\circ}$ can be written as

$$ R_s^{\alpha^\circ} = \int_0^s f(t, X_t, \alpha_t)dt + \mathbf{E}^\beta \left\{ \int_s^T f(t, X_t, \beta_t)dt + \phi(T, X_T) | \mathcal{F}_s \right\} \qquad \mathbf{P}_\beta \quad \text{a.s.,} $$

(9.9)

where the conditional expectation in (9.9) does depend, due to the properties of the probability measure $\mathbf{P}_\beta$ only on the values of the control process from the time s on. This observation enables us to write (9.9) as

$$ R_s^{\alpha^\circ} = \int_0^s f(t, X_t, \alpha_t)dt + v^\beta(s, x) \qquad \mathbf{P}_\beta \quad \text{a.s.,} \tag{9.10} $$

which together with (9.8) and (9.5) suggests that any control process switching before time s to the optimal control β does better than α°. Hence, let us use the right-hand side of (9.10) to define the functional $M : (s, \alpha) \longrightarrow M_s^\alpha$ with M_s^α given by means of

$$ M_s^\alpha = \int_0^s f(t, X_t, \alpha_t)dt + v(s, x) \qquad \mathbf{P}_\beta \quad \text{a.s.} \tag{9.11} $$

The process $M = (M_s^\alpha)_{s \geq 0}$ represents the *conditional optimal reward* resulting from steering the process on the basis of an arbitrary control α up to time s and from this time on switching to the optimal control β, under the assumption that at time s the relationship $X_s = x$ holds.

The process representing the conditional optimal reward exhibits an interesting dynamic behaviour known as the supermartingale property. The key feature of the supermartingale property is the fact that whenever a control action drives a system optimally from an observation time s to the final state on the basis of the information $\mathcal{F}_s$, then this control action needs to be optimal along the time interval preceding s. Thus, given the information $\mathcal{F}_s$, any control which does not use the optimal control all along $[0, s]$ cannot do as good as one that does it. The practical meaning of this backward rationale is that it is necessary to let the system evolve under an arbitrary control before one collects the information from which the optimal control action results.

P 9.4 *Supermartingale property of the process* $M = (M_s^\alpha)_{s \geq 0}$

- For any $s \leq t$ and any control process $\alpha \in \mathcal{A}$, the process $M = (M_t^\alpha)_{t \geq 0}$ satisfies the inequality

$$\mathsf{E}^\beta [M_t^\alpha | \mathcal{F}_s] \leq M_s^\alpha \qquad \mathsf{P}_\beta \quad \text{a.s.}, \tag{9.12}$$

  almost surely with the probability measure P_β.

A glance at (9.12) suggests, optimality obtains only when the control process α switches from the very beginning to the optimal control β and this is precisely what Bellman's Principle of Optimality says, a rationale on which dynamic programming relies. Eq. (9.12) has the following natural interpretation: Failure of switching at the time s to the optimal control β entails a reward loss given by

$$M_s^\alpha - \mathsf{E}^\beta [M_t^\alpha | \mathcal{F}_s],$$

if one persists in using the non-optimal control α all along the time interval $[s, t]$. That means, any control policy that does not update as new information becomes available, cannot be efficient.

This formulation of the control problem is deeply related to various measurability requirements and concepts of semimartingale calculus which are presented in [59, 60, 83] to which we refer.

Theorem 9.4 (Martingale Principle of Optimality) *For any admissible control policy* $\alpha \in \mathcal{A}$, *the process* $M = (M_t^\alpha)_{t \geq 0}$ *is a supermartingale and the control* α *is optimal if and only if* $M = (M_t^\alpha)_{t \geq 0}$ *is a martingale.*

In addition to the interpretation associated with (9.12), the Martingale Principle of Optimality says that whenever one has followed an optimal control policy α until time s, it remains optimal as long as one takes into account only the information available at the time s, i.e.

$$\mathsf{E}^\beta [M_t^\alpha | \mathcal{F}_s] = M_s^\beta \qquad \mathsf{P}_\alpha \quad \text{a.s.}$$

As we shall later, this amounts to requiring the control action α to be (s, α)-conditionally optimal for any $s \in \mathcal{T}$. The problem of optimisation is then linked to the construction of a $\mathcal{O}$-measurable supermartingale $J = (J_t^\alpha)_{t \geq 0}$ with α ranging over $\mathcal{A}$ so that for any $s \in \mathcal{T}$ and $\beta \in \mathcal{A}$ the optimal control, one has $J_s^\beta - J(s, \beta)$ P_β-a.s.. The supermartingale J is said to aggregate the conditional maximal reward in a framework, where all controlled objects P_α are dominated by the probability measure P which acts as the reference (probability) measure. Viewing the process J as an envelope of a $(\mathcal{T}, \mathcal{A})$-system of conditional rewards provides an intuitive feeling of the actual meaning of the characterisation of J as a supermartingale.

9.2   Remarks on controlled objects

In this section we shall make some comments and remarks on the controlled objects introduced in the foregoing section with the purpose of elucidating some links between the structure underlying control actions and their associate rewards. In this connection it is extremely important to recognise how the mentioned links contribute to understanding the mechanisms on which the structure underlying $\mathfrak{P}$ rest. The central idea is introducing a martingale process, for which the optimal control acts as a parameter, in a manner rendering this martingale a version of the density of the probability measure with respect to the reference measure; this martingale is said to be steered by the optimal control. The martingale process is, as already stated, linked to the controlled process first directly by means of its dynamics and second by means of a functional of its optimal path. To see how this is, let us survey the main steps the martingale approach follows solving a control problem.

9.2.1   Main steps of the control problem

The increasing use of the theory of martingales in stochastic control may be attributed to the transparency and intuition it adheres to the theoretical framework, and to the fact the application of martingale techniques provides a natural setting for the formulation and analysis of various underlying processes and optimisation issues. We shall draw attention to the main control theoretical steps one follows solving control problems in accordance with the martingale philosophy.

P 9.5 *Basic theoretical steps of the martingale approach*

1. Define precisely various processes related to the performance index of the basic process like the conditional optimal reward $M = (M_s^\alpha)_{s \geq 0}$, the value function $v = (v(s,x))_{s \geq 0}$ and others, see (9.5)–(9.8) and (9.10)–(9.11) as a way of example.

2. Establish an equivalence relationship between the family of probability measures $\mathbf{P}_\alpha \in \mathfrak{P}$ associated with admissible controls $\alpha \in \mathcal{A}$ and a fixed measure $\mathbf{P}$, and introduce moreover the corresponding probability density using the exponential formula.

3. Check the fulfilment of conditions for the application of the Doob-Meyer decomposition to the $\mathbf{P}_\alpha$-martingale $M = (M_t^\alpha)_{t \geq 0}$ and to the existence of the adjoint process $p = (p_t)_{t \geq 0}$. This provides a sufficient condition for the existence of the value function, the adjoint state and the optimal control process among those maximising certain functionals.

4. Show that the Martingale Principle of Optimality stated in Theorem 9.4 holds. Construct the Hamiltonian $H = (H_t(x,p))_{t \geq 0}$ with $H_t(x,p) = H_t(x,p,\beta)$ given by

$$H_t(x,p,\beta) = p_t \, b(t,x,\beta) + f(t,x,\beta), \qquad (9.13)$$

where the control process β is the optimal one.

5. Conclude any control policy β satisfying, for any arbitrary but fixed $t \in [0, T]$, the relationships

$$H_t(x, \beta) = \operatorname*{ess\,sup}_{\alpha \in \mathcal{A}(\beta, s)} \{p_t\, b(t, x, \alpha) + f(t, x, \alpha)\} \qquad \mathbf{P}_\beta \quad \text{a.s.,} \tag{9.14}$$

$$H(s, \beta) = \operatorname*{ess\,sup}_{\alpha \in \mathcal{A}(\beta, s)} H_t(x, p, \alpha) \qquad \mathbf{P}_\beta \quad \text{a.s.,} \tag{9.15}$$

is also an optimal control according to the Martingale Principle of Optimality.

6. Derive an equation for the adjoint process $p = (p_t)_{t \geq 0}$, verify that $p_t = \nabla_x v(t, x)$ and prove the differentiability of the Hamiltonian with respect to the state variable.

A striking feature of the martingale approach is that optimal controls are constructed by maximising a process leading to the *Hamiltonian*. This is possible since the adjoint process arises without having to recur to the HJB equation. The value process emerges only from probabilistic considerations.

Since we are here mainly concerned with martingales which are functionals of a Wiener process, the question of how to represent martingales usefully and explicitly in terms of a Wiener process is of great relevance. Therefore, martingale representation theorems act as a bridge between abstract theory and applications. In some instances this representation parallels the role of matrix representation of linear operators as a link between abstract theory and linear algebra.

9.2.2 Changes of measure induced by control operations

As we already know, an admissible control $\alpha \in \mathcal{A}$ determines a measure $\mathbf{P}_\alpha$ on the sample space $\widetilde{\Omega}$ and this in turn induces a process $X^\alpha = (X_t)_{t \geq 0}$ which enables us to define the value function v and the conditional optimal reward M corresponding to the control process β, see eqs. (9.8) and (9.11).

Let $\mathbf{P}$ denote the measure on the basis of which one defines the process X without any control intervention, a general situation corresponding to a purely random process which can be here visualised as the case of a SDE lacking drift and a control problem with a reward function lacking the instantaneous rate of reward, i.e. the case where $b \equiv f \equiv 0$; compare with (5.6) and (5.8). Further, let $\mathbf{P}_\alpha$ be as before the probability measure associated with the control $\alpha \in \mathcal{A}$ which generates the process X^α characterised by the presence of a drift term representing the intensity impact of the control action α upon the original process X. Thus, the control action has two consequences: first, transforming the probability law of X into $\mathbf{P}_\alpha$ and second, inducing a drift term which turns X into X^α.

It turns out, the probability measures $\mathbf{P}_\alpha$ are absolutely continuous with respect to the measure $\mathbf{P}$. This fact leads to the problem of absolutely continuous changes of measures associated with switching of control actions due mainly to the operation of concatenation, a feature which stresses that control interventions act through the probabilistic law governing the dynamics of the system, as reflected in the process determined by (5.8). We proceed to outline briefly some of these issues .

The fact that for any admissible $\alpha \in \mathcal{A}$ the associated probability measure P_α on $\mathcal{F}_\zeta$ is equivalent to the (reference) measure P enables us to introduce for any $\alpha \in \mathcal{A}$ a martingale $m(\alpha) = (m_t(\alpha))_{t \geq 0}$ given by the relationships

$$m_T(\alpha) = \frac{\mathsf{P}_\alpha(d\omega)}{\mathsf{P}(d\omega)}, \tag{9.16}$$

$$m_t(\alpha) = \mathsf{E}\left[m_T(\alpha)|\mathcal{F}_t\right], \tag{9.17}$$

where $m_t(\alpha)$ stands for a right-continuous version with left-limits of the process of probability densities with respect to P. These probability densities correspond precisely to the restrictions of the measures P_α to the σ-algebra $\mathcal{F}_t$, $t \in [0,T]$. In other words, for any $t \in [0,T]$ and $\alpha \in \mathcal{A}$, the $\mathcal{F}_t$-measurable $m_t(\alpha)$ satisfies for any $A \in \mathcal{F}_t$ the relationship $\mathsf{P}^t_\alpha(A) = \mathsf{E}\{\mathbf{1}_A \, m_t(\alpha)\}$ by means of which the process P_α characterised by

$$\mathsf{P}_\alpha = (\mathsf{P}^\alpha_t)_{t \geq 0} \quad \text{and} \quad \mathsf{P}^\alpha_t = \mathsf{P}_\alpha|_{\mathcal{F}_t}$$

comes about. At this point let us mention, admissible controls need to be so as to ensure the stochastic integrals involved are well defined. That means, control actions have to be chosen in a way which renders the resulting integrands predictible and the probability densities a strictly positive uniformly integrable martingale; see (5.7) and its accompanying discussion.

Moreover, let us point out the martingale $(m_t(\alpha))_{t \geq 0}$ possesses a right continuous modification with left hand limits, i.e. with $m_{t-}(\alpha) = \lim_{s \uparrow t} m_s(\alpha)$ a.s. The process $m(\alpha)$ defined by (9.17) is indeed described probabilistically by the following lemma, see [139] pp. 207–208.

Lemma 9.5 *Let the process $m(\alpha) = (m_t(\alpha), \mathcal{F}_t)_{t \leq T}$ satisfy*

$$\mathsf{P}\left(\int_0^T m_s^2(\alpha)ds < \infty\right) = 1$$

and let $m_t(\alpha) \geq 0$ P a.s., $0 \leq t \leq T$. Then the process $m(\alpha) = (m_t(\alpha), \mathcal{F}_t)$ is a (nonnegative) supermartingale. In particular, $\mathsf{E}\{m_t(\alpha)\} \leq 1$ and $m(\alpha)$ is a martingale if and only if $\mathsf{E}\{m_t(\alpha)\} = 1$.

We shall assume these densities build an exponential martingale characterised as follows.

A 9.1 *Assumptions on the probability densities $m_t(\alpha)$*

- With any admissible control $\alpha \in \mathcal{A}$, there exists an associated local P_α-martingale $X = (X^\alpha_t)_{t \geq 0}$ such that the relationship $X_{0-} = 0$ holds P_α a.s. and the inequality $\Lambda X_s + 1 > 0$ for all s holds P_α a.s., where $\Lambda X^\alpha_s = X^\alpha_s - X^\alpha_{s-}$ represents a jump of the process X at the time s.

- Let $m(\alpha)$ denote the exponential martingale attached to X^{α}, i.e. the process $m(\alpha) = \mathcal{E}(X)$ known as the Doléans-Dade exponential. The process $m(\alpha)$ is the unique solution of Doléans' stochastic integral equation

$$m_t(\alpha) = m_{0-}(\alpha) + \int_o^t m_{s-}(\alpha)dX_s^{\alpha},$$

where m_{0-} is a given initial value. Usually $m_{0-}(\alpha) = m_0(\alpha) = 1$ and $X_0 = 0$, a situation to which we adhere.

- The process $m(\alpha)$ is a local $\mathbf{P}_{\alpha}$-martingale strictly positive, stopped at time ζ. Furthermore, it is a uniformly integrable $\mathbf{P}_{\alpha}$-martingale such that the terminal random variable $m_{\zeta}(\alpha)$ is the density of $\mathbf{P}_{\alpha}$ with respect to $\mathbf{P}$.

The Doléans-Dade exponential martingale $\mathcal{E}(X) = (\mathcal{E}_t(X^{\alpha}))_{t \geq 0}$ associated with the process $X = (X_t^{\alpha})_{t \geq 0}$ is defined by the formula

$$\mathcal{E}_t(X^{\alpha}) = \exp\left\{ X_t^{\alpha} - \frac{1}{2}\langle X^c, X^c \rangle_t \right\} \prod_{s \leq t} (1 + \Lambda X_s^{\alpha}) e^{-\Lambda X_s^{\alpha}}$$

where X^c is the continuous part of local martingale $X = (X_t^{\alpha})_{t \geq 0}$. Here it is necessary to emphasise that a martingale emerges only when α becomes the optimal control β. The countable product presented above is $\mathbf{P}_{\alpha}$ a.s. absolutely convergent and defines a purely discontinuous process of bounded variation, see [59], pp. 178–180. The properties of exponential martingales have been object of a great deal of investigation as we already mentioned in Chapter 8. The presence of control actions can be accounted viewing exponential martingales just as processes depending on an exogeneously given parameter. The relevance of these ideas in connection with the change of measure attached to the change of control is due to a great extent to the following Lemma own to van Schuppen and Wong, see [60], Ch. 13, pp. 165–166.

Lemma 9.6 *Suppose the exponential martingale $m(\alpha) = \mathcal{E}(X)$ is given as in A 9.1 and measure $\mathbf{P}_{\alpha}$ on $(\Omega, \mathcal{F}_T)$ is defined by (9.16). Further, let X be a local $\mathbf{P}$-martingale such that the cross-variation process $\langle X, m(\alpha) \rangle$ exists and set $\tilde{X} = X - \langle X, m(\alpha) \rangle$. Then, it turns out that $\tilde{X}$ is a local $\mathbf{P}_{\alpha}$-martingale.*

Proof: See [60], pp. 165–166   ∎

Let us point out that from the general formula connecting Radon-Nikodym derivatives and conditional expectations one obtains using the results just mentioned the following enlightening relationship

$$\mathbf{E}^{\alpha}\left[\tilde{X}_t | \mathcal{F}_s\right] = \mathbf{E}\left[m_t(\alpha)\tilde{X}_t | \mathcal{F}_s\right] m_s^{-1}(\alpha),$$

which tells us that $\tilde{X}$ is a local $\mathbf{P}_{\alpha}$-martingale if and only if $m_t(\alpha)\tilde{X}$ is a local $\mathbf{P}$-martingale. Concerning the special situation we have just introduced, i.e. when $b \equiv f \equiv 0$, one easily notes from (9.3) and (9.4) that in this particular case we get

$$v(x) = \int_{\widetilde{\Omega}} \phi(T, \omega_T)\mathbf{P}_{\alpha}(d\omega) = \mathbf{E}^{\alpha}\left\{\phi(T, \omega_T)\right\}$$

where $\widetilde{\Omega}$ denotes the sample space, i.e. $\mathbb{C}([0,T] \longrightarrow \mathbb{R}^n)$ of real-valued continuous functions and $\mathcal{F}_t$ the algebra generated by the ω_t coordinate functions. As we pointed out already, under these simplifying assumptions we have that $\mathbf{P}_\alpha \ll \mathbf{P}$ holds.

Thus, assuming our process lives under these conditions we have that $m(\alpha) = (m_t(\alpha))_{t \geq 0}$ is a positive martingale, $\mathbf{E}\{m_t(\alpha)\} = 1$ with $m_0(\alpha) = 1$ a.s., since $\mathcal{F}_0$ is the completion of the trivial σ-field $\{\sigma, \Omega\}$.

Moreover, Lemma 9.6 delivers the simple relationship for this special case of our control problem

$$\mathbf{E}^\alpha\{\phi\} = \mathbf{E}\{m_t(\alpha)\phi(T, \omega_T)\}. \tag{9.18}$$

Therefore, a look at eq. (8.10) suggests that our optimal control problem is equivalent to choosing $\alpha \in \mathcal{A}$ such that the mean reward attains an optimum. Afterwards, a drift transformation switches us back to the original problem, i.e. one with $b \neq 0$ and $f \neq 0$; in Chapter 5 we deal in detail with the cases of control adapted to the state in a martingale framework. Hence, it remains to say that in order to apply these ideas to our control problem one needs to make sure that operations with admissible controls like concatenation, restriction, etc. deliver again an admissible control.

9.2.3 Probabilistic representation of the adjoint state

Let us sketch Hausmann's way of obtaining $\nabla_x v$. Considering the fundamental adjoint solution Ψ, one gets an equation the adjoint satisfies, i.e.

$$
\begin{aligned}
-dp_s &= \{p_s\left[\nabla_x b^\beta(s, Y_s) - \nabla_x \sigma^k(s, Y_s)(\nabla_x \sigma^k(s, Y_s))^*\right] + \\
&\quad + \nabla_x f^\beta(s, Y_s) + (\nabla_x \sigma^k(s, Y_s))^* \gamma^k \Psi_s\} \, ds - \\
&\quad - \left[\Psi_s \gamma^k - p_s \nabla_x \sigma^k(s, Y_s)\right] dB_x^{\beta(k)}. \\
p_T &= \nabla_x \phi(T, Y_T)
\end{aligned}
$$

The unknown functions γ can be identified in a way similar as in Theorem 5.14. Thus, we shall concentrate on some aspects related to the (BHJ) partial differential equation, under the assumption it is smooth and does not degenerate. For the sake of simplicity, for any function $\phi(t, x, \alpha)$ we shall write $\widehat{\phi}(t, x)$ for $\phi(t, x, \widehat{\alpha}(t, x))$. However, to avoid confusion we shall notice $\nabla_x \widehat{\phi}(t, x)$ is not the same as $\widehat{\phi}_x(t, x) + \widehat{\phi}_\alpha(t, x)\widehat{\alpha}_x(t, x)$.

Theorem 9.7 *Let us assume A 5.9. Then we have the following representation for $p_t(x)$:*

$$
\begin{aligned}
p_t(x) &= \nabla_x v(t, x) \\
&= \mathbf{E}^\beta\left[\int_t^T \nabla_x \widehat{f}(s, Y_s)\Phi_{s,t} \, ds + \nabla_x \phi(T, Y_T)\Phi_{T,t}\Big|\mathcal{F}_t\right]. \tag{9.19}
\end{aligned}
$$

Proof: Differentiating (5.39) and (5.40) as a Schwartz distribution, with respect to x, [103] obtains for $z = \nabla_x^{(k)} v$, $k = 1, 2, \cdots, n$, the equation

$$0 = \partial_s z + \sum_{i=1}^{n} \nabla_x^{(i)} \left(\frac{1}{2} \sum_{i=1}^{n} (a(s, x)^*)^{(ij)} \nabla_x^{(j)} z \right) -$$

$$-\frac{1}{2} \sum_{i=1}^{n} \nabla_x^{(i)} (a(s, x)^*)^{(ij)} \nabla_x^{(j)} z + Z, \tag{9.20}$$

$$(s, x) \in (0, T) \times \mathbb{R}^n$$

$$z_T(x) = \nabla_x^{(k)} v(T, x) = \nabla_x^{(k)} \phi(T, Y_T), \tag{9.21}$$

where

$$Z(s, Y_s) = \frac{1}{2} \sum_{i,j=1}^{n} \nabla_x^{(k)} ((a(t, x)^*)^{(ij)} \nabla_x^{(i)} \nabla_x^{(j)} v + \sum_{i=1}^{n} \widehat{b}^{(i)}(t, x) \nabla_x^{(i)} \nabla_x^{(k)} v +$$

$$+ \nabla_x^{(k)} \widehat{b}(t, x) \nabla_x^{(k)} v + \nabla_x^{(k)} \widehat{f}(t, x).$$

Knowing that (9.20) and (9.21) have a unique solution z, we can obtain also a differential for z with the purpose of computing $d(\Phi_{s,t} z_s)$. The procedure is already known, one applies Itô to $z_s(Y_s(x))$ and uses the foregoing PDE for z. Thus we get for z the differential

$$-dz_s = \left[z_s \nabla_x \widehat{b}(s, Y_s) - \nabla_x \sigma^k(s, Y_s) \nabla_x z_s(Y_s) \sigma^k(s, Y_s) \right.$$

$$\left. + (\nabla_x f^\beta(s, Y_s))^* \right] ds - \nabla_x z_s(Y_s) \sigma(s, Y_s) dB_s^\beta, \tag{9.22}$$

$$z_T(Y_T) = \nabla_x \phi(T, Y_T) \tag{9.23}$$

where $\nabla_x z$ is the Hessian of the value function v. Applying Itô's lemma to the product $\eta_s z_s(Y_s)$ yields

$$- d\eta_s z_s = \left(\nabla_x \widehat{b}(s, Y_s) \eta_s ds + \nabla_x \sigma^{(k)}(s, Y_s) \eta_s dB^{\beta(k)} \right) z +$$

$$\eta_s \left[\left(z_s \nabla_x \widehat{b}(s, Y_s) - \nabla_x \sigma^{(k)}(s, Y_s) \nabla_x z_s(Y_s) \sigma^{(k)}(s, Y_s) \right. \right.$$

$$\left. + (\nabla_x f^\beta(s, Y_s))^* \right) ds - \nabla_x z_s(Y_s) \sigma^{(k)}(s, Y_s) dB_s^{\beta(k)} \right]$$

$$+ d\eta_s dz. \tag{9.24}$$

Integrating from t to T, using (9.20) and (9.21), and (9.22) and (9.23) gives

$$-\nabla_x \phi(Y_T) \eta_T + \eta_t z_t(x) = \nabla_x z_s \sigma^k(s, Y_s)) dB_s^{\beta(k)} +$$

$$+ \int_t^T \eta_s \nabla_x f^\beta(s, Y_s) ds + \int_t^T \eta_s (\nabla_x \sigma^k(s, Y_s) z_s$$

Recalling $\eta_t = I_n$ and taking conditional expectation at both sides of the foregoing equation, one obtains

$$z_t(x) = \nabla_x v(t, x) = \mathsf{E}^\beta \left[\nabla_x \phi(T, Y_T) \eta_T + \int_t^T \nabla_x f^\beta(s, Y_s) \eta_s \, ds \middle| \mathcal{F}_t \right] \tag{9.25}$$

which is the representation we look for.   ∎

A rigorous proof of this theorem is given in [45] and [103]. It approximates the function z and its corresponding partial derivatives by $\mathbb{C}^{\infty}$ functions and obtained our equations above on the basis of uniform convergence on appropriate spaces.

Bibliography

[1] M. Aglietta. *A Theory of Capitalist Regulation.* Verso, London, 1987.

[2] M. Aglietta and A. Orléans. *La violence de la monnaie.* Presses Universitaires de France, Paris, 1982.

[3] A. Aguilar. *Origenes del Subdesarollo.* Plaza & Janes, Bogotá, 1982.

[4] A. N. Al-Hussaini. A simplified proof of the Representation of Functionals of Diffusions. *Applied Mathematics and Optimization,* 20:63–69, 1989.

[5] S. Albeverio and G. L. Gómez M. Equations of motion in economics and mechanics: Elements of a stochastic theory of value. *Institute of Mathematics, University of Erlangen-Nürnberg,* Erlangen, FRG, 1991.

[6] S. Albeverio and G.L. Gomez M. *Systems Interaction and Entropic Processes in Physics and Economic Dynamics: Toward an Evolutionary Theory.* Work in progress.

[7] D. Aldous. *Probability Approximations via the Poisson Clumping Heuristic.* Springer Verlag, New York, 1989.

[8] L. Althusser. *Essays on Ideology.* Verso, London, 1971.

[9] K. Arrow and M. Kurz. *Public Investment, the Rate of Return and Optimal Fiscal Policy.* The John Hopkins University Press, Baltimore, 1970.

[10] A. K. Bagchi. *The Political Economy of Underdevelopment.* Cambridge University Press, Cambridge, 1982.

[11] P. A. Baran. *The Political Economy of Growth.* Monthly Review, New York, 1957.

[12] J. Bather. Optimal Stopping Problems for Brownian Motion. *Advanced Applied Probability,* 2:259–286, 1970.

[13] M. Beaud. *A History of Capitalism 1500-1980.* Monthly Review Press, New York, 1983.

[14] D.R. Bell. *The Malliavin Calculus.* Longman Scientific & Technical, Essex, 1987.

422

[15] A. Bensoussan. Control of Stochastic Partial Differential Equations. In W. H. Ray and D. G. Lainiotis, editors, *Distributed Parameter Systems*. Marcel Dekker, New York, 1978.

[16] A. Bensoussan. Lectures on stochastic control. In *Lecture Notes in Mathematics, Vol. 972*. Springer Verlag, Berlin, 1982.

[17] A. Bensoussan. *Stochastic Control by Functional Analysis Methods*. North Holland, Amsterdam, 1982.

[18] A. Bensoussan. Maximum principle and dynamic programming approaches of the optimal control of partially observed diffusions. *Stochastics*, 9:169–222, 1983.

[19] A. Bensoussan. Stochastic Maximum Principle for Distributed Parameter Systems. *Journal of the Franklin Institute*, 315(5/6):387–406, 1983.

[20] A. Bensoussan and J. Lions. *Applications des Inéquations Variationnelles en Contrôle Stochastique*. Dunod, Paris, 1978.

[21] A. Bensoussan and J. Lions. *Contrôle Impulsionnel et Inéquations Quasi-Variationnelles*. Dunod, Paris, 1982.

[22] H. Bernstein, B. Crow, M. Mackintosh, and C. Martin. *The Food Question: Profit versus People*. Monthly Review Press, New York, 1990.

[23] A. Bhaduri. *Macroeconomics: The Dynamics of Commodity Production*. The Macmillan Press, London, 1986.

[24] Ph. Blanchard, Ph. Combe, and W. Zheng. *Mathematical and Physical Aspects of Stochastic Mechanics*. Springer Verlag, Berlin, 1987.

[25] C. J. Bliss. *Capital Theory and the Distribution of Income*. North-Holland, Amsterdam, 1975.

[26] T. Bottomore. *A Dictionary of Marxist Thought*. Harvard University Press, Cambridge, Mass., 1983.

[27] K.E. Boulding. *Ecodynamics*. Sage Publications, Beverly Hills, 1981.

[28] K.E. Boulding. *Evolutionary economics*. Sage Publications, Beverly Hills, 1981.

[29] F. Bourguignon. A particular class of continuous–time stochastic growth-models. *Journal of Economic Theory*, 9:141–158, 1974.

[30] F. Braudel. *Civilization and Capitalism. 15th–18th Century*, volume 1–3. Collins/Fontana Press, London, 1988.

[31] H. Braverman. *Labor and Monopoly Capital*. Monthly Review Press, New York, 1974.

[32] R. G. Brockett. *Finite Dimensional linear systems*. Wiley, New York, 1970.

[33] N. Bukharin. *Economic Theory of the Leisure Class*. Monthly Review Press, New York, 1972.

[34] E. Burmeister. *Capital Theory and Dynamics*. Cambridge University Press, Cambridge, 1980.

[35] E. Burmeister and A. R. Dobell. *Mathematical Theories of Economic Growth*. The Macmillan Press, London, 1970.

[36] S. Chakravarty. *Capital and Development Planning*. The MIT Press, Cambridge, Mass., 1969.

[37] F. R. Chang and A. G. Malliaris. Asymptotic growth under uncertainty: existence and uniqueness. *Review of Economic Studies*, 54:169–174, 1987.

[38] H. Chernoff. Optimal Stochastic Control. *Sankhiā, Series A*, 30:221–252, 1968.

[39] N. Chomsky. *Deterring Democracy*. Verso, London, 1991.

[40] K.L. Chung. *Lectures from Markov Processes to Brownian Motion*. Springer Verlag, New York, 1982.

[41] L. Colletti. *From Rousseau to Lenin*. Monthly Review Press, New York, 1972.

[42] M. Cornforth. *The Theory of Knowledge*. Lawrence & Wishart Ltd., London, 1963.

[43] J. Crank. *The Mathematics of Diffusion*. Oxford University Press, Oxford, 1975.

[44] P. Dasgupta, A. Sen, and S. Marglin. *Guidelines for Project Evaluation*. Unido Publications, New York, 1972.

[45] M. H. A. Davis. Functionals of diffusion processes as stochastic integrals. *Mathematical Proceedings of the Cambridge Philosophical Society*, 87:157–166, 1980.

[46] M. H. A. Davis and G. L. Gómez M. The martingale maximum principle and the allocation of labour surplus. *Journal of Economic Dynamics and Control*, 11:241–247, 1987.

[47] M. H. A. Davis and G. L. Gómez M. The semi-martingale approach to the optimal resource allocation in the controlled labour-surplus economy. In S. Albeverio et al., editors, *Lecture Notes in Mathematics, Vol. 1250*, pages 36–74. Springer Verlag, Berlin, 1987.

[48] M. H. A. Davis and P. Varaiya. Dynamic Programming conditions for partially observable stochastic systems. *SIAM Journal on Control and Optimization*, 11:226–261, 1973.

[49] J. de Castro. *The Geopolitics of Hunger*. Monthly Review Press, New York, 1977.

[50] C. Dellacherie and P.-A. Meyer. *Probabiltés et potentiel: Théorie des martingales*. Hermann, Paris, 1980.

[51] M. Dobb. *Studies in the Development of Capitalism*. Routledge & Kegan Paul, London, 1963.

[52] M. Dunford and D. Perrons. *The Arena of Capital*. The Macmillan Press, London, 1983.

[53] A.K. Dutt. *Growth, distribution and uneven development*. Cambridge University Press, Cambridge, 1990.

[54] E. B. Dynkin. *Markov Processes*, volume I and II. Springer Verlag, Berlin, 1965.

[55] E. B. Dynkin. Harmonic functions associated with several Markov processes. *Advances in Applied Mathematics*, 2:260–283, 1981.

[56] E. B. Dynkin and R.J. Vanderbei. Stochastic waves. *Transactions of the American Mathematical Society*, 275:771–779, 1983.

[57] A. Eichner, editor. *A Guide to Post-Keynesian Economics*. M. E. Sharpe, 1979.

[58] A. Eichner. *Toward a New Economics*. The Macmillan Press, 1985.

[59] N. El Karoui. Les Aspects Probabilistes du Contrôle Stochastique. In *Lecture Notes in Mathematics, Vol. 876*, pages 73–238. Springer Verlag, Berlin, 1981.

[60] R. J. Elliott. *Stochastic Calculus and Applications*. Springer Verlag, Berlin, 1982.

[61] R. J. Elliott and M. Kollmann. Martingale representation and the Malliavin Calculus. *Applied Mathematics and Optimization*, 20:105–112, 1989.

[62] M. Emery. *Stochastic Calculus in Manifolds*. Springer Verlag, Berlin, 1989.

[63] E. Eshag. *Fiscal and Monetary Policies and Problems in Developing Countries*. Cambridge University Press, Cambridge, 1983.

[64] A.G. Fakeev. Optimal stopping rules for stochastic processes with continuous parameters. *Theory of Probability and its Applications*, 15:324–331, 1970.

[65] F. Fanon. *The Wretched of the Earth*. Penguin Books, London, 1985.

[66] E. Farjourn and M. Machover. *Laws of Chaos: A Probabilistic Approach to Political Economy*. Verso, London, 1983.

[67] I. Fisher. *The Theory of Interest.* Porcupine, Philadelphia, 1977.

[68] W. Fleming. Non–linear semigroups for controlled partially observed diffusions. *SIAM Journal of Control and Optimization*, 20, 1982.

[69] W. Fleming and E. Pardoux. Optimal control of partially observed diffusions. *SIAM Journal of Control and Optimization*, 20, 1982.

[70] W. Fleming and R. Rishel. *Deterministic and Stochastic Optimal Control.* Springer Verlag, New York, 1975.

[71] D. K. Foley. *Understanding Capital: Marx's Economic Theory.* Harvard University Press, Cambridge, Mass., 1986.

[72] I. Forbes. *Marx and the New Individual.* Unwin Hyman, London, 1990.

[73] J. Foster. *Evolutionary Macroeconomics.* Unwin Hyman, London, 1987.

[74] A. G. Frank. *World Accumulation 1492-1780.* Monthly Review Press, New York, 1978.

[75] A Friedman. *Stochastic Differential Equations and Applications.* Academic Press, New York, 1975.

[76] M. Fujisaki, G. Kallianpur, and H. Kunita. Stochastic differential equations for the nonlinear filtering problem. *Osaka Journal of Mathematics*, 9:19–40, 1972.

[77] C. Furtado. *Accumulation and Development.* St. Martin's Press, New York, 1983.

[78] E. Galeano. *Open Veins of Latin America: Five Centuries of the Pillage of a Continent.* Monthly Review Press, New York, 1973.

[79] A. Garcia. *Bases de Economia Contemporanea.* Plaza & Janes, Bogotá, 1984.

[80] S. George. *How the other half dies: The Real Reasons for World Hunger.* Penguin Books, London, 1986.

[81] N. Geras. *Marx & Human Nature.* Verso, London, 1983.

[82] I. I. Gihman and A. V. Skorohod. *Stochastic Differential Equations.* Springer Verlag, Berlin, 1972.

[83] I. I. Gihman and A. V. Skorohod. *Controlled Stochastic Processes.* Springer Verlag, Berlin, 1979.

[84] M. Godelier. *Rationality and Irrationality in Economics.* Monthly Review Press, New York, 1972.

[85] G. L. Gómez M. Modelling the economic development by means of impulsive control techniques. In X. J. Avula and R. E. Kalman, editors, *Mathematical Modelling in Sciences and Technology*, pages 802–806. Pergamon Press, New York, 1984.

[86] G. L. Gómez M. On the Markov Stopping Rule Associated with the Problem of Controlling a Dual Economy. In T. Basar and L. F. Pau, editors, *Dynamic Modelling and Control of National Economies*, pages 197–204. Pergamon Press, New York, 1984.

[87] G. L. Gómez M. The intertemporal labour allocation inherent in the optimal stopping of the dual economy: the dynamic case. *Methods of Operations Research*, 49:523–543, 1985.

[88] G. L. Gómez M. Discounted values and stochastic rates arising in control theory. *Methods of Operations Research*, 57:379–392, 1987.

[89] G. L. Gómez M. Attainability and Reversibility of a Golden Age for the Labour Surplus Economy: A Stochastic Variational Approach. In S. Alberverio et al., editors, *Stochastic Processes in Physics and Engineering*, pages 107–148. D. Reidel Publishing Co., Dordrecht, 1988.

[90] G. L. Gómez M. Intertemporal issues associated with the control of macroeconomic systems. *Computers & Mathematics with Applications*, in press, 1991.

[91] G. L. Gómez M. *Lectures on Economic Development: Theory and Models*. Work in progress, 1991.

[92] G. L. Gómez M. A mathematical dynamic model of the dual economy emphasising unemployment, migration and structural change. *Institute of Mathematics, University of Erlangen–Nürnberg*, Erlangen, FRG., 1991.

[93] G. L. Gómez M. On the assessment of social benefits derived from employment policies. *Institute of Mathematics, University of Erlangen-Nürnberg*, Erlangen, FRG., 1991.

[94] G.L. Gomez M. *Probabilistic Analysis of Disorder and Structural Change: Essays in Socioeconomic Discontinuity*. Submitted to D. Reidel Publishing Co., Dordrecht. Theory and Decision Library. Series B: Mathematical and Statistical Methods.

[95] A. Gorz. *Critique of Economic Reason*. Verso, London, 1989.

[96] A. Gramsci. *Selection from Prison Notebooks*. Lawrence and Wishart, London, 1986.

[97] K. Griffin. *Alternative Strategies for Economic Development*. The Macmillan Press, London, 1989.

[98] G. Haag and W. Weidlich. Concepts of the dynamic migration model. In W. Weidlich and G. Haag, editors, *Interregional Migration*, pages 9–20. Springer Verlag, Berlin, 1988.

[99] J. Habermas. *Knowledge and Human Interests*. Beacon Press, Boston, 1971.

[100] J. Habermas. *Theorie des kommunikativen Handelns, Band 1 und 2.* Suhrkamp Verlag, Frankfurt a.M., 1988.

[101] U. Hausmann. On the stochastic maximum principle. *SIAM Journal on Control*, 16:236–251, 1978.

[102] U. Hausmann. On the integral representation of functionals of Itô process. *Stochastics*, 3:17–27, 1979.

[103] U. Hausmann. On the adjoint process for optimal control of diffusion processes. *SIAM Journal of Control and Optimization*, 19:221–243, 1981.

[104] S.W. Hawking. *A Brief History of Time.* Bantam Press, London, 1989.

[105] T. Hayter. *The Creation of World Poverty.* Pluto Press, London, 1990.

[106] G. Heal. *The Theory of Economic Planning.* North-Holland, Amsterdam, 1973.

[107] J.F. Henry. *The Making of Neoclassical Economics.* Unwin Hyman, London, 1990.

[108] A. O. Hirschman. *The Strategy of Economic Development.* Yale University Press, New Haven, 1958.

[109] A. O. Hirschman. *The Passions and the Interests: The Political Arguments for Capitalism before its Triumph.* Princeton University Press, Princeton, 1977.

[110] G. Hodgson. *Economics and Institutions.* Polity Press, London, 1988.

[111] R. A. Howard. *Dynamic Probabilistic Systems*, volume I and II. Wiley, New York, 1971.

[112] N. Ikeda and S. Watanabe. *Stochastic Differential Equations and Diffusion Processses.* North–Holland, Amsterdam, 1981.

[113] B. Ingrao and G. Israel. *The Invisible Hand. Economic Equilibrium in the History of Science.* The MIT Press, Cambridge, Mass., 1990.

[114] M. D. Intriligator. *Mathematical Optimization and Economic Theory.* Prentice–Hall, London, 1971.

[115] J. Jacod and A.N. Shiryayev. *Limit Theorems for Stochastic Processes.* Springer Verlag, Berlin, 1987.

[116] M. Kalecki. *Essays on Devoloping Economies.* The Harvester Press, Sussex, 1976.

[117] S. Kalmanovitz. *Ensayos sobre el desarollo del capitalismo dependiente.* Editorial La Oveja Negra, Bogotá, 1980.

[118] S. Kalmanovitz. *El desarollo tardio del capitalismo: Un enfoque crítico de la teoría de la dependencia.* Siglo veintiuno editores, Bogotá, 1983.

[119] C. Kay. *Latin American Theories of Development and Underdevelopment.* Routledge, London, 1989.

[120] G. Kay. *Development and Underdevelopment: A Marxist Analysis.* The Macmillan Press, London, 1975.

[121] N.M. Kay. *The Emergent Firm.* The Macmillan Press, London, 1984.

[122] J.M. Keynes. *A Treatise on Money. The Applied Theory of Money.* The Macmillan Press, London, 1973.

[123] H. Koning. *Columbus: His Enterprise.* Monthly Review Press, New York, 1991.

[124] T. Koopmans. On the Concept of Optimal Economic Growth. In *The Econometric Approach to Planning.* Rand McNally, Chicago, 1966.

[125] P. Körner, G. Maass, T. Siebold, and R. Tetzlaff. *The IMF and the Debt Crisis.* Zed Books Ltd., London, 1986.

[126] J.A. Kregel. *The Reconstruction of Political Economy.* The Macmillan Press, London, 1978.

[127] L. Krüger, L.J. Daston, and M. Heidelberger, editors. *The Probabilistic Revolution.* MIT Press, Cambridge, Mass., 1987.

[128] N. V. Krylov. *Controlled Diffusion Processes.* Springer Verlag, Berlin, 1980.

[129] H. Kunita and S. Watanabe. On square integrable martingales. *Nagoya Mathematical Journal*, 30:209–245, 1967.

[130] H. Kurz. *Capital, Distribution and Effective Demand.* Polity Press, Cambridge, 1990.

[131] H. J. Kushner. *An Introduction to Stochastic Control.* Holt, Rinehart and Winston, New York, 1973.

[132] H. J. Kushner. *Probability Methods for Approximations in Stochastic Control and for Elliptic Equations.* Academic Press, New York, 1977.

[133] H. J. Kushner. *Weak Convergence Methods and Singularity Perturbed Stochastic Control and Filterin Problems.* Birkhäuser, Boston, 1990.

[134] P. Lax. The flowering of applied mathematics in America. *SIAM Review*, 31:533–541, 1989.

[135] E. Lenglart. Transformation des martingales locales par changement absolument continu de probabilités. *Zeitschrift für Wahrscheinlichkeitstheorie*, 39:65–70, 1977.

[136] P. M. Lichtenstein. *An Introduction to Post–Keynsian and Marxian Theories of Value and Price.* The Macmillan Press, London, 1983.

[137] A. Lipietz. *Le capital et son espace*. La Decouverte/Maspero, Paris, 1983.

[138] A. Lipietz. *Mirages and Miracles*. Verso, London, 1987.

[139] R. S. Liptser and A. N. Shiryayev. *Statistics of Random Processes I*. Springer Verlag, Berlin, 1977.

[140] B. Malinowski. *A Scientific Theory of Culture*. Galaxy Books, New York, 1944.

[141] E. Mandel. *Marxist Economic Theory*, volume I and II. Monthly Review Press, New York, 1968.

[142] P. Marcuse. A letter from the German Democratic Republic. *Monthly Review*, 42(3):30–62, 1990.

[143] P. Marcuse. *Missing Marx*. Monthly Review Press, New York, 1991.

[144] H.J. Marcussen and J.E. Torp. *Internationalization of Capital*. Zed Books Ltd., London, 1982.

[145] S. A. Marglin. *Value and Price in the Labour–Surplus Economy*. Oxford University Press, London, 1976.

[146] S. A. Marglin. *Growth, Distribution and Prices*. Harvard University Press, Cambridge, Mass, 1984.

[147] S.A. Marglin. What do bosses do? part i. *Review of Radical Political Economics*, 7:20–37, 1974.

[148] R. Merton. An asymptotic theory of growth under uncertainty. *Review of Economic Studies*, 42:375–393, 1975.

[149] I. Mészáros. *The Power of Ideology*. New York University Press, New York, 1989.

[150] P. Mirowski. *More heat than light*. Cambridge University Press, Cambridge, 1989.

[151] M Morishima. *The Theory of Economic Growth*. Clarendon Press, Oxford, 1969.

[152] M Morishima. *The Economic Theory of Modern Society*. Cambridge University Press, Cambridge, 1976.

[153] T. Negishi. *History of Economic Theory*. North-Holland, Amsterdam, 1989.

[154] E. Nelson. *Dynamical Theories of Brownian Motion*. Princeton University Press, Princeton, 1967.

[155] C. Nwoke. *The Third World Minerals and Global Pricing*. Zed Books, London, 1987.

430

[156] J. Nyilas, editor. *The Theory and Practice of Development*. A. W. Sijthoff, Leyden, 1977.

[157] J. Nyilas, editor. *The Changing Face of the Third World*. A. W. Sijthoff, Leyden, 1978.

[158] J. Nyilas. *World Economy and its Main Development Tendencies*. Martinus Nijoff Publishers, The Hague, 1982.

[159] B. Øksendal. The high contact principle in optimal stopping and stochastic waves. In *Seminar on stochastic processes 1989*, pages 177–192. 1989.

[160] L.I. Pasinetti. *Structural Change and Economic Growth*. Cambridge University Press, Cambridge, 1981.

[161] Ch. Payer. *The Debt Trap: the International Monetary Fund and The Third World*. Monthly Review Press, New York, 1974.

[162] Ch. Payer. *The World Bank: A Critical Analysis*. Monthly Review Press, New York, 1982.

[163] M. Peschel. *Modellbildung für Signale und Systeme*. VEB Verlag Technik, Berlin, 1978.

[164] E. S. Phelps. *Inflation Policy and Unemployment Theory*. The Macmillan Press, London, 1972.

[165] C. A. Pissarides. *Labour market adjustment: Microeconomic foundations of short-run neoclassical and Keynesian dynamics*. Cambridge University Press, Cambridge, 1976.

[166] D. Pitchford. *Population in Economic Growth*. North Holland, Amsterdam, 1974.

[167] J. Polkinghorne. *One World*. Princeton University Press, Princeton, 1986.

[168] R. Prebisch. *Capitalismo Periférico. Crisis y Transformación*. Fondo de Cultura Económica, Mexico, 1981.

[169] R. Prebisch. Five stages in my thinking on development. In G.M. Meier and D. Seers, editors, *Pioneers in Development*. Oxford University Press, Oxford, 1984.

[170] I. Prigogine. *From Being to Becoming – Time and Complexity in Physical Sciences*. W. H. Freeman and Company, San Francisco, 1980.

[171] I. Prigogine and I. Stengers. *Order out of Chaos*. Bantam Books, New York, 1984.

[172] P. Protter. *Stochastic Integration and Differential Equations*. Springer Verlag, Berlin, 1990.

[173] K. Raffer. *Unequal Exchange and the Evolution of the World System.* The Macmillan Press, London, 1987.

[174] C. Raghavan. *Recolonization.* Zed Books Ltd., London, 1990.

[175] S. A. Resnick and R. D. Wolff. *Knowledge and Class: A Marxian Critique of Political Economy.* The University of Chicago Press, Chicago, 1987.

[176] D. Revuz and M. Yor. *Continuous Martingales and Brownian Motion.* Springer Verlag, Berlin, 1991.

[177] R.J. Richards. *Darwin and the Emergence of Evolutionary Theories of Mind and Behaviour.* The University of Chicago Press, Chicago, 1987.

[178] J. Rifkin. *Entropy: A New World View.* Viking Press, New York, 1980.

[179] R. Rishel. Necessary and sufficient conditions for continuous-time stochastic optimal control. *SIAM Journal on Control,* 8:559–571, 1970.

[180] J. Robinson. *The Accumulation of Capital.* The Macmillan Press, London, 1956.

[181] J. Robinson. *Essays in the Theory of Economic Growth.* The Macmillan Press, London, 1962.

[182] J. Robinson. *The Economics of Imperfect Competion.* The Macmillan Press, London, 1969.

[183] J. Robinson. *Aspects of development and underdevelopment.* Cambridge University Press, Cambridge, 1979.

[184] J. Robinson. *The Generalization of the General Theory and other Essays.* The Macmillan Press, London, 1979.

[185] J. Roddick. *The Dance of the Millions.* Betrand Russell House, London, 1988.

[186] W. Rodney. *How Europe Underdeveloped Africa.* Bogle L'Ouverture Publications, London, 1988.

[187] L. C. G. Rogers and D. Williams. *Diffusions, Markov Processes, and Martingales,* volume 2. Wiley, Chichester, 1987.

[188] D. Rosenberg. The colonization of East Germany. *Monthly Review,* 43(4):14–33, 1991.

[189] K. W. Rotschild. *Arbeitslose: Gibt's die?* Metropolis-Verlag, Marburg, 1990.

[190] G. Ryle. *The Concept of Mind.* The University of Chicago Press, Chicago, 1984.

[191] P. Samuelson. Maximum principles in analytical economics. *The American Economy Review,* 62:249–262, 1972.

[192] J.P. Sartre. *Critique of Dialectical Reason*, volume 1. Verso, London, 1991.

[193] R. Sau. *Unequal Exchange, Imperialism and Underdevelopment.* Oxford University Press, Calcutta, 1978.

[194] R. Sau. *Trade, Capital and Underdevelopment.* Oxford University Press, Calcutta, 1982.

[195] E. Schrödinger. *What is Life?* Cambridge University Press, Cambridge, 1944.

[196] J. A. Schumpeter. *History of Economic Analysis.* Oxford University Press, New York, 1954.

[197] J.T. Schwartz. *Lectures on the Mathematical Method in Analytical Economics.* Gordon and Breach, Science Publishers, New York, 1961.

[198] M. F. Scott. *A New View of Economic Growth.* Oxford University Press, Oxford, 1989.

[199] A. Sen. *Collective Choice and Social Welfare.* North–Holland, Amsterdam, 1970.

[200] A. Sen. *Choice, Welfare and Measurement.* Basil Blackwell, Oxford, 1982.

[201] A. Sen. *Resources, Values and Development.* Harvard University Press, Cambridge, Mass., 1984.

[202] A. Sen. *Commodities and Capabilities.* North–Holland, Amsterdam, 1985.

[203] A. Sen. *The Standard of Living.* Cambridge University Press, Cambridge, 1987.

[204] A. Sen. Development as capability expansion. In *Human Development and the International Development Strategy for the 1990's*, pages 41–58. The Macmillan Press, London, 1990.

[205] A. N. Shiryayev. *Optimal Stopping Rules.* Springer Verlag, Berlin, 1978.

[206] J. L. Simon. *The Ultimate Resource.* Princeton University Press, Princeton, N. J., 1981.

[207] J. L. Simon. *The Theory of Population and Economic Growth.* Basil Blackwell, Oxford, 1986.

[208] A. Smith. *The Theory of Moral Sentiments.* Oxford University Press, 1976.

[209] W. Sombart. *Der Moderne Kapitalismus.* Deutscher Taschenbuch Verlag, München, 1987. Bd. 1, 2 und 3.

[210] J. Steindl. *Economic Papers 1941–1988*. The Macmillan Press, London, 1990.

[211] C. Striebel. *Optimal Control of Discrete Time Stochastic Systems*. Springer Verlag, Berlin, 1975.

[212] C. Striebel. Martingale conditions for the optimal control of continuous time stochastic systems. *Stochastic Processes and their Applications*, 18:329–347, 1984.

[213] D.W. Stroock and S.R.S. Varadhan. *Multidimensional Diffusion Processes*. Springer Verlag, Berlin, 1979.

[214] P. M. Sweezy. *The Theory of Capitalist Development*. Monthly Review Press, New York, 1970.

[215] T. Szentes. *The Political Economy of Underdevelopment*. Akademiai Kiado, Budapest, 1976.

[216] M. Tanzer. *The Race for Resources*. Monthly Review Press, New York, 1980.

[217] C. S. Tapiero. *Managerial Planning: An Optimum and Stochastic Control Approach*, volume 1 and 2. Gordon and Breach, New York, 1977.

[218] C. S. Tapiero. *Applied Stochastic Models and Control in Management*. North–Holland, Amsterdam, 1988.

[219] L. Taylor. *Macro Models for Developing Countries*. McGraw-Hill, New York, 1979.

[220] L. Taylor. *Structuralist Macroeconomics*. Basic Books, New York, 1983.

[221] G. Tintner and J. Sengupta. *Stochastic Economics*. Academic Press, New York, 1972.

[222] T. Veblen. *The Theory of the Leisure Class: An Economic Study of Institutions*. Modern Library, 1934.

[223] J. Viner. *The Role of Providence in the Social Order*. Princeton University Press, Princeton, 1972.

[224] J. von Neuman and O. Morgenstern. *Theory of Games and Economic Behaviour*. Princeton University Press, Princeton, 1944.

[225] I. Wallerstein. *The Capitalist World-Economy*. Cambridge University Press, 1989.

[226] M. Weber. *Economy and Society: An Outline of Interpretative Sociology*, volume 1 and 2. University of California Press, Berkeley, 1978.

[227] M. Weber. *Die protestantische Ethik*. Gütersloher Verlagshaus, Gütersloh, 1984. Bd. I und II.

[228] C. West. *The Ethical Dimensions of Marxist Thought*. Monthly Review Press, New York, 1991.

[229] N. Wiener. *Cybernetics: or Control and Communication in the Animal and the Machine*. MIT Press, Cambridge, Mass., 1964.

[230] R. D. Wolff and S. A. Resnick. *Economics: Marxian Versus Neoclassical*. The Johns Hopkins University Press, Baltimore, 1987.

[231] J. Zabczyk. Introduction to the theory of optimal stopping. In *Stochastic Control Theory and Stochastic Differential Systems*, pages 227–250. Springer Verlag, Berlin, 1979.

[232] P. Zarembka. *Towards a Theory of Economic Development*. Holden–Day, San Francisco, 1972.

[233] A. Zimbalist, editor. *Case Studies on the Labour Process*. Monthly Review Press, New York, 1979.

Index

438

distributional relationship, 79
Doléans-Dade exponential, 378
drift, 95, 390
    forward, 305
    vector, 181
drive to accumulate, 38
dual
    problem, 37
    space, 387
dynamic
    economising problem, 230
    efficiency condition, 270
    efficiency conditions, 333
    system, 119
dynamics, system, 309, 353

earning
    potentials, 303
    power, 295, 314, 337
economic
    growth, fundamental equation
      of, 182
    policy, 112
    policy practices, common, 27
    surplus, 73
      actual, 74
      potential, 74
      social, 73
    surplus of society, 71
economics
    neo-Marxian, 70
    neoclassical, 70
    post-Keynesian, 70
effective
    demand, 183, 299
efficiency
    condition
      dynamic, 270, 333
efficient performance, 111
employment
    relevant range of, 108
energy
    potential, 220, 282
enrichment of human life, 109
entitlement, 85, 197
    individual, 331
entropic

nature, 282
    process, 352
entropy
    law of, 337
    principle of, 176
envelope
    of a $(\mathcal{T},\mathcal{A})$-system, 412
    Snell's, 191, 192
environment, 351
    political, 111
    socioeconomic, 111
equation
    deterministic adjoint, 254
    of economic growth, fundamen-
      tal, 182
    of motion, 127
    price, 79
    production, 79
    quantity, 79
    state, 361
equations
    Kolmogorov's forward, 126
    of motion, Hamilton, 326
escape, probability of immediate, 190
essential supremum, 409
essentialism, 162
evanescent set, 368
evolution, 198
    of temporal processes, 129
excessive
    function, 190
    functionals, 189
    majorant, 190
      smallest, 190
exchange, 112
existence
    social, 287
expected
    conditional reward, 408, 411
    income streams, 197
    total reward, 407
explosion
    time, 374, 391
exponential
    decay, 194
extended Poisson process, 373
extension, capability of, 355

440

442

property, 187, 411

surplus
    economic, 73
    of society
      economic, 71
    social economic, 73

syntactic
    features of a model, 356
    purpose, 355

system, 351, 354
    dynamic, 119
    dynamics, 309, 353
    socioeconomic, 352
    structure, 353
    without an after-effect, 126

$(\mathcal{T},\mathcal{A})$-system, 409

system-environment interaction, 352

systems
    characterisation, 352
    theory, 351, 353
    theory of, 15

$(\mathcal{T},\mathcal{A})$-system, 409

tacit knowledge, 154

tangency condition, 266

technical
    coefficients, 78, 180, 294

technique
    choice of, 175

technology
    of the unified field, 18

temporal
    character of decisions, 129
    character of information, 129
    processes, evolution of, 129

terminal
    reward function, 318
    set, 362
    state, 111
    variable, 369

terminating time, 403

theory
    control, 353
    of knowledge, 357
    of systems, 15
    systems, 353

time

    change of, 368
    decentralisation, 335
    evolution, 351
    explosion, 374
    historical, 123
    hitting, 368
    impulsion, 112
    life, 403
    Markov, 190
    mechanistic notion of, 119
    of discontinuity, 372
    of observation, 403
    problems, stopping, 183
    shape of income streams, 198
    stopping, 367
    structure of individual needs, 129
    structure of social needs, 129
    structure of transformation processes, 129
    terminating, 403

timing decisions, 129

total
    reward, 406
    reward, expected, 407
    semimartingale, 370
    variation, 369

trade-offs, 79

tranquillity, state of, 283

transformation
    marginal rate of, 262
    processes, time structure of, 129

transient phenomena, 16

transition probability, 126

transversality
    condition, 269, 273

turnover
    period of, 101, 135

uncertainty, 92, 154
    principle of, 360

underdeveloped countries, 26

underdevelopment, 10
    development of, 25
    roots of, 26

uniqueness in probability law, 392

unity and separation, 155

THEORY AND DECISION LIBRARY

SERIES B: MATHEMATICAL AND STATISTICAL METHODS
Editor: H. J. Skala, *University of Paderborn, Germany*

1. D. Rasch and M.L. Tiku (eds.): *Robustness of Statistical Methods and Nonparametric Statistics*. 1984   ISBN 90-277-2076-2

2. J.K. Sengupta: *Stochastic Optimization and Economic Models*. 1986
          ISBN 90-277-2301-X

3. J. Aczél: *A Short Course on Functional Equations*. Based upon Recent Applications to the Social Behavioral Sciences. 1987
          ISBN Hb 90-277-2376-1; Pb 90-277-2377-X

4. J. Kacprzyk and S.A. Orlovski (eds.): *Optimization Models Using Fuzzy Sets and Possibility Theory*. 1987   ISBN 90-277-2492-X

5. A.K. Gupta (ed.): *Advances in Multivariate Statistical Analysis*. Pillai Memorial Volume. 1987   ISBN 90-277-2531-4

6. R. Kruse and K.D. Meyer: *Statistics with Vague Data*. 1987
          ISBN 90-277-2562-4

7. J.K. Sengupta: *Applied Mathematics for Economics*. 1987
          ISBN 90-277-2588-8

8. H. Bozdogan and A.K. Gupta (eds.): *Multivariate Statistical Modeling and Data Analysis*. 1987   ISBN 90-277-2592-6

9. B.R. Munier (ed.): *Risk, Decision and Rationality*. 1988
          ISBN 90-277-2624-8

10. F. Seo and M. Sakawa: *Multiple Criteria Decision Analysis in Regional Planning*. Concepts, Methods and Applications. 1988   ISBN 90-277-2641-8

11. I. Vajda: *Theory of Statistical Inference and Information*. 1989
          ISBN 90-277-2781-3

12. J.K. Sengupta: *Efficiency Analysis by Production Frontiers*. The Non-parametric Approach. 1989   ISBN 0-7923-0028-9

13. A. Chikán (ed.): *Progress in Decision, Utility and Risk Theory*. 1991
          ISBN 0-7923-1211-2

14. S.E. Rodabaugh, E.P. Klement and U. Höhle (eds.): *Applications of Category Theory to Fuzzy Subsets*. 1992   ISBN 0-7923-1511-1

15. A. Rapoport: *Decision Theory and Decision Behaviour*. Normative and Descriptive Approaches. 1989   ISBN 0-7923-0297-4

16. A. Chikán (ed.): *Inventory Models*. 1990   ISBN 0-7923-0494-2

17. T. Bromek and E. Pleszczyńska (eds.): *Statistical Inference*. Theory and Practice. 1991   ISBN 0-7923-0718-6

18. J. Kacprzyk and M. Fedrizzi (eds.): *Multiperson Decision Making Models Using Fuzzy Sets and Possibility Theory*. 1990   ISBN 0-7923-0884-0

19. G.L. Gómez M.: *Dynamic Probabilistic Models and Social Structure*. Essays on Socioeconomic Continuity. 1992   ISBN 0-7923-1713-0

20. H. Bandemer and W. Näther: *Fuzzy Data Analysis*. 1992
ISBN 0-7923-1772-6

KLUWER ACADEMIC PUBLISHERS – DORDRECHT / BOSTON / LONDON